*Edited by*
*Dennis G. Hall*

**Boronic Acids**

## Related Titles

Blaser, H.-U., Federsel, H.-J. (eds.)

**Asymmetric Catalysis on Industrial Scale**

Challenges, Approaches and Solutions

580 pages in 1 volumes with 131 figures and 94 tables
2010
Hardcover
ISBN: 978-3-527-32489-7

Bolm, C., Hahn, F. E. (eds.)

**Activating Unreactive Substrates**

The Role of Secondary Interactions

481 pages with approx. 485 figures
2009
Hardcover
ISBN: 978-3-527-31823-0

Carreira, E. M., Kvaerno, L.

**Classics in Stereoselective Synthesis**

651 pages with 600 figures and 3 tables
2009
Hardcover
ISBN: 978-3-527-32452-1

Dupont, J., Pfeffer, M. (eds.)

**Palladacycles**

Synthesis, Characterization and Applications

431 pages
2008
Hardcover
ISBN: 978-3-527-31781-3

Toru, T., Bolm, C. (eds.)

**Organosulfur Chemistry in Asymmetric Synthesis**

448 pages with 494 figures and 34 tables
2008
Hardcover
ISBN: 978-3-527-31854-4

Plietker, B. (ed.)

**Iron Catalysis in Organic Chemistry**

Reactions and Applications

295 pages with 340 figures and 29 tables
2008
Hardcover
ISBN: 978-3-52 7-31927-5

Warren, S., Wyatt, P.

**Workbook for Organic Synthesis**

Strategy and Control

500 pages
2008
Softcover
ISBN: 978-0-471-92964-2

Ribas Gispert, J.

**Coordination Chemistry**

640 pages with 427 figures and 74 tables
2008
Softcover
ISBN: 978-3-527-31802-5

Maruoka, K. (ed.)

**Asymmetric Phase Transfer Catalysis**

228 pages with 213 figures and 34 tables
2008
Hardcover
ISBN: 978-3-527-31842-1

Christmann, M., Bräse, S. (eds.)

**Asymmetric Synthesis - The Essentials**

395 pages with 85 figures and 3 tables
2008
Softcover
ISBN: 978-3-527-32093-6

Edited by Dennis G. Hall

# Boronic Acids

Volume 1
Preparation and Applications in Organic Synthesis,
Medicine and Materials

Second Completely Revised Edition

With a Foreword by Akira Suzuki

WILEY-VCH Verlag GmbH & Co. KGaA

**The Editors**

*Prof. Dennis G. Hall*
4-010 Centennial Centre for
Interdisciplinary Science
Department of Chemistry
University of Alberta
Edmonton, Alberta, T6G 2G2
Canada

**Cover Graphics**
Displayed at the forefront is the space-filling structure of ortho-iodophenylboronic acid, a novel type of highly active catalyst for direct amidation reactions between carboxylic acids and amines (Al-Zoubi, Marion, Hall; Angew. Chem. Int. Ed. **2008**, *47*, 2876). It is shown over a background representation of the X-ray crystallographic structure of AN2690, a benzoxaborole drug bound to the editing site of its bacterial protein target, the isoleucyl tRNA synthetase (Anacor Pharmaceuticals Laboratory; Science **2007**, *316*, 1759). These two graphics components illustrate the breadth of novel conceptual applications that are continually emerging using boronic acid derivatives.

All books published by **Wiley-VCH** are carefully produced. Nevertheless, authors, editors, and publisher do not warrant the information contained in these books, including this book, to be free of errors. Readers are advised to keep in mind that statements, data, illustrations, procedural details or other items may inadvertently be inaccurate.

**Library of Congress Card No.:** applied for

**British Library Cataloguing-in-Publication Data**
A catalogue record for this book is available from the British Library.

**Bibliographic information published by the Deutsche Nationalbibliothek**
The Deutsche Nationalbibliothek lists this publication in the Deutsche Nationalbibliografie; detailed bibliographic data are available on the Internet at http://dnb.d-nb.de

© 2011 Wiley-VCH Verlag & Co. KGaA, Boschstr. 12, 69469 Weinheim, Germany

All rights reserved (including those of translation into other languages). No part of this book may be reproduced in any form – by photoprinting, microfilm, or any other means – nor transmitted or translated into a machine language without written permission from the publishers. Registered names, trademarks, etc. used in this book, even when not specifically marked as such, are not to be considered unprotected by law.

**Typesetting**  Thomson Digital, Noida, India
**Printing and Binding**  Fabulous Printers Pte Ltd, Singapore
**Cover Design**  Grafik-Design Schulz, Fußgönheim

Printed in Singapore
Printed on acid-free paper

**Print ISBN:** 978-3-527-32598-6
**ePDF ISBN:** 978-3-527-63934-2
**oBook ISBN:** 978-3-527-63932-8
**ePub ISBN:** 978-3-527-33214-4
**Mobi ISBN:** 978-3-527-63935-9

# Foreword

Hydroboration, discovered in 1956, has made organoboranes readily available. This discovery opened the gate to a new continent for the chemical community to explore, develop, and exploit. Mainly in the late-1960s and 1970s, many novel types of carbon–carbon and carbon–heteroatom bond forming reactions of organoboranes were discovered and developed for use in organic synthesis.

The palladium-catalyzed cross-coupling of organoboron compounds with organic electrophiles such as organic halides in the presence of base was developed in 1979. Over the past 30 years or so, transition metal-catalyzed cross-coupling reactions of boronic acid derivatives have emerged as one of the most important and widely used organometallic reactions for carbon–carbon and carbon–heteroatom bond formation, and is now regarded as an integral part of any synthetic route toward building complex organic chemicals. The coupling reaction has many advantages: the reactants are readily available, nontoxic, and air- and water-stable, and they react under mild conditions and are amenable to a variety of reaction conditions, including the use of aqueous solvents. Moreover, the inorganic boron byproduct can be easily removed after the reaction. Most important of all, the coupling proceeds with high regio- and stereoselectivity, and is little affected by steric hindrance. The process does not affect other functional groups in the molecule and can thus be used in one-pot strategies. In addition, the reaction has proved to be extremely versatile. Consequently, these coupling reactions have been actively utilized not only in academic laboratories but also in industrial processes, like in pharmaceutical and agrochemical industries as well as other industries for the production of liquid crystals and organic LEDs.

Today, the use of boronic acid derivatives and their applications continue to evolve with many new findings reported during the past decade. For example, new reactions, catalysts and ligands have been developed. Increasingly, industry is seeking to use more environment-friendly processes. These often require ingenious solutions to which Suzuki coupling is well suited. We can expect to see many more interesting versions of the coupling reactions and other applications of boronic acids in the future.

Hokkaido University, Sapporo, Japan                              *Akira Suzuki*

# Contents to Volume 1

**Foreword**  V

**Contents to Volume 2**  XIII

**Preface**  XV

**List of Contributors**  XIX

| | |
|---|---|
| **1** | **Structure, Properties, and Preparation of Boronic Acid Derivatives: Overview of Their Reactions and Applications**  *1* |
| | *Dennis G. Hall* |
| 1.1 | Introduction and Historical Background  *1* |
| 1.2 | Structure and Properties of Boronic Acid Derivatives  *2* |
| 1.2.1 | General Types and Nomenclature of Boronic Acid Derivatives  *2* |
| 1.2.2 | Boronic Acids  *3* |
| 1.2.2.1 | Structure and Bonding  *3* |
| 1.2.2.2 | Physical Properties and Handling  *8* |
| 1.2.2.3 | Safety Considerations  *9* |
| 1.2.2.4 | Acidic Character  *9* |
| 1.2.2.5 | Chemical Stability  *12* |
| 1.2.3 | Boronic Acid Derivatives  *15* |
| 1.2.3.1 | Boroxines (Cyclic Anhydrides)  *15* |
| 1.2.3.2 | Boronic Esters  *16* |
| 1.2.3.3 | Acyloxy- and Diacyloxyboronates  *25* |
| 1.2.3.4 | Dialkoxyboranes and Other Heterocyclic Boranes  *25* |
| 1.2.3.5 | Diboronyl Esters  *26* |
| 1.2.3.6 | Azaborolidines and Other Boron–Nitrogen Heterocycles  *27* |
| 1.2.3.7 | Dihaloboranes and Dihydroalkylboranes  *29* |
| 1.2.3.8 | Trifluoro- and Trihydroxyborate Salts  *30* |
| 1.3 | Preparation of Boronic Acids and Their Esters  *31* |
| 1.3.1 | Arylboronic Acids  *31* |
| 1.3.1.1 | Electrophilic Trapping of Arylmetal Intermediates with Borates  *31* |
| 1.3.1.2 | Transmetalation of Aryl Silanes and Stannanes  *41* |

| | | |
|---|---|---|
| 1.3.1.3 | Coupling of Aryl Halides with Diboronyl Reagents | 42 |
| 1.3.1.4 | Direct Boronation by Transition Metal-Catalyzed Aromatic C–H Functionalization | 43 |
| 1.3.1.5 | Cycloadditions of Alkynylboronates | 43 |
| 1.3.1.6 | Other Methods | 43 |
| 1.3.2 | Diboronic Acids | 44 |
| 1.3.3 | Heterocyclic Boronic Acids | 44 |
| 1.3.4 | Alkenylboronic Acids | 45 |
| 1.3.4.1 | Electrophilic Trapping of Alkenylmetal Intermediates with Borates | 45 |
| 1.3.4.2 | Transmetalation Methods | 45 |
| 1.3.4.3 | Transition Metal-Catalyzed Coupling between Alkenyl Halides/Triflates and Diboronyl Reagents | 45 |
| 1.3.4.4 | Hydroboration of Alkynes | 55 |
| 1.3.4.5 | Alkene Metathesis | 58 |
| 1.3.4.6 | Diboronylation and Silaboration of Unsaturated Compounds | 59 |
| 1.3.4.7 | Other Methods | 60 |
| 1.3.5 | Alkynylboronic Acids | 60 |
| 1.3.6 | Alkylboronic Acids | 61 |
| 1.3.7 | Allylic Boronic Acids | 63 |
| 1.3.8 | Chemoselective Transformations of Compounds Containing a Boronic Acid (Ester) Substituent | 63 |
| 1.3.8.1 | Oxidative Methods | 64 |
| 1.3.8.2 | Reductive Methods | 64 |
| 1.3.8.3 | Generation and Reactions of $\alpha$-Boronyl-Substituted Carbanions and Radicals | 66 |
| 1.3.8.4 | Reactions of $\alpha$-Haloalkylboronic Esters | 68 |
| 1.3.8.5 | Other Transformations | 70 |
| 1.3.8.6 | Protection of Boronic Acids for Orthogonal Transformations | 72 |
| 1.4 | Isolation and Characterization | 73 |
| 1.4.1 | Recrystallization and Chromatography | 74 |
| 1.4.2 | Solid Supports for Boronic Acid Immobilization and Purification | 75 |
| 1.4.2.1 | Diethanolaminomethyl Polystyrene | 75 |
| 1.4.2.2 | Other Solid-Supported Diol Resins | 76 |
| 1.4.2.3 | Soluble Diol Approaches | 76 |
| 1.4.3 | Analytical and Spectroscopic Methods for Boronic Acid Derivatives | 76 |
| 1.4.3.1 | Melting Points, Combustion Analysis, and HPLC | 76 |
| 1.4.3.2 | Mass Spectrometry | 77 |
| 1.4.3.3 | Nuclear Magnetic Resonance Spectroscopy | 77 |
| 1.4.3.4 | Other Spectroscopic Methods | 78 |
| 1.5 | Overview of the Reactions of Boronic Acid Derivatives | 78 |
| 1.5.1 | Metalation and Metal-Catalyzed Protodeboronation | 78 |
| 1.5.2 | Oxidative Replacement of Boron | 79 |
| 1.5.2.1 | Oxygenation | 79 |
| 1.5.2.2 | Amination and Amidation | 81 |
| 1.5.2.3 | Halodeboronation | 81 |
| 1.5.3 | Carbon–Carbon Bond Forming Processes | 85 |

| | | |
|---|---|---|
| 1.5.3.1 | Transition Metal-Catalyzed Cross-Coupling with Carbon Halides and Surrogates (Suzuki–Miyaura Cross-Coupling) | 85 |
| 1.5.3.2 | Transition Metal-Catalyzed Insertions, Cycloisomerizations, and C–H Functionalizations Based on Transmetalation of Boronic Acids | 88 |
| 1.5.3.3 | Heck-Type Coupling to Alkenes and Alkynes | 90 |
| 1.5.3.4 | Rhodium- and Other Transition Metal-Catalyzed Additions to Alkenes, Carbonyl Compounds, and Imine Derivatives | 90 |
| 1.5.3.5 | Diol-Catalyzed Additions of Boronic Esters to Unsaturated Carbonyl Compounds and Acetals | 92 |
| 1.5.3.6 | Allylation of Carbonyl Compounds and Imine Derivatives | 93 |
| 1.5.3.7 | Uncatalyzed Additions of Boronic Acids to Imines and Iminiums | 93 |
| 1.5.4 | Carbon–Heteroatom Bond Forming Processes | 94 |
| 1.5.4.1 | Copper-Catalyzed Coupling with Nucleophilic Oxygen and Nitrogen Compounds | 94 |
| 1.5.5 | Other Reactions | 94 |
| 1.6 | Overview of Other Applications of Boronic Acid Derivatives | 97 |
| 1.6.1 | Use as Reaction Promoters and Catalysts | 97 |
| 1.6.2 | Use as Protecting Groups for Diols and Diamines | 99 |
| 1.6.3 | Use as Supports for Immobilization, Derivatization, Affinity Purification, Analysis of Diols, Sugars, and Glycosylated Proteins and Cells | 100 |
| 1.6.4 | Use as Receptors and Sensors for Carbohydrates and Other Small Molecules | 102 |
| 1.6.5 | Use as Antimicrobial Agents and Enzyme Inhibitors | 103 |
| 1.6.6 | Use in Neutron Capture Therapy for Cancer | 105 |
| 1.6.7 | Use in Transmembrane Transport | 105 |
| 1.6.8 | Use in Bioconjugation and Labeling of Proteins and Cell Surface | 106 |
| 1.6.9 | Use in Chemical Biology | 107 |
| 1.6.10 | Use in Materials Science and Self-Assembly | 108 |
| | References | 109 |
| | | |
| 2 | **Metal-Catalyzed Borylation of C–H and C–Halogen Bonds of Alkanes, Alkenes, and Arenes for the Synthesis of Boronic Esters** | **135** |
| | *Tatsuo Ishiyama and Norio Miyaura* | |
| 2.1 | Introduction | 135 |
| 2.2 | Borylation of Halides and Triflates via Coupling of H–B and B–B Compounds | 137 |
| 2.2.1 | Borylation of Aryl Halides and Triflates | 138 |
| 2.2.2 | Alkenyl Halides and Triflates | 143 |
| 2.2.3 | Allylic Halides, Allylic Acetates, and Allylic Alcohols | 145 |
| 2.2.4 | Benzylic Halides | 148 |
| 2.3 | Borylation via C–H Activation | 148 |
| 2.3.1 | Aliphatic C–H Bonds | 148 |
| 2.3.2 | Alkenyl C–H Bonds | 151 |
| 2.3.3 | Aromatic C–H Bonds | 153 |
| 2.4 | Catalytic Cycle | 159 |

2.5   Summary   *161*
      References   *161*

**3**   **Transition Metal-Catalyzed Element-Boryl Additions to Unsaturated Organic Compounds**   171
        *Michinori Suginome and Toshimichi Ohmura*
3.1     Introduction   *171*
3.2     Diboration   *172*
3.2.1   Diboron Reagents for Diboration   *172*
3.2.2   Diboration of Alkynes   *173*
3.2.3   Diboration of Alkenes, Allenes, 1,3-Dienes, and Methylenecyclopropanes   *176*
3.2.4   Synthetic Applications of Diboration Products   *183*
3.3     Silaboration   *185*
3.3.1   Silylborane Reagents for Silaboration   *185*
3.3.2   Silaboration of Alkynes   *187*
3.3.3   Silaboration of Alkenes, Allenes, 1,3-Dienes, and Methylenecyclopropanes   *191*
3.3.4   Synthetic Application of Silaboration Products   *200*
3.4     Carboboration   *202*
3.4.1   Direct Addition: Cyanoboration and Alkynylboration   *203*
3.4.2   Transmetalative Carboboration   *205*
3.5     Miscellaneous Element-Boryl Additions   *207*
3.6     Conclusion   *208*
        References   *208*

**4**   **The Contemporary Suzuki–Miyaura Reaction**   213
        *Cory Valente and Michael G. Organ*
4.1       Introduction   *213*
4.1.1     Preamble and Outlook   *213*
4.1.2     A Brief History   *214*
4.1.3     Mechanistic Aspects   *214*
4.2       Developments Made in the Coupling of Nontrivial Substrates   *215*
4.2.1     Rational Design of Ligands for Use in the Suzuki–Miyaura Reaction   *215*
4.2.1.1   Organophosphine Ligands and Properties   *217*
4.2.1.2   N-Heterocyclic Carbene Ligands and their Properties   *219*
4.2.2     The Suzuki–Miyaura Cross-Coupling of Challenging Aryl Halides   *220*
4.2.2.1   Overview of Challenges   *220*
4.2.2.2   Organophosphine-Derived Catalysts   *221*
4.2.2.3   NHC-Derived Catalysts   *228*
4.2.3     The Suzuki–Miyaura Reaction Involving Unactivated Alkyl Halides   *234*
4.2.3.1   Associated Difficulties   *234*
4.2.3.2   Cross-Couplings Promoted by Phosphines and Amine-Based Ligands   *235*

| 4.2.3.3 | Cross-Coupling-Promoted NHC Ligands 240 |
| 4.3 | Asymmetric Suzuki–Miyaura Cross-Couplings 241 |
| 4.3.1 | Achieving Axial Chirality in the Suzuki–Miyaura Reaction 241 |
| 4.3.1.1 | Axial Chirality Induced by Chiral Ligands/Catalysts 241 |
| 4.3.1.2 | Axial Chirality Induced by Point Chirality 244 |
| 4.3.1.3 | Axial Chirality Induced by Planar Chirality 246 |
| 4.3.2 | Achieving Point Chirality in the Suzuki–Miyaura Reaction 246 |
| 4.4 | Iterative Suzuki–Miyaura Cross-Couplings 248 |
| 4.4.1 | ortho Metalation–Cross-coupling Iterations 248 |
| 4.4.2 | Triflating–Cross-Coupling Iterations 248 |
| 4.4.3 | Iterative Cross-Couplings via Orthogonal Reactivity 249 |
| 4.4.3.1 | Bifunctional Electrophiles 249 |
| 4.4.3.2 | Bifunctional Organoboranes 252 |
| 4.5 | Conclusions and Future Outlook 256 |
| | References 257 |

| 5 | **Rhodium- and Palladium-Catalyzed Asymmetric Conjugate Additions of Organoboronic Acids** 263 |
| | *Guillaume Berthon-Gelloz and Tamio Hayashi* |
| 5.1 | Introduction 263 |
| 5.2 | Rh-Catalyzed Enantioselective Conjugate Addition of Organoboron Reagents 263 |
| 5.2.1 | α,β-Unsaturated Unsaturated Ketones 264 |
| 5.2.1.1 | A Short History 264 |
| 5.2.1.2 | Mechanism 264 |
| 5.2.1.3 | Model for Enantioselection 266 |
| 5.2.1.4 | Organoboron Sources Other Than Boronic Acids 266 |
| 5.2.1.5 | Rh Precatalysts 269 |
| 5.2.1.6 | Ligand Systems 269 |
| 5.2.1.7 | α,β-Unsaturated Aldehydes 278 |
| 5.2.2 | Enantioselective Addition to α,β-Unsaturated Esters and Amides 279 |
| 5.2.2.1 | Diastereoselective Conjugate Addition 282 |
| 5.2.2.2 | Fumarate and Maleimides 284 |
| 5.2.2.3 | Synthetically Useful Acceptors 286 |
| 5.2.2.4 | Conjugate Additions of Boryl and Silyl Groups 286 |
| 5.2.3 | Addition to Other Electron-Deficient Alkenes 288 |
| 5.2.3.1 | Arylmethylene Cyanoacetates 288 |
| 5.2.3.2 | Alkenylphosphonates 288 |
| 5.2.3.3 | Nitroalkene 288 |
| 5.2.3.4 | Sulfones 289 |
| 5.2.3.5 | Addition to *cis*-Allylic Alcohols 291 |
| 5.2.3.6 | 1,4-Addition/Enantioselective Protonation 291 |
| 5.2.4 | 1,6-Conjugate Additions 294 |
| 5.2.5 | Rh-Catalyzed Enantioselective Conjugate Addition with Other Organometallic Reagents 296 |

| 5.2.6 | Rh-Catalyzed Tandem Processes  297 |
| --- | --- |
| 5.2.6.1 | Tandem Enantioselective Conjugate Addition/Aldol Reaction  297 |
| 5.2.6.2 | Tandem Carborhodation/Conjugate Addition  298 |
| 5.3 | Pd-Catalyzed Enantioselective Conjugate Addition of Organoboron Reagents  299 |
| 5.3.1 | Introduction  299 |
| 5.3.2 | Addition to α,β-Unsaturated Ketones  300 |
| 5.3.3 | Addition to α,β-Unsaturated Esters, Amides, and Aldehydes  304 |
| 5.3.4 | Palladium-Catalyzed Tandem Processes  305 |
| 5.4 | Conclusions  306 |
| | References  307 |

| 6 | **Recent Advances in Chan–Lam Coupling Reaction: Copper-Promoted C–Heteroatom Bond Cross-Coupling Reactions with Boronic Acids and Derivatives**  *315* |
| --- | --- |
| | *Jennifer X. Qiao and Patrick Y.S. Lam* |
| 6.1 | General Introduction  315 |
| 6.2 | C–O Cross-Coupling with Arylboronic Acids  316 |
| 6.2.1 | Intermolecular C–O Cross-Coupling  316 |
| 6.2.2 | Intramolecular C–O Cross-Coupling  320 |
| 6.3 | C–N Cross-Coupling with Arylboronic Acids  321 |
| 6.3.1 | C–N (Nonheteroarene NH) Cross-Coupling  321 |
| 6.3.1.1 | Application of Chan–Lam Cross-Coupling in Solid-Phase Synthesis  324 |
| 6.3.2 | C–N (Heteroarene) Cross-Coupling  324 |
| 6.3.2.1 | Factor Xa Inhibitors  326 |
| 6.3.2.2 | Purines  326 |
| 6.3.2.3 | Heteroarene–Heteroarene Cross-Coupling  329 |
| 6.3.3 | Intramolecular C–N Cross-Coupling  330 |
| 6.3.4 | Catalytic Copper-Mediated C–N Cross-Coupling  331 |
| 6.3.5 | Additional N-Containing Substrates in Chan–Lam Cross-Coupling  332 |
| 6.4 | Substrate Selectivity and Reactivity in Chan–Lam Cross-Coupling Reaction  335 |
| 6.5 | C–N and C–O Cross-Coupling with Alkenylboronic Acids  336 |
| 6.6 | C–N and C–O Cross-Coupling with Boronic Acid Derivatives  338 |
| 6.6.1 | Boroxines, Boronic Esters, and Trifluoroborate Salts  338 |
| 6.6.2 | Alkylboronic Acids  343 |
| 6.7 | C–S and C–Se/C–Te Cross-Coupling  346 |
| 6.8 | Mechanistic Considerations  349 |
| 6.8.1 | Empirical Observations  349 |
| 6.8.2 | General Mechanistic Observations  351 |
| 6.8.3 | Mechanistic Study of the Catalytic Reaction  352 |
| 6.9 | Other Organometalloids  354 |
| 6.10 | Conclusion  355 |
| 6.11 | Note Added in Proof  355 |
| | References  357 |

## Contents to Volume 2

**7** Transition Metal-Catalyzed Desulfitative Coupling of Thioorganic Compounds with Boronic Acids  *363*
Ethel C. Garnier-Amblard and Lanny S. Liebeskind

**8** Catalytic Additions of Allylic Boronates to Carbonyl and Imine Derivatives  *393*
Tim G. Elford and Dennis G. Hall

**9** Recent Advances in Nucleophilic Addition Reactions of Organoboronic Acids and Their Derivatives to Unsaturated C–N Functionalities  *427*
Timothy R. Ramadhar and Robert A. Batey

**10** Asymmetric Homologation of Boronic Esters with Lithiated Carbamates, Epoxides, and Aziridines  *479*
Matthew P. Webster and Varinder K. Aggarwal

**11** Organotrifluoroborates: Organoboron Reagents for the Twenty-First Century  *507*
Gary A. Molander and Ludivine Jean-Gérard

**12** Borate and Boronic Acid Derivatives as Catalysts in Organic Synthesis  *551*
Joshua N. Payette and Hisashi Yamamoto

**13** Applications of Boronic Acids in Chemical Biology and Medicinal Chemistry  *591*
Nanting Ni and Binghe Wang

**14** Boronic Acids in Materials Chemistry  *621*
Jie Liu and John J. Lavigne

Index  *677*

# Preface

From the first isolation of a boronic acid by Frankland in 1860 to the report of their palladium-catalyzed cross-coupling with carbon halides by Suzuki in 1979, advances in the chemistry and biology of boronic acids have been few and far between. The early 1980s announced a drastic turn. In the past two decades alone, numerous breakthroughs have been reported. From Miyaura's discovery of rhodium-catalyzed couplings to alkenes and aldehydes to the commercialization of Velcade$^{TM}$, the first boronic acid drug used in human therapy, new reactions and applications of boronic acids have been reported at a spectacular rate. As seen in Figure P.1, the number of publications focused on boronic acid derivatives has increased exponentially, elevating boronic acids to a new status, that of an essential class of organic compounds. The attribution of the 2010 Chemistry Nobel Prize for palladium-catalyzed cross-coupling reactions, shared by Professor Akira Suzuki, cements the importance of boronic acids in this revolutionary class of C–C bond forming processes.

This sudden rise in the usefulness and popularity of boronic acids necessitated a comprehensive book on their synthetic and biological applications. In just a few years working in the field of boronic acid chemistry, I had quickly come to regret the absence of a specialized book on this topic. Thus, I could not turn down an opportunity to help fulfill this need and lead such a project that led to the first edition of *Boronic Acids* in 2005. I was most fortunate to assemble a select group of experts, who literally included legends in the field. The successful result of this project was a popular handbook containing 13 chapters that covered all modern aspects of boronic acid derivatives. All efforts were made to achieve a comprehensive coverage of the field, with particular emphasis on topics of great interest to a large audience of synthetic organic, organometallic, and medicinal chemists. A quick look at Figure P.1 is sufficient to justify the need for an expanded, two-volume second edition only 6 years later. The recent period of 2005–2009 following the publication of the first edition of *Boronic Acids* shows a continuous, exponential burst of research activity around boronic acids, with the year 2010 showing no signs of stagnation. Clearly, it has become difficult to keep up with the literature on boronic acids, and it is anticipated that the second edition of *Boronic Acids* will be of invaluable assistance. In the past 5 years alone, impressive new advances have been made in the use of boronic acids in molecular recognition, chemical biology, materials science, and catalysis. Compared to the first edition, this second edition consists of chapters containing entirely new material, replacing pre-

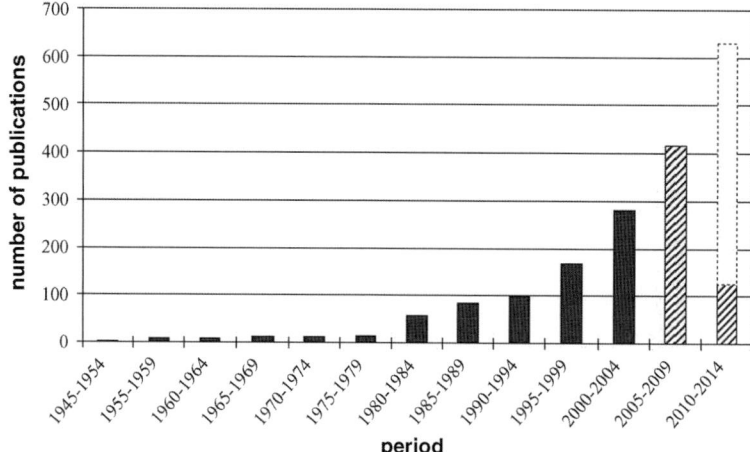

**Figure P.1** Number of publications focused on boronic acids over time (note that only those publications carrying the word "boronic" in their titles were included). Hatched bars indicate the volume of publications reported after publication of the first edition of *Boronic Acids*.

vious chapters that have seen less advances in recent years, while other chapters have been completely revised and updated.

Our understanding of the structure and properties of boronic acids, their important ester derivatives, and other parent compounds such as trifluoroborate salts, is described in Chapter 1. In the past, the limited number of methods for the preparation of boronic acid derivatives had long impeded their use as synthetic reagents. This has changed drastically, and Chapter 1 describes modern methods for the preparation of all types of boronic acid derivatives, including several useful tables of examples. It also provides an overview of their synthetic, biological, and medicinal applications. One of the latest advances in the preparation of boronic acids, the use of transition metal-catalyzed borylation of C–X and C–H bonds, is discussed in Chapter 2. In the same manner, Chapter 3 describes metalloborylation reactions of unsaturated compounds. Much has happened in the development of new conditions and catalysts to expand the scope of transition metal-catalyzed C–C bond formation processes using boronic acids. Chapter 4 describes the most recent advances in the Suzuki–Miyaura cross-coupling reaction. A few years ago, rhodium(I) complexes were found to catalyze the addition of boronic acids to enones and aldehydes. These discoveries have now flourished into highly efficient catalytic enantioselective processes that can afford functionalized products with over 99% optical purity. These impressive advances are reviewed in Chapter 5. The copper-catalyzed coupling of boronic acids with heteroatom functionalities, such as phenols, amines, and amides, is yet another recent synthetic application that has contributed to the emergence of boronic acids as a popular class of reagents. This new and useful process, described in Chapter 6, has already become firmly established in the synthesis of natural products and in medicinal chemistry research. Chapter 7 describes the Liebeskind–Srogl cross-coupling reaction between boronic acids and thioorganic compounds.

Already a workhorse in the synthesis of polypropionate compounds, the addition of allylboronates to carbonyl compounds and imine derivatives is still getting increasing attention as a result of new modes of catalytic activation highlighted in Chapter 8. The important discovery that boronic acids add to imine derivatives and iminium ions, even in a three-component fashion, has been exploited in a number of synthetic applications and progress in this area is reviewed in Chapter 9. Described in Chapter 10 is a new twist to the seminal Matteson homologation of boronic esters, using novel reagents such as lithiated carbamates, epoxides, and aziridines. Chapter 11 presents an overview of the synthetic applications of organotrifluoroborates, boronic acid derivatives that have become a very popular class of stable and efficient "go-to" reagents in cross-coupling chemistry. Boronic acids and several of their ester derivatives can serve as stable and mild Lewis acids, and this unique property has inspired the development of catalysts for several reaction processes, including asymmetric transformations; this topic is reviewed in Chapter 12. Boronic acids have long been known to bind and inhibit the action of certain classes of proteolytic enzymes. This important topic, as well as the applications of boronic acids in chemical biology, is discussed in Chapter 13 along with other emerging therapeutic applications. Finally, Chapter 14 presents an impressive overview of all the recently uncovered potential of boronic acids in materials science. From the rich contents of this book, it is clear that the spectacular rise of boronic acids as a class of compounds may have just begun. It is hoped that the second edition will contribute to generating more work and continue to attract more researchers to the field.

The success of a book project relies heavily on the involvement of several dedicated individuals. I would like to thank all authors and coauthors who have generously agreed to contribute a chapter. Their expertise and professionalism were invaluable assets to this ambitious project. Grateful acknowledgements are also offered to the Wiley-VCH editorial staff, in particular to Elke Maase and Renate Doetzer in the first edition, and Bernadette Gmeiner for the newer, second edition. For their valued support in various stages of editing this book I am also indebted to thanking Jack Lee, Jin-Yong Lu, Ho-Yan Sun, Hongchao Zheng, and Nitin Vashisht.

Edmonton (Alberta, Canada)  *Dennis Hall*
May 2011

# List of Contributors

**Varinder K. Aggarwal**
University of Bristol
School of Chemistry
Cantock's Close
Bristol BS8 1TS
UK

**Robert A. Batey**
University of Toronto
Department of Chemistry
Davenport Research Laboratories
80 St. George Street
Toronto, Ontario, M5S 3H6
Canada

**Guillaume Berthon-Gelloz**
Syngenta Crop Protection
Münchwilen AG
Schaffhauserstrasse
4332 Stein
Switzerland

**Tim G. Elford**
Department of Chemistry
University of Alberta
Edmonton, Alberta, T6G 2G2
Canada

**Ethel C. Garnier-Amblard**
Emory University
Sanford S. Atwood Chemistry Center
1515 Dickey Drive
Atlanta, GA 30322
USA

**Dennis G. Hall**
4-010 Centennial Centre for
Interdisciplinary Science
Department of Chemistry
University of Alberta
Edmonton, Alberta, T6G 2G2
Canada

**Tamio Hayashi**
Kyoto University
Graduate School of Science
Department of Chemistry
Sakyo
Kyoto 606-8502
Japan

**Tatsuo Ishiyama**
Hokkaido University
Graduate School of Engineering
Division of Chemical Process
Engineering
Kita 13, Nishi 8
Sapporo, Hokkaido 060-8628
Japan

**Ludivine Jean-Gérard**
University of Pennsylvania
Department of Chemistry
231 South 34th Street
Philadelphia, PA 19104-6323
USA

**Patrick Y.S. Lam**
Bristol Myers Squibb
Pharmaceutical Co.
Discovery Chemistry
Princeton, NJ 08543-5400
USA

**John J. Lavigne**
University of South Carolina
Department of Chemistry and
Biochemistry
631 Sumter Street
Columbia, SC 29208
USA

**Lanny S. Liebeskind**
Emory University
Sanford S. Atwood Chemistry Center
1515 Dickey Drive
Atlanta, GA 30322
USA

**Jie Liu**
University of South Carolina
Department of Chemistry and
Biochemistry
631 Sumter Street
Columbia, SC 29208
USA

**Norio Miyaura**
Hokkaido University
Graduate School of Engineering
Division of Chemical Process
Engineering
Kita 13, Nishi 8
Sapporo, Hokkaido 060-8628
Japan

**Gary A. Molander**
University of Pennsylvania
Department of Chemistry
231 South 34th Street
Philadelphia, PA 19104-6323
USA

**Nanting Ni**
Georgia State University
Department of Chemistry and Center
for Biotechnology and Drug Design
Atlanta, GA 30302-4098
USA

**Toshimichi Ohmura**
Kyoto University
Graduate School of Engineering
Department of Synthetic Chemistry and
Biological Chemistry
Katsura, Nishikyo-ku
Kyoto 615-8510
Japan

**Michael G. Organ**
York University
Department of Chemistry
4700 Keele Street
Toronto, Ontario, M3J 1P3
Canada

**Joshua N. Payette**
The University of Chicago
Department of Chemistry
5735 South Ellis Avenue
Chicago, IL 60637
USA

**Jennifer X. Qiao**
Bristol Myers Squibb
Pharmaceutical Co.
Discovery Chemistry
Princeton, NJ 08543-5400
USA

# List of Contributors

**Timothy R. Ramadhar**
University of Toronto
Department of Chemistry
Davenport Research Laboratories
80 St. George Street
Toronto, Ontario, M5S 3H6
Canada

**Michinori Suginome**
Kyoto University
Graduate School of Engineering
Department of Synthetic Chemistry
and Biological Chemistry
Katsura, Nishikyo-ku
Kyoto 615-8510
Japan

**Cory Valente**
Northwestern University
Department of Chemistry
2145 Sheridan Road
Evanston, IL 60208
USA

**Binghe Wang**
Georgia State University
Department of Chemistry and Center
for Biotechnology and Drug Design
Atlanta, GA 30302-4098
USA

**Matthew P. Webster**
University of Illinois
School of Chemical Sciences Chemistry
Roger Adams Lab, Room 237, Box 90-5
600 S Mathews
Urbana, IL 61801
USA

**Hisashi Yamamoto**
The University of Chicago
Department of Chemistry
5735 South Ellis Avenue
Chicago, IL 60637
USA

# 1
## Structure, Properties, and Preparation of Boronic Acid Derivatives
Overview of Their Reactions and Applications

*Dennis G. Hall*

### 1.1
### Introduction and Historical Background

Structurally, boronic acids are trivalent boron-containing organic compounds that possess one carbon-based substituent (i.e., a C−B bond) and two hydroxyl groups to fill the remaining valences on the boron atom (Figure 1.1). With only six valence electrons and a consequent deficiency of two electrons, the $sp^2$-hybridized boron atom possesses a vacant p-orbital. This low-energy orbital is orthogonal to the three substituents, which are oriented in a trigonal planar geometry. Unlike carboxylic acids, their carbon analogues, boronic acids, are not found in nature. These abiotic compounds are derived synthetically from primary sources of boron such as boric acid, which is made by the acidification of borax with carbon dioxide. Borate esters, one of the key precursors of boronic acid derivatives, are made by simple dehydration of boric acid with alcohols. The first preparation and isolation of a boronic acid was reported by Frankland in 1860 [1]. By treating diethylzinc with triethylborate, the highly air-sensitive triethylborane was obtained, and its slow oxidation in ambient air eventually provided ethylboronic acid. Boronic acids are the products of a twofold oxidation of boranes. Their stability to atmospheric oxidation is considerably superior to that of borinic acids, which result from the first oxidation of boranes. The product of a third oxidation of boranes, boric acid, is a very stable and relatively benign compound to humans (Section 1.2.2.3).

Their unique properties and reactivity as mild organic Lewis acids, coupled with their stability and ease of handling, are what make boronic acids a particularly attractive class of synthetic intermediates. Moreover, because of their low toxicity and their ultimate degradation into boric acid, boronic acids can be regarded as "green" (environment-friendly) compounds. They are solids, and tend to exist as mixtures of oligomeric anhydrides, in particular the cyclic six-membered boroxines (Figure 1.1). For this reason and other considerations outlined later in this chapter, the corresponding boronic esters are often preferred as synthetic intermediates. Although other classes of organoboron compounds have found tremendous utility

---

*Boronic Acids: Preparation and Applications in Organic Synthesis, Medicine and Materials*, Second Edition.
Edited by Dennis G. Hall.
© 2011 Wiley-VCH Verlag GmbH & Co. KGaA. Published 2011 by Wiley-VCH Verlag GmbH & Co. KGaA.

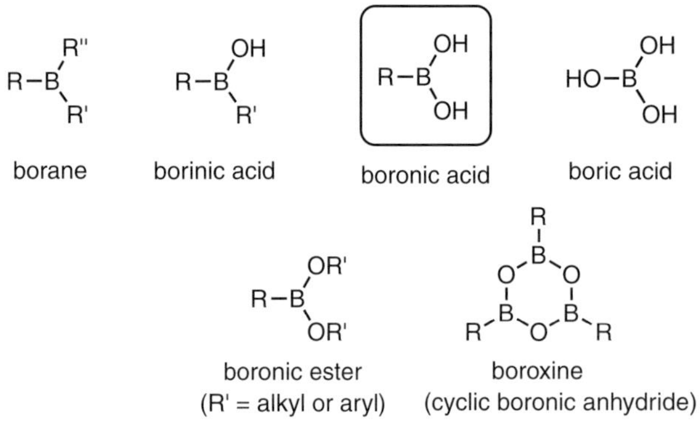

**Figure 1.1** Oxygen-containing organoboron compounds.

in organic synthesis, this book focuses on the most recent applications of the convenient boronic acid derivatives. For a comprehensive description of the properties and reactivity of other classes of organoboron compounds, interested readers may refer to a selection of excellent monographs and reviews by Brown [2], Matteson [3], and others [4–8]. In the past two decades, the status of boronic acids in chemistry has gone from that of peculiar and rather neglected compounds to that of a prime class of synthetic intermediates in their own right. The attribution of the 2010 Chemistry Nobel Prize for palladium-catalyzed cross-coupling reactions to Professor Akira Suzuki and other pioneers recognized the great importance of boronic acids in this revolutionary class of C−C bond forming processes. In the past 5 years, impressive advances have been made in the use of boronic acids in molecular recognition, materials science, and catalysis. The approval of the anticancer agent Velcade®, the first boronic acid-containing drug to be commercialized (Section 1.6.5), further confirms the growing status of boronic acids as an important class of compounds in chemistry and medicine. This chapter describes the structural and physicochemical properties of boronic acids and their many derivatives, as well as modern methods for their preparation. A brief overview of their synthetic and biological applications is presented, with an emphasis on topics that are not covered in other chapters of this book.

## 1.2
### Structure and Properties of Boronic Acid Derivatives

#### 1.2.1
#### General Types and Nomenclature of Boronic Acid Derivatives

The reactivity and properties of boronic acids highly depend upon the nature of their single variable substituent, more specifically, on the type of carbon group (R, Figure 1.1) directly bonded to boron. In the same customary way employed for other functional groups, it is convenient to classify boronic acids into subtypes such as alkyl-, alkenyl-, alkynyl-, and arylboronic acids.

When treated as an independent substituent, the prefix borono is employed to name the boronyl group (e.g., 3-boronoacrolein). For cyclic derivatives such as boronic esters, the IUPAC RB-1-1 rules for small heterocycles (i.e., the Hantzsch–Widman system) are employed along with the prefix "boro." Thus, saturated five- and six-membered cyclic boronic esters are, respectively, named as dioxaborolanes and dioxaborinanes. For example, the formal name of the pinacol ester of phenylboronic acid is 2-phenyl-4,4,5,5-tetramethyl-1,3,2-dioxaborolane. The corresponding nitrogen analogues are called diazaborolidines and diazaborinanes, and the mixed nitrogen–oxygen heterocycles are denoted by the prefix oxaza. Unsaturated heterocycles wherein the R group and the boron atom are part of the same ring are named as boroles.

### 1.2.2
### Boronic Acids

#### 1.2.2.1 Structure and Bonding

The X-ray crystal structure of phenylboronic acid (**1**, Figure 1.2) was reported in 1977 by Rettig and Trotter [9]. The crystals are orthorhombic, and each asymmetric unit was found to consist of two distinct molecules, bound together through a pair of O–H–O hydrogen bonds (Figure 1.3a and b). The $CBO_2$ plane is quite coplanar with the benzene ring, with a respective twist around the C–B bond of 6.6° and 21.4° for the two independent molecules of $PhB(OH)_2$. Each dimeric ensemble is also linked with hydrogen bonds to four other similar units to give an infinite array of layers

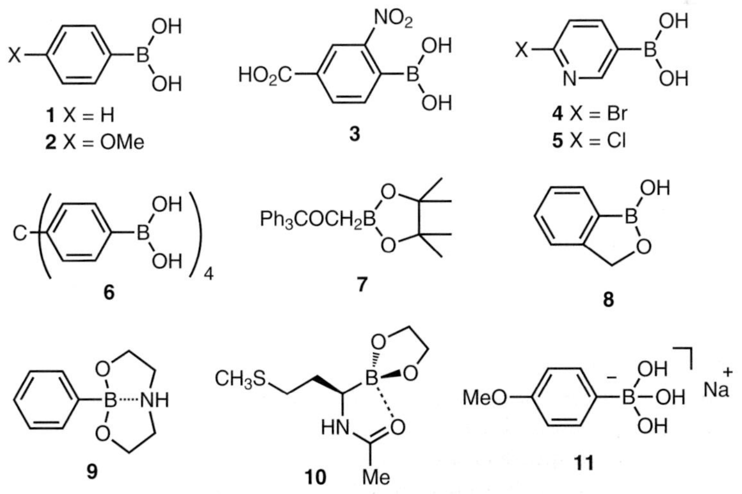

**Figure 1.2** Boronic acid derivatives analyzed by X-ray crystallography.

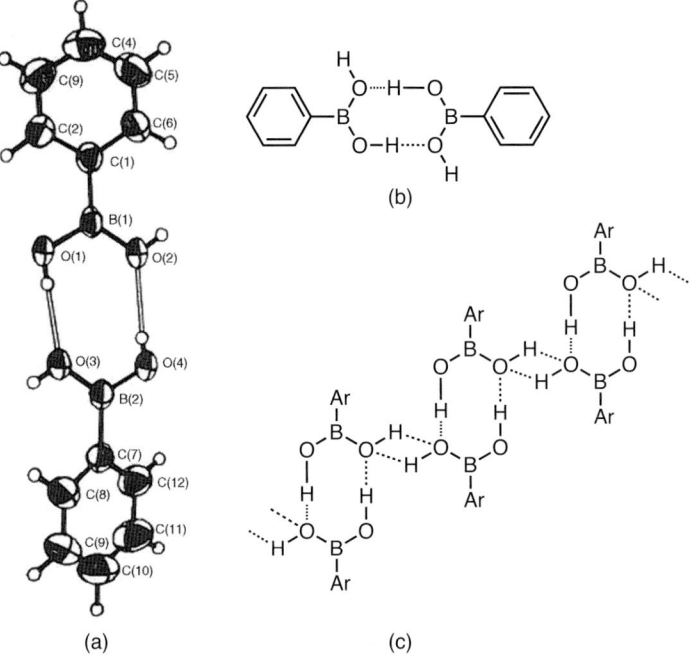

**Figure 1.3** Representations of the X-ray crystallographic structure of phenylboronic acid. (a) ORTEP view of a dimeric unit. (b) Structure of the dimeric unit showing hydrogen bonds. (c) Structure of the extended hydrogen-bonded network.

(Figure 1.3c). The X-ray crystallographic analysis of other arylboronic acids like *p*-methoxyphenyl boronic acid (**2**) [10] and 4-carboxy-2-nitrophenylboronic acid (**3**, Figure 1.2) [11] is consistent with this pattern. The structures of heterocyclic boronic acids such as 2-bromo- and 2-chloro 5-pyridylboronic acids (**4** and **5**) were reported [12]. Although the boronate group has a trigonal geometry and is fairly coplanar with the benzene ring in structures **1**, **2**, **4**, and **5**, it is almost perpendicular to the ring in structure **3**. This observation is likely due to a combination of two factors: minimization of steric strain with the *ortho*-nitro group, and because of a possible interaction between one oxygen of the nitro group and the trigonal boron atom. Based on the structural behavior of phenylboronic acid and its propensity to form hydrogen-bonded dimers, diamond-like porous solids were designed and prepared by the crystallization of tetrahedral-shaped tetraboronic acid **6** (Figure 1.2) [13]. With a range of approximately 1.55–1.59 Å, the C–B bond of boronic acids and esters is slightly longer than typical C–C single bonds (Table 1.1). The average C–B bond energy is also slightly smaller than that of C–C bonds (323 versus 358 kJ/mol) [14]. Consistent with strong B–O bonds, the B–O distances of tricoordinate boronic acids such as phenylboronic acid are fairly short, and lie in the range of 1.35–1.38 Å (Table 1.1). These values are slightly larger than those observed in boronic esters. For example, the B–O bond distances observed in the X-ray crystal-

## 1.2 Structure and Properties of Boronic Acid Derivatives

**Table 1.1** Bond distances from X-ray crystallographic data for selected boronic acid derivatives (Figure 1.2).

| Compound | B—C (Å) | B—O$^1$ (Å) | B—O$^2$ (Å) | B—X (Å) | Reference |
|---|---|---|---|---|---|
| 1  | 1.568 | 1.378 | 1.362 |       | [9]  |
| 2  | 1.556 |       |       |       | [10] |
| 3  | 1.588 | 1.365 | 1.346 |       | [11] |
| 4  | 1.573 | 1.363 | 1.357 |       | [12] |
| 5  | 1.573 | 1.362 | 1.352 |       | [12] |
| 7  | 1.560 | 1.316 | 1.314 |       | [15] |
| 8  | 1.494 | 1.408 | 1.372 |       | [16] |
| 9  | 1.613 | 1.474 | 1.460 | 1.666 | [18] |
| 10 | 1.613 | 1.438 | 1.431 | 1.641 | [22] |
| 11 | 1.631 | 1.492 | 1.487 | 1.471 | [23] |

lographic structures of the trityloxymethyl pinacolate boronic esters (e.g., **7** in Figure 1.2) are in the range of 1.31–1.35 Å (Table 1.1), and the dioxaborolane unit of these derivatives is nearly planar [15]. The X-ray crystallographic structure of cyclic hemiester **8** (Figure 1.2) was described [16]. Like phenylboronic acid, this benzoxaborole also crystallizes as a hydrogen-bonded dimer, however without the extended network due to the absence of a second hydroxyl group. The cyclic nature of this derivative induces a slight deviation from planarity for the tricoordinate boronate unit, as well as a distortion of the bond angles. The endocyclic B—O bond in **8** is slightly longer than the B—OH bond. This observation was attributed to the geometrical constraints of the ring, which prevents effective lone pair conjugation between the endocyclic oxygen and the vacant orbital of boron. The unique properties and reactivity of benzoxaboroles along with their preparation were recently reviewed [17].

In order to complete boron's octet, boronic acids and their esters may also coordinate basic molecules and exist as stable tetracoordinated adducts. For example, the X-ray crystallographic structure of the diethanolamine adduct of phenylboronic acid (**9**, Figure 1.2) [18] confirmed the transannular B—N bridge long suspected from other spectroscopic evidence such as NMR [19, 20]. This dative B—N bond has a length of 1.67 Å (Table 1.1), and it induces a strong $N^{\delta+}-B^{\delta-}$ dipole that points away from the plane of the aryl ring. This effect was elegantly exploited in the design of a diboronate receptor for paraquat [21]. Chelated boronic ester **10** presents characteristics similar to that of **9** [22]. Trihydroxyborate salts of boronic acids are discrete, isolable derivatives that had not been characterized until recently [23]. The sodium salt of *p*-methoxyphenyl boronic acid (**11**) was recrystallized in water and its X-ray structural elucidation showed the borate unit in the expected hydrogen bonding network accompanied with the sodium cation coordinated with six molecules of water. In principle, the boron atom in tetrahedral complexes can be stereogenic if it is bonded to four different ligands. Hutton and coworkers recently reported the first example of one such optically pure complex stereogenic at boron only [24]. Stable complex **12** (Figure 1.4) was made through a chirality transfer process described in

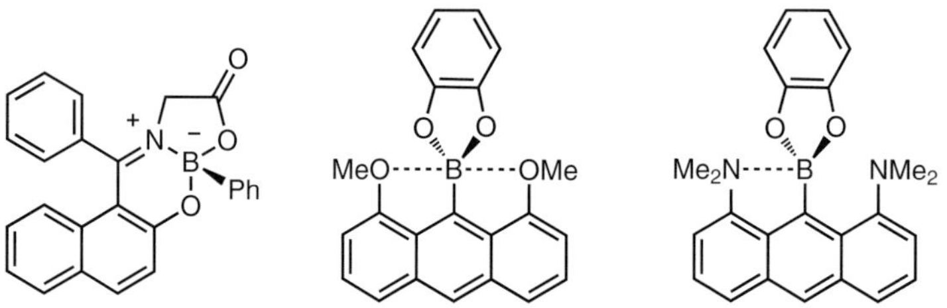

**Figure 1.4** B-Chiral tetrahedral boronate **12** and model compounds for boron hypercoordination.

Section 1.2.3.6. When tetracoordinated such as in structures **9–11** [23] (Figure 1.2), the B–O bond length of boronic acids and esters increases to about 1.43–1.48 Å, which is as much as 0.10 Å longer than the corresponding tricoordinate analogues (Table 1.1). These markedly longer B–O bonds are comparable to normal C–O ether bonds (~1.43 Å). These comparisons further emphasize the considerable strength of B–O bonds in trigonal boronic acid derivatives. Not surprisingly, trigonal B–O bonds are much stronger than the average C–O bonds of ethers (519 versus 384 kJ/mol) [14]. This bond strength is believed to originate from the conjugation between the lone pairs on the oxygens and boron's vacant orbital, which confers partial double bond character to the B–O linkage. In fact, it was estimated that formation of tetrahedral adducts (e.g., with $NH_3$) may result in a loss of as much as 50 kJ/mol of B–O bond energy compared to the tricoordinate boronate [25].

In rare instances where geometrical factors allow it, boronic acid derivatives may become hypervalent. For example, the catechol ester **13** (Figure 1.4) was found by X-ray crystallographic analysis to be pentacoordinated in a highly symmetrical fashion as a result of the rigidly held ether groups, which are perfectly positioned to each donate lone pair electrons to both lobes of the vacant p-orbital of boron [26]. The boronyl group of this two electron–three atom center is planar, in a $sp^2$ hybridization state, and the resulting structure possesses a slightly distorted trigonal bipyramidal geometry. According to DFT calculations, the bonding is weak and ionic in nature [26b]. The corresponding diamine **14**, however, behaved quite differently and demonstrated coordination with only one of the two $NMe_2$ groups [27].

Due to electronegativity differences (B = 2.05, C = 2.55) and notwithstanding the electronic deficiency of boron, which is compensated by the two electron-donating oxygen atoms (see above), the inductive effect of a boronate group should be that of a weak electron donor. The $^{13}C$ NMR alpha effect of a boronate group is in fact very small [28]. On the other hand, the deficient valency of boron and its size relatively similar to that of carbon have long raised the intriguing question of possible pi-bonding between carbon and boron in aryl- and alkenylboronic acids and esters [29]. NMR data and other evidence, such as UV and photoelectron spectroscopy and LCAO-MO calculations, suggest that B–C pi-conjugation occurs to a moderate extent in alkenylboranes [30–32], and is even smaller in the case of the considerably less acidic boronate derivatives. A thorough comparative study of $^{13}C$ NMR shift effects,

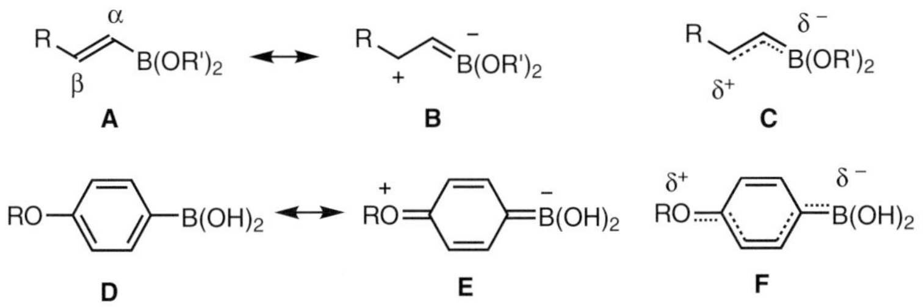

## 1.2 Structure and Properties of Boronic Acid Derivatives

**Figure 1.5** Limit mesomeric forms involving B–C pi-overlap.

in particular the deshielding of the beta-carbon, concluded to a certain degree of mesomeric pi-bonding in the case of boranes and catechol boronates [28]. For example, compared to analogous aliphatic boronates, the beta-carbons of a dialkyl alkenylboronate and the corresponding catechol ester are deshielded by 8.6 and 18.1 ppm, respectively. In all cases, the beta-carbon is more affected by the boronate substituent than the alpha-carbon, which is consistent with some contribution from the B–C pi-bonded form **B** to give resonance hybrid **C** (Figure 1.5). X-ray crystallography may also provide insights into the extent of B–C pi-bonding. The difference in B–C bond distances for arylboronic acids (Table 1.1) is significant enough to suggest a small degree of B–C pi-bonding. The B–C bond distance (1.588 Å) in the electron-poor boronic acid **3**, which is incapable of pi-conjugation because it has its vacant p-orbital placed orthogonally to the pi-system of the phenyl ring, is expectedly longer than that of phenylboronic acid (1.568 Å). Interestingly, the B–C bond of **2** stands at 1.556 Å, suggesting only a minimal contribution from the mesomeric form **E** (Figure 1.5). On the other hand, the B–C bond distance of 1.613 Å in the diethanolamine adduct **9** (Table 1.1), where the boron vacant orbital is also incapacitated from B–C pi-bonding, is 0.045 Å longer than that of free phenylboronic acid (**1**). In so far as bond length data correlate with the degree of pi-bonding [33], this comparison is consistent with a small B–C pi-bonding effect in arylboronic acids and esters (i.e., hybrid form **F** in Figure 1.5). This view is further supported by chemical properties such as substituent effects on the acidity of arylboronic acids (see Section 1.3.8.3) and $^{11}$B chemical shifts correlations [34]. Likewise, B–C pi-bonding is also present in alkenylboronic acids and esters, but this effect must be weak in comparison to the electron-withdrawing effect of a carbonyl or a carboxyl group. For instance, alkenylboronic esters do not readily act as Michael acceptors with organometallic reagents in the same way as the unsaturated carbonyl compounds do [35]. On the other hand, the formal electron-withdrawing behavior of the boronate group manifests itself in cycloadditions of dibutylethylene boronate with ethyldiazoacetate [36] and in Diels–Alder reactions where it provides cycloadducts with dienes like cyclopentadiene [37] and cyclohexadiene, albeit only at elevated temperatures (about 130 and 200 °C, respectively) [38, 39]. The higher reactivity of ethylene boronates as dienophiles compared to ethylene has been rationalized by MO calculations [29], but their reactivity stands far from that of acrylates in the same cycloadditions. In fact, more recent high-level calculations suggest that the reactivity of alkenylboronates may be mainly due to a three-atom–two-electron center stabilization of the transition state rather than a true LUMO-lowering electron-withdrawing mesomeric effect from the boronate substituent [40]. Another evidence for the rather weak electron-withdrawing character of boronic esters comes from their modest stabilizing effect on boronyl-substituted carbanions, where their effect has been compared to that of a phenyl group (see Section 1.3.8.3).

#### 1.2.2.2 Physical Properties and Handling

Most boronic acids exist as white crystalline solids that can be handled in air without special precautions. At the ambient temperature, boronic acids are chemically stable and most display shelf stability for long periods of time (Section 1.2.2.5). Alkyl-substituted and some heteroaromatic boronic acids, however, were shown to have a limited shelf stability under aerobic conditions [41]. Boronic acids normally do not tend to disproportionate into their corresponding borinic acid and boric acid even at high temperatures. To minimize atmospheric oxidation and autoxidation, however, they should be stored under an inert atmosphere. When dehydrated, either with a water-trapping agent or through coevaporation or high vacuum, boronic acids form cyclic and linear oligomeric anhydrides such as the trimeric boroxines already mentioned (Figure 1.1). Fortunately, this behavior is usually inconsequential when boronic acids are employed as synthetic intermediates. Many of their most useful reactions (Section 1.5), including the Suzuki-Miyaura cross-coupling, proceed regardless of the hydrated state (i.e., free boronic acid or anhydride). Anhydride formation, however, may complicate analysis, quantitation, and characterization efforts (Section 1.4.3). Furthermore, upon exposure to air, dry samples of boronic acids may be prone to decompose rapidly, and it has been proposed that boronic anhydrides may be initiators of the autoxidation process [42]. For this reason, it is often better to store boronic acids in a slightly moist state. Presumably, coordination of water or hydroxide ions to boron protects boronic acids from the action of oxygen [42, 43]. Incidentally, commercial samples tend to contain a small percentage of water that may help in their long-term preservation. Due to their facile dehydration, boronic acids tend to provide somewhat unreliable values of melting points (Section 1.4.3.1). This inconvenience and the other above-mentioned problems associated with anhydride formation explain in large part the popularity of boronic esters and other derivatives as surrogates of boronic acids (Section 1.2.3.2).

The Lewis acidity of boron in boronic acids and the hydrogen bond donor capability of their hydroxyl groups combine to lend a polar character to most of these compounds. Although the polarity of the boronic acid head can be mitigated by a relatively hydrophobic tail as the boron substituent, most small boronic acids are amphiphilic. Phenylboronic acid, for instance, was found to have a benzene–water partition ratio of 6 [44]. The partial solubility of boronic acids in both neutral water and polar organic solvents often complicates isolation and purification efforts (Section 1.4). Evidently, boronic acids are more water soluble in their ionized form in high-pH aqueous solutions and can be extracted more readily into organic solvents from aqueous solutions of low pH (see Section 1.2.2.4).

### 1.2.2.3 Safety Considerations

As evidenced by their application in medicine (Chapter 13), most boronic acids present no particular toxicity compared to other organic compounds [45]. Small water-soluble boronic acids demonstrate low toxicity levels, and are excreted largely unchanged by the kidney [46]. Larger fat-soluble boronic acids were found to be moderately toxic [46–48]. At high doses, boronic acids may interact promiscuously with nucleophilic enzymes and complex weakly to biological diols (Section 1.2.3.2.3). Boronic acids present no particular environmental threat, and the ultimate fate of all boronic acids in air and aqueous media is their slow oxidation into boric acid. The latter is a relatively innocuous compound, and may be toxic only under high daily doses [49]. A single acute ingestion of boric acid does not even pose a threatening poisoning effect to humans [50] unless it is accompanied by other health malfunctions such as dehydration [51].

### 1.2.2.4 Acidic Character

By virtue of their deficient valence, boronic acids possess a vacant p-orbital. This characteristic confers them unique properties as a mild class of organic Lewis acids capable of coordinating basic molecules. When doing so, the resulting tetrahedral adducts acquire a carbon-like configuration. Thus, despite the presence of two hydroxyl groups, the acidic character of most boronic acids is not that of a Brønsted acid (i.e., oxyacid) (Equation 1.1, Figure 1.6) but usually that of a Lewis acid (Equation 1.2). When coordinated with an anionic ligand, the resulting negative charge is formally drawn on the boron atom, but it is in fact spread out on the three heteroatoms.

#### 1.2.2.4.1 Complexation Equilibrium in Water and Structure of the Boronate Anion

Boronic acids are more soluble in aqueous solutions of high pH (>8). Although the acidic character of boronic acids in water had been known for several decades, it is only in 1959 that the structure of the boronate ion, the conjugate base, was elucidated. In their classical paper on polyol complexes of boronic acids [52], Lorand and Edwards demonstrated that the trivalent neutral form, likely hydrated, is in equilibrium with the anionic tetrahedral species (Equation 1.2, Figure 1.6) and not with the structurally related Brønsted base (i.e., the trivalent ion shown in Equation 1.1). The first X-ray crystallographic structure of a trihydroxyboronate salt has been reported recently (**11** in Figure 1.2) [23]. It is this ability to ionize water and form hydronium ions by "indirect" proton transfer that characterizes the acidity of most boronic acids in water. Hence, the most acidic boronic acids possess the most

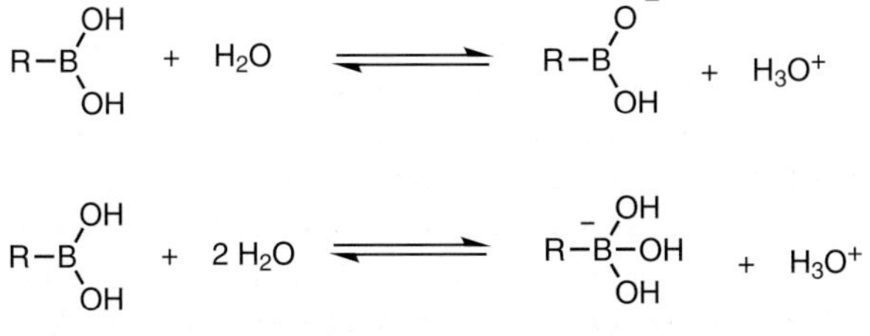

$$R-B(OH)_2 + H_2O \rightleftharpoons R-B(O^-)(OH) + H_3O^+ \quad (1.1)$$

$$R-B(OH)_2 + 2 H_2O \rightleftharpoons R-B(OH)_3^- + H_3O^+ \quad (1.2)$$

**Figure 1.6** Ionization equilibrium of boronic acids in water.

**Table 1.2** Ionization constant ($pK_a$) for selected boronic acids.

| Boronic acid, RB(OH)$_2$ | $pK_a$ | Reference |
| --- | --- | --- |
| Boric acid, B(OH)$_3$ | 9.0 | [58] |
| Methyl | 10.4 | [58] |
| Phenyl | 8.9 | [59] |
| 3,5-Dichlorophenyl | 7.4 | [59] |
| 3,5-Bis(trifluoromethyl)phenyl | 7.2 | [59] |
| 2-Methoxyphenyl | 9.0 | [57] |
| 3-Methoxyphenyl | 8.7 | [59] |
| 4-Methoxyphenyl | 9.3 | [60] |
| 4-Carboxyphenyl | 8.4 | [56] |
| 2-Nitrophenyl | 9.2 | [61] |
| 4-Nitrophenyl | 7.1 | [60] |
| 4-Bromophenyl | 8.6 | [59] |
| 4-Fluorophenyl | 9.1 | [59] |
| 2-Methylphenyl | 9.7 | [62] |
| 3-Methylphenyl | 9.0 | [62] |
| 4-Methylphenyl | 9.3 | [62] |
| 3,5-Dimethylphenyl | 9.1 | [59] |
| 3-Methoxycarbonyl-5-nitrophenyl | 6.9 | [63] |
| 2-Fluoro-5-nitrophenyl | 6.0 | [57] |
| 3-Pyridyl (**15**) | 4.0, 8.2 | [64] |
| 3-Benzyl-3-pyridylium | 4.2 | [57] |
| 8-Quinolinyl | 4.0, 10 | [65] |
| 2-($R^1R^2NCH_2$)phenyl (e.g., **16**) | 5.2–5.8 | [66] |

electrophilic boron atom that can best form and stabilize a hydroxyboronate anion. The acidity of boronic acids in water has been measured using electrochemical methods as early as the 1930s [53–55]. Values of $pK_a$ are now measured more conveniently by UV spectrophotometry [56] and $^{11}B$ NMR spectroscopy. Phenylboronic acid, with a $pK_a$ value of 8.9 in water, has an acidity comparable to a phenol (Table 1.2). It is slightly more acidic than boric acid ($pK_a$ 9.2). With the $pK_a$ values as shown in Table 1.2, the relative order of acidity for the different types of boronic acids is aryl > alkyl. More values can be found elsewhere [57]. For *para*-monosubstituted aromatic boronic acids, the relationship between the $pK_a$ and the electronic nature of the substituent can be described with a Hammet plot [57]. Bulky substituents proximal to the boronyl group can decrease the acid strength due to steric inhibition in the formation of the tetrahedral boronate ion. For example, *ortho*-tolylboronic acid is slightly less acidic than its *para*-isomer ($pK_a$ 9.7 versus 9.3, Table 1.2) [62]. This difference was explained in terms of F-strain in the resulting ion (Equation 1.3, Figure 1.7) [67]. As expected, the presence of electron-withdrawing substituents in the aryl group of arylboronic acids increases the acid strength by a fairly significant measure [53, 55, 60, 68]. For example, the highly electron-poor 3-methoxycarbonyl-5-nitrophenyl boronic acid was attributed a $pK_a$ value of 6.9 [63]. Exceptionally, *ortho*-nitrobenzeneboronic acid [61] is much less acidic than its *para*-isomer [60] ($pK_a$ 9.2 versus 7.1, Table 1.2) presumably due to internal coordination of

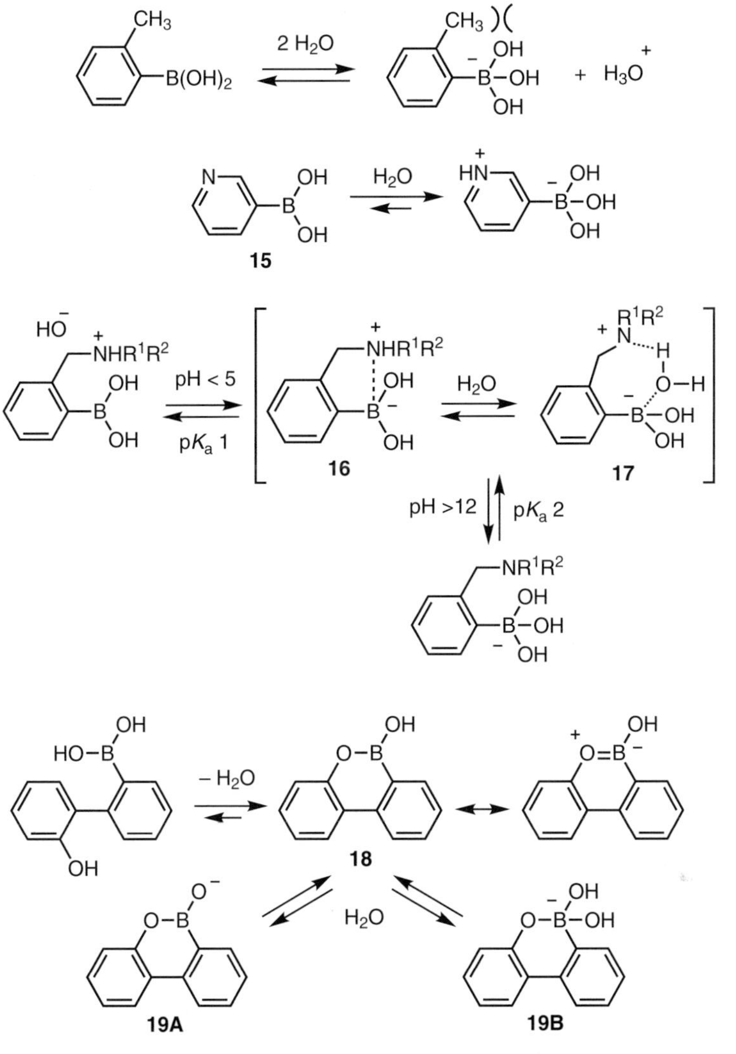

**Figure 1.7** Ionization equilibrium of special boronic acids.

one of the nitro oxygens that prevents the complexation of a hydroxyl anion [55]. Perhaps one of the most acidic of all known boronic acids, with a p$K_a$ of approximately 4.0, 3-pyridylboronic acid (**15**) exists mainly as a zwitterion in water (Equation 1.4, Figure 1.7) [64]. Similarly, arylboronic acids of type **16** (Equation 1.5), which benefit from anchimeric participation of the *ortho*-dialkylaminomethyl group, display a relatively low value of p$K_a$ of about 5.2 [66]. In this case, the actual first p$K_a$ is that of ammonium ion deprotonation and formation of the putative tetrahedral B−N ate adduct **16**. The latter form was shown to exist in organic solvents, but in water and other hydroxylic solvents, complex **17** forms through a water-insertion mechanism [69]. The application of boronic acids of type **16** in the aqueous recognition of saccharides is briefly discussed in Chapter 13. Fluoride ions also form strong dative bonds with boron, and it has been noted long ago that boronic acids dissolved in aqueous solutions of hydrofluoric acid are very difficult to extract into organic solvents unless the fluoride is precipitated out [70].

Boronic acids display Brønsted acidity (cf. Equation 1.1, Figure 1.6) only in exceptional cases where the formation of a tetrahedral boronate adduct is highly unfavorable. For example, coordination of hydroxide ion to boron in heterocyclic boronic acid derivative **18**, to form **19B**, would break the partial aromatic character of the central ring (Equation 1.6, Figure 1.7). Indeed, based on $^{11}$B NMR and UV spectroscopic evidence, it was suggested that **18** acts as a Brønsted acid in water and forms conjugate base **19A** through direct proton transfer [71]. A small number of other boronic acids are suspected of behaving as Brønsted acids due to the same reasons [72].

#### 1.2.2.4.2 Bimolecular Lewis Acid–Base Complexation under Nonaqueous Conditions

As evidenced by the high pH required in the formation of boronate anions, boronic acids and most dialkyl esters are weak Lewis acids. This behavior is in sharp contrast with trialkylboranes, which form strong adducts with phosphines, amines, and other Lewis bases [73]. Apart from the formation of boronate anions, discussed in the previous section, very few examples of stable intermolecular acid–base adducts of boronic acids (esters) exist. It has been known for a long time that aliphatic amines and pyridine can form complexes in a 1 : 3 amine:boronic acid stoichiometry [74]. Combustion analyses of these air-stable solids suggested that two molecules of water are lost in the process, which led the authors to propose structure **20** (Equation 1.7, Figure 1.8). Much later, Snyder *et al.* used IR spectroscopy to demonstrate that these 1 : 3 complexes rather involved the fully dehydrated boroxine (**21**) [75]. Boronic esters are generally weak Lewis acids but catechol boronates are quite acidic, and provided that cooperative effects are exploited, bimolecular complexes with fluoride anions and amines have been reported [76–78]. The B−F bond strength is a key factor in these complexes as other halide salts do not form similar adducts. As suggested by $^1$H NMR spectroscopic studies, an *ortho*-phenyldiboronic ester (**22**) showed cooperative binding of two amine molecules in putative complex **24** (Equation 1.8, Figure 1.8) [79]. Other diboronate receptors were found to bind to diamines selectively using the two boron centers for B−N coordination [80–82]. Catechol esters and other cyclic five-membered boronic esters with $sp^2$ centers are more acidic as complexation to form a tetrahedral boron atom relieves strain. The concept of strain has recently been exploited in the design of a receptor with photoswitchable Lewis acidity [83]. Pyridine complexation studies by $^1$H NMR spectroscopy showed that bisthiophene boronate receptor **25** is more acidic in its closed cross-conjugated form **26** compared to the less strained, open form **25** (Equation 1.9).

### 1.2.2.5 Chemical Stability

#### 1.2.2.5.1 Ligand Exchange and Disproportionation
Several favorable factors contribute to the stability of boronic acids and their esters. Substitution of the carbon-containing group of boronic acids with other substituents is a slow process, and B−C/B−O bond metatheses to give the corresponding disproportionation products

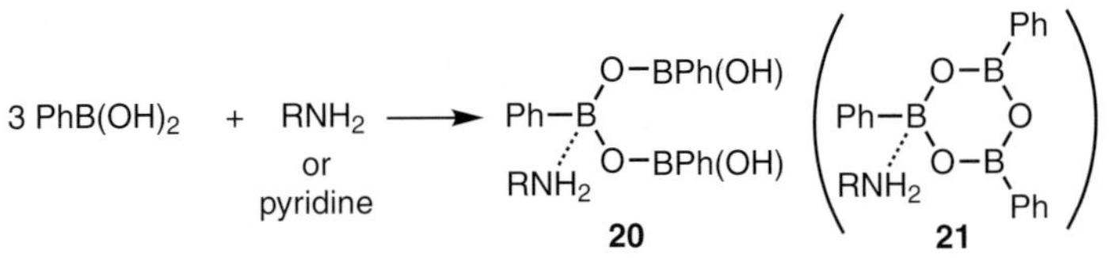

**Figure 1.8** Bimolecular Lewis acid–base complexes with boronic esters.

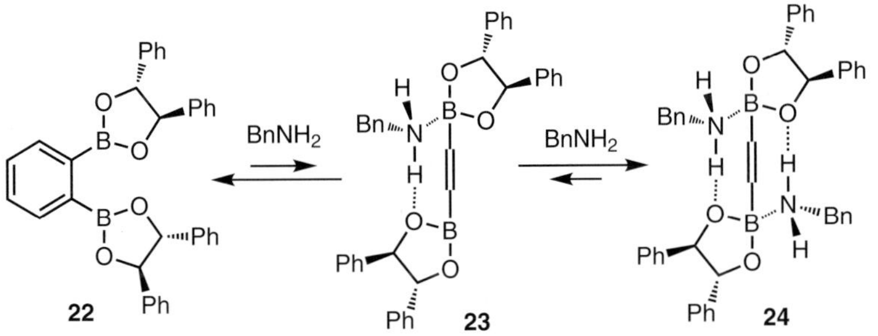

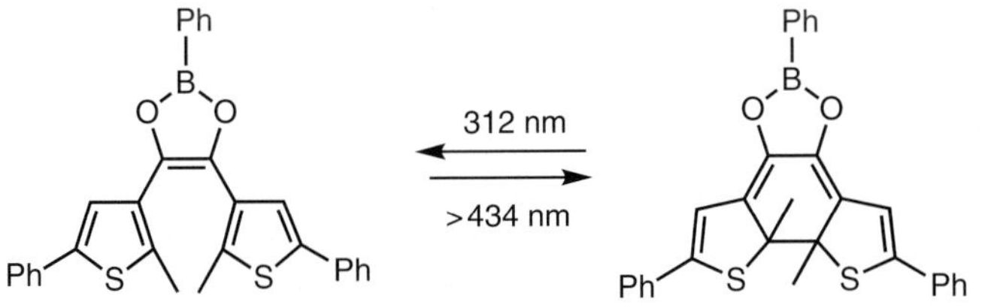

(trialkylborane, borinic acid, or boric acid) are thermodynamically unfavorable [25]. This redox disproportionation is rather used to transform borinic esters into boronic esters [84]. Similarly, thermodynamic considerations make the exchange of the hydroxyl substituents of boronic acids with other ligands quite unfavorable. Substitution with most alcohols or diols to form boronic esters usually requires dehydration techniques in order to drive the reaction forward (Section 1.2.3.2.1). In general, from the B–X bond energy values of all possible boronic acid derivatives (RBX$_2$), it can be said that free boronic acids remain unchanged when dissolved in solutions containing other potential anionic ligands [24]. The only type of B–X bond stronger than a B–O bond is the B–F bond. Chemical methods to accomplish this type of exchange and other B–O bond derivatizations are described in Sections 1.2.3.7 and 1.2.3.8.

### 1.2.2.5.2 Atmospheric Oxidation

A significant thermodynamic drive for C–B bond oxidation results as a direct consequence of the large difference between B–O and B–C bond energies (Section 1.2.2.1). Heats of reaction for the oxidative cleavage of methylboronic acid with water and hydrogen peroxide are −112 and −345 kJ/mol, respectively [25]. Yet, fortunately for synthetic chemists, oxidative cleavage of the B–C bond of boronic acid derivatives with water or oxygen is a kinetically slow process, and most boronic acids can be manipulated in ambient air and are stable in water in a wide range of pH. This is particularly true for aryl- and alkenylboronic acids, and in general, samples of all types of boronic acids tend to be significantly more stable when moist (coordination of water to boron likely acts as a protection) (Section 1.2.2.2) [42, 43, 85]. Exceptionally, the highly electron-poor arylboronic acid 4-carboxy-2-nitrophenylboronic acid was reported to undergo slow oxidation to the corresponding phenol when left in aqueous basic solutions (pH 9) [11]. On the other hand, basic aqueous solutions of alkylboronate ions were claimed to be highly tolerant of air oxidation [42]. Free alkylboronic acids, however, are quite prone to a slow atmospheric oxidation and variable amounts of the corresponding alcohols may form readily when dried samples are left under ambient air. Likewise, solutions of arylboronic acids in tetrahydrofuran devoid of stabilizer may turn rapidly into the corresponding phenols. The propensity of alkylboronic acids to undergo autoxidation depends on the degree of substitution, with primary alkyl substituents being less reactive than the secondary and tertiary alkyl substituents, respectively [85]. More potent oxidants such as peroxides readily oxidize all types of boronic acids and their corresponding esters (Section 1.5.2.1). This propensity for oxidation must be kept in mind while handling boronic acids.

1.2.2.5.3 **Protolytic Deboronation** Most types of boronic acids are highly resistant to protolysis of the C−B bond in neutral aqueous solutions even at high temperatures. For example, *p*-tolylboronic acid was recovered unchanged after 28 h in boiling water [86]. Aqueous protodeboronation can become problematic at higher temperatures; *p*-tolylboronic acid was completely deboronated to toluene after 6 h under pressure at 130–150 °C [86]. Deboronation of arylboronic acids can be effected quite readily in highly acidic or basic aqueous solutions [87]. In particular, *ortho*-substituted and especially electron-poor arylboronic acids are notorious for their propensity to protodeboronate under basic aqueous conditions, a process that can be exacerbated by exposure to light [64]. Consequently, competitive deboronation may plague some reactions employing boronic acids as reagents like the Suzuki–Miyaura cross-coupling reaction (Section 1.5.3.1), which requires basic conditions often at high temperatures. Under acidic aqueous conditions, however, the more electron-rich arylboronic acids tend to deboronate faster [87, 88]. For example, *p*-carboxyphenylboronic acid was found to be more tolerant than phenylboronic acid to the highly acidic conditions of ring nitration under fuming nitric acid and concentrated sulfuric acid [89]. Certain heteroaromatic boronic acids with the boronyl group next to the heteroatom (α-substituted) are notoriously prone to protodeboronation, but they can be stabilized as tetrahedral adducts (Section 1.2.3.3) [41, 90]. The effect of acid, temperature, and ring substitution of arylboronic acids on the kinetics of electrophilic protolytic deboronation with strong aqueous acid has been studied by Kuivila and Nahabedian [91]. A relatively complex behavior was found, and at least two possible pH-dependent mechanisms were proposed. In contrast to their behavior with aqueous acids, most arylboronic acids and esters appear to be very resistant to nonaqueous acids, as evidenced by their recovery from reaction processes using strong organic acids. For example, a phenolic methoxymethyl ether was deprotected with a 2 : 1 $CH_2Cl_2/CF_3CO_2H$ (TFA) mixture that left intact a pinacol boronic ester functionality [92]. Likewise, free arylboronic acids have been shown to tolerate, at ambient temperature, similar organic acid conditions that effect cleavage of *t*-butoxycarbonyl groups (Equation 1.10) [93]. On the other hand, a report emphasized that arylboronic acids can be protodeboronated thermally without added acid by prolonged heating in refluxing ethereal solvents [94].

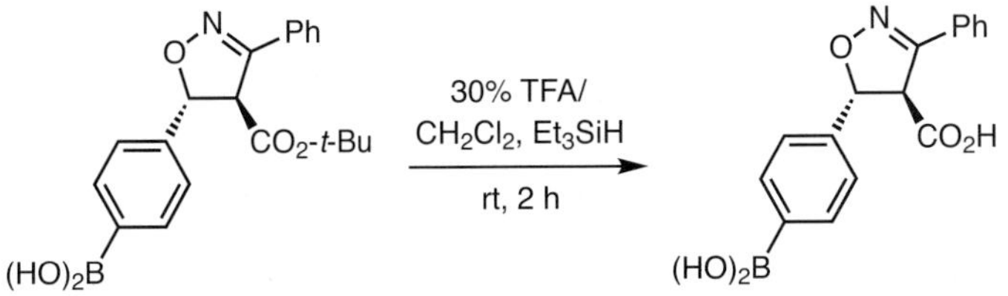

$$\text{(1.10)}$$

In contrast to arylboronic acids, early reports document the great stability of alkylboronic acids under aqueous acidic solutions. For example, a variety of simple alkylboronic acids were unaffected by prolonged heating in 40% aqueous HBr or HI [42]. Like arylboronic acids, however, deboronation is observed in hot basic aqueous solutions [85]. Alkenylboronic esters undergo protonolysis in refluxing AcOH [95], and alkynylboronic acids were reported to be quite unstable in basic aqueous solutions (Section 1.3.5).

All types of boronic acids can be protodeboronated by means of metal-promoted C−B bond cleavage, and these methods are described separately later in this chapter (Section 1.5.1).

## 1.2.3
## Boronic Acid Derivatives

For the sake of convenience in their purification and characterization, boronic acids are often best handled as ester derivatives where the two hydroxyl groups are masked. On the other hand, transformation of the hydroxyl groups into other substituents such as halides or borate salts may also provide an increase in reactivity necessary for a number of synthetic applications. The next sections describe the most important classes of boronic acid derivatives.

### 1.2.3.1 Boroxines (Cyclic Anhydrides)
Boroxines are the cyclotrimeric anhydrides of boronic acids. Their properties and applications have been reviewed recently [96]. By virtue of boron's vacant orbital, boroxines are isoelectronic to benzene, but it is generally accepted that they possess little aromatic character [97]. Several theoretical and experimental studies have addressed the nature and structure of these derivatives [96]; in particular, the X-ray crystallographic analysis of triphenylboroxine confirmed that it is virtually flat [98]. Boroxines are easily produced by the simple dehydration of boronic acids, either thermally through azeotropic removal of water or by exhaustive drying over sulfuric acid or phosphorus pentoxide [42]. These compounds can be employed invariably as substrates in many of the same synthetic transformations known to affect boronic acids. Interest in the applications of boroxines as end products has increased in the past decade. Their use has been proposed as flame retardants [99] and as functional materials (see Chapter 14) [100]. The formation of boroxine cross-linkages has been employed as a means to immobilize blue light-emitting oligofluorene diboronic

acids [101]. Samples of boroxines, which may also contain oligomeric acyclic analogues, were found to be sensitive to autoxidation when dried exhaustively (Sections 1.2.2.2 and 1.2.2.5.2). A study examined the thermodynamic parameters of boroxine formation in water (Equation 1.11) [102]. Using $^1$H NMR spectroscopy, the reaction was found to be reversible at room temperature, and the equilibrium constants, relatively small ones, were found to be subject to substituent effects. For example, boroxines with a *para*-electron-withdrawing group have smaller equilibrium constants. This observation was interpreted as an outcome of a back-reaction (i.e., boroxine hydrolysis) that is facilitated by the increased electrophilicity of boron. Steric effects also come into play, as indicated by a smaller K-value for *ortho*-tolylboronic acid compared to the *para*-isomer. Variable temperature studies provided useful thermodynamic information, which was found consistent with a significant entropic drive for boroxine formation due to the release of three molecules of water.

$$3\ R{-}C_6H_4{-}B(OH)_2 \rightleftharpoons (R{-}C_6H_4{-}BO)_3 + 3\ H_2O \quad (1.11)$$

$$K = \frac{[\text{boroxine}][H_2O]^3}{[\text{boronic acid}]^3}$$

#### 1.2.3.2 Boronic Esters

By analogy with carboxylic acids, the replacement of the hydroxyl groups of boronic acids by alkoxy or aryloxy groups provides esters. By losing the hydrogen bond donor capability of the hydroxyl groups, boronic esters are less polar and easier to handle. They also serve as protecting groups that can mitigate the particular reactivity of boron–carbon bonds. Most boronic esters with a low molecular weight are liquid at room temperature and can be conveniently purified by distillation. Exceptionally, the trityloxymethyl esters described above are crystalline solids [15]. A selection of the most commonly encountered boronic esters is shown in Figure 1.9. Many of these esters are chiral and have also been used as inducers in stereoselective reactions discussed in Section 1.3.8.4. In addition, a number of macrocyclic oligomeric esters have been described [103].

##### 1.2.3.2.1 Stoichiometric Formation in Nonaqueous Conditions

The preparation of boronic esters from boronic acids and alcohols or diols is straightforward (Equation 1.12, Figure 1.9). The overall process is an equilibrium and the forward reaction is fast with preorganized diols, and particularly favorable when the boronate product is insoluble in the reaction solvent. The backward process (hydrolysis) can be slowed to a practical extent by using bulky diols such as pinanediol or pinacol. Otherwise, ester formation can be driven by azeotropic distillation of the water produced using a Dean–Stark apparatus or, alternatively, with the use of a dehydrating agent (e.g.,

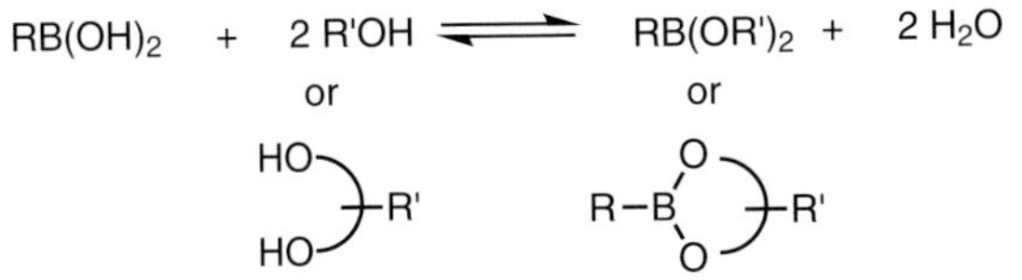

## 1.2 Structure and Properties of Boronic Acid Derivatives

$$RB(OH)_2 + 2\ R'OH \rightleftharpoons RB(OR')_2 + 2\ H_2O \quad (1.12)$$

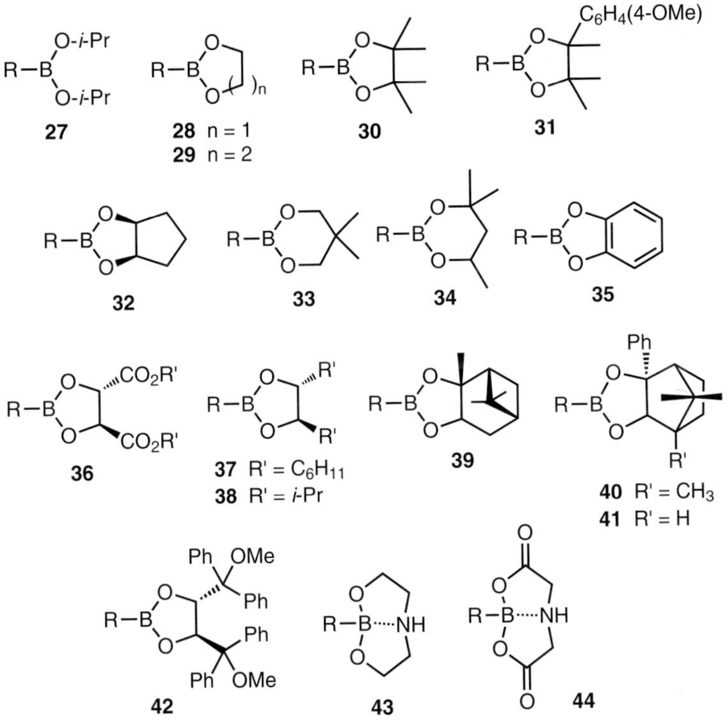

Figure 1.9 Common types of boronic esters.

MgSO$_4$, molecular sieves, etc.). The use of mechanochemistry (i.e., solvent-less grinding) has been reported for the preparation of cyclic esters by condensation of certain diols with aliphatic and aromatic boronic acids [104]. Boronic esters can also be made by transesterification of smaller dialkyl esters like the diisopropyl boronates, with distillation of the volatile alcohol by-product driving the exchange process. In the case of cyclic esters made from the more air-sensitive alkylboronic acids, an alternate method involves treatment of a diol with lithium trialkylborohydrides [105]. Likewise, cyclic ethylboronates were prepared by reaction of polyols with triethylborane at elevated temperatures [106]. One of the first reports on the formation of boronic esters from diols and polyols, by Kuivila et al., described the preparation of several esters of phenylboronic acid by reaction of the latter, in warm water, with sugars like mannitol and sorbitol and 1,2-diols like catechol and pinacol [107]. The desired nonpolar boronic esters precipitated upon cooling the solution. Interestingly, cis-1,2-cyclohexanediol failed to provide the corresponding cyclic ester and the authors rationalized this observation on the basis of the unfavorable diol geometry of

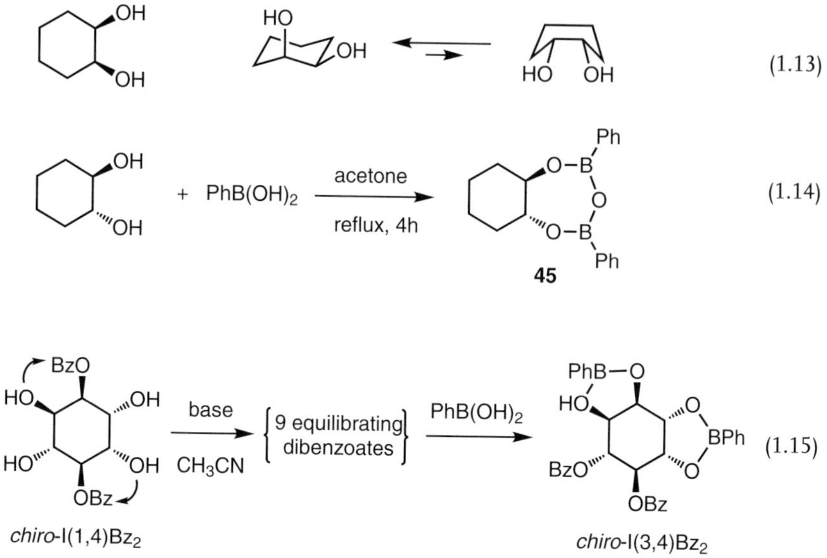

# 1 Structure, Properties, and Preparation of Boronic Acid Derivatives

(1.13)

(1.14)

(1.15)

**Figure 1.10** Specific examples of boronic ester formation with cyclic diols.

the substrate. Thus, although the two diols are not oriented in the same plane in the chair conformation (Equation 1.13, Figure 1.10), they can adopt such a favorable orientation only in the boat conformer, which is thermodynamically unfavorable [107]. Under anhydrous conditions (i.e., refluxing acetone), phenylboronic esters of cis-1,2-cyclopentanol and cis-1,2-cyclohexanol can be isolated [108]. The trans-isomers, however, failed to give a 1 : 1 adduct, and based on elemental analysis and molecular weight determinations, rather gave 1 : 2 adducts such as **45** (Equation 1.14). The existence of a seven-membered trans 1 : 2 adduct of a glucopyranoside was recently demonstrated by NMR spectroscopy [109]. This behavior can be explained in terms of the large energy required for the trans-diol to adopt a coplanar orientation, which would increase ring strain and steric interactions between axial atoms. The marked preference for the formation of boronic esters from cyclic cis-diols was exploited in the concept of dynamical combinatorial chemistry, using phenylboronic acid as a selector to amplify and accumulate one out of nine possible dibenzoate isomers of chiro-inositol that exist under equilibrating conditions through base-promoted intramolecular acyl migration (Equation 1.15) [110]. The relative thermodynamic stability of several boronic esters was examined by comparing the equilibrium composition of products in the transesterification of 2-phenyl-1,3,2-dioxaborolane with various diols by NMR spectroscopy in deuterated chloroform (Figure 1.11) [111]. Rigid, preorganized diols like pinanediol (**39**) provide the most robust esters and it was also found that six-membered esters are generally more stable than the corresponding five-membered boronates (i.e., **29** versus **28**). Presumably, the stabilizing effect of B—O conjugation via overlap of boron with oxygen lone pairs is geometrically optimal in the larger rings. Diethanolamine boronic esters (**43**, Figure 1.9) represent a useful class of boronic acid derivatives [112].

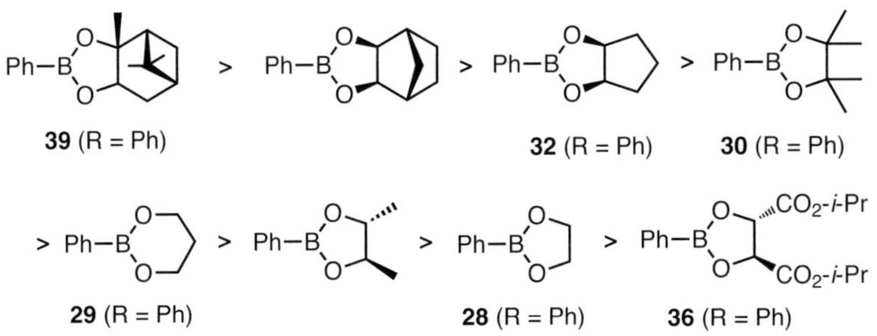

**Figure 1.11** Relative thermodynamic stability in a series of boronic esters.

Other N-substituted derivatives were characterized [113]. The presence of internal coordination between the nitrogen lone pair and boron's vacant orbital constitutes a unique structural characteristic of these tetrahedral derivatives. This coordination makes the hydrolysis reaction less favorable and even stabilizes the boron atom against atmospheric oxidation. Diethanolamine boronic esters can be conveniently formed in high yields, often without any need for dehydration techniques, as they tend to crystallize out of solution. These adducts are solids, often crystalline, with sharp melting points, and can thus be used for purifying and characterizing boronic acids, as well as in the chemical protection of the boronyl group toward various transformations (see Section 1.3.8.6). The concept of internal coordination in diethanolamine esters has been exploited in the development of the DEAM-PS resin for immobilization and derivatization of boronic acids (Section 1.4.2.1).

1.2.3.2.2 **Hydrolysis and Cleavage** From a thermodynamic standpoint, the stability of B–O bonds in boronic acids and their ester derivatives is comparable (Section 1.2.2.1). Consequently, hydrolysis, in bulk water or even by simple exposure to atmospheric moisture, is a threatening process when handling boronic esters that are kinetically vulnerable to the attack of water. In fact, hydrolysis is very rapid for all acyclic boronic esters such as **27** (Figure 1.9) and for small unhindered cyclic ones such as those made from ethylene or propylene glycol (**28** and **29**) and tartrate derivatives (**36**) [114]. Catechol esters (**35**) are another class of popular derivatives as they are the direct products of hydroboration reactions with catecholborane (Section 1.3.4.4). Due to the opposing conjugation between the phenolic oxygens and the benzene ring, these derivatives are more Lewis acidic and are quite sensitive to hydrolysis. They are stable only in nonhydroxylic solvents and are not compatible with silica chromatography [115]. In the hydrolytic cleavage of catechol boronic esters, it is often necessary to carefully monitor the pH and buffer the acidity of the released catechol.

In contrast, hydrolysis can be slowed down considerably in the case of hindered cyclic aliphatic esters such as the C2-symmetrical derivatives **37** [116] and **38** [117], pinacol (**30**) [107], pinanediol (**39**) [118], Hoffmann's camphor-derived diols (**40** and **41**) [119], and the newer **42** [120] (Figure 1.9). Indeed, many of these boronic esters tend to be stable to aqueous workups and silica gel chromatography. The robustness of the esters of trans-1,4-dimethoxy-1,1,4,4-tetraphenyl-2,3-butanediol (**42**) was demonstrated in its applications as a protecting group for alkenylboronic acids [120].

The resulting alkenylboronic esters are tolerant of a wide variety of reaction conditions (Section 1.3.8.6). Unfortunately, the bulky boronic esters **39–42** are very robust to hydrolysis, and their conversion back to boronic acids is notoriously difficult. The removal of the bulky pinanedioxy group in boronates **39** exemplifies the magnitude of this particular problem. It is generally not possible to cleave a pinanediol ester quantitatively in water even under extreme pH conditions. It can be released slowly (over several days) and rather ineffectively by treatment with other rigid diols in chloroform [121]. Cleavage of various pinanedioxy boronates has been achieved by transborylation with boron trichloride [22, 121–125], which destructs the pinanediol unit, or by reduction to the corresponding borane using lithium aluminum hydride (Equations 1.16 and 1.17, Figure 1.12) [126]. Both of these derivatives can be subsequently hydrolyzed to afford the desired boronic acid. More recently, mild approaches have been developed to convert the robust DICHED, pinacol, and pinanediol esters into difluoroboranes or trifluoroborate salts (Equation 1.18, Figure 1.12) [127, 128]. The latter can then be hydrolyzed to the corresponding boronic acids using various methods (Section 1.2.3.8) [128, 129]. Two-phase transesterification procedures with polystyrylboronic acid [130] or with phenylboronic acid have been described, but the latter is only applicable to small, water-soluble boronic acids [131]. Most of these procedures, such as the BCl$_3$-promoted method, were applied to the particular case of pinanediol esters of α-acylaminoalkylboronic acids [22, 125]. Using such a substrate, **46**, an oxidative method allowed the recovery of free boronic acid **47** in good yield from a periodate-promoted cleavage that destructs the pinanediol unit or by using the biphasic transesterification method in hexanes/water (pH 3) (Equations 1.19 and 1.20, Figure 1.12) [132]. The cleavage of methoxyphenyl-substituted pinacol-like boronates **31** (Figure 1.9) can be effected under oxidative conditions, providing an orthogonal strategy to protect boronic acid compounds in various transformations [133].

Long ago, the hydrolysis of a series of five-, six-, and seven-membered phenylboronic esters was studied by measuring the weight increase of samples subjected to air saturated with water vapor (i.e., under neutral conditions) [134]. The occurrence of hydrolysis was confirmed by the observation of phenylboronic acid deposits. This early study confirmed that hindered esters such as phenylboron pinacolate (PhBpin) hydrolyze at a much slower rate, and that six-membered boronates are more resistant to hydrolysis than the corresponding five-membered analogues. These results were interpreted in terms of the relative facility of boron–water complexation to form a tetracoordinate intermediate. Two factors were proposed: (1) the increase of steric effects on neighboring atoms upon formation of the hydrated complex, and (2) the release of angle strain, which is optimal in the five-membered boronates due to the decrease of the O—B—O and B—O—C bond angles from about 120° to 109° upon going from a planar configuration to the tetracoordinate hydrated form with tetrahedral B and O atoms. Propanediol derivative **34** emphasizes the importance of steric hindrance to the coordination of water in order to minimize kinetic hydrolysis. The hydrolysis of **34** is considerably slower compared to the unsubstituted 1,3-propanediol ester (**29**). The superior stability of ester **34** toward hydrolysis was attributed to the axial

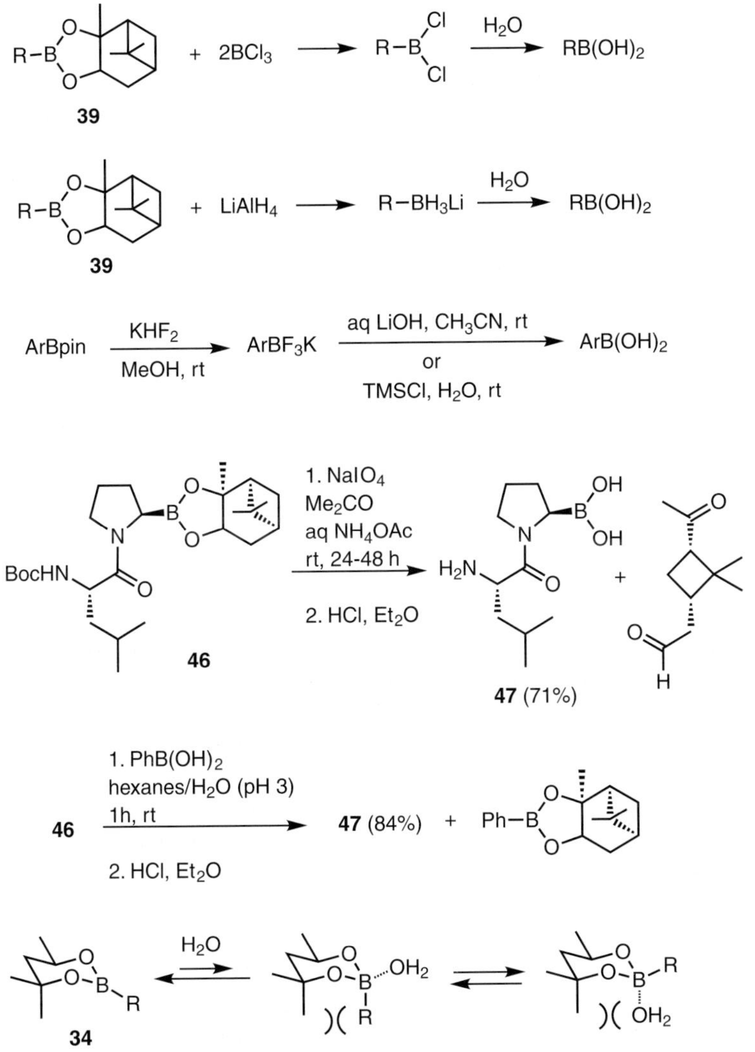

$$R-B\begin{smallmatrix}O\\O\end{smallmatrix}\hspace{-2pt}\text{(pinanediol)} + 2BCl_3 \longrightarrow R-B\begin{smallmatrix}Cl\\Cl\end{smallmatrix} \xrightarrow{H_2O} RB(OH)_2 \quad (1.16)$$

**39**

$$R-B\begin{smallmatrix}O\\O\end{smallmatrix}\hspace{-2pt}\text{(pinanediol)} + LiAlH_4 \longrightarrow R-BH_3Li \xrightarrow{H_2O} RB(OH)_2 \quad (1.17)$$

**39**

$$ArBpin \xrightarrow[MeOH, rt]{KHF_2} ArBF_3K \xrightarrow[\substack{\text{or}\\TMSCl, H_2O, rt}]{aq\ LiOH, CH_3CN, rt} ArB(OH)_2 \quad (1.18)$$

**46** → **47 (71%)** + [cyclobutanone byproduct] (1.19)

Reagents: 1. NaIO$_4$, Me$_2$CO, aq NH$_4$OAc, rt, 24–48 h; 2. HCl, Et$_2$O

**46** → **47 (84%)** + Ph–B(pinanediol) (1.20)

Reagents: 1. PhB(OH)$_2$, hexanes/H$_2$O (pH 3), 1h, rt; 2. HCl, Et$_2$O

[Equilibrium scheme showing hydrolysis of **34** via boronate with OH$_2$ intermediates] (1.21)

**Figure 1.12** Cleavage of pinanediol boronic esters.

methyl groups, which develop a 1,3-diaxial interaction with the boron center in the approach of water from either face (Equation 1.21, Figure 1.12).

While developing a novel two-phase system for the basic hydrolysis of DICHED esters, **37**, Matteson proposed a useful generalization on the process of thermodynamic hydrolysis of boronic esters (Scheme 1.1) [135]. Using a relatively dilute nonmiscible mixture of 1 M aqueous sodium hydroxide and diethyl ether (required to avoid precipitation of boronate salt **48**), the equilibrium ratio of 42 : 1 of **49** to **37** in the ether phase was reached slowly only after 18 h by using a large excess of sodium hydroxide with respect to the boronic ester **37**. By using soluble triols like

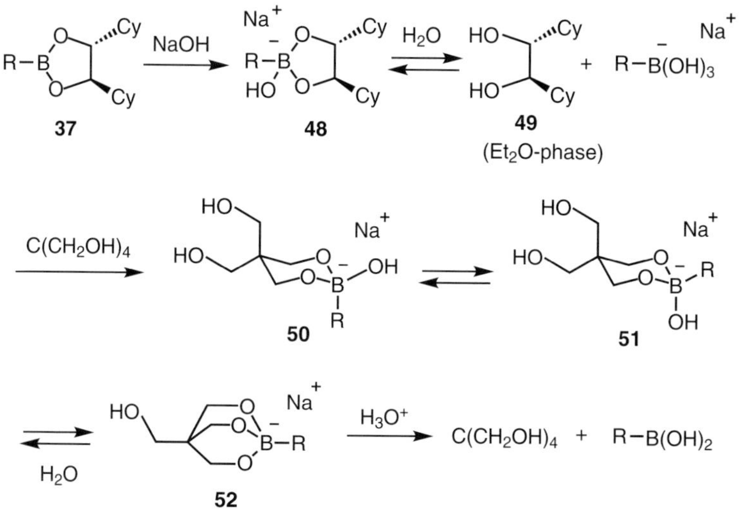

**Scheme 1.1** Hydrolysis of boronic esters in a two-phase system.

pentaerythrol to transesterify salt **48** into a more water-soluble salt (i.e., **50/51/52**) and thus facilitate the liberation of DICHED, a higher ratio of 242 : 1 was obtained. The free boronic acid could then be recovered by acidification of the aqueous phase containing a mixture of **50–52**, followed by extraction with ethyl acetate. This new procedure, however, was not successful for the complete hydrolysis of pinanediol phenylboronic ester, providing the optimal pinanediol:boronic ester ratio of only 3.5 : 1 in the ether phase. These results were interpreted in terms of the determining thermodynamic factors controlling the reversible hydrolysis or transesterification of boronic esters. Entropic factors in the hydrolysis of cyclic esters are unfavorable as three molecules are converted into only two. In this view, transesterification with a diol, instead of hydrolysis, is entropically even and thus more favorable. More important factors affecting the equilibrium are the effect of steric repulsions on enthalpy and the entropies of internal rotation of the free diols. For example, *trans*-4,5-disubstituted dioxaborolanes such as DICHED esters present a minimal extent of steric repulsions as the two cyclohexyl substituents eclipse C−H bonds. On the contrary, pinacol esters present a significant amount of steric repulsion from the four eclipsing methyl groups. Consequently, it is not surprising that pinacol esters can be transesterified easily with *trans*-DICHED so as to relieve these eclipsing interactions [15, 136]. In this scenario, the exceptional resistance of pinanediol esters to thermodynamic hydrolysis would be due to the rigid cyclic arrangement, whereby the two hydroxyls are preorganized in a coplanar fashion to form a boronic ester with essentially no loss of entropy from internal rotation compared to the free pinanediol. Other types of esters, including DICHED [137] and the robust pinacol esters of peptidyl boronates [138], have also been converted to the boronic acids through transesterification with diethanolamine in organic solvent, followed by acidic aqueous hydrolysis. This method, however, is effective only if the resulting diethanolamine ester crystallizes from the solution so as to drive the equilibrium forward. As stated above, the transesterification of cyclic boronic esters with diols is often slow, and particularly so in organic solvents. Wulff *et al.* found that a number of boronic acids possessing proximal basic atoms or substituents (e.g., **16**, Figure 1.7) lead to a large neighboring group effect, and the transesterification equilibriums are reached much faster with these boronic acids as a result of a rapid proton transfer [139].

#### 1.2.3.2.3 Boronic Acid–Diol (Sugar) Equilibrium in Water and Protic Solvents

The reversible formation of boronic esters by the interaction of boronic acids and polyols in water was first examined in the seminal study of Lorand and Edwards [52]. This work followed an equally important study on the elucidation of structure of the borate ion [140]. By measuring the complexation equilibrium between several model diols and monosaccharides using the method of pH depression, it was shown that ester formation is more favorable in solutions of high pH where the boronate ion exists in high concentrations (Equation 1.22, Figure 1.13). This study also confirmed the Lewis acid behavior of boronic acids and the tetracoordinate structure of their conjugate base, that is, the hydroxyboronate anion (Section 1.2.2.4). Another conclusion made from this study is the lower Lewis acid strength of free boronic acids compared to that of their neutral complexes with 1,2-diols. For example, the p$K_a$ of PhB(OH)$_2$ decreases from 8.8 to 6.8 and 4.5 upon formation of cyclic esters with glucose and fructose, respectively [141]. To explain the favorable thermodynamic effect observed at high pH (Equation 1.22, Figure 1.13) in comparison to neutral pH (Equation 1.23), it was hypothesized that the formation of hydroxyboronate complexes of 1,2-diols is accompanied by a significant release of angle strain resulting from the rehybridization of the boron from sp$^2$ to sp$^3$ (i.e., 120° versus 109° bond angles) [52]. A series of investigations on the equilibria and mechanism of complexation between boric acid or boronic acids with polyols and other ligands in water were reported by Pizer and coworker. Early work by this group [58] and others [142] showed that the stability constants of complexes increase when the aryl substituent on the boronic acid is electron poor, which is consistent with the view that the formation of anionic hydroxyboronate complexes is the drive for release of angle strain. Using methylboronic acid and simple 1,2- and 1,3-diols, equilibrium

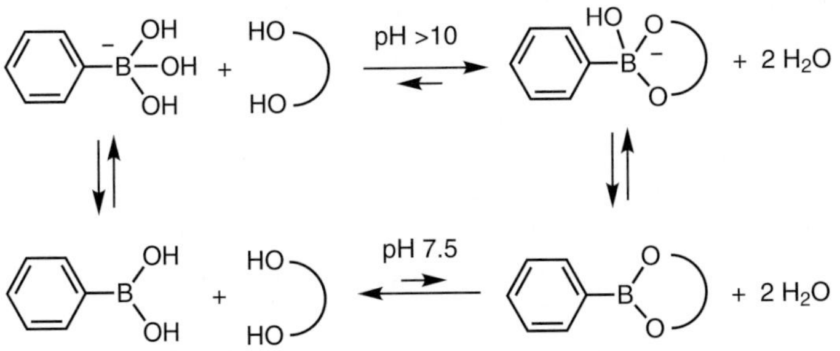

**Figure 1.13** Equilibrium formation of boronic esters from diols at high (Equation 1.22) and neutral (Equation 1.23) pH in water.

constants were measured both by pH titration and $^{11}$B NMR spectroscopy [143]. Constants of 2.5, 5.5, and 38 were found for 1,3-propanediol, 1,2-ethanediol, and 1,2,3-propanetriol respectively, with the latter binding much preferentially with a 1,2-diol unit. The results of this work also suggested that the tetracoordinate hydroxyboronate anion is much more reactive than the trigonal neutral boronic acid in forming esters with diols (at least $10^4$ times faster), with forward rate constants in the range of $10^3$–$10^4$ M/s. It was suggested that the high reactivity of the boronate anion could be interpreted in terms of an associative transition state involving proton transfer and hydroxide displacement within a pentacoordinated boron. This fundamental view has been challenged in a recent experimental study claiming that boronate formation with aliphatic diols occurs through trigonal boronic acids, with a high pH needed only to provide a small but sufficient concentration of the anionic, monodeprotonated diol [144]. In the past decade, interest in the interaction between boronic acids and *cis*-diols has developed tremendously due to its applications in the development of receptors and sensors for saccharides and in the design of new materials (Sections 1.6.4 and 1.6.10 and Chapters 13 and 14). For instance, the reversibility of boronic ester formation in hydroxylic solvents has been exploited in the crystallization-induced dynamical self-assembly between tetraol **53** and *p*-phenyldiboronic acid (Figure 1.14) [145]. Different inclusion complexes are observed depending on the solvent composition and the presence or absence of methanol is utilized as an *on/off* switch. For example, the [2 + 2] boxed toluene complex **54**, structurally characterized by X-ray crystallography, is formed in toluene, whereas a [3 + 3] box is formed in benzene. Similar to the case of simple polyols discussed above, the binding of carbohydrates to boronic acids is subject to the same geometrical preference for a preorganized, coplanar diol unit. In fact, it was demonstrated that in water, boronic acid receptors bind to glucose in the furanose form, which presents a very favorable, coplanar 1,2-diol [146]. X-ray crystallographic structures of 2 : 1 complexes between phenylboronic acid and D-fructose and D-glucose (in its furanose form), respectively, have been obtained [147, 148]. All these observations

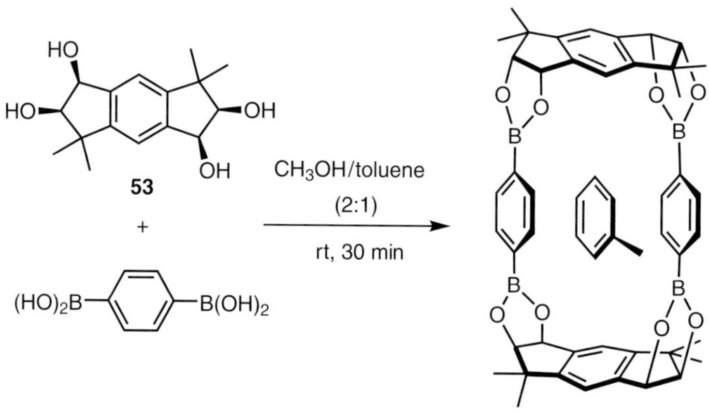

**Figure 1.14** Self-assembled, reversible tetraboronic ester cages.

concur with the absence of appreciable complexation between normal boronic acids and nonreducing sugars (glycosides) and the low affinity of 1–4 linked oligosaccharides such as lactose [149, 150]. Recently, however, benzoboroxoles such as **8** (Figure 1.2) were demonstrated to complex glycopyranosides weakly [151] and these units have been employed in the design of "synthetic lectins" (Chapter 13). Fluorescent catechol derivatives such as the dye alizarin red S (ARS) also form covalent adducts with boronic acids in water, and this equilibrium has been used as a competitive color- and fluorescence-based assay for both qualitative and quantitative determination of saccharide binding [152]. Using the ingenious ARS assay, Springsteen and Wang presented an interesting cautionary tale from discrepancies found in the measurements of boronic acid–diol binding constants based on the above-mentioned method of pH depression [141]. The latter method may not always be reliable at providing the true overall equilibrium constants due to the multiple states of ionization of the boronic acid and the resulting ester (neutral trigonal or tetrahedral hydroxyboronate), which is further complicated by the pronounced effect of the solvent, pH, and buffer components and the concentration of these species on the equilibrium [141, 153]. A follow-up study further concluded that despite some accepted generalizations, exceptions exist and the optimal pH for diol–boronic acid complexation is not always above the $pK_a$ of the boronic acid [57]. Likewise, boronic acids with a lower $pK_a$ do not always show greater binding affinity to diols.

#### 1.2.3.3 Acyloxy- and Diacyloxyboronates

Acyloxyboronates have seldom been employed as boronic acid derivatives compared to diacyloxyboronates [154]. N-Alkyliminodiacetate complexes of boronic acids homologous to **44** (Figure 1.9) were found to be even more robust than diethanolamine complexes (B–N $\Delta G^{\neq} > 90$ versus 60 kJ/mol for **43**) [20]. Compared to the alkoxy groups of **43**, the electronic effect of the carboxyl groups leads to a more acidic boron atom, hence a stronger B–N interaction. The N-methyl derivatives **55** (Equation 1.24), termed MIDA boronates, form easily in benzene–DMSO mixtures with a Dean–Stark apparatus and can be cleaved relatively easily in basic media [155]. MIDA boronates tolerate various reaction conditions and have recently been exploited as a means to mask boronic acids in iterative cross-coupling strategies (Section 1.3.8.6) [156].

$$RB(OH)_2 + \underset{R = \text{alkenyl, aryl, heteroaryl}}{\text{HOOC-CH}_2\text{-N(CH}_3\text{)-CH}_2\text{-COOH}} \xrightarrow{\text{Dean–Stark conditions}} \underset{\mathbf{55}}{\text{MIDA boronate}} \quad (1.24)$$

#### 1.2.3.4 Dialkoxyboranes and Other Heterocyclic Boranes

Several cyclic dialkoxyboranes such as 4,4,6-trimethyl-1,3,2-dioxaborinane **56** [157], 1,3,2-benzodioxaborole (catecholborane) **57** [158], and pinacolborane **58** [159] have been described (Figure 1.15). Dialkoxyboranes can be synthesized simply by the

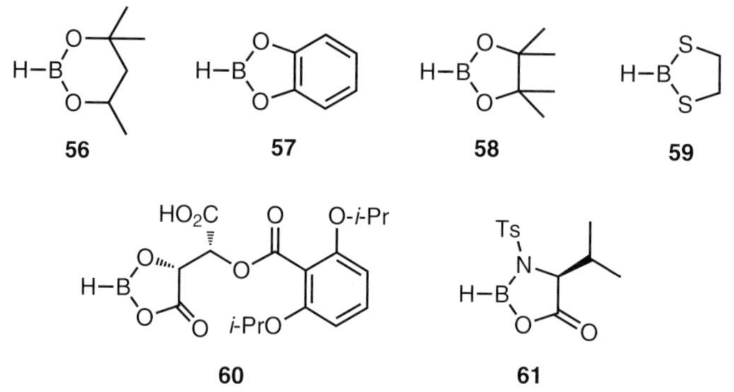

**Figure 1.15** Common dialkoxyboranes and heterocyclic analogues.

reaction between equimolar amounts of borane and the corresponding diols. These borohydride reagents have been employed as hydroborating agents, in carbonyl reduction and more recently as boronyl donors in cross-coupling reactions. Dialkoxyboranes have also been invoked as intermediates in the intramolecular, alkoxy-directed hydroboration of β,γ-unsaturated esters [160]. Sulfur-based heterocyclic boranes were reported, including 1,3,2-dithiaborolane **59** [161]. Acyloxyboranes such as Yamamoto's tartaric acid-derived CAB catalyst **60** [162] and related oxazaborolidinones such as **61**, derived from N-sulfonylated amino acids, have been used as chiral promoters for cycloadditions and aldol reactions of silyl enol ethers [163]. Synthetic applications of these catalysts are described in Chapter 12.

#### 1.2.3.5 Diboronyl Esters

A number of synthetically useful diboronyl esters such as B$_2$cat$_2$ **62** and particularly B$_2$pin$_2$ **63** have been described (Figure 1.16) [164]. The mixed reagent **64** has been reported recently and employed in regioselective alkyne diborations [165]. Reagent **63** is commercially available at a relatively low cost. Diboronyl esters can be prepared by condensation of a diol with tetrakis(dimethylamino)diboron precursor, which can be made in three steps from boron tribromide [166]. A shorter and more practical synthesis of B$_2$cat$_2$ was described [167]. The discovery that diboronyl compounds can be employed with transition metal catalysts in a variety of efficient cross-coupling and direct addition reactions to unsaturated compounds and C–H bonds can be considered one of the most significant advances in boronic acid chemistry in the past 15 years. The chemistry of diboronyl compounds has been reviewed regularly [164] and is discussed in several sections of this chapter and also in Chapters 2 and 3.

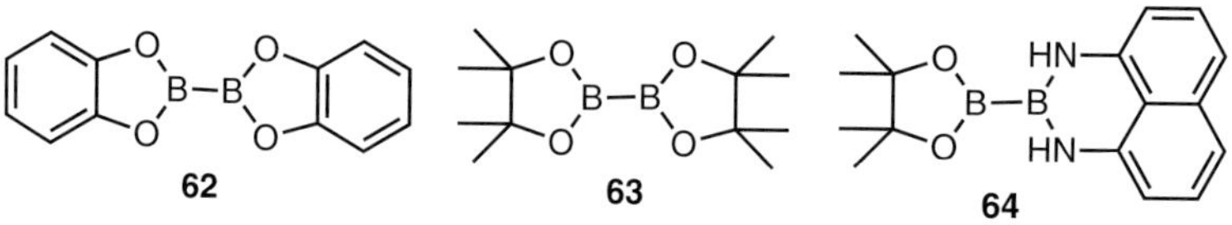

**Figure 1.16** Common diboronyl reagents.

### 1.2.3.6 Azaborolidines and Other Boron–Nitrogen Heterocycles

A large number of heterocyclic derivatives of boronic acids have been described, and useful X-ray crystallographic data were obtained for many of these compounds. It is beyond the scope of this chapter to present a comprehensive account of these derivatives; thus, only representative examples will be described in this section (Figure 1.17). The benzodiazaborole products (65) of 1,2-phenylenediamine and free boronic acids form readily in refluxing toluene [168, 169]. Both aliphatic and aromatic acids are applicable, and it was claimed that the resulting adducts are easier to recrystallize than the diethanolamine boronates. An intramolecular adduct was also reported [170]. These benzodiazaboroles are air-stable, and the adduct of phenylboronic acid was found to hydrolyze only slowly in aqueous solutions. With anhydrous hydrogen chloride in toluene, a dihydrochloride salt was formed [168, 169]. The unusual stability of adducts 65 was further supported by the fact that

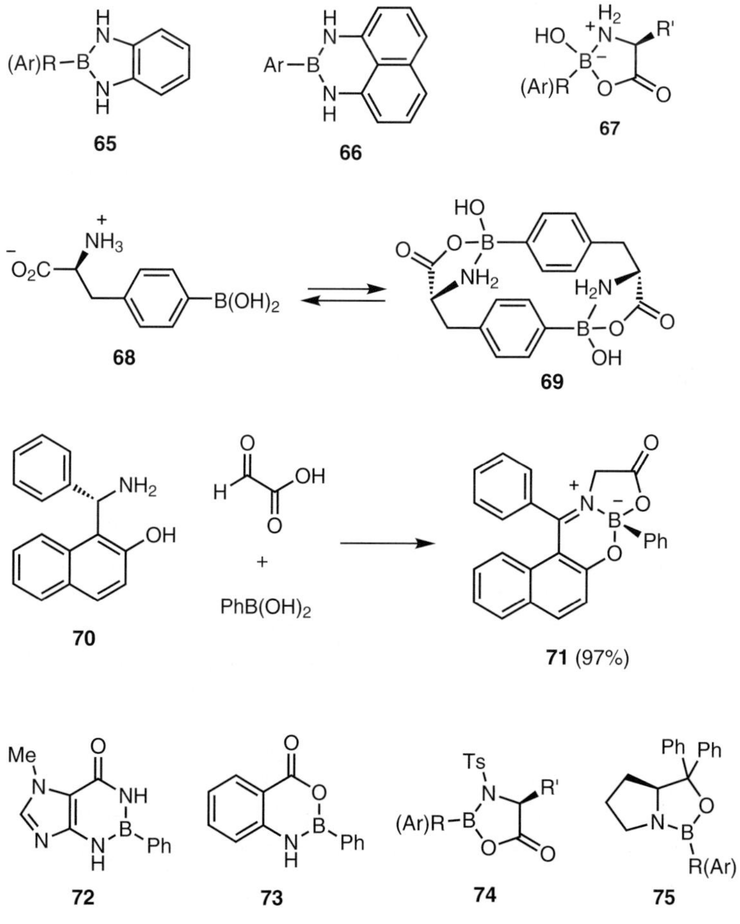

**Figure 1.17** Examples of azaborolidines and other heterocyclic analogues.

they even form by exchange of tartrate esters with 1,2-phenylenediamine at room temperature in benzene. Control studies showed that the position of the equilibrium lies much toward the diazaborole, which is surprising with respect to thermodynamic factors such as the much higher energy of covalent B–O bonds compared to B–N bonds (see Section 1.2.2.1). As both ethylenediamine and aniline itself did not form similar covalent adducts under the same conditions, it was suggested that the favorable geometry of 1,2-phenylenediamine and the stability of the resulting five-membered ring and its partial aromatic character were responsible for the highly favorable formation of adducts **65** [168]. The 1,8-diaminonaphthalene adducts **66** form readily in refluxing toluene, are cleaved with aqueous acid, and have been exploited recently as boronic acid masking groups in iterative cross-coupling (see Section 1.3.8.6) [171]. Diazaborolidines from aliphatic 1,2-diamines, on the other hand, are not prepared with such ease. For example, a number of chiral ones evaluated as chiral proton sources were prepared from dichloroboranes [172].

Amino acids can condense with boronic acids to form 1 : 1 chelates of type **67** [173]. The tetracoordinate structure of these adducts is very apparent by NMR due to the formation of a new stereocenter at boron. Interestingly, 4-boronophenylalanine (**68**), a potential BNCT agent, was shown to dimerize to form head-to-tail paracyclophane derivative **69** in a reversible fashion in DMSO (Equation 1.25, Figure 1.17) [174]. This dimer is prevalent at low concentrations (<50 mM), while oligomeric mixtures predominate at higher concentrations. Amino acid adducts of boronic acids are hydrolytically unstable, and **69** was indeed found to revert to free **68** upon addition of water to the solution. In contrast, the optically pure internally coordinated monomer **71** (Equation 1.26, Figure 1.17) is stable [24]. It was prepared in optically pure form from amino alcohol **70** through a remarkable C–to–B chirality transfer process via a 1,3 H-shift. Purine analogue **72** was found to hydrolyze readily in aqueous ethanolic solutions [175]. The addition product **73** between anthranilic acid and phenylboronic acid was also reported [176]. Salicylhydroxamic acid adducts of arylboronic acids are more resistant to hydrolysis and were proposed as components of an affinity system for bioconjugation (Section 1.6.8) [177]. Both B-alkyl and B-aryl oxazaborolidinones **74**, made from *N*-sulfonylated amino acids such as tryptophan, have been employed as chiral Lewis acids in several synthetic transformations (Chapter 12) [178] and in crystallization-induced asymmetric transformations [179]. Amino alcohols can form oxazaborolidines by condensation with boronic acids under anhydrous conditions. Chiral oxazaborolidines derived from reduced amino acids (e.g., **75**) have been a popular class of Lewis acids for cycloadditions [180] and as catalysts and reagents for the enantioselective reduction of ketones and imine derivatives [181]. The analogous cationic oxazaborolidinium catalysts are even more efficient (see Chapter 12) [182].

*ortho*-Aminophenylboronic acid exists as a hydrogen-bonded heterodimer in anhydrous aprotic solutions (Equation 1.27, Figure 1.18) [183]. In addition to the benzoboroxole described in Section 1.2.2.1 (**8**, Figure 1.2) [16, 184, 185], there are several other examples of internal heteroaromatic boronic acid derivatives where an *ortho*-substituent closes onto the boron atom with either a dative or a covalent bond [186]. For example, *ortho*-anilide derivatives **76** and the corresponding ureas (**77**), of putative internally chelated form **A**, were shown to exist mainly in their cyclic

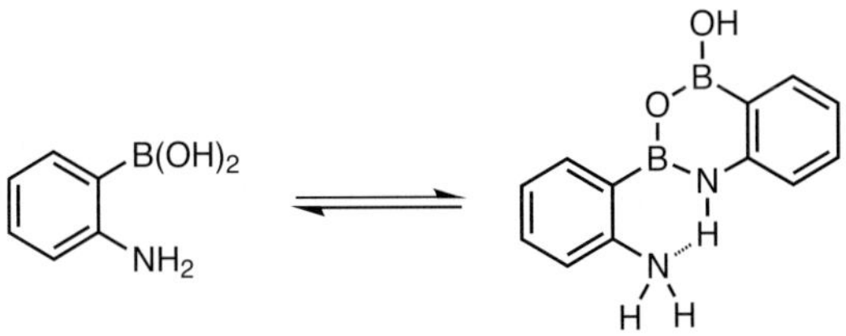

## 1.2 Structure and Properties of Boronic Acid Derivatives

(1.27)

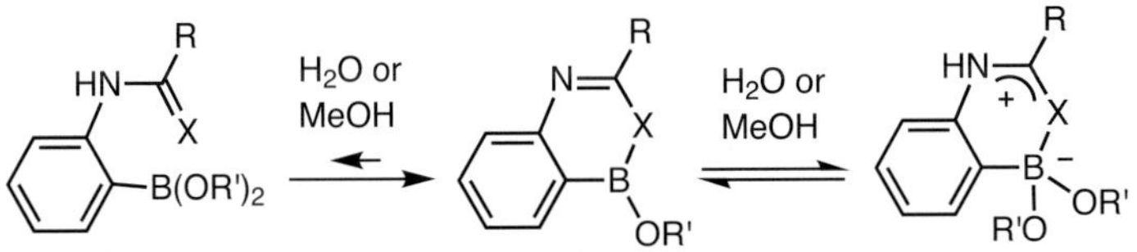

(1.28)

**A**
**76** X = O,
**77** X = NH, NR
R = H, NH$_2$, alkyl or aryl
R' = H, CH$_3$, or CMe$_2$CMe$_2$

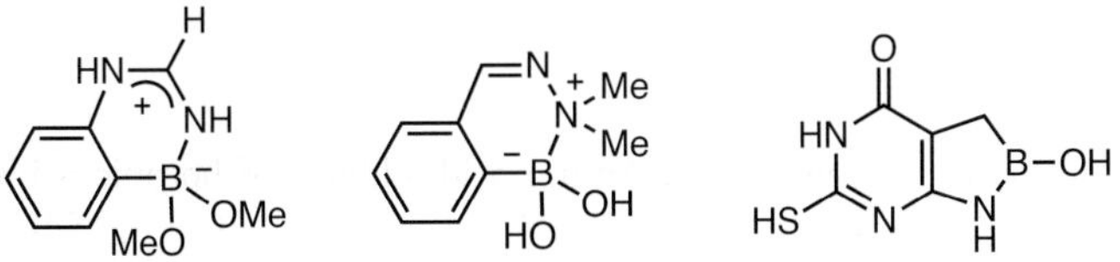

**B**

**C**

**78**

**79**

**80**

**Figure 1.18** Hemi-heterocyclic boronic ester derivatives.

monodehydrated form **B** (Equation 1.28, Figure 1.18) [183]. This is probably true even in aqueous or alcohol solutions owing to the partial aromatic character of these boron-containing analogues of purine heterocycles. In fact, it has even been shown that these and similar compounds can add one molecule of water or alcohol by 1,4-addition and thus exist in equilibrium with form **C**. One such derivative **78** was obtained from recrystallization in methanol, and the X-ray crystallographic analysis proved its zwitterionic structure with a tetrahedral boronate anion. A class of related derivatives made from 2-formylboronic acid and hydrazines was also characterized [186], and the boroxine of one internally chelated derivative **79** was studied by X-ray crystallography [187]. Other examples of heterocyclic derivatives include pyrimidine analogue **80** [188].

### 1.2.3.7 Dihaloboranes and Dihydroalkylboranes

The highly electrophilic dihaloboranes can undergo reactions that do not affect boronic acids and esters. For example, to achieve an oxidative amination of the B–C bonds in boronate derivatives, it is necessary to transform boronic esters into the corresponding dichlorides (Section 1.5.2.2). Several methods have been described for the preparation of alkyl- and aryl- dichloroboranes, but only a few of those conveniently employ boronic acids and esters as substrates. They can be

$$RB(OR')_2 + 2BCl_3 \xrightarrow{FeCl_3} RBCl_2 + 2BCl_2OR' \qquad (1.29)$$

$$RB(OR')_2 + LiAlH_4 \longrightarrow RBH_3Li \xrightarrow[TMSCl]{3HCl\ or} RBCl_2 \qquad (1.30)$$

$$RB(OH)_2 \xrightarrow{KHF_2} RBF_3K \qquad (1.31)$$

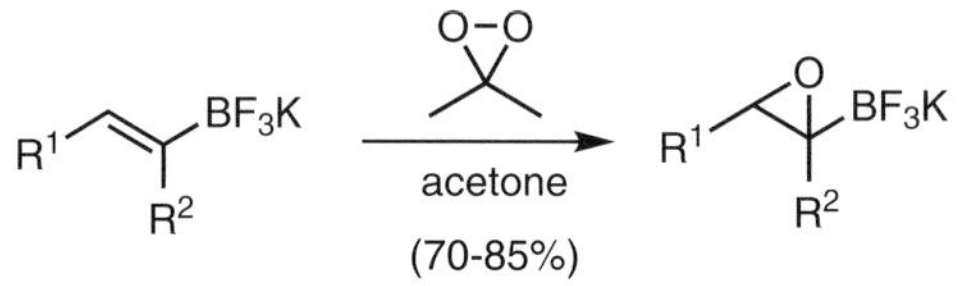

$$\underset{R^2}{\overset{R^1}{\diagup}}\!\!\!=\!\!\!\diagdown BF_3K \xrightarrow[\text{acetone} \atop (70\text{-}85\%)]{O-O} R^1\text{-}\underset{R^2}{\overset{O}{\diagdown}}\!\!\text{-}BF_3K \qquad (1.32)$$

**Figure 1.19** Synthesis of dichloroboranes, monoalkylboranes, and trifluoroborate salts.

accessed either by iron trichloride-catalyzed exchange of the boronic ester with BCl$_3$ (Equation 1.29, Figure 1.19) [189] or by treatment of the corresponding monoalkylborane with TMSCl [190] or acidification with anhydrous HCl in dimethyl sulfide (Equation 1.30) [191]. The requisite monoalkyl and monoaryl borohydride salts can be made by treating boronic esters with LiAlH$_4$ [192], and the use of HCl in dimethyl sulfide leads to the isolation of the stable RBCl$_2$–SMe$_2$ adducts (Equation 1.30) [191]. Both of these methods can be performed without any detectable epimerization when using chiral boronic esters originating from the asymmetric hydroboration of alkenes [189, 191].

### 1.2.3.8 Trifluoro- and Trihydroxyborate Salts

The organotrifluoroborate salts discussed in Section 1.2.3.2.2 are a class of air-stable boronic acid derivatives that can be easily prepared according to a procedure described by Vedejs et al. (Equation 1.31, Figure 1.19) [193]. Boronic esters also react to give the desired salts [127]. These crystalline derivatives are easy to handle, are competent substrates in many of the same reactions that employ free boronic acids, and often outperform boronic acids as cross-coupling reagents. Chapter 11 provides an overview of their synthetic applications in transition metal cross-coupling reactions and other transformations [194]. They have also been evaluated as tracer molecules for positron emission tomography (PET) applications [195]. There are several methods to hydrolyze trifluoroborate salts back to boronic acids (see Chapter 11) and they can also be conveniently transformed into dichlororoboranes by treatment with SiCl$_4$ in THF [196]. The incompatibility of boron–carbon bonds with several oxidants limits the ability to further transform compounds containing a boronic acid (ester) functionality. Taking advantage of the strong B−F bonds, the use of organotrifluoroborate salts may be viewed as a way to protect boron's vacant orbital against an electrophilic reaction with a strong oxidant. Thus, as described in Chapter 11, organotrifluoroborate salts have been shown to tolerate oxidations, including the remarkable epoxidation of 1-alkenyltrifluoroborate salts with preservation of the carbon–boron bonds in good yields with dimethyldioxirane (Equation 1.32, Figure 1.19) [197]. It is significant that under the same conditions, 1-alkenylboronic acids and the corresponding pinacol esters rather lead to the corresponding aldehyde resulting from B–C bond oxidation. Owing to their unique properties, interest in the chemistry of trifluoroborate salts has grown tremendously and several hundreds have become commercially available.

Although they have long been postulated to exist in basic aqueous solutions, trihydroxyborate salts of boronics are discrete, isolable derivatives that had not been characterized until recently [23]. The sodium salt of $p$-methoxyphenyl boronic acid (**11**, Figure 1.2) was recrystallized in water and its X-ray crystallographic structure elucidated. When concentrated NaOH is added, these borate salts precipitate from organic solutions and they can be isolated through a simple filtration. Not surprisingly, they can undergo Suzuki–Miyaura cross-coupling in the absence of an added base [23]. Cyclic trialkoxyborate salts (made from triols) behave similarly [198].

## 1.3
## Preparation of Boronic Acids and Their Esters

The increasing importance of boronic acids as synthetic intermediates in the past decades has motivated the development of new, mild, and efficient methods to provide access to these important compounds. Of particular interest is the synthesis of arylboronic acids substituted with a wide range of other functional groups. As a consequence of their growing popularity and advances in methods available for their preparation, a few thousands of functionalized boronic acids have become available from several commercial sources. Although a number of methods like the oxidation or hydrolysis of trialkylboranes bear a significant historical and fundamental relevance, this section is devoted mainly to modern methods of practical value for synthetic chemists.

### 1.3.1
### Arylboronic Acids

Arylboronic acids remain the most popular class of boronic acids. Their popularity in medicinal chemistry is in large part due to their role as cross-coupling partners for the synthesis of biaryl units (Section 1.5.3.1), which are present in the structure of several pharmaceutical drugs. Several methods, summarized in a generic way in Figure 1.20, are now available for the synthesis of complex arylboronic acids and the following section provides an overview of these methods with selected examples highlighted in Table 1.3.

#### 1.3.1.1 Electrophilic Trapping of Arylmetal Intermediates with Borates
One of the first and probably still the most common way of synthesizing arylboronic acids involves the reaction of a hard organometallic intermediate (i.e., lithium or

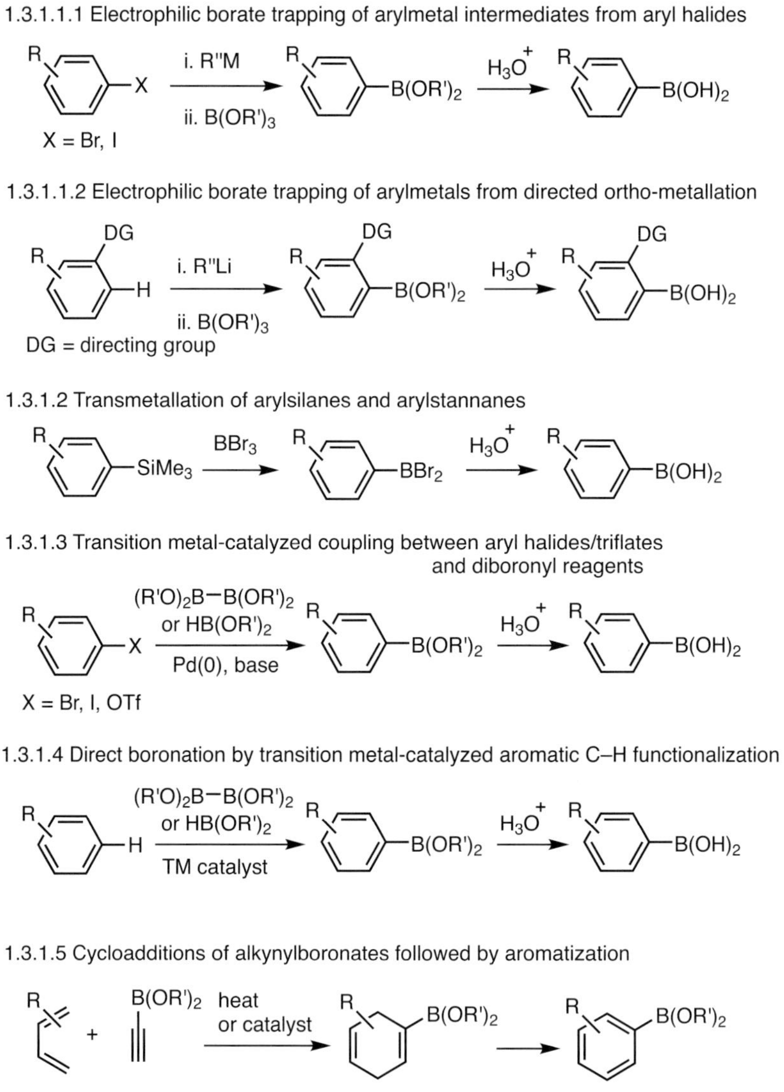

### 1.3.1.1.1 Electrophilic borate trapping of arylmetal intermediates from aryl halides

$$\text{R-Ar-X} \xrightarrow[\text{ii. B(OR')}_3]{\text{i. R''M}} \text{R-Ar-B(OR')}_2 \xrightarrow{H_3O^+} \text{R-Ar-B(OH)}_2$$

X = Br, I

### 1.3.1.1.2 Electrophilic borate trapping of arylmetals from directed ortho-metallation

$$\text{R-Ar(DG)-H} \xrightarrow[\text{ii. B(OR')}_3]{\text{i. R''Li}} \text{R-Ar(DG)-B(OR')}_2 \xrightarrow{H_3O^+} \text{R-Ar(DG)-B(OH)}_2$$

DG = directing group

### 1.3.1.2 Transmetallation of arylsilanes and arylstannanes

$$\text{R-Ar-SiMe}_3 \xrightarrow{BBr_3} \text{R-Ar-BBr}_2 \xrightarrow{H_3O^+} \text{R-Ar-B(OH)}_2$$

### 1.3.1.3 Transition metal-catalyzed coupling between aryl halides/triflates and diboronyl reagents

$$\text{R-Ar-X} \xrightarrow[\text{Pd(0), base}]{(R'O)_2B-B(OR')_2 \text{ or } HB(OR')_2} \text{R-Ar-B(OR')}_2 \xrightarrow{H_3O^+} \text{R-Ar-B(OH)}_2$$

X = Br, I, OTf

### 1.3.1.4 Direct boronation by transition metal-catalyzed aromatic C–H functionalization

$$\text{R-Ar-H} \xrightarrow[\text{TM catalyst}]{(R'O)_2B-B(OR')_2 \text{ or } HB(OR')_2} \text{R-Ar-B(OR')}_2 \xrightarrow{H_3O^+} \text{R-Ar-B(OH)}_2$$

### 1.3.1.5 Cycloadditions of alkynylboronates followed by aromatization

diene + alkynyl-B(OR')$_2$ $\xrightarrow{\text{heat or catalyst}}$ cyclohexadiene-B(OR')$_2$ $\longrightarrow$ aryl-B(OR')$_2$

**Figure 1.20** Common methods for the synthesis of arylboronic acids (esters).

magnesium) with a borate ester at a low temperature, which is necessary to minimize double addition leading to borinate side product. The corresponding zinc and cadmium species are much less effective [199].

1.3.1.1.1 **By Metal–Halogen Exchange with Aryl Halides** Provided that the aryl halide substrate is compatible with its transformation into a strongly basic and nucleophilic arylmetal reagent, relatively simple aryl, alkenyl, and even alkylboronic acids can be made from a sequence of metal–halogen exchange followed by electrophilic trapping with a trialkylborate. The first such methods for preparing phenylboronic acid, which

Table 1.3 Selected examples of preparative methods for arylboronic acids and esters.

| Entry | Substrate | Conditions | Product | Reference |
|---|---|---|---|---|
| 1 | 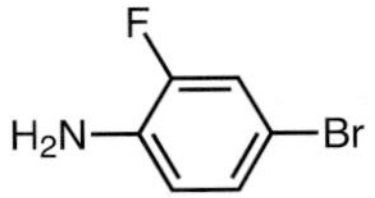 4-bromo-2-fluoroaniline | 1. i. n-BuLi (2 equiv), THF, 0 °C<br>ii. TMSCl (2 equiv)<br>2. i. t-BuLi (2.2 equiv) Et₂O, −78 °C<br>ii. B(OMe)₃ (xs), −78 °C<br>iii. 0.1N aq HCl | 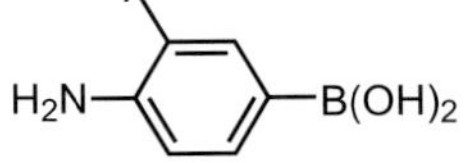 4-amino-3-fluorophenylboronic acid (45%) | [210] |
| 2 | 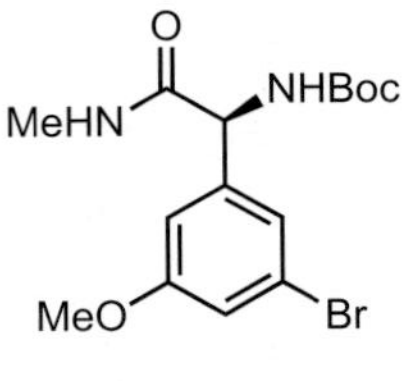 aryl bromide with NHBoc and MeO | i. MeMgCl (5 equiv), THF, 0 °C<br>ii. t-BuLi (5 equiv), −78 °C<br>iii. B(OMe)₃ (10 equiv), 0 °C | 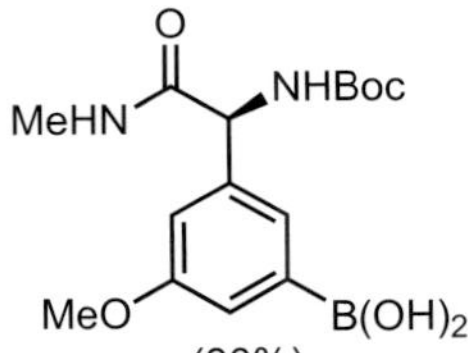 arylboronic acid with NHBoc and MeO (80%) | [211] |
| 3 | 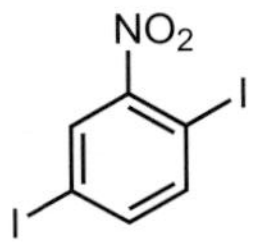 2,5-diiodo-nitrobenzene | i. PhMgCl (1.1 equiv), THF, −60 °C<br>ii. B(OMe)₃, 0.5 h, −60 °C<br>iii. 2M HCl, −20 °C | 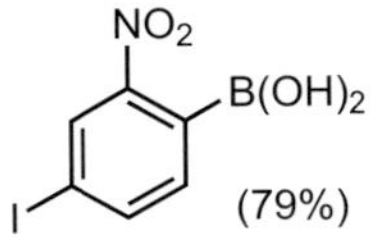 2-nitro-4-iodophenylboronic acid (79%) | [214] |
| 4 | 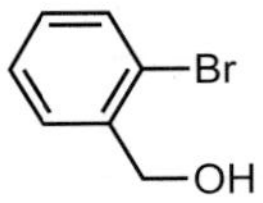 2-bromobenzyl alcohol | i. n-BuLi (2 equiv) Et₂O, 0 °C, 2 h; −78 °C<br>ii. B(OMe)₃ (1 equiv)<br>iii. aq HCl | 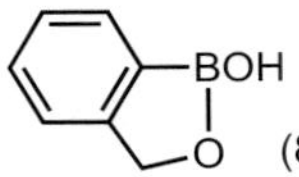 benzoxaborole (86%) | [16] |

(Continued)

**34** | *1 Structure, Properties, and Preparation of Boronic Acid Derivatives*

**Table 1.3** (Continued)

| Entry | Substrate | Conditions | Product | Reference |
|---|---|---|---|---|
| 5 | 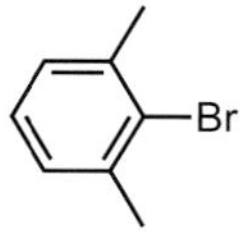 2,6-dimethylbromobenzene | i. i-PrMgBr, THF, −40 °C<br>ii. B(OMe)₃, THF, −78 °C<br>iii. HOCH₂CH₂OH, toluene | 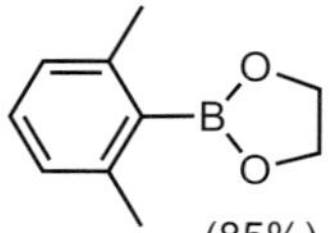 2,6-dimethylphenyl-1,3,2-dioxaborolane (85%) | [215] |
| 6 | 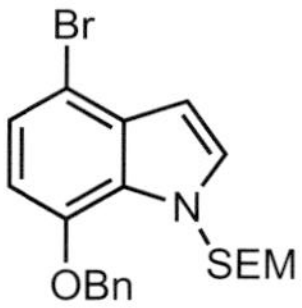 4-Br-7-OBn-N-SEM-indole | i. t-BuLi, THF, −78 °C<br>ii. 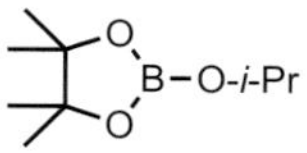 isopropyl pinacol boronate | 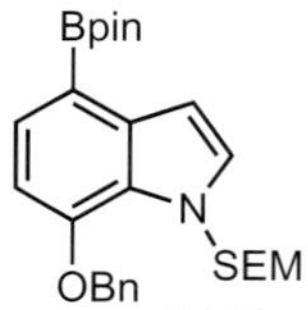 4-Bpin-7-OBn-N-SEM-indole (68%) | [216] |
| 7 | 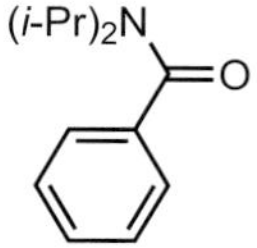 (i-Pr)₂N-C(O)-phenyl | i. s-BuLi, TMEDA THF, −78 °C<br>ii. B(OMe)₃<br>iii. 5% aq HCl | 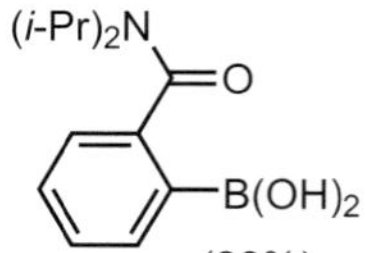 2-[(i-Pr)₂NC(O)]-phenylboronic acid (80%) | [221] |
| 8 | 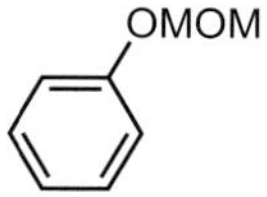 OMOM-phenyl | i. s-BuLi, TMEDA THF, −78 °C<br>ii. B(OMe)₃<br>iii. 5% aq HCl | 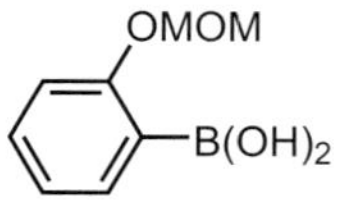 2-OMOM-phenylboronic acid | [222] |

| | | | | |
|---|---|---|---|---|
| 9 | 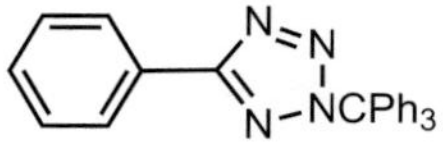 ![phenyl-tetrazole-NCPh3] | i. *n*-BuLi (1 equiv) THF, ≤20 °C ii. B(O-*i*-Pr)$_3$ (1.3 equiv) ii. *i*-PrOH·NH$_4$Cl·H$_2$O | 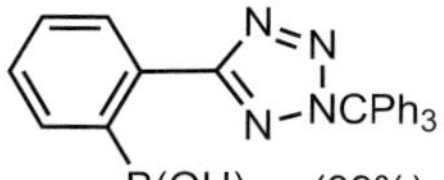 2-(tetrazolyl)phenyl B(OH)$_2$ (89%) | [224] |
| 10 | 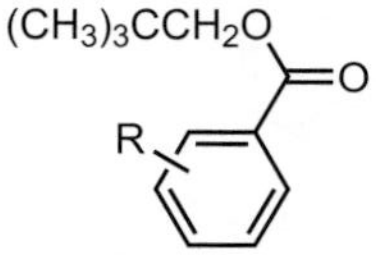 (CH$_3$)$_3$CCH$_2$O-C(=O)-Ar, R = *p*-Br or *o*-Br | i. LDA (1.2 equiv) B(O-*i*-Pr)$_3$ (2.6 equiv), THF ii. Diethanolamine (1.1 equiv) | 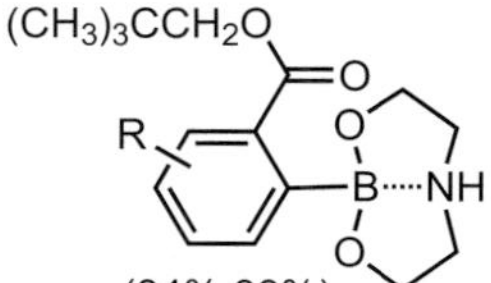 diethanolamine boronate (84%, 88%) | [226] |
| 11 | 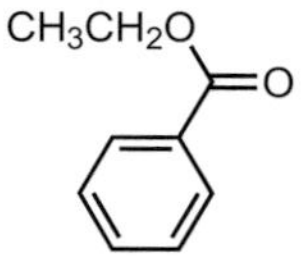 CH$_3$CH$_2$O-C(=O)-Ph | i. LTMP (1.5 equiv) B(O-*i*-Pr)$_3$ (2 equiv) THF, −78 °C ii. HOCH$_2$CMe$_2$CH$_2$OH | 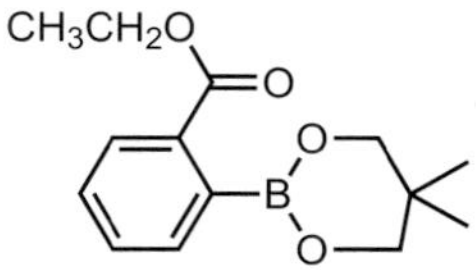 neopentyl glycol boronate (92%) | [227] |
| 12 | 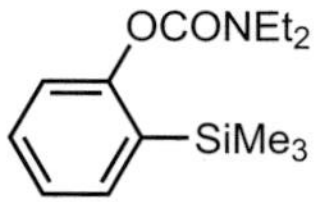 2-(OCONEt$_2$)-C$_6$H$_4$-SiMe$_3$ | i. BBr$_3$ (1.5 equiv) CH$_2$Cl$_2$, −78 °C to rt ii. 5% aq HCl | 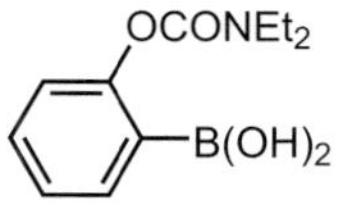 2-(OCONEt$_2$)C$_6$H$_4$B(OH)$_2$ (>85%) | [222] |
| 13 | 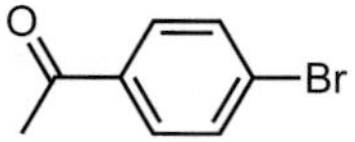 4-Br-C$_6$H$_4$-C(=O)CH$_3$ | B$_2$pin$_2$ (1.1 equiv) PdCl$_2$(dppf) (3 mol%) KOAc (3 equiv) DMSO, 80 °C, 1 h | 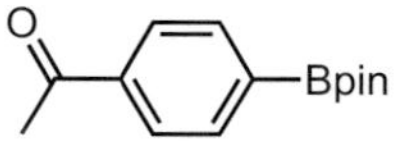 4-Bpin-C$_6$H$_4$-C(=O)CH$_3$ (80%) | [232] |

*(Continued)*

1.3 Preparation of Boronic Acids and Their Esters | 35

**36** | *1 Structure, Properties, and Preparation of Boronic Acid Derivatives*

**Table 1.3** *(Continued)*

| Entry | Substrate | Conditions | Product | Reference |
|---|---|---|---|---|
| 14 | 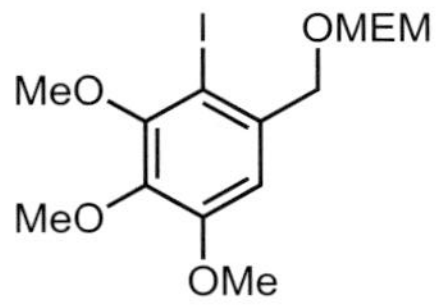 (aryl with OMEM, MeO, MeO, OMe) | 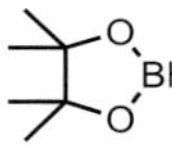 O-BH-O (2 equiv)<br>Et₃N (3 equiv)<br>Pd(OAc)₂ (5 mol%)<br>PCy₂(o-biph) (10 mol%)<br>80 °C, 0.5 h | 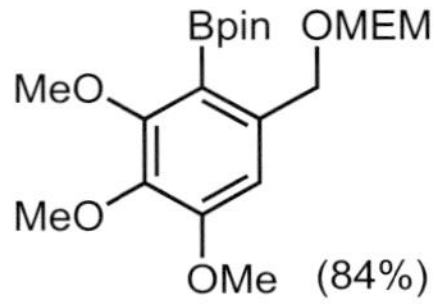 Bpin-aryl with OMEM, MeO, MeO, OMe (84%) | [234] |
| 15 | 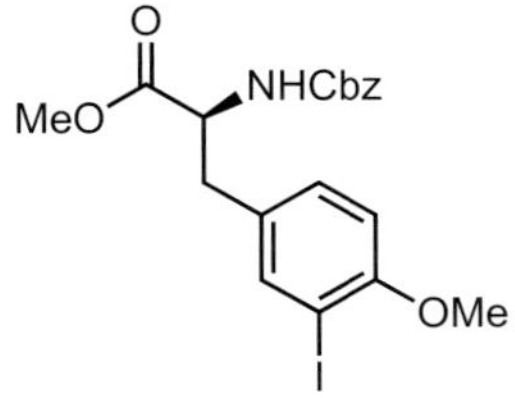 (phenylalanine derivative with NHCbz, OMe) | B₂pin₂<br>PdCl₂(dppf) (3 mol%)<br>KOAc (3 equiv)<br>DMSO, 80 °C, 3 h | 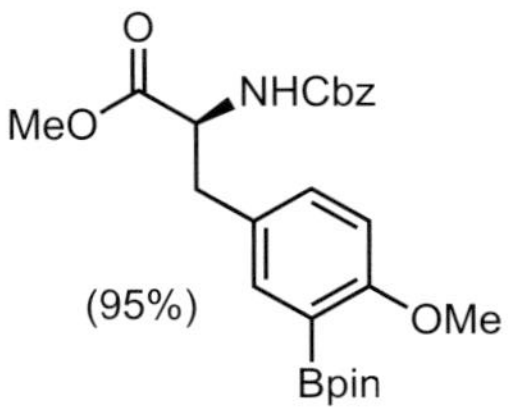 Bpin product with NHCbz, OMe (95%) | [237] |
| 16 | 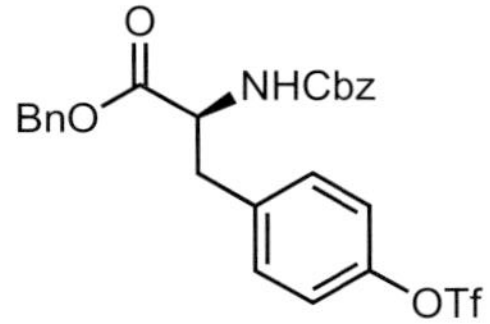 (tyrosine derivative BnO, NHCbz, OTf) | 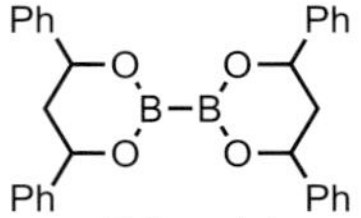 Ph-substituted diol boronate (1.1 equiv)<br>PdCl₂(dppf) (8 mol%)<br>KOAc, DMF, 100 °C, 3 h | 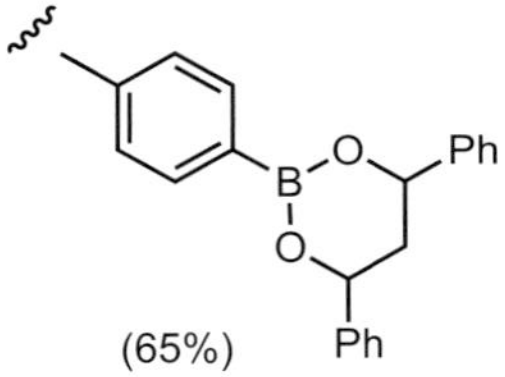 aryl-B(OR)₂ with Ph,Ph diol (65%) | [240] |
| 17 |  MeO-C₆H₄-Cl | B₂(OH)₄ (1.5 equiv)<br>Pd(0) (1 mol%)<br>XPhos (2 mol%)<br>NaOt-Bu (1 mol%)<br>KOAc (3 equiv)<br>EtOH, 80 °C, 18 h | 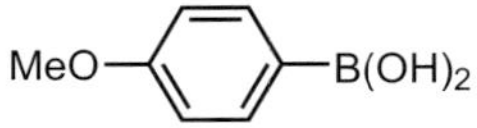 MeO-C₆H₄-B(OH)₂ (>90%) | [241] |

| # | Substrate | Conditions | Product | Ref |
|---|---|---|---|---|
| 18 | 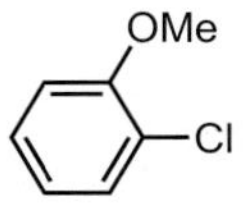 2-chloroanisole | B₂pin₂ (1.1 equiv), Pd(dba)₂ (3 mol%), PCy₃ (7.2 equiv), KOAc (1.5 equiv), dioxane, 80 °C, 48 h | 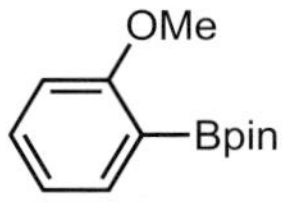 2-Bpin anisole (70%) | [242] |
| 19 | 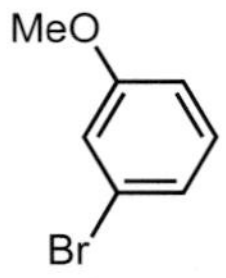 3-bromoanisole | B₂pin₂ (1.1 equiv), 1/2[IrCl(COD)]₂ + bpy (3 mol%), benzene, 80 °C, 16 h | 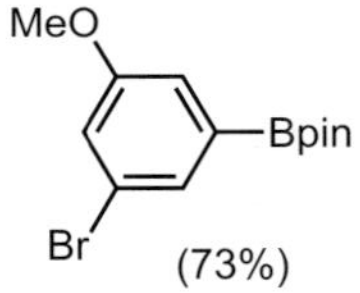 3-MeO-5-Br-phenyl-Bpin (73%) | [253] |
| 20 |  4-iodobenzonitrile | B₂pin₂ (1.0 equiv), [Ir(OMe)(COD)]₂ (1.5 mol%), dtbpy (3 mol%), THF, 25 °C, 8–48 h | 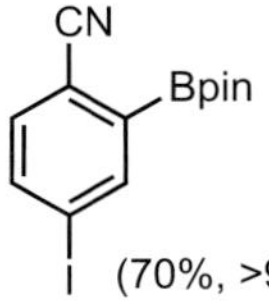 (70%, >99:1 regio) | [255] |
| 21 | 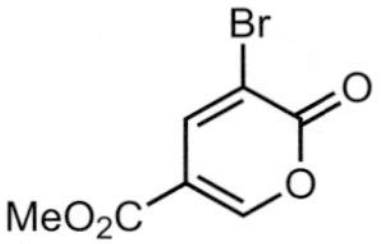 3-bromo-5-(methoxycarbonyl)-pyranone + n-Bu—≡—Bpin | o-Cl₂C₆H₄, 180 °C, 18 h (−CO₂) | 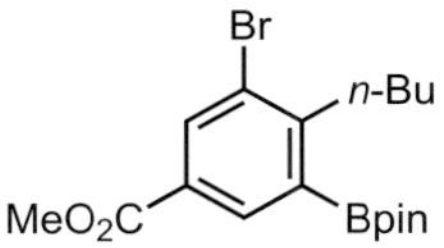 n-Bu/Bpin arene with Br, MeO₂C (82%, 10:1 regio) | [260] |
| 22 | 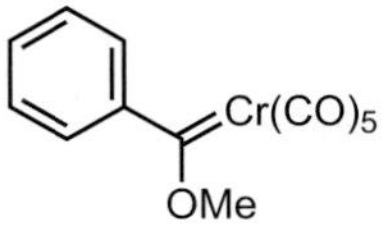 Ph-C(=Cr(CO)₅)OMe + n-Bu—≡—Bpin(pinacolboronate alkyne) | 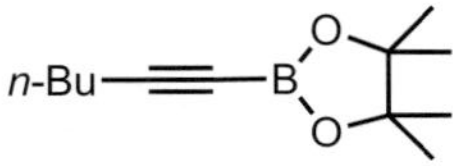 THF, 45 °C, 16 h | 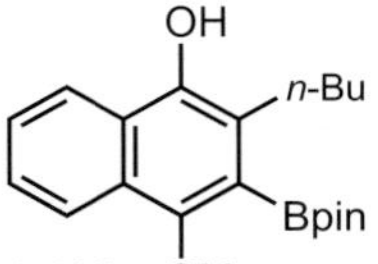 n-Bu, Bpin naphthol with OH, OMe (73%) | [261] |

(Continued)

**Table 1.3** (Continued)

| Entry | Substrate | Conditions | Product | Reference |
|---|---|---|---|---|
| 23 | 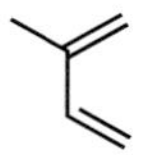 (isoprene structure) | 1. 1. n-Bu—≡—Bpin<br>CoBr₂(dppe) (10 mol%)<br>ZnI₂ (20 mol%)<br>Zn (20 mol%) CH₂Cl₂, rt<br>2. DDQ, C₆H₆ | 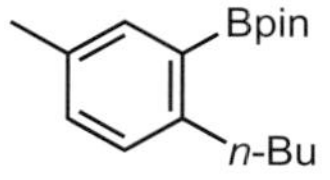 (aryl with Bpin, Me, n-Bu)<br>(80%, >95:5 regio) | [263] |
| 24 | 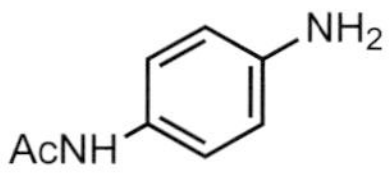 AcNH–C₆H₄–NH₂ | B₂pin₂<br>t-BuONO<br>(PhCO₂)₂O (2 mol%)<br>MeCN, rt | 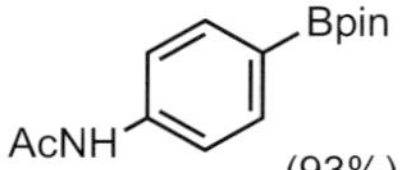 AcNH–C₆H₄–Bpin<br>(93%) | [264] |
| 25 | 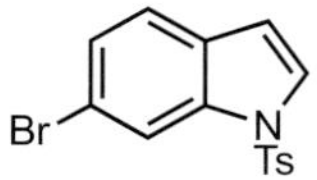 6-Br-N-Ts-indole | 1. Hg(OAc)₂, AcOH<br>H₂O, HClO₄<br>2. BH₃·THF<br>3. H₂O | 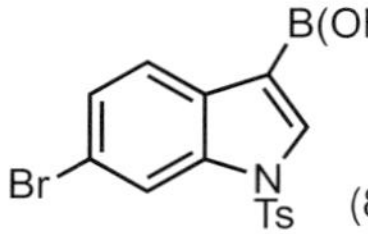 6-Br-N-Ts-indol-3-yl-B(OH)₂<br>(85%) | [216] |
| 26 | 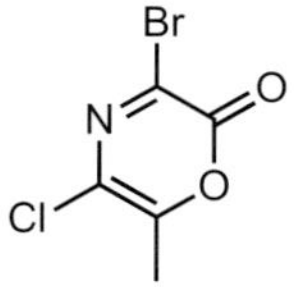 3-Br-5-Cl-6-methyl-2H-1,3-oxazin-2-one | Ph—≡—Bpin<br>toluene, reflux,<br>48 h (–CO₂) | 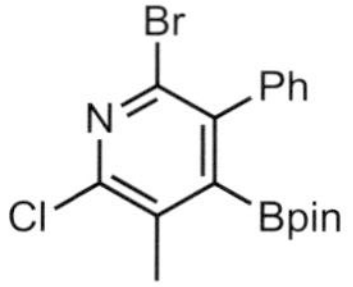 2-Br-3-Ph-4-Bpin-5-Me-6-Cl-pyridine<br>(74%, 20:1 regio) | [271a] |

involved the addition of methylborate to an ethereal solution of phenylmagnesium bromide at −15 °C, became notorious for providing a low yield of desired product [200]. Boron trifluoride was also employed in place of borates [201]. In the early 1930s, Johnson and coworkers developed the first practical and popular method for preparing phenylboronic acid and other arylboronic acids with an inverse addition procedure meant to minimize the undesired formation of borinic acid by-product [202, 203]. In this variant, phenylmagnesium bromide is slowly added to a solution of n-butylborate at −70 °C. In the reaction between an arylmagnesium bromide and a trialkylborate, the exhaustive formation of undesired borinic acid and borane via a second and third displacement on the intermediate boronic ester is prevented by the precipitation of the magnesium trialkoxyphenylborate salt (**81**, M = MgX, in Equation 1.33, Figure 1.21). The latter salt is also thought not to dissociate into the corresponding boronic ester and metal alkoxide at low temperatures, which is key to protecting the desired boronate from a second displacement by the Grignard reagent (Equation 1.34). Then, the free boronic acid is obtained following a standard aqueous workup to hydrolize the labile boronic ester substituents. These types of procedures have been used successfully in the kilogram-scale preparation of arylboronic acids [204–206]. Borinic esters may form in significant amounts at higher temperatures or when using electron-rich arylmagnesium reagents. Equilibration of mixtures leading to enrichment in boronic esters is practically useful in some limited cases [84].

The isolation of free boronic acids using an aqueous workup may lead to low yields especially in the case of small or polar ones, which tend to be water soluble even at a low pH (Sections 1.2.2.2 and 1.4). In such cases, it is often better to isolate the desired compound as a boronic ester. Using an improved procedure that does not involve an aqueous workup, Brown and Cole reported that the reaction of several types of organolithium intermediates with triisopropylborate was found to be very effective for the synthesis of arylboronic esters [207]. To help minimize the possible formation of borinic acids and boranes by multiple displacements (i.e., Equation 1.34 in Figure 1.21), the Brown–Cole protocol involves the slow addition of the organolithium to a solution of triisopropylborate in diethyl ether cooled to −78 °C. The use of smaller borate esters such as trimethylborate gave large proportions of multiple addition products (i.e., borinic acid and borane). With the use of triisopropylborate, however, the clean formation of lithium alkoxyboronate salt (**81**, M = Li, R = i-Pr, Figure 1.21) was demonstrated by NMR spectroscopy, and the boronic ester can be

$$ArM + B(OR)_3 \longrightarrow \underset{\mathbf{81}}{M[ArB(OR)_3]} \rightleftarrows ArB(OR)_2 + ROM \quad (1.33)$$

$$ArB(OR)_2 + ArM \longrightarrow M[Ar_2B(OR)_2] \rightleftarrows Ar_2B(OR) + ROM \quad (1.34)$$

**Figure 1.21** Equilibrium involved in the reaction between arylmetal intermediates (Li or Mg) and borates.

obtained in high purity as the final product upon addition of anhydrous hydrogen chloride at 0 °C. The use of pinacol borates leads directly to pinacol boronates. An improvement to this procedure involves pyrolysis or the use of acid chlorides to breakdown the lithium triisopropylboronate salt, thereby avoiding the generation of free isopropanol and lithium chloride and facilitating the isolation of the boronic ester [208]. An *in situ* quench variant whereby triisopropylborate is present in the flask prior to the addition of butyllithium was described, and in many cases this simpler procedure afforded superior yields of arylboronic and heteroaryl boronic acids compared to the sequential addition procedure [209]. In addition to arylboronic esters, alkenyl, alkynyl, alkyl, and even α-haloalkylboronic esters were made in this way [207]. If desired, the free boronic acid may be obtained by hydrolysis of the ester. The metal–halogen exchange route, both from aryllithium and arylmagnesium intermediates, can even be applied to functionalized substrates containing acidic hydrogen atoms provided that temporary protection via silylation is effected (entry 1, Table 1.3) or a suitable excess of organometallic reagent is employed (entries 2 and 4). All isomers of hydroxybenzeneboronic acid were synthesized from the corresponding bromophenols using this method [212]. An efficient, low-temperature I-to-Mg exchange protocol [213] compatible with esters, nitriles, and benzylic bromide functionalities was employed in a facile synthesis of *o*-nitro arylboronic acids (entry 3, Table 1.3) [214a]. A variant using in situ borate quench provides a noncryogenic preparation of other arylboronic esters [214b].

A new convenient procedure to synthesize arylboronic esters from Grignard reagents and trimethylborate was described [215]. This method involves a nonaqueous workup procedure where the resulting solution of aryldimethoxyboronate is evaporated to eliminate the excess $B(OMe)_3$ and the residual solid is refluxed overnight in a solution of diol in toluene. In particular, several ethylene glycol arylboronic esters were prepared using this method (e.g., entry 5, Table 1.3). Alternatively, the robust pinacol ester can be obtained directly by electrophilic quench of the aryllithium intermediate with a pinacol borate ester (entry 6). The use of bis(diisopropylamino)boron chloride as trapping agent in the reaction of both organolithium and magnesium compounds provides the corresponding bis(diisopropylamino)boranes, which can be easily transformed into the corresponding boronic esters and oxazaborolidines by exchange with a diol or an aminodiol [217].

1.3.1.1.2 **By Directed ortho-Metalation** The metalation of arenes functionalized with coordinating *ortho*-directing groups such as amines, ethers, anilides, esters, amides, and carbamates is yet another popular way to access arylmetal intermediates that can be trapped with boric esters. Early work showed that the *ortho*-lithiation of N,N-dialkylated benzylamines was a suitable method for the synthesis of *ortho*-methylamino-benzeneboronic acids [218–220]. Sharp and Snieckus further demonstrated the efficiency of this method in the preparation of *ortho*-carboxamido phenylboronic acids (entry 7, Table 1.3) [221]. This protocol was then generalized to many other substrates. For example, methoxymethoxybenzene (entry 8) and pivaloylaniline can be treated with *s*-BuLi in the presence of TMEDA in THF at

−78 °C, and the resulting *ortho*-lithiated intermediates are quenched with trimethyl borate followed by an aqueous acidic workup described above (Section 1.3.1.1.1), giving the corresponding arylboronic acids in good yields [222, 223]. Although the crude boronic acids could be used directly in Suzuki cross-coupling reactions, they were characterized as their stable diethanolamine adducts. The *ortho*-metalation route to arylboronic acids constitutes a reliable process in pharmaceutical chemistry, where it can be applied to heterocyclic intermediates such as a tetrazole required in the synthesis of the antihypertensive drug losartan (entry 9, Table 1.3) [224]. N,N-Diethyl *O*-carbamates are particularly valuable directors for the introduction of *ortho*-boronyl groups, as they can also be employed as orthogonal partners in Suzuki–Miyaura cross-couplings [225]. The use of carboxyesters as directing groups is more problematic as the metalated intermediate can undergo condensation with the benzoate substrate, giving a benzophenone. In a newer protocol, the metalation step is performed in the presence of the borate electrophile [226]. This *in situ* metalation–boronylation procedure employs LDA as base, and neopentyl esters were found particularly suitable because of their stability in the presence of this base. Most importantly, the LDA is compatible with boric esters under the conditions employed, and its inertness to bromide-substituted benzoates provides another significant advantage over the use of BuLi for the deprotonation step. Thus, treatment of a solution of bromo-substituted neopentyl benzoate esters and excess triisopropylborate with LDA (1.1–1.5 equiv) in THF led to the isolation of crude *ortho*-carboxy arylboronic acids, which were isolated as diethanolamine adducts in high yields (entry 10, Table 1.3). A limitation of this method, using LDA as the base, is the requirement for an electron-withdrawing substituent to activate the arene substrate. Neopentyl benzoate, for example, does not undergo directed metalation and rather gives the corresponding diisopropyl carboxamide. A recent variant of this *in situ* trapping procedure using 2,2,6,6-tetramethylpiperidide (LTMP) as the base led to a more general methodology allowing the presence of other substituents normally incompatible with standard *ortho*-metalation procedures with alkyllithium bases [227]. For example, ethyl benzoate, benzonitrile, and fluoro- and chlorobenzene were transformed in high yield into the corresponding *ortho*-substituted boronic acids as neopentylglycol esters. As demonstrated in the case of ethyl benzoate (entry 11), the use of LTMP as the base is particularly advantageous because LDA fails to metalate this substrate and rather provides the carboxamide product of addition to the ester.

#### 1.3.1.2 Transmetalation of Aryl Silanes and Stannanes

One of the earliest methods for preparing aromatic boronic acids involved the reaction between diaryl mercury compounds and boron trichloride followed by hydrolysis [228]. Borane can also be employed [229]. As organomercurial compounds are to be avoided for safety and environmental reasons, this approach has remained unpopular. In this respect, trialkylaryl silanes and stannanes are more suitable and both can be transmetalated efficiently with a hard boron halide such as boron tribromide [230]. The apparent thermodynamic drive for this reaction is the higher stability of B−C and Si(Sn)−Br bonds of product compared to the respective B−Br and Si(Sn)−C bonds in the substrates. It is noteworthy that the resulting dibromoboranes can be cross-coupled *in situ* with aryl halides [231]. Using this ipso-desilylation method, functionalized arylboronic acids, including indolylboronic acids, can be made following an aqueous acidic workup to hydrolyze the arylboron dibromide intermediate [222, 231]. For example, some boronic acids were synthesized more conveniently from the trimethylsilyl derivative than by a standard low-temperature *ortho*-metalation procedure (entry 12, Table 1.3). The use of pinacol in the workup leads to the corresponding boronic esters [231].

#### 1.3.1.3 Coupling of Aryl Halides with Diboronyl Reagents

The traditional method involving the trapping of aryllithium or arylmagnesium reagents with boric esters is limited by the functional group compatibility of these hard organometallic species as well as the rigorously anhydrous conditions required. In search for milder conditions amenable to a wider scope of substrates and functionalities, Miyaura and coworkers found that diboronyl esters such as $B_2pin_2$ (**63**, Figure 1.16) undergo a smooth cross-coupling reaction with aryl bromides, iodides, and triflates under palladium catalysis [232]. This modern reaction process is described in Chapter 2; thus, only a brief summary is presented in this section. A detailed mechanism has been proposed [164b, 232], and a number of diboronyl reagents are now commercially available, including the common diborylpinacolate ($B_2pin_2$). Standard conditions for the coupling reaction involve $PdCl_2(dppf)$ as catalyst, with potassium acetate as the base in a polar aprotic solvent [232]. The mildness of these conditions is apparent in the use of carbonyl-containing substrates such as benzophenones (entry 13, Table 1.3) or benzaldehydes [92], which would be incompatible with the metal–halogen exchange procedures described in Section 1.3.1.1. Pinacolborane (**58**, Figure 1.15) can also serve as an efficient boronyl donor in this methodology (entry 14) [233]. The use of cedranediolborane has also been proposed as an alternative to pinacolborane, which gives pinacol esters that are notoriously difficult to hydrolyze [235]. The scope of arene substrates in coupling reactions with diboronyl esters or pinacolborane is very broad. The preparation of peptide dimers has been described using a one-pot borylation/Suzuki coupling [236]. Hindered [238], electron-rich aryl halides (entries 14 and 15), and even pyridinyl halides [239], may also be used with high efficiency. Of particular significance is the use of pinacolborane with aryltriflates, which can be made with ease from phenols [233]. For instance, 4-borono-phenylalanine is now easily accessible from tyrosine using this approach (entry 16). As shown with this example and others [238], the use of diboronyl reagents with hydrolytically labile substituents is advantageous if the desired product is the free boronic acid. Alternatively, a recent procedure employs $B_2(OH)_4$ directly with aryl halides (entry 17) [241]. Aryl chlorides are more attractive substrates compared to bromides and iodides due to their low cost and wider commercial availability. In this regard, modified conditions with $Pd(dba)_2$ and tricyclohexylphosphine as catalyst system provide pinacolates from aryl chlorides, even electron-rich ones (entry 18, Table 1.3) [242]. Other palladium ligand systems are efficient [243]. Alternatively, a microwave-promoted procedure for aryl chlorides using a palladium/imidazolium system has been described [244]. A similar procedure employed aryldiazonium salts as substrates [245]. Recent procedures for nickel-catalyzed borylations for aryl halides [246] and sulfonates [247] and copper-catalyzed variants have been reported [248].

#### 1.3.1.4 Direct Boronation by Transition Metal-Catalyzed Aromatic C–H Functionalization

In terms of atom economy, a very attractive strategy for preparing boronic acids and esters is the direct boronation of arenes through a transition metal-catalyzed C–H functionalization [249]. In addition to the catalyst, a suitable boron donor is required, and both diboronyl esters and dialkoxyboranes were found to be very appropriate in this role. The concept of this type of direct borylation was first demonstrated on alkanes using photochemical conditions [250]. For arene substrates, several research groups including those of Smith [251], Hartwig [252], Miyaura/Hartwig [253], and Marder [254] have pioneered a number of efficient procedures using iridium and rhodium catalysts (entry 19, Table 1.3). The most active catalyst for this chemistry is an iridium di-*t*-butylbipyridine (dtbpy) complex [253], and room-temperature borylations of *para*-substituted cyanoarenes with this system tend to occur mainly *ortho* to the nitrile (entry 20) [255]. Otherwise, regioselectivity is under steric control and it remains a major challenge except for monosubstituted and 1,3-disubstituted arenes where *meta*-borylation is the main pathway, thus complementing the directed *ortho*-metalation/borylation approach described in Section 1.3.1.2 [256]. Exceptionally, a heterogeneous, silica-supported Ir-phosphine catalyst provides directed *ortho*-borylation of a wide variety of monosubstituted arenes in very high selectivity [257]. The scope and mechanism of this contemporary approach to the synthesis of boronic acid derivatives are discussed in detail in Chapter 2.

#### 1.3.1.5 Cycloadditions of Alkynylboronates

Alkynylboronates are versatile cycloaddition partners [258]. Thermal [4 + 2] cycloadditions are possible with activated dienes [259, 260]. For example, 2-pyrones provide polysubstituted arylboronates in variable regioselectivities following $CO_2$ extrusion (entry 21, Table 1.3) [260]. Harrity and coworkers also described the application of 2-substituted 1-alkynylboronic esters in the Dötz cycloaddition of Fisher chromium carbene complexes, affording in a highly regioselective manner a novel class of hydroxy-naphthyl boron pinacolates (entry 22, Table 1.3) [261]. These reaction products also provided, upon treatment with ceric ammonium nitrate, the corresponding quinone boronic esters. A one-pot 3-component ruthenium-catalyzed process between alkynylboronates, propargylic alcohols, and terminal alkynes provides benzoboroxoles similar to **8** [262], whereas a recent cobalt-catalyzed formal [4+2] cycloaddition leads to arylboronates after an elimination or a chemoselective oxidative treatment that leaves the C–B bond intact (entry 23, Table 1.3) [263].

#### 1.3.1.6 Other Methods

An intriguing deaminoborylation of aniline was recently reported using benzoyl peroxide [264]. This process can provide simple, monofunctionalized arylboron pinacolates in variable yields (entry 24, Table 1.3).

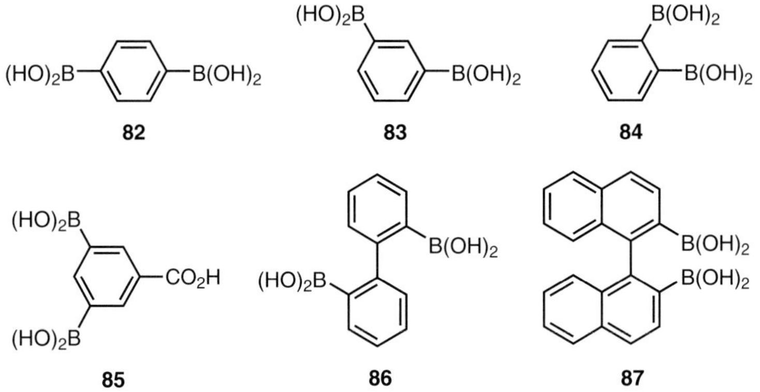

**Figure 1.22** Selected examples of diboronic acids.

## 1.3.2
### Diboronic Acids

The preparation of all three substitution patterns of benzenediboronic acid has been reported (Figure 1.22). Although the preparation of the 1,4- and 1,3-benzenediboronic acids **82** and **83** from the corresponding dibromides was well described [186a,265], the preparation of the *ortho*-isomer **84** is more tedious [79, 266]. A number of other mono- and polycyclic aromatic diboronic acids such as **85** [177], **86** [267], and the binaphthyl derivative **87** [268] were described. Tetraboronic acid **6** (Figure 1.2) is a popular building block in materials chemistry (see Chapter 14).

## 1.3.3
### Heterocyclic Boronic Acids

Heterocyclic aromatic boronic acids, in particular pyridinyl, pyrrolyl, indolyl, thienyl, and furyl derivatives, are popular cross-coupling intermediates in natural product synthesis and medicinal chemistry. The synthesis of heterocyclic boronic acids has been reviewed in 2004 [269]. In general, these compounds can be synthesized using methods similar to those described in the above section for arylboronic acids. Of particular note, all three isomers of pyridineboronic acid have been described, including protected forms of the unstable and hitherto elusive 2-substituted isomer, notorious for its tendency to protodeboronate [41, 270]. Many 2-boronyl heteroarenes are sensitive, but they can be stabilized significantly as MIDA adducts (Section 1.2.3.3) [41]. Improvements and variants of the established methods for synthesizing heterocyclic boronic acids have been constantly reported [12, 209]. For example, a Hg–B transmetalation procedure was employed to synthesize a highly functionalized indolylboronic acid (entry 25, Table 1.3) [216]. Recent advances in the preparation of nitrogen-containing heteroaromatic boronic acids and esters include various thermal and catalyzed cycloadditions of alkynylboronates [271] (entry 26).

## 1.3.4
## Alkenylboronic Acids

Alkenylboronic acids constitute another class of highly useful synthetic intermediates. They are particularly popular as partners in the Suzuki–Miyaura cross-coupling reaction for the synthesis of dienes and other unsaturated units present in a large number of natural products (see Section 1.5.3.1). Several methods are available for the preparation of a wide range of alkenylboronic acids with different substitution patterns. These approaches are summarized in a generic way in Figure 1.23 and are described in the following sections.

### 1.3.4.1 Electrophilic Trapping of Alkenylmetal Intermediates with Borates

Alkenylboronic acids can be synthesized from reactive alkenylmetal species in a way similar to that described above for arylboronic acids (Section 1.3.1.1.1) [272]. Typically, alkenyl bromides or iodides are treated sequentially with $n$-BuLi and a borate (entry 1, Table 1.4). A nonpolar trienylboronic acid was synthesized using this approach [274]. As described in Section 1.2.2.2, small boronic acids tend to be highly soluble in water and may be difficult to isolate when made using the traditional approach involving an aqueous workup. In such cases, exemplified with the polymerization-prone ethyleneboronic acid prepared from vinylmagnesium bromide, it has proved more convenient to rather isolate it as a dibutyl ester by extraction of the acidic aqueous phase with butanol [275]. Alkoxy-functionalized butadienyl- and styrenylboronic esters were synthesized from $\alpha,\beta$-unsaturated acetals by treatment with Schlosser's base and subsequent trapping with triisopropylborate (entry 2) [276]. Both alkenyl- [277] and cycloalkenylboronates [278] can be made from ketones using a Shapiro reaction with trapping of the alkenyllithium intermediate with a borate (entry 3). Terminal 2-alkenylboronates ($\alpha$-vinylboronates) are not easily accessible, but a regioselective Ni-catalyzed approach to alkenylalanes followed by borate trapping was recently reported (entry 4) [279].

### 1.3.4.2 Transmetalation Methods

The treatment of trialkylsilyl derivatives with boron halides described in Section 1.3.1.2 is applicable to alkenyltrimethylsilanes [280]. It was employed as a method for preparing ethylene boronic esters [281]. Recently, isomerically pure tetrasubstituted alkenylboronic esters were synthesized by this approach, following an esterification of the intermediate dichloroborane with pinacol (entry 5, Table 1.4) [282]. *trans*-Alkenylboronic acids can also be synthesized from zirconocene intermediates obtained from the hydrozirconation of terminal alkynes (entry 6) [283].

### 1.3.4.3 Transition Metal-Catalyzed Coupling between Alkenyl Halides/Triflates and Diboronyl Reagents

Alkenyl halides and triflates are suitable substrates in the palladium-catalyzed borylation reaction described for aromatic substrates in Section 1.3.1.3. In this reaction, the geometry of the starting alkenyl halide is preserved in the product, and several functionalities are tolerated in the substrate. At the outset, however, Miyaura

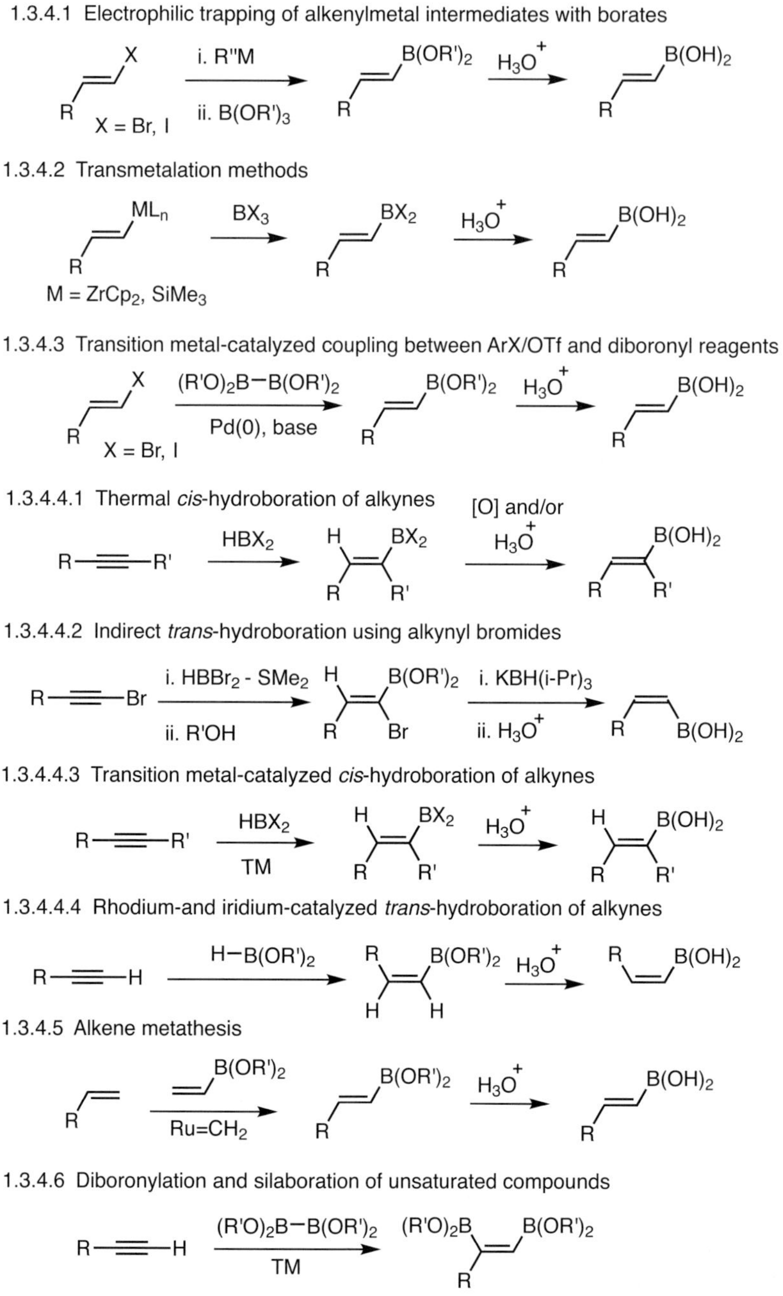
1.3.4.1 Electrophilic trapping of alkenylmetal intermediates with borates
X
i. R''M
ii. B(OR')3
X = Br, I
B(OR')2
H3O+
B(OH)2
1.3.4.2 Transmetalation methods
MLn
BX3
M = ZrCp2, SiMe3
BX2
H3O+
B(OH)2
1.3.4.3 Transition metal-catalyzed coupling between ArX/OTf and diboronyl reagents
(R'O)2B−B(OR')2
Pd(0), base
X = Br, I
B(OR')2
H3O+
B(OH)2
1.3.4.4.1 Thermal cis-hydroboration of alkynes
R−≡−R'
HBX2
BX2
[O] and/or
H3O+
B(OH)2
1.3.4.4.2 Indirect trans-hydroboration using alkynyl bromides
R−≡−Br
i. HBBr2 - SMe2
ii. R'OH
B(OR')2
Br
i. KBH(i-Pr)3
ii. H3O+
B(OH)2
1.3.4.4.3 Transition metal-catalyzed cis-hydroboration of alkynes
R−≡−R'
HBX2
TM
BX2
H3O+
B(OH)2
1.3.4.4.4 Rhodium-and iridium-catalyzed trans-hydroboration of alkynes
R−≡−H
H−B(OR')2
B(OR')2
H3O+
B(OH)2
1.3.4.5 Alkene metathesis
B(OR')2
Ru=CH2
B(OR')2
H3O+
B(OH)2
1.3.4.6 Diboronylation and silaboration of unsaturated compounds
R−≡−H
(R'O)2B−B(OR')2
TM
(R'O)2B
B(OR')2

**Figure 1.23** Common methods for the synthesis of alkenylboronic acids (esters).

**Table 1.4** Selected examples of preparative methods for alkenylboronic acids and esters.

| Entry | Substrate | Conditions | Product | Reference |
|---|---|---|---|---|
| 1 | 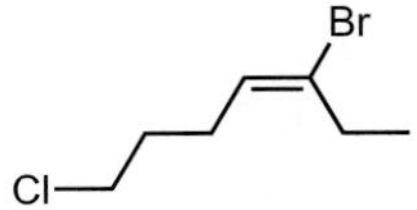 (bromoalkenyl chloride) | i. s-BuLi, THF, −78 °C<br>ii. B(OR')$_3$, −78 °C, 1 h<br>iii. HCl/Et$_2$O, −78 °C to rt<br>iv. H$_2$O<br>v. HO(CH$_2$)$_3$OH | 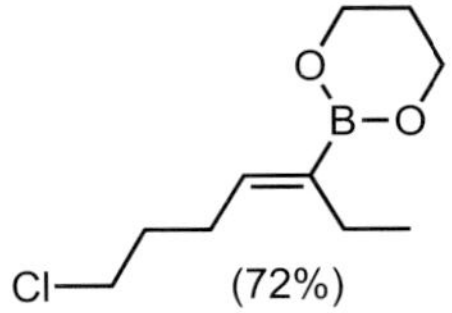 (dioxaborinane product) (72%) | [273] |
| 2 | 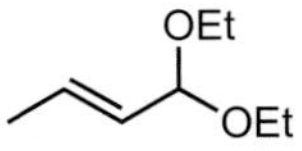 (diethoxy alkene) | i. n-BuLi/KO-t-Bu (2.5 equiv) THF, −95 °C, 2 h<br>ii. B(O-i-Pr)$_3$ (2 equiv) −95 °C to rt<br>iii. H$_2$O, extraction<br>iv. HOCH$_2$CMe$_2$CH$_2$OH toluene, rt, 12 h | 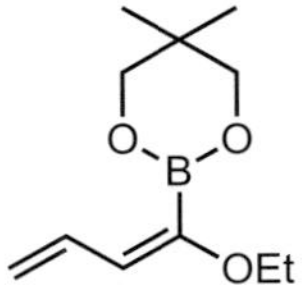 (dioxaborinane OEt product) (93%) | [276] |
| 3 | 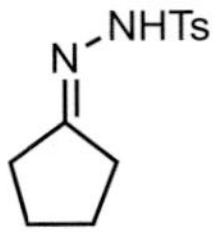 (cyclopentanone N-Ts hydrazone) | i. n-BuLi (4 equiv) TMEDA, hexane −78 °C, 1 h; rt, 2 h<br>ii. i-PrOBpin, −78 °C to rt | 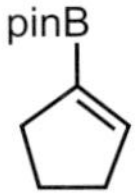 pinB-cyclopentenyl (80%) | [278] |
| 4 | 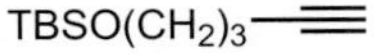 TBSO(CH$_2$)$_3$-≡ (1.1 equiv) | i. DIBALH (1.3 equiv) Ni(dppp)Cl$_2$ (3 mol%) THF, 0 °C, 2 h<br>ii. MeOBpin (3 equiv) 0–80 °C, 24 h | 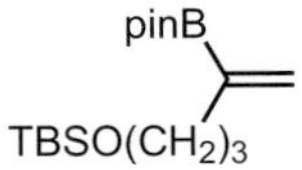 pinB / TBSO(CH$_2$)$_3$ (78%) | [279] |

(Continued)

## 1 Structure, Properties, and Preparation of Boronic Acid Derivatives

**Table 1.4** (Continued)

| Entry | Substrate | Conditions | Product | Reference |
|---|---|---|---|---|
| 5 | 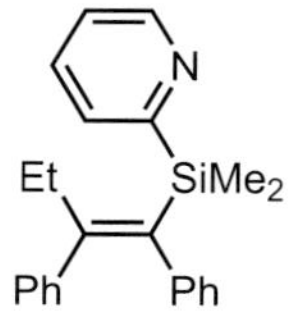 Et–C(Py-SiMe₂)=C(Ph)–CH(Ph)–Et (pyridyl-silyl substituted alkene with Ph groups) | 1. BCl₃ (2.2 equiv) CH₂Cl₂, −40 °C, 5 h  2. Pinacol, Et₃N | 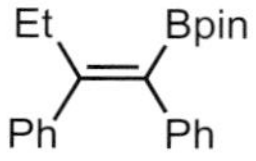 Et(Bpin)C=C(Ph)(Ph) (82%, Z/E 98:2) | [282] |
| 6 | 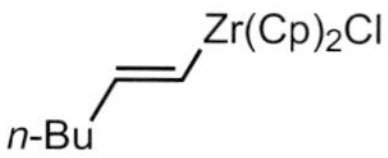 n-Bu–CH=CH–Zr(Cp)₂Cl | catBCl CH₂Cl₂, 0 °C | 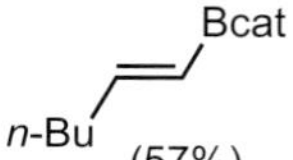 n-Bu–CH=CH–Bcat (57%) | [283] |
| 7 | 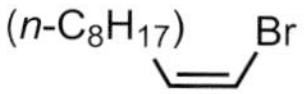 (n-C₈H₁₇)CH=CH–Br (Z) | B₂pin₂ (1.1 equiv) PdCl₂(dppf) (3 mol%) PPh₃ (6 mol%) KOPh (1.5 equiv) toluene, 50 °C, 5 h | 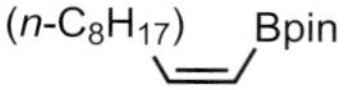 (n-C₈H₁₇)CH=CH–Bpin (74%) | [285] |
| 8 | 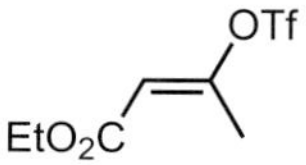 EtO₂C–CH=C(Me)–OTf | B₂pin₂ (1.1 equiv) PdCl₂(PPh₃)₂ (3 mol%) PPh₃ (6 mol%) KOPh (1.5 equiv) toluene, 50 °C, 1 h | 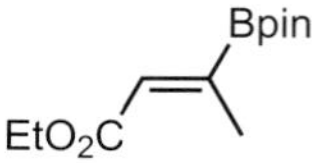 EtO₂C–CH=C(Me)–Bpin (93%, >99% Z:E) | [286] |

| | | | | |
|---|---|---|---|---|
| 9 | 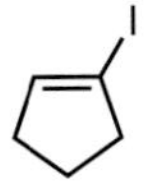 (iodocyclopentene) | HBpin (1.5 equiv)<br>PdCl$_2$(dppf) (3 mol%)<br>AsPh$_3$ (12 mol%)<br>Et$_3$N (3 equiv)<br>dioxane, 80 °C, 16 h | 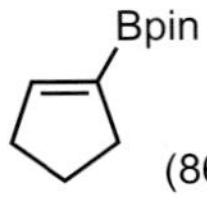 cyclopentenyl-Bpin (86%) | [287] |
| 10 | 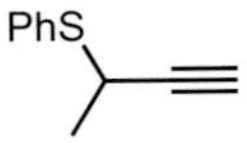 PhS–C(Me)–C≡CH | i. Cy$_2$BH (1 equiv)<br>DME, rt, 1 h<br>ii. Me$_3$NO (2 equiv), reflux<br>iii. HOCMe$_2$CMe$_2$OH<br>rt, 12 h | 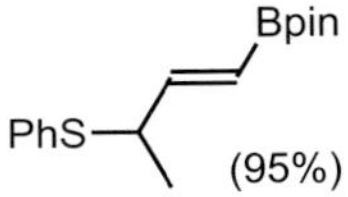 PhS–CH(Me)–CH=CH–Bpin (95%) | [295] |
| 11 | 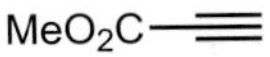 MeO$_2$C–C≡CH | i. Ipc$_2$BH, THF<br>−35 °C to 0 °C<br>ii. CH$_3$CHO (10 equiv)<br>0–40 °C<br>iii. HOCMe$_2$CMe$_2$OH (1 equiv)<br>rt, 12 h | 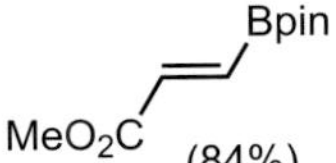 MeO$_2$C–CH=CH–Bpin (84%) | [297, 299] |
| 12 | 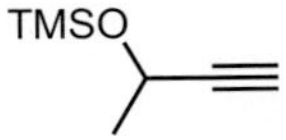 TMSO–CH(Me)–C≡CH | i. Ipc$_2$BH, THF<br>−35 °C to rt, 5 h<br>ii. CH$_3$CHO (xs)<br>0 °C; reflux, 12 h<br>iii. HO(CH$_2$)$_3$OH | 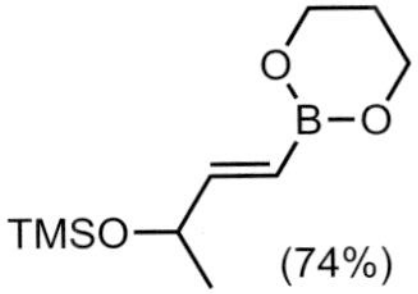 TMSO–CH(Me)–CH=CH–B(OCH$_2$CH$_2$CH$_2$O) (74%) | [298] |
| 13 | 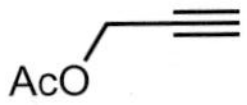 AcO–CH$_2$–C≡CH | i. **93** (1 equiv)<br>ii. H$_2$O, rt, 0.5 h<br>iii. aq CH$_2$O (1 equiv)<br>rt, 1 h<br>iv. HOCMe$_2$CMe$_2$OH (1.1 equiv)<br>rt, 12 h | 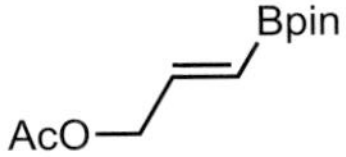 AcO–CH$_2$–CH=CH–Bpin (55%, 97:3 regio) | [300] |

(*Continued*)

**Table 1.4** (Continued)

| Entry | Substrate | Conditions | Product | Reference |
|---|---|---|---|---|
| 14 | 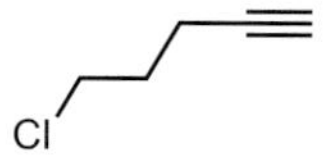 Cl-(CH₂)₃-C≡CH | i. CBH 57 (1 equiv), 70 °C, 1 h<br>ii. H₂O, 25 °C, 1 h<br>iii. Filtration | 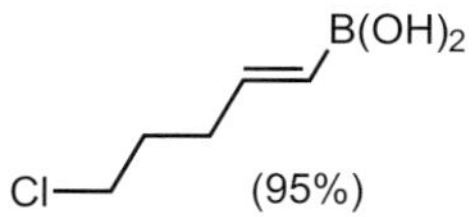 Cl-(CH₂)₃-CH=CH-B(OH)₂ (95%) | [301b] |
| 15 | 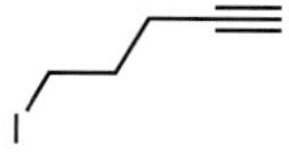 I-(CH₂)₃-C≡CH | HBpin (2 equiv)<br>CH₂Cl₂, 25 °C, 6 h | 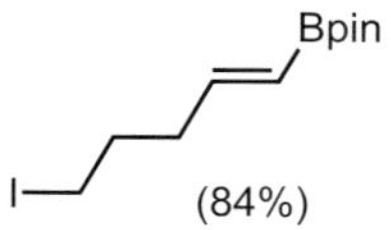 I-(CH₂)₃-CH=CH-Bpin (84%) | [159] |
| 16 | 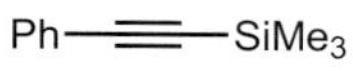 Ph-C≡C-SiMe₃ | i. HBCl₂ (1 equiv)<br>BCl₃ (1 equiv), pentane<br>−78 °C; rt, 12 h<br>ii. MeOH, Et₃N, 0 °C | 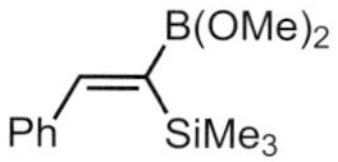 Ph-C(B(OMe)₂)=CH-SiMe₃ (46%) | [305] |
| 17 | 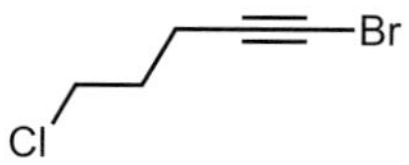 Cl-(CH₂)₃-C≡C-Br | i. HBBr₂·SMe₂, CH₂Cl₂<br>ii. MeOH, pentane<br>iii. K(i-PrO)₃BH, Et₂O<br>0 °C to rt, 0.5 h<br>iv. H₂O, 0 °C<br>v. HO(CH₂)₃OH | 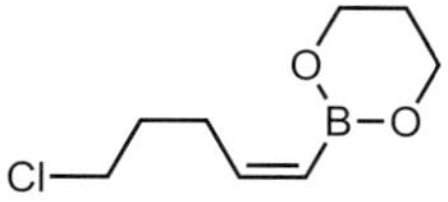 Cl-(CH₂)₃-CH=CH-B(OCH₂CH₂CH₂O) (89%) | [308] |
| 18 | 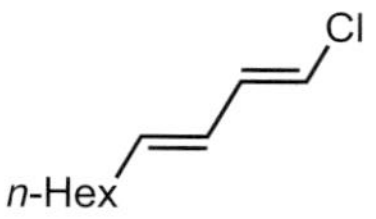 Cl-CH=CH-CH=CH-n-Hex | i. n-BuLi (1.05 equiv)<br>THF, −90 °C, 15 min<br>ii. PhMe₂SiB(OCMe₂)₂<br>warm up to rt, 12 h | 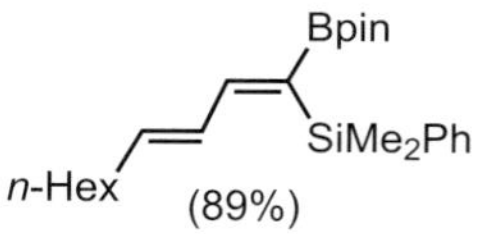 (Bpin)(SiMe₂Ph)C=CH-CH=CH-n-Hex (89%) | [310] |

| | Substrate | Conditions | Product | Ref. |
|---|---|---|---|---|
| 19 | 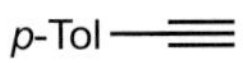 p-Tol—≡ | HBcat (1 equiv) Cp$_2$Ti(CO)$_2$ (4 mol%) C$_6$H$_6$, 25 °C, 2 h | 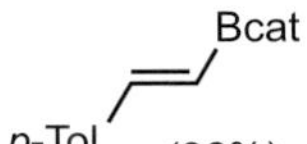 p-Tol—CH=CH—Bcat (96%) | [313] |
| 20 | 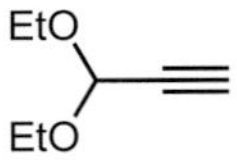 EtO—C(H)(OEt)—≡ | HBpin (1.05 equiv) HZrCp$_2$Cl (5 mol%) CH$_2$Cl$_2$, 25 °C, 24 h | 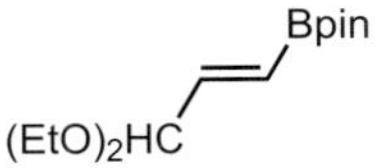 (EtO)$_2$HC—CH=CH—Bpin (82%) | [314] |
| 21 | 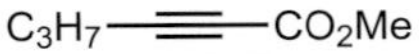 C$_3$H$_7$—≡—CO$_2$Me | pinBH (1.1 equiv) [(Ph$_3$P)CuH] (2 mol%) Ph$_3$P (3 mol%) | 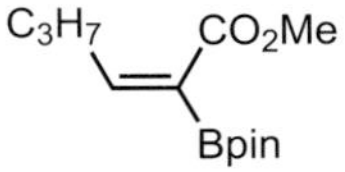 C$_3$H$_7$(Bpin)C=CH—CO$_2$Me (85%, >25:1 Z/E) | [319] |
| 22 | 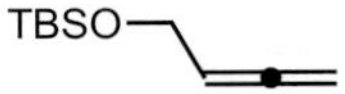 TBSO—CH$_2$—CH=C=CH$_2$ | HBpin (1.5 equiv) Pt(dba)$_2$ (3 mol%) P(t-Bu)$_3$ (6 mol%) toluene, 50 °C, 2 h | 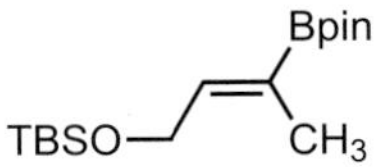 TBSO—CH$_2$—C(Bpin)=C(CH$_3$)H (82%) | [321] |
| 23 | 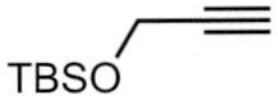 TBSO—CH$_2$—C≡CH (1.2 equiv) | HBcat (1 equiv) [Rh(cod)Cl]$_2$ (1.5 mol%) PPr$_3$ (6 mol%) Et$_3$N (1 equiv) cyclohexane, rt, 2 h |  TBSO—CH=CH—Bcat (71%, 99% Z) | [322] |
| 24 | 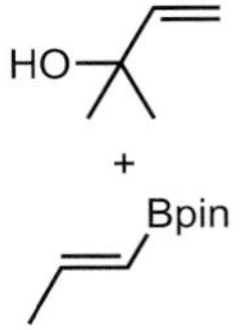 HO—C(CH$_3$)$_2$—CH=CH$_2$ + CH$_2$=CH—Bpin (1 equiv each) | 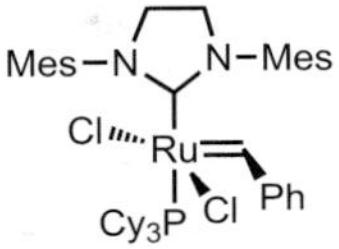 Mes—N⏜N—Mes, Cl···Ru=CHPh, Cy$_3$P, Cl (5 mol%) CH$_2$Cl$_2$, reflux | 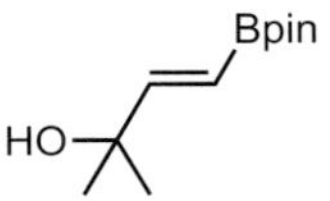 HO—C(CH$_3$)$_2$—CH=CH—Bpin (61%, >20:1 E:Z) | [327] |

(Continued)

**52** | *1 Structure, Properties, and Preparation of Boronic Acid Derivatives*

**Table 1.4** (Continued)

| Entry | Substrate | Conditions | Product | Reference |
|---|---|---|---|---|
| 25 | 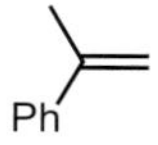 CH₂=C(CH₃)Ph | B₂pin₂ (0.67 equiv)<br>[RhCl(CO)(PPh₃)₂] (5 mol%)<br>3:1 toluene–CH₃CN<br>80 °C, 3 d | 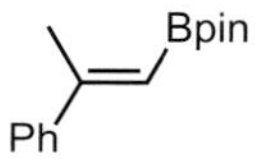 (E)-Ph-C(CH₃)=CH-Bpin<br>(90%) | [331] |
| 26 | 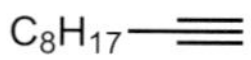 C₈H₁₇–C≡CH<br>(1.1 equiv) | B₂pin₂ (1 equiv)<br>Pt(PPh₃)₄ (3 mol%)<br>DMF, 80 °C, 24 h | 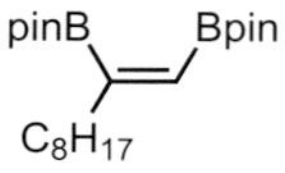 pinB-CH=C(C₈H₁₇)-Bpin<br>(86%) | [332] |
| 27 | 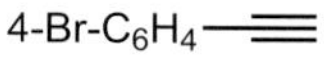 4-Br-C₆H₄–C≡CH | (dan)BBpin **64** (0.67 equiv)<br>[IrCl(COD)]₂ (1.5 mol%)<br>toluene, 80 °C, 24 h | 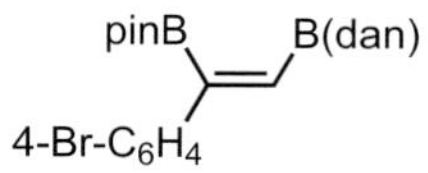 pinB-CH=C(4-Br-C₆H₄)-B(dan)<br>(83% 98:2 *E/Z*) | [165] |
| 28 | 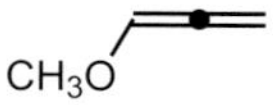 CH₃O-CH=C=CH₂<br>(1.5 equiv) | B₂pin₂ (1 equiv)<br>Pt(dba)₂ (10 mol%)<br>PCy₃ (10 mol%)<br>toluene, 50 °C, 18 h | 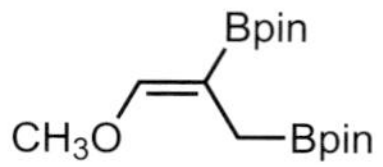 CH₃O-CH=C(Bpin)-CH₂-Bpin<br>(85%) | [334] |
| 29 | 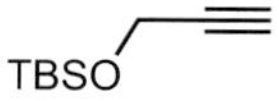 TBSO-CH₂-C≡CH | B₂pin₂ (1.1 equiv)<br>CuCl (1.1 equiv)<br>KOAc (1.1 equiv)<br>P(*t*-Bu)₃ (1.1 equiv)<br>DMF, rt, 16 h | 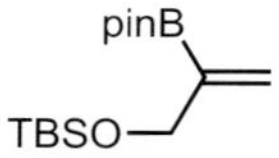 CH₂=C(pinB)-CH₂-OTBS<br>(62%, 91:9 regioselectivity) | [335] |

| | | | | |
|---|---|---|---|---|
| 30 | 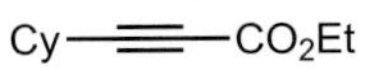 Cy—≡—CO₂Et | B₂pin₂ (1.1 equiv)<br>CuCl (3 mol%)<br>NaO-t-Bu (6 mol%)<br>Xantphos (3 mol%)<br>MeOH (2 equiv)<br>THF, rt | 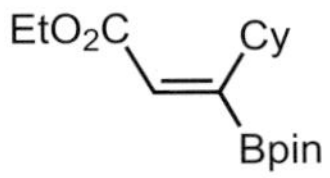 EtO₂C\_/Cy with Bpin (99%) | [336] |
| 31 | 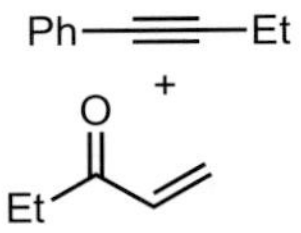 Ph—≡—Et  +  CH₂=CH–C(O)Et | B₂pin₂ (1.5 equiv)<br>[Ni(COD)₂] (5 mol%)<br>n-Bu₃P (10 mol%)<br>tol/MeOH 3:1<br>40 °C, 10 h | 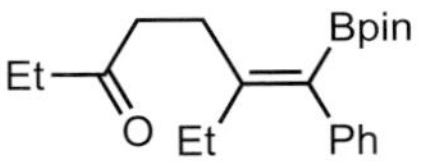 Bpin, Ph, Et, C(O)Et (80%) | [339] |
| 32 | 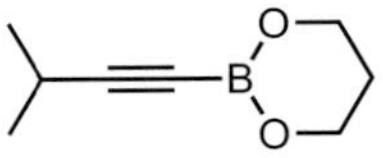 isobutyl-C≡C-B(OCH₂CH₂CH₂O) | i. H₂, Lindlar, pyridine<br>1,4-dioxane, rt, 1.5 h<br>ii. H₂O<br>iii. HO(CH₂)₃OH, pentane | 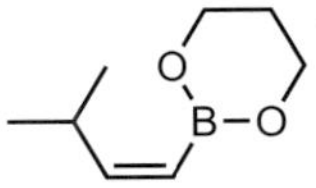 cis-alkenyl boronate (83%, 95% Z) | [341] |
| 33 | 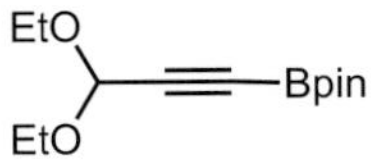 EtO–CH(OEt)–C≡C–Bpin | i. HZrCp₂Cl (1.2 equiv)<br>THF, 25 °C, 0.5 h<br>ii. H₂O, 0.5 h | 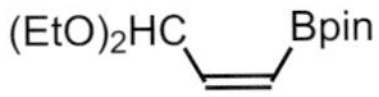 (EtO)₂HC\_/Bpin (82%) | [342] |
| 34 | 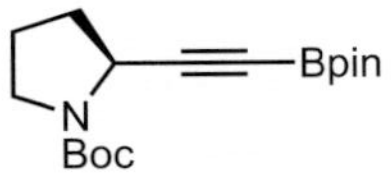 N-Boc-pyrrolidinyl-C≡C–Bpin | i. Cy₂BH (1.0 equiv)<br>Et₂O, 0 °C<br>ii. AcOH (2.2 equiv), 0 °C<br>iii. HOCH₂CH₂NH₂, 0 °C to rt | 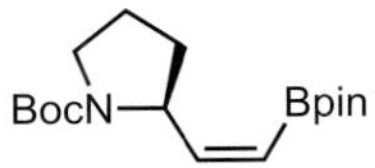 BocN-pyrrolidinyl-CH=CH-Bpin (57%) | [343] |

(*Continued*)

**54** | *1 Structure, Properties, and Preparation of Boronic Acid Derivatives*

**Table 1.4** (Continued)

| Entry | Substrate | Conditions | Product | Reference |
|---|---|---|---|---|
| 35 | 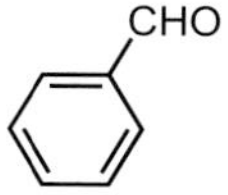 PhCHO | 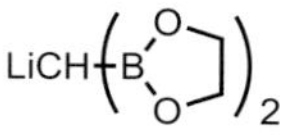 LiCH(Bpin-dioxolane)₂<br>THF/CH₂Cl₂<br>−78 °C, 3 h | 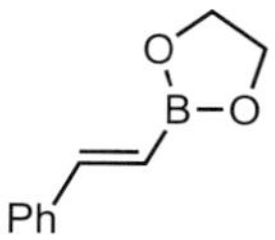 styryl-dioxaborolane, Ph<br>(87%, >93% *E*) | [348] |
| 36 | 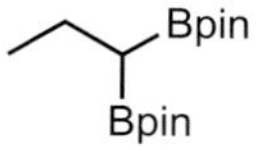 sec-butyl bis(Bpin) | i. LTMP<br>THF, 0 °C, 5 min<br>ii. MeCOPh | 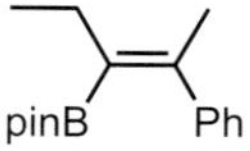 pinB−C(Et)=C(Me)Ph<br>(94%, >99:1) | [352] |
| 37 | 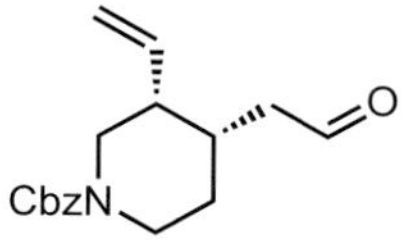 3-vinyl-4-(CH₂CHO)-CbzN-piperidine | Cl₂CHBpin (2 equiv)<br>CrCl₂ (8 equiv)<br>LiI (4 equiv)<br>THF, 25 °C | 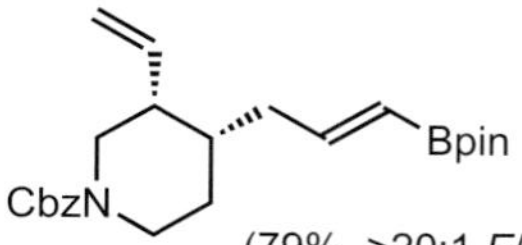 3-vinyl-4-(CH=CH-Bpin)-CbzN-piperidine<br>(79%, >20:1 *E/Z*) | [354] |
| 38 | 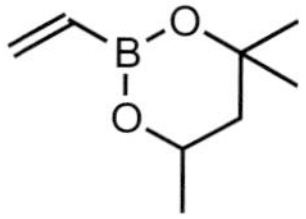 vinyl-Bneop | 4-MeC₆H₄I<br>Pd(OAc)₂ (5 mol%)<br>PPh₃ (12 mol%)<br>*n*Bu₃N (1.2 eq)<br>toluene, reflux, 8 h | 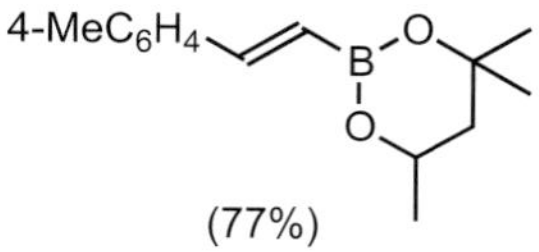 4-MeC₆H₄−CH=CH−Bneop<br>(77%) | [355] |

and coworkers found that the conditions utilized for aryl halide substrates led to low yields of the desired alkenylboronate due to competing reactions such as the formation of homocoupled product of Suzuki cross-coupling [284]. To improve the rate of transmetalation between the diboronyl reagent ($B_2Pin_2$) and the oxidative addition Pd(II) intermediate, stronger bases were evaluated. In the optimal procedure, potassium phenoxide was found to be the most effective base, with a less polar solvent (toluene) than that used with aryl halides, and triphenylphosphine as ligand in place of dppf. Alkenyl bromides and triflates were found to be superior over iodides, and generally afforded good yields in the 70–90% range. The mildness of these conditions opened up a rather impressive scope of suitable substrates [285], including Z-alkenes (entry 7, Table 1.4), and both acyclic and cyclic ones with functionalities such as alkyl halides, silyl-protected alcohols, and carboxylic esters (entry 8) [286]. Pinacolborane was found to be effective in the borylation of alkenyl halides under a new set of optimal conditions (entry 9) [287]. No competing hydroboration was observed, but acyclic Z-configured substrates are inverted under these reaction conditions. Likewise, borylation of alkenyltriflates within a pyran ring can lead to isomerization to the corresponding allylboronate [287].

#### 1.3.4.4 Hydroboration of Alkynes

*1.3.4.4.1 Thermal cis-Hydroboration* Since its discovery by Brown and Rao in 1956 [288], hydroboration chemistry has been a central reaction in the preparation of organoboron compounds [289]. The *cis*-hydroboration of terminal alkynes provides ready access to *trans*-2-substituted alkenylboronic acids [290], and several borane reagents have been used for this purpose (Figure 1.24). Unsymmetrical internal alkynes usually give mixtures of regioisomeric alkenylboron compounds. With terminal alkynes, however, the hydroboration is highly regioselective and adds boron at the terminal carbon. Likewise, whereas small borane reagents tend to undergo a double hydroboration with alkyne substrates, more hindered boranes allow the hydroboration process to stop with ease after one addition, avoiding further hydroboration of the desired product into a diboroalkane [290]. Thus, the bulky dialkylborane reagents disiamylborane (**88**) [289], thexylborane (**89**) [291], dicyclohexylborane (**90**) [292], and 9-BBN (**91**) [293] all react with terminal alkynes to provide

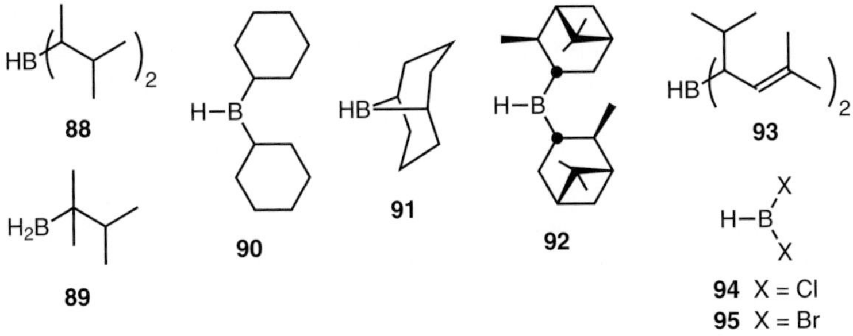

**Figure 1.24** Common hydroborating agents for alkynes.

2-substituted dialkylalkenylboranes in a very high regioselectivity. The corresponding alkenylboronic acid may be obtained after an appropriate oxidative workup, which is generally performed with a mild and selective oxidant for the two $sp^3$ C–B bonds. Toward this end, trimethylamine oxide was found most suitable [294], leaving not only the alkenyl boron–carbon bond intact but also a selenide and a sulfide substituent (entry 10, Table 1.4) [295]. In the hydrolysis of the resulting alkenylboronate, the ensuing separation of the desired boronic acid from the alcohol by-product originating from the oxidation of the dialkylborane is not always straightforward. Hoffmann and Dresely described a procedure with dicyclohexylborane where the boronic acid is esterified *in situ* as a pinacolate after the oxidation step and then purified by distillation to eliminate the residual cyclohexanol [295]. This way, several functionalized (*E*)-1-alkenylboronates were isolated, and it was found that the use of DME, a polar coordinating solvent, was essential when using a propargylic ether as substrate. For substrates that may be sensitive to the oxidative workup or to avoid the cyclohexanol by-product, diisopinocampheylborane (**92**, Figure 1.24) [296] offers a milder alternative. With this reagent, the alkyne is hydroborated and then subjected to a gentle oxidative dealkylation using acetaldehyde to afford a diethyl alkenylboronic ester along with 2 equiv of pinene [297–299]. The crude diethyl alkenylboronate can be transesterified with diols such as pinacol to yield the corresponding pinacol ester, which in most cases must be purified by distillation or chromatography. Although the synthesis of several highly functionalized alkenylboronates was reported using this method (entries 11 and 12), it is often difficult to completely eliminate the pinene by-product by distillation. The newer reagent di(isopropylprenyl)borane, **93**, was described [300]. Much like reagent **92**, it features a mild neutral workup with aqueous formaldehyde or water (entry 13).

The use of 4,4,6-trimethyl-1,3,2-dioxaborinane (**56**, Figure 1.15) [157], catecholborane (**57**) [301], pinacolborane (**58**) [159], or the more reactive 1,3,2-dithiaborolane (**59**) [161] provides the boronic acid derivative directly after a nonoxidative hydrolytic workup. Yet, these methods are not without disadvantages. Dialkoxyboranes are less reactive than the dialkylboranes described above. For example, alkyne hydroborations with catecholborane are often performed at temperatures as high as 100 °C, whereas dialkylboranes such as $Cy_2BH$ were found to catalyze these hydroborations at ambient temperature [302]. Although catecholborane was employed with highly functionalized substrates [303], it was reported that it does not tolerate acetal or ether functionalities at the propargylic carbon [295, 298], and the acidic catechol released in the aqueous workup needs to be neutralized and removed from the mixture (entry 14). By producing the robust pinacolate ester in a single operation, the use of pinacolborane (**58**) is quite advantageous, although the addition also tends to be sluggish (entry 15). Dibromoborane (**95**, Figure 1.24), in the form of a methyl sulfide complex, conveniently gives access to 1-alkenylboronic acids bearing alkyl or aryl substituents at the 2-position following alcoholysis of the intermediate alkenyldibromoborane [304]. Several other functionalities, however, are not well tolerated by this reagent. The related dichloroborane (**94**) was found to undergo a regioselective hydroboration with silylacetylenes, giving the (*E*)-1-trimethylsilyl-1-alkenylboronic ester after methanolysis (entry 16) [305]. Dichloroborane is difficult to handle, but a simple variant presumed to generate it *in situ* by reaction of trimethylsilane with boron trichloride was also shown to hydroborate alkynes [306]. Alternatively, a more recent report demonstrated the suitability of the stable and commercially available $Cl_2BH$–dioxane complex for the preparation of 1-alkenylboronic acids [307].

#### 1.3.4.4.2 Indirect trans-Hydroboration Using Alkynyl Bromides

All the above hydroboration methods provide terminal *trans*-alkenylboronic acids by a highly regioselective *syn*-addition of the B–H bond across the terminal alkyne. To provide the *cis*-alkenylboronic acids, Brown and Imai developed an ingenious two-step method based on the regioselective hydroboration of bromoalkynes with dibromoborane (Figure 1.23) [308]. In this procedure, the resulting (Z)-1-bromo-alkenyldibromoboranes are transformed into the corresponding esters through simple alcoholysis. The isolated boronates are then treated with potassium triisopropoxyborohydride (KIPBH) to effect a stereospecific bromide substitution by inversion of configuration, thereby affording the *cis*-alkenylboronic esters. Although dibromoborane presents a limited scope of chemoselectivity, KIPBH is relatively mild. For example, it tolerates a primary alkyl chloride on the substrate (entry 17, Table 1.4). Furthermore, an extension of this approach employing organolithium or Grignard reagents in place of KIPBH leads to the stereoselective preparation of (*E*)-1-substituted-1-alkenylboronic esters that could not be obtained via the hydroboration of alkynes [309]. Recently, a similar nucleophilic substitution mechanism has also been proposed in a new method involving the addition of alkenyllithium intermediates to the diboronyl reagent $B_2pin_2$ or the related dimethylphenylsilyl(pinacolato)borane [310]. In this reaction, which accomplishes a geminal difunctionalization of formal alkenylidene-type carbenoids, 1,1-diboronylalkenes or 1-silyl-1-alkenylboronates are produced (entry 18).

#### 1.3.4.4.3 Transition Metal-Catalyzed cis-Hydroboration

Since the discovery of the rhodium-catalyzed hydroboration of alkenes by Männig and Nöth in 1985 [311], the application of this method to alkynes has generally not provided satisfactory results [312]. He and Hartwig, however, found that dicarbonyltitanocene effectively catalyzes the hydroboration of alkynes with catecholborane without the contamination of by-products of catecholborane decomposition usually observed under rhodium catalysis (entry 19, Table 1.4) [313]. By taking advantage of the superior stability of pinacolborane over catecholborane, Pereira and Srebnik developed a very convenient zirconocene-catalyzed procedure for the pinacolboration of terminal alkynes (entry 20) [314]. This method, which features lower reaction temperature and times compared to the noncatalyzed variant of Knochel and coworkers [159], provides the (*E*)-1-alkenylboronates as their convenient pinacolate ester in high yields and high regioselectivity. A modified procedure affords improved *E/Z* selectivities with oxygen-containing alkynes [315], and the efficient use of reagent **56** was also reported [316]. Other transition metal catalysts such as $Rh(CO)(Ph_3P)_2Cl$ and $NiCp(Ph_3P)Cl$ were also found to be effective in conjunction with pinacolborane as the hydroboration agent [317]. Like the noncatalyzed hydroboration, internal alkynes tend to give mixtures of regioisomers. Using thioalkynes, however, a nickel-catalyzed catecholboration method provides 2-alkylthio-1-alkenylboronates in a high regioselectivity [318]. A copper hydride-catalyzed copper–to–boron transmetalation procedure with pinacolborane affords 1-carboalkoxy alkenylboronates regioselectively (entry 21) [319].

A $Pd(PPh_3)_4$-catalyzed catecholboration of an enyne afforded an allenylboronate [320]. Miyaura and coworkers also reported the $Pt(dba)_2$-catalyzed pinacolboration of terminal allenes, and the regioselectivity was found to be highly dependent on the nature of the added phosphine ligand [321]. For example, whereas the bulky tris (2,4,6-trimethoxyphenyl)phosphine often led to substantial amounts of the external Markovnikov product, the use of tris($t$-butylphosphine) provided the internal hydroboration product as single isomer (entry 22, Table 1.4). It is noteworthy that the resulting 1-substituted-1-alkenylboronate is not accessible regioselectively using the uncatalyzed hydroboration of terminal allenes or terminal alkynes.

#### 1.3.4.4.4 Rhodium- and Iridium-Catalyzed trans-Hydroboration

Direct alkyne hydroboration methods, whether catalyzed or not, afford *trans*-alkenylboronic acids by a highly regioselective *syn*-addition of the reagent's B—H bond across the terminal alkyne. The indirect Brown method to effect formal *trans*-hydroboration (Section 1.3.4.4.2) is limited by the need for a bromoalkyne and the harshness of the dibromoborane reagent employed. To fill this important methodological void and allow a direct and mild formation of *cis*-alkenylboronic acids, a true "*trans*-hydroboration" method was developed by Miyaura and coworkers. It was found that the hydroboration of alkynes with either catecholborane or pinacolborane in the presence of triethylamine and catalytic amounts of rhodium or iridium phosphine complex provides good to high yields of (*Z*)-1-alkenylboronic esters in a very high selectivity (entry 23, Table 1.4) [322]. Interestingly, deuterium labeling experiments showed that the *cis*-hydrogen substituent does not originate from the borane, it comes from the terminal alkyne instead. Based on this information, a mechanism involving migration of the acetylenic hydrogen and proceeding through a metal–vinylidene complex was proposed [322] to explain the selectivity of this unique "*trans*-hydroboration" method that has been employed in complex natural product synthesis [323].

### 1.3.4.5 Alkene Metathesis

Recently, the advent of efficient catalysts for alkene metathesis has opened up new opportunities for the synthesis of alkenylboronic acids. For example, it was shown that ring-closing metathesis of dienylboronic acids provides cyclic alkenylboronic acids that would be difficult to obtain otherwise [324]. Chemoselectivity in cross-metathesis chemistry is a significant issue that tends to pose strict limits to the synthesis of acyclic alkenes using these novel catalysts [325]. With most terminal alkenes, mixtures of disubstituted alkene products are obtained, and often with a low *E/Z* selectivity. Exceptionally, a number of alkene substrates are prone to undergo a highly chemoselective cross-metathesis with other terminal alkenes [325]. Fortunately, ethylene and 1-propenyl pinacol boronic esters rank among those favorable partners [326, 327]. Morrill and Grubbs discovered that they undergo a clean cross-metathesis with terminal alkenes, catalyzed by a ruthenium alkylidene, to provide the (E)-1-alkenylboronic ester in high selectivity (entry 24, Table 1.4) [327]. This methodology was tested in the synthesis of complex molecules such as epothilone analogues [328]. Ene–yne metathesis reactions based on alkynylboronic ester annulation strategies provide polysubstituted 2-butadienyl boronic esters [329, 330].

### 1.3.4.6 Diboronylation and Silaboration of Unsaturated Compounds

Diboronyl reagents such as $B_2pin_2$ (63) can be employed in various ways to access mono- or diboronyl alkenes depending on the reaction conditions [249]. Marder and coworkers developed a dehydrogenative borylation of vinylarenes to access 2,2-disubstituted-1-alkenylboronates that are not accessible by standard alkyne hydroboration chemistry [331]. By using the catalyst precursor $RhCl(CO)(PPh_3)_2$ and $B_2pin_2$ or $B_2neop_2$, the authors found conditions that prevent any significant competitive hydrogenation or hydroboration of the product. For example, (E)-Ph(Me)C=CH(Bpin) was obtained from α-methylstyrene in high yield and high geometrical selectivity (entry 25, Table 1.4). A mechanism that accounts for the beneficial role of acetonitrile as cosolvent was proposed. Diboronyl compounds add onto terminal and internal alkynes under platinum catalysis to provide cis-1,2-diboronylalkenes [332]. For example, $Pt(PPh_3)_4$ catalyzes the addition of bis(pinacolato)diboron (63) to 1-decyne, affording the corresponding alkenylbisboronate (entry 26, Table 1.4). Several other metal complexes tested, including palladium, rhodium and nickel complexes failed to promote the same reaction. Recently, the use of reagent 64 (Figure 1.16) was found to give high selectivities for the Bpin group on the internal position (entry 27) [165]. Mechanistically, these reactions' catalytic cycle is thought to be initiated by the oxidative addition of Pt(0) into the B−B bond, followed by a cis-boro-platination of the alkyne, and the cycle is terminated by the reductive elimination of the alkenyl-Pt(II)-Bpin unit to regenerate the Pt(0) catalyst [333]. Allenes also react similarly (entry 28) [334]. In a related process, $B_2pin_2$ was found to add to terminal alkynes at room temperature in the presence of stoichiometric copper (I) chloride and potassium acetate as the base [335]. It was proposed that a boron–copper transmetalation is involved, giving a putative boryl-copper species (CuBpin). The reaction provides a variable ratio of 1-boronyl and 2-boronyl alkenes depending on the additive employed, which can either be a phosphine or LiCl (entry 29). With α,β-ethylenic esters, Z-configured β-boronyl enoates are obtained [336] (entry 30), thus complementing the above formal hydroboration approach [299]. Murakami and coworkers reported a palladium-catalyzed silaboration of allenes, affording 2-boronyl-allylsilanes [337]. The same group also described a palladium- and nickel-catalyzed intramolecular cyanoboronation of homopropargylic alkynes [338]. An interesting nickel-catalyzed borylative coupling of alkynes and enones provides tri- and tetrasubstituted alkenylboronates (entry 31) [339]. Many more diboronylation and silaboration processes have been developed and are described in detail in Chapter 3. It should be noted that unlike the direct aromatic borylations discussed in Section 1.3.1.4, direct transition metal-catalyzed borylations of alkenes' C−H bonds with diboronyl reagents are complicated by competitive allylic borylation [340].

### 1.3.4.7 Other Methods

The conceptually simple, photochemical E–Z isomerization of double bonds is not an efficient approach for accessing geometrically pure alkenylboronic esters [305, 309b]. Alkynylboronic esters, however, are very useful precursors of alkenylboronates. For instance, they can be selectively hydrogenated over Lindlar's catalyst with 1,4-dioxane as the optimal solvent for providing (Z)-1-alkenylboronates with stereochemical purity over 95% (entry 32, Table 1.4) [341]. Likewise, highly pure (Z)-1-alkenylboron pinacolates were isolated from the corresponding alkynylboronates and from a sequence of regioselective hydrozirconation and aqueous protonolysis (entry 33) [342]. A similar hydroboration/protodeboronation approach was recently reported, giving functionalized (Z)-1-alkenylboronates (entry 34) [343]. In the past few years, various transition metal-catalyzed alkylative insertions and cycloadditions of alkynylboronates [344] and allenylboron pinacolate [345] have emerged, affording tri- and tetrasubstituted alkenylboronates usually with limited scope.

Addition of a α-silylallylboronate to aldehydes gives 4-alkoxy (E)-1-alkenylboronates [346]. Matteson and Majumdar have reported a Peterson-type olefination of the anion derived from an α-trimethylsilylmethylboronic ester (LiCH(SiMe$_3$)Bpin) [347]. Addition of the latter onto aldehydes provided the corresponding alkenylboronic ester as a mixture of geometrical isomers (∼70 : 30 Z/E). No further optimization was reported toward controlling the E/Z selectivity in this potentially useful and unique method for synthesizing alkenylboronic esters from aldehydes. The corresponding lithiomethylenediboronic esters tend to provide mixtures favoring the E-isomer (entry 35) [348, 349], and this approach to access alkenylboronic acids from aldehydes was employed in the total syntheses of natural products such as palytoxin [350] and the macrolide antibiotic rutamycin B [351]. An extension of this method using lithium 2,2,6,6-tetramethylpiperazide (LTMP) as base and ketones as electrophiles produces tetrasubstituted alkenylboronates (entry 36) [352]. A variant of the traditional Takai reaction using Cl$_2$CHBpin provides trans-1-alkenylboronic esters from aldehydes [353], and this procedure was recently employed in a synthesis of quinine (entry 37) [354]. The pinacol and 2-methyl-2,4-pentanediol esters of ethylene boronic acid are efficient substrates for Heck couplings with aryl and alkenyl halides, giving 2-aryl- and 2-butadienylboronates, respectively, with minimal side product from Suzuki–Miyaura cross-coupling (entry 38) [355]. A radical promoted variant employs xanthates to produce 3-oxo-(E)-1-alkenylboronates in low yields [356]. To access 2,2-disubstituted-1-alkenylboranes, a two-step sequence of bromoboration/Negishi coupling was described [357]. Advances in the Pd-catalyzed intramolecular carboboration of alkynes give access to tetrasubstituted alkenylboronates [358]. The synthesis of alkenylboronates using other types of additions and cycloadditions to alkynylboronates is described elsewhere (Chapter 3) [258, 359].

### 1.3.5
### Alkynylboronic Acids

Like their aryl and alkenyl counterparts, alkynylboronic acids can be made by displacement of magnesium or lithium acetylides with borate esters. For example,

Matteson and Peacock described the preparation of dibutyl acetyleneboronate from ethynylmagnesium bromide and methyl borate [360]. It was observed that the C—B linkage is stable in neutral or acidic hydroxylic solvents, but readily hydrolyzes in basic media such as aqueous sodium bicarbonate. Brown and coworkers eventually applied their organolithium procedure toward the preparation of alkynylboronic esters, and in this way provided a fairly general access to this class of compounds [361].

## 1.3.6
### Alkylboronic Acids

Compared to aryl- and alkenylboronic acids, alkylboronic acids and esters have not found widespread use as synthetic intermediates aside for their oxidation into alcohols (Section 1.5.2.1). This is due in part to their limited shelf stability. In addition, their transmetalation with transition metal catalysts such as palladium is presumed to be more difficult compared to unsaturated and aromatic boronic acid derivatives [362]. For example, alkylboronic acids have long been known to be reluctant substrates in the Suzuki–Miyaura cross-coupling reaction, and they have become suitable only very recently with the use of special bases and the advent of new and highly active catalyst systems (Section 1.5.3.1 and Chapter 4). Arguably, the most synthetically useful class of alkylboronic acids are the α-haloalkyl derivatives popularized by Matteson (Section 1.3.8.4). Specifically, the Matteson asymmetric homologation of α-haloalkylboronic esters provides a general access to functionalized, chiral alkylboronic esters in high enantioselectivities. Recent applications of this elegant chemistry and variants thereof are also described in Chapter 10.

Alkylboronic acids and esters can also be synthesized from the trapping of organomagnesium and organolithium intermediates with borates. Methylboronic esters, for example, are made using the condensation of methyllithium and triisopropylborate [207]. Likewise, the useful α-chloromethylboronate reagents **96** (Figure 1.25) can be made with the *in situ* trapping variant whereby butyllithium is added to a mixture of $ICH_2Cl$ and triisopropylborate [363]. The corresponding bromide (**97**) [364] and iodides (**98**) [365] were also reported. Recently, a method for the preparation of benzylic boronates was devised using a catalytic amount of magnesium [366]. Both catalyzed and uncatalyzed hydroboration of alkenes serve as powerful methods to access enantiopure alkylboronic esters. Because a selective oxidation of two of the resulting three B—C bonds following hydroboration with dialkylboranes is difficult, a hydroboration route to alkylboronic acids and esters is limited to reagents such as $ipc_2BH$ (**92**), dihaloboranes, and dialkoxyboranes (e.g., catechol- and pinacolborane). The asymmetric hydroboration of alkenes with $ipc_2BH$ or $ipcBH_2$ (Equation 1.35, Figure 1.25) [367, 368], or using chiral rhodium catalysts [369, 370], constitutes well-established routes to access chiral alkylboronic esters or the corresponding alcohols or amines after a stereospecific oxidation of the B—C bond (Sections 1.5.2.1 and 1.5.2.2). A remarkable NHC-Cu(I)-catalyzed formal hydroboration of aryl-substituted alkenes was recently reported (Equation 1.36) [371]. Chiral cyclopropylboronic esters were obtained by catalytic enantioselective pinacolboration of cyclopropenes (Equation 1.37) [372], and other methods to

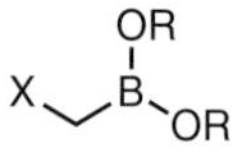

**62** | *1 Structure, Properties, and Preparation of Boronic Acid Derivatives*

$$X\diagdown B(OR)_2 \quad \begin{array}{l} \textbf{96 } X = Cl \\ \textbf{97 } X = Br \\ \textbf{98 } X = I \end{array}$$

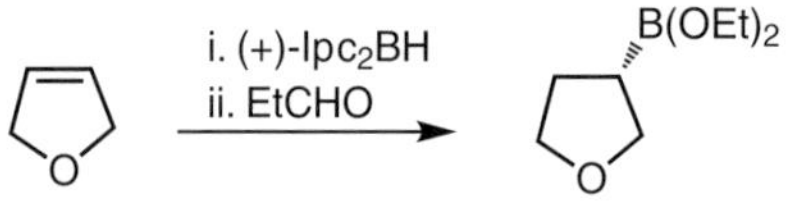

(1.35) dihydrofuran → i. (+)-Ipc$_2$BH; ii. EtCHO → tetrahydrofuran-B(OEt)$_2$

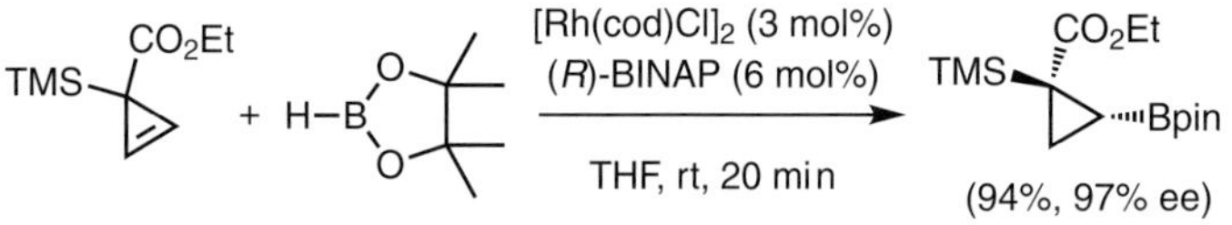

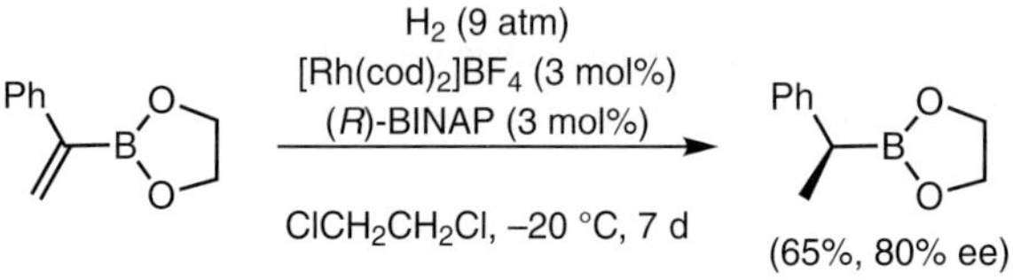

(1.36) PhCH=CHMe (styrene) + B$_2$pin$_2$ (1.1 equiv), chiral NHC (7.5 mol%), CuCl (7.5 mol%), KO-*t*-Bu (30 mol%), MeOH (2 equiv), THF, −50 °C, 48 h → PhCH$_2$CH(Me)Bpin (80%, 98% ee)

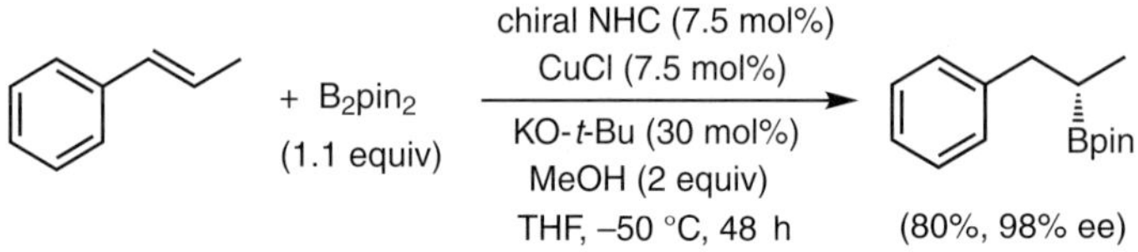

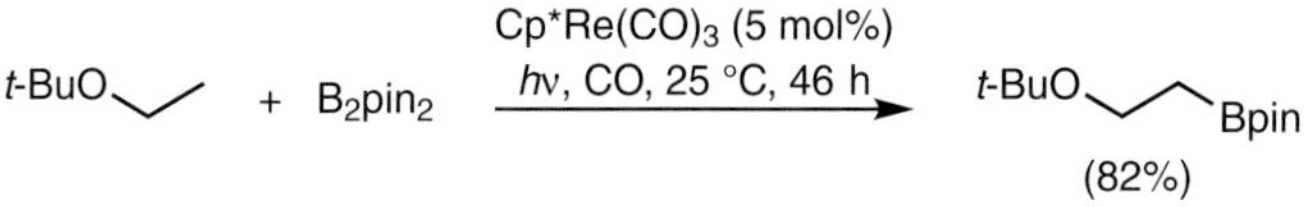

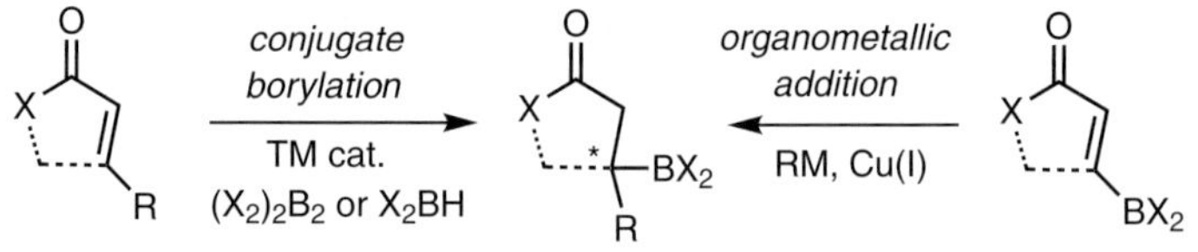

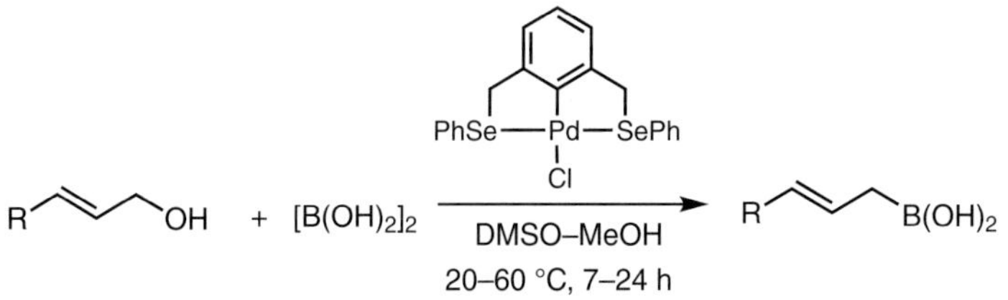

(1.37) TMS-cyclopropane-CO$_2$Et + H−Bpin, [Rh(cod)Cl]$_2$ (3 mol%), (*R*)-BINAP (6 mol%), THF, rt, 20 min → TMS-cyclopropane(CO$_2$Et)(Bpin) (94%, 97% ee)

(1.38) Ph-C(=CH$_2$)-B(OCH$_2$CH$_2$O), H$_2$ (9 atm), [Rh(cod)$_2$]BF$_4$ (3 mol%), (*R*)-BINAP (3 mol%), ClCH$_2$CH$_2$Cl, −20 °C, 7 d → Ph-CH(Me)-B(OCH$_2$CH$_2$O) (65%, 80% ee)

(1.39) *t*-BuO-CH=CH$_2$ + B$_2$pin$_2$, Cp*Re(CO)$_3$ (5 mol%), hν, CO, 25 °C, 46 h → *t*-BuO-CH$_2$CH$_2$-Bpin (82%)

(1.40) enone (X = O/N, R) → conjugate borylation, TM cat., (X$_2$)$_2$B$_2$ or X$_2$BH → β-boryl ketone −BX$_2$ ← organometallic addition, RM, Cu(I) ← enone-BX$_2$

(1.41) R-CH=CH-CH$_2$OH + [B(OH)$_2$]$_2$, PhSe−Pd(Cl)−SePh (pincer), DMSO−MeOH, 20–60 °C, 7–24 h → R-CH=CH-CH$_2$-B(OH)$_2$

**Figure 1.25** Alkylboronic acids (esters): selected examples of enantioselective preparative methods.

access vinylcyclopropylboronic esters are known [373]. Enantiomerically enriched alkylboronic esters can also be obtained through less common methods such as the hydrogenation of chiral alkenylboronic esters [374] and even with enantioselective variants using chiral catalysts (Equation 1.38) [375a]. Though not a general method, alkylboronic acids have also been isolated via a regioselective rhenium-catalyzed C–H activation/boronylation reaction (Equation 1.39) [250b]. Several other transition metal-catalyzed mono- and diboration reactions of aldehydes [376], unactivated alkenes [377], alkynes, [378], and dienes [379] provide new ways to chiral alkylboronic esters. The transition metal-catalyzed asymmetric conjugate borylation of $\alpha,\beta$-unsaturated carbonyl compounds [380] delivers alkylboronates with high enantioselectivities (Equation 1.40) [381]. Recently, metal-free, carbene-catalyzed and chiral phosphine-catalyzed variants have appeared [382]. An efficient alternative is the asymmetric conjugate addition to $\beta$-boronyl acceptors (Equation 1.40) [383].

## 1.3.7
**Allylic Boronic Acids**

Because of their tremendous utility as carbonyl and imine allylation agents (Section 1.5.3.6 and Chapter 8), several methodologies have been developed for synthesizing allylic boronic acids and their various esters. The preparation and reactions of allylboronic esters have been reviewed in the past 5 years [384], but new methods appear constantly, including asymmetric variants [385]. Among others, metal-catalyzed diborylation of allylic electrophiles, including free alcohols (Equation 1.41) [386], and even direct allylic borylation of alkenes [340] provide mild approaches to allylic boronates. Recently, efficient methods to produce $\alpha$-substituted allylic boronates using catalytic regio- and stereoselective hydroborations of 1,3-dienes have appeared [387].

## 1.3.8
**Chemoselective Transformations of Compounds Containing a Boronic Acid (Ester) Substituent**

New boronic acid derivatives can be made by the derivatization of compounds that already contain a boronic acid (ester) functionality. The scope of possible transformations, however, relies on the compatibility of these reaction conditions with the boronate group and, in particular, on the oxidatively labile C–B bond. One seminal example that best illustrates the limitations imposed by the intrinsic reactivity of boronic acids is that of $\alpha$-aminoalkylboronic acids, the boron analogues of amino acids (Section 1.3.8.4) [388]. The synthesis of these important derivatives remained an elusive goal for several years. The reason for the instability of compounds of type **99** is the incompatibility of free $\alpha$-amino groups possessing hydrogen substituents, which undergo a spontaneous 1,3-rearrangement to give the homologated amine **101** following hydrolysis of the transposed intermediate **100** (Equation 1.42) [124]. It was eventually found that this undesired process could be prevented through a rapid acylation of the amino group or its neutralization as a salt [124]. This undesirable rearrangement was later exploited in a method for mono-*N*-methylation of primary amines [389]. Also of note is the lability of alkylboronic acids with a leaving group in the β-position, which, as exemplified with the formation of ethylene by debromoboronation of 2-bromoethaneboronic acid, are unstable under basic conditions [390]. A review by Matteson provides a detailed overview of the chemical compatibility of boronic acids and esters, and can undoubtedly be of great advice for evading trouble when derivatizing a boronic acid-containing compound [391]. Therefore, only selected examples of boronate-compatible transformations will be discussed in the following sections. It is noteworthy that accrued information on the chemical compatibility of free boronic acids has recently made it possible to perform multistep syntheses on boronic acid-containing compounds. This is realized in conjunction with a new liquid–liquid, pH-driven phase switching strategy employing sorbitol as a phase transfer agent [392]. For example, the antilipidemic drug ezetimibe (**102**) was synthesized in five steps from *p*-boronobenzaldehyde, making use of ubiquitous transformations such as [2 + 2] cycloaddition, cross-metathesis, and hydrogenation without a need for chromatographic purifications of intermediates up until the final productive detagging operation that unmasked the desired phenol via C–B bond oxidation (Figure 1.26) [392].

$$
\underset{\mathbf{99}}{R\!\!\begin{array}{c}\text{OR'}\\|\\ \text{B}\\ \end{array}\!\!\text{OR'}}\quad\longrightarrow\quad \underset{\mathbf{100}}{R\!\!\begin{array}{c}\text{OR'}\\|\\ \text{N}-\text{B}\\ |\\ \text{H}\end{array}\!\!\text{OR'}}\quad\longrightarrow\quad \underset{\mathbf{101}}{R\!\!\frown\!\!\text{NH}_2} \qquad (1.42)
$$

#### 1.3.8.1 Oxidative Methods

The sensitivity of the B–C bond of boronic acids and esters to oxidation was discussed in Section 1.2.2.5.2. Although basic hydrogen peroxide and other strong oxidants rapidly oxidize B–C bonds, a certain degree of selectivity is possible. For example, sulfide and alcohol functionalities can be selectively oxidized without affecting a pinacol boronate (Equations 1.43 and 1.44, Figure 1.27) [295]. The reagent 2-iodoxybenzoic acid (IBX) oxidizes alcohols selectively on substrates containing a free arylboronic acid moiety [392]. On the other hand, the epoxidation of alkenylboronic esters is known to fail, but it can be achieved indirectly from trifluoroborate salts (Equation 1.32, Figure 1.19) [197]. The permanganate oxidation method is commonly employed to access carboxy-substituted arylboronic acids from methyl-substituted precursors [393]. Radical bromination of methyl-substituted arylboronic acids provides a route to the corresponding hydroxymethyl and formyl derivatives (Equations 1.45 and 1.46) [184]. The bromination of *p*-tolylboronic acid, followed by alkylation of acetaminomalonic ester, hydrolysis, and decarboxylation, affords 4-borono-phenylalanine [184].

#### 1.3.8.2 Reductive Methods

Care must be taken in using strong hydride reagents as they can transform boronic esters into dihydridoboranes (Section 1.2.3.7). A subsequent hydrolysis, however,

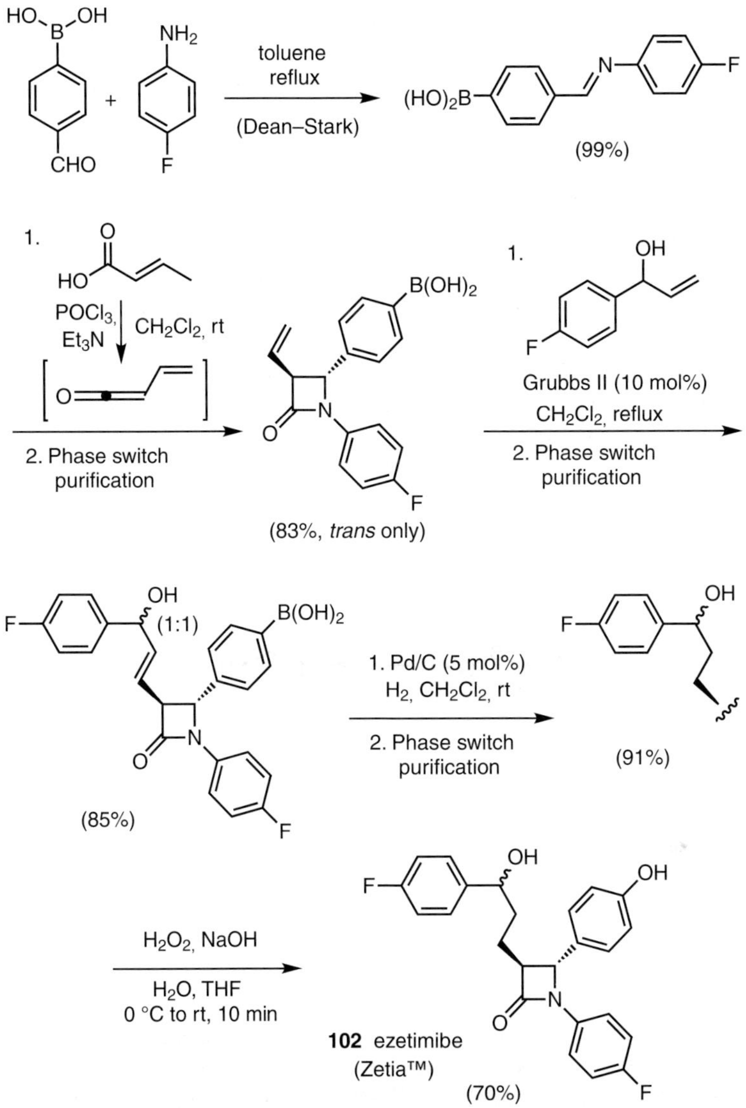

**Figure 1.26** Chemoselective, multistep phase switch synthesis of ezetimibe using boronic acid as a phase tag.

can restore the boronic acid. DIBALH reduced a carboxyester selectively on a substrate containing a free arylboronic acid moiety [392]. Catalytic hydrogenation methods appear to be quite compatible with boronate groups and even with free boronic acids, as shown by the examples of Figure 1.26 and Figure 1.28 (Equations 1.47 and 1.48) [394, 395].

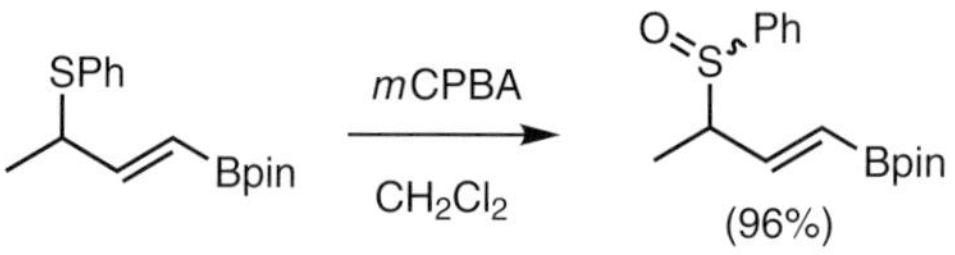

(1.43)

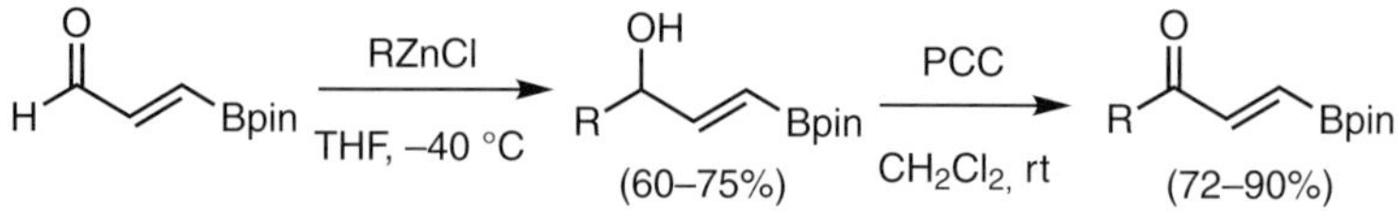

(1.44)

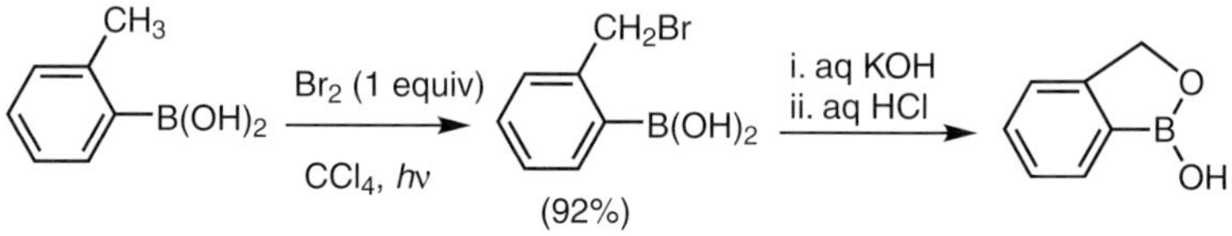

(1.45)

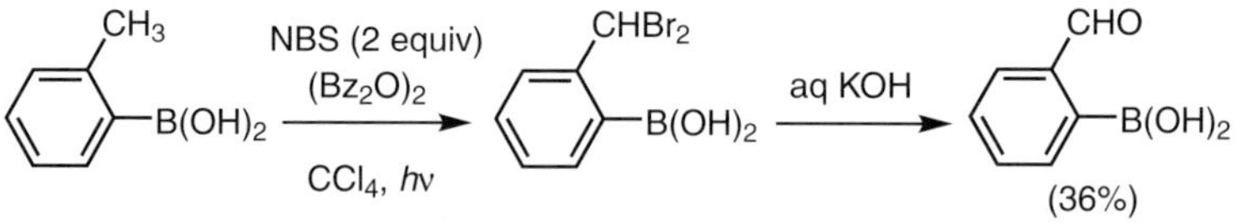

(1.46)

**Figure 1.27** Chemoselective oxidation reactions involving boronic acid derivatives.

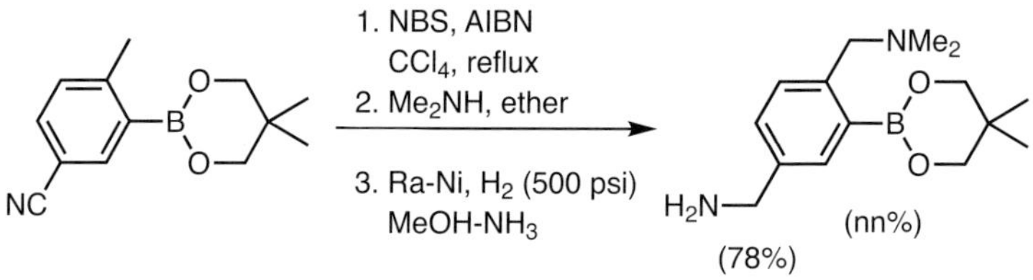

(1.47)

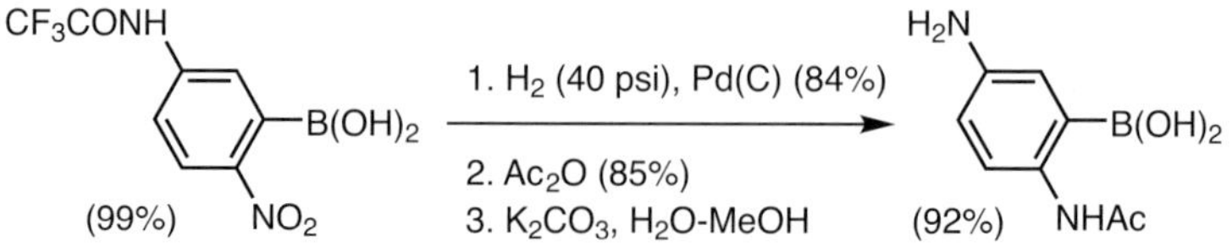

(1.48)

**Figure 1.28** Chemoselective reduction reactions involving boronic acid derivatives.

#### 1.3.8.3 Generation and Reactions of α-Boronyl-Substituted Carbanions and Radicals

Carbanions adjacent to a boronate group can be generated by two general approaches: direct deprotonation, or metalation by replacement of an α-substituent. Direct deprotonation of simple alkylboronic esters like 2,4,4,5,5-pentamethyl-1,3,2-dioxaborolane (**103** with $(RO)_2 = OCMe_2CMe_2O$, Equation 1.49 in Figure 1.29) is not possible even with strong bases like LDA or lithium 2,2,6,6-tetramethylpiperidide (LiTMP) [349]. An activating group must be present next to the boronate, and it has

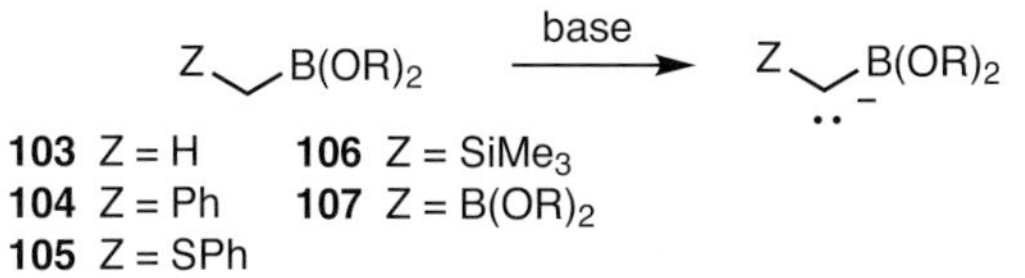

## 1.3 Preparation of Boronic Acids and Their Esters | 67

$$Z\diagdown B(OR)_2 \xrightarrow{\text{base}} Z\diagdown \underset{..}{B(OR)_2} \qquad (1.49)$$

**103** Z = H     **106** Z = SiMe$_3$
**104** Z = Ph   **107** Z = B(OR)$_2$
**105** Z = SPh

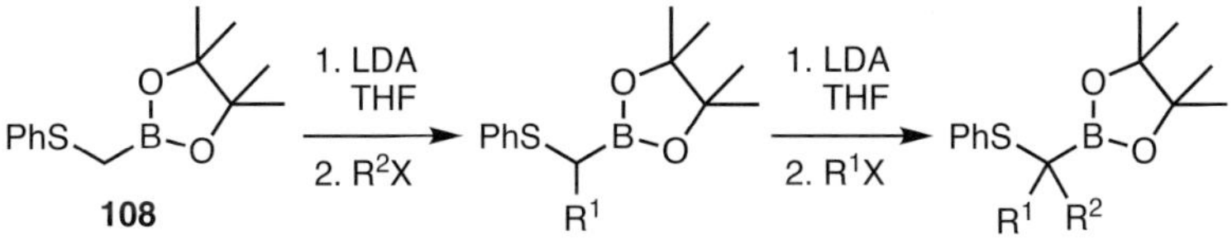

(1.50)

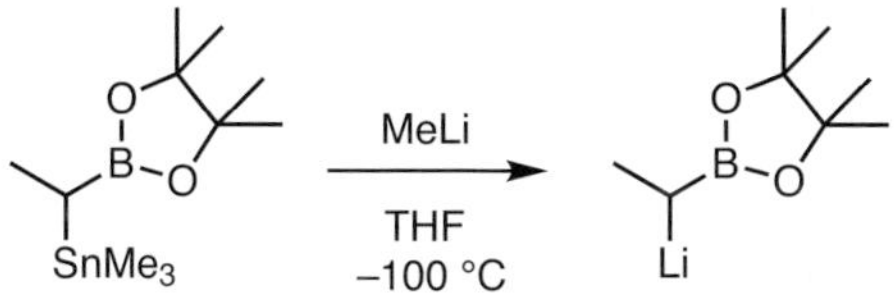

(1.51)

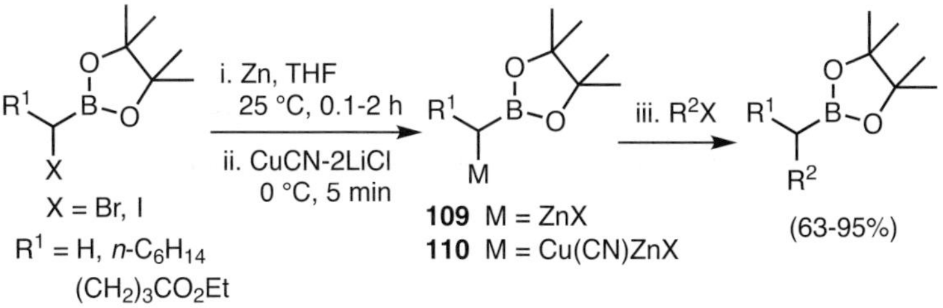

(1.52)

X = Br, I
R$^1$ = H, n-C$_6$H$_{14}$
(CH$_2$)$_3$CO$_2$Et

**109** M = ZnX
**110** M = Cu(CN)ZnX

(63–95%)

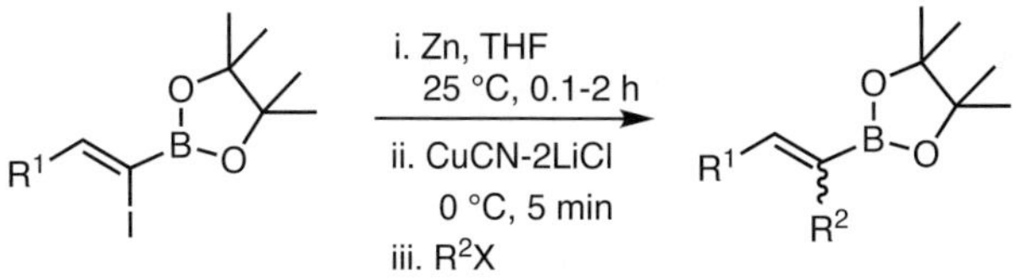

(1.53)

**Figure 1.29** Formation and reactions of boronyl-substituted carbanions.

been shown that phenyl [349], thioether [396], trimethylsilyl [347, 397], triphenylphosphonium [397], and another boronate group [349] are all suitable in this role (**104–107**, Equation 1.49). Relatively hindered bases and a large boronic ester are preferable in order to favor C–H abstraction over the formation of a B–N ate adduct. For example, the carbanion of bis(1,3,2-dioxaborin-2-yl)methane (**107** with (RO)$_2$ = O(CH$_2$)$_3$O) can be generated by treatment with LiTMP (1 equiv) and 1 equiv of the additive tetramethylethylenediamine (TMEDA) in tetrahydrofuran (−78 to 0 °C) [349]. Some of these species can be alkylated efficiently with primary halides and tosylates. Propanediol bisboronate **107** ((RO)$_2$ = O(CH$_2$)$_3$O) and the useful α-phenylthio derivative **108**, deprotonated with LDA, can even be alkylated twice in a sequential manner (Equation 1.50) [396]. The anion of **108** was also reacted with epoxides and lactones, and more recently it was used in the synthesis of functionalized boronic acid analogues of α-amino acids [398]. The carbanions of *gem*-diboronic esters **107** and α-trimethylsilyl pinacolboronate (**106** with $(RO)_2 = O(CH_2)_3O$) undergo other transformations and also behave as substituted Wittig-like reagents by adding to aldehydes or ketones to provide alkenylboronates (e.g., entries 35 and 36, Table 1.4) [348], which can also be oxidized and hydrolyzed to provide the homologated aldehydes [348, 399]. One drawback to the use of **107** is its low-yielding preparation. The corresponding carbanion can also be accessed by reaction of tris(dialkoxyboryl)methanes with an alkyllithium, but this approach lacks generality [400]. Substituted *gem*-diboronates can be made via a sequential hydroboration of 1-alkynes [401], and their anions are generated by treatment with LiTMP (see entry 36, Table 1.4, for addition to ketones). It has been suggested that bis(1,3,2-dioxaborin-2-yl)methane (**107** with $(RO)_2 = O(CH_2)_3O$) is slightly more acidic than triphenylmethane ($pK_a$ 30.6 in DMSO) [349], which confirms the rather weak stabilizing effect of a boronate group compared to a carboxyester ($pK_a$ of dimethylmalonate ~13). The calculation of Huckel delocalization energies confirmed that a boronate group is indeed slightly more stabilizing than a phenyl group ($pK_a$ of diphenylmethane = 32.6 in DMSO), and the calculation of B–C pi-bond orders indicated a very high degree of B–C conjugation in the carbanion [349]. This suggestion appears to be in contradiction with the apparently modest degree of B–C pi-overlap in alkenyl and aryl boronates discussed in Section 1.2.2.1; however, those cases concern neutral species.

Other methods for the generation of α-boronyl carbanions include transmetalations such as the lithiation of an α-trimethylstannyl derivative (Equation 1.51, Figure 1.29) [402] and the formation of the corresponding organozinc or organocopper species from α-bromo or α-iodo alkylboronates (Equation 1.52) [403]. In the latter example, the mildness of the zinc and copper organometallic intermediates expands the range of compatible functional groups compared to the corresponding organolithium intermediates described above. Thus, reagents **109** and **110**, even with a carboxyester-containing side chain as $R^1$ substituent, were reacted with a variety of electrophiles such as allylic halides, aldehydes, and Michael acceptors in good to excellent yields (Equation 1.52) [403]. Likewise, the related $sp^2$ 1,1-bimetallics can be generated from 1-iodoalkenylboronic pinacol esters albeit with loss of stereochemical integrity of the olefin geometry (Equation 1.53) [404]. In one example, the Negishi coupling of a 1-iodozincalkenylboronate with an alkenyl iodide partner led to the formation of a 2-boronylbutadiene.

### 1.3.8.4 Reactions of α-Haloalkylboronic Esters

One of the most powerful methods for modifying alkylboronic esters involves the nucleophilic attack and 1,2-rearrangement on α-haloalkylboronic esters (**111**) (Figure 1.30). The addition of organometallic species to these boronic esters induces a facile boron-promoted displacement (Equation 1.54). Heteroatom-containing nucleophiles as well as organometallic reagents can be employed in this substitution reaction. Conversely, the addition of α-haloalkyl carbanions to alkyl- and alkenylboronic esters leads to the same type of intermediates and constitutes a formal one-carbon homologation of boronic esters (Equation 1.55). Sulfides from the addition of carbanions of α-thioethers can also undergo this rearrangement in the presence of

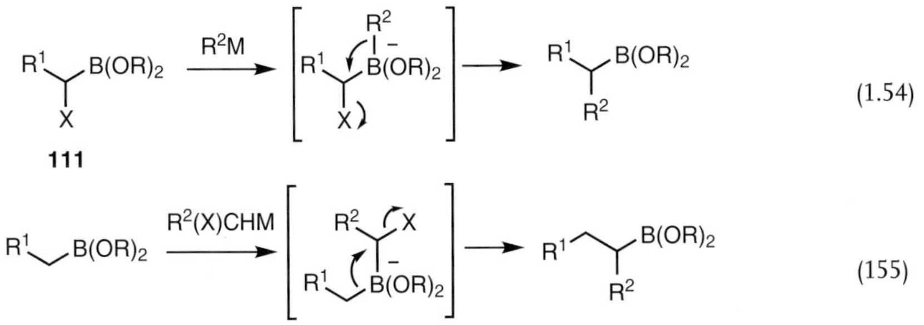

## 1.3 Preparation of Boronic Acids and Their Esters

$$R^1\underset{X}{\overset{}{\diagdown}}B(OR)_2 \xrightarrow{R^2M} \left[ R^1\underset{X)}{\overset{R^2}{\diagdown}}B(OR)_2^- \right] \longrightarrow R^1\underset{R^2}{\overset{}{\diagdown}}B(OR)_2 \quad (1.54)$$

**111**

$$R^1\diagdown B(OR)_2 \xrightarrow{R^2(X)CHM} \left[ R^1\underset{}{\overset{R^2\diagdown X}{\diagdown}}B(OR)_2^- \right] \longrightarrow R^1\underset{R^2}{\overset{}{\diagdown}}B(OR)_2 \quad (155)$$

**Figure 1.30** Substitution reactions of α-haloalkylboronic esters.

mercuric salts [405]. A very efficient asymmetric variant of this chemistry was developed to allow the synthesis of chiral α-chloroalkylboronates, which can further undergo substitution reactions with a broad range of nucleophiles [406]. These α-chloroboronates are obtained in very high enantiomeric purity through the Matteson asymmetric homologation reaction, which features the $ZnCl_2$-promoted addition of dichloromethyllithium to the boronates of pinanediol and certain C2-symmetrical 1,2-diols. This elegant methodology was used in the synthesis of complex natural products, and is at the cornerstone of the design and preparation of α-acylaminoalkylboronic acid enzyme inhibitors. As exemplified with the synthesis of **112** (Scheme 1.2), α-aminoalkylboronic esters are obtained via the displacement of α-chloroalkylboronates with hexamethyldisilazide anion. This example also emphasizes the powerful neighboring group effect of boron, which allows selectivity in the addition of $Cl_2CHLi$ in the presence of a primary alkyl bromide [407]. More recently, this chemistry was applied to Hoppe's chiral lithiated carbamates (X = $OCONMe_2$ in Equation 1.55) [408], and the applications of these methods in stereoselective synthesis are described in detail in Chapter 10.

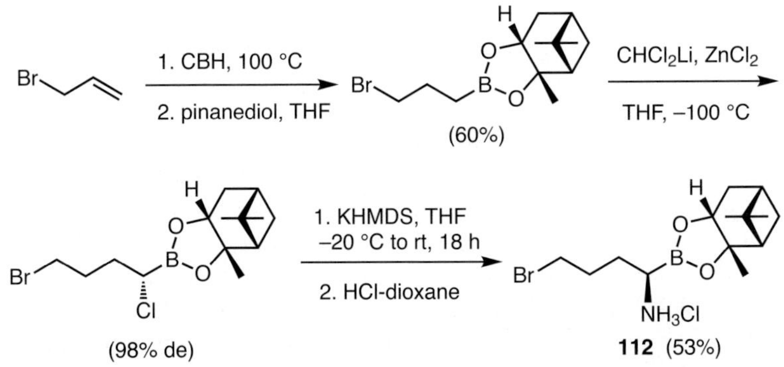

**Scheme 1.2** Application of the Matteson asymmetric homologation to the synthesis of a chiral α-aminoboronic ester.

### 1.3.8.5 Other Transformations

Several other reactions can be performed on free boronic acid compounds and the corresponding esters while preserving the boronyl group. The nitration of free arylboronic acids under fuming nitric acid and concentrated sulfuric acid has been known since the 1930s [70]. The use of low temperatures (e.g., Equation 1.56, Figure 1.31) is recommended in order to minimize protodeboronation (Section 1.2.2.5.3) [395, 409]. Other successful transformations of arylboronic acids that preserve the boronyl group include diazotization/hydrolysis [203], bromination [410], iodination (Equation 1.57) [411], and nucleophilic aromatic substitutions [203]. Pinacol arylboronates can be halogenated using a gold-catalyzed halogenation with halosuccinimides [412]. In the context of developing the phase switch synthesis concept already described (Section 1.3.8), reactions such as reductions, oxidations, amidations, Wittig olefinations, cross-metathesis, and others were also found to be compatible with boronic acids [392]. Azide-alkyne cycloadditions require

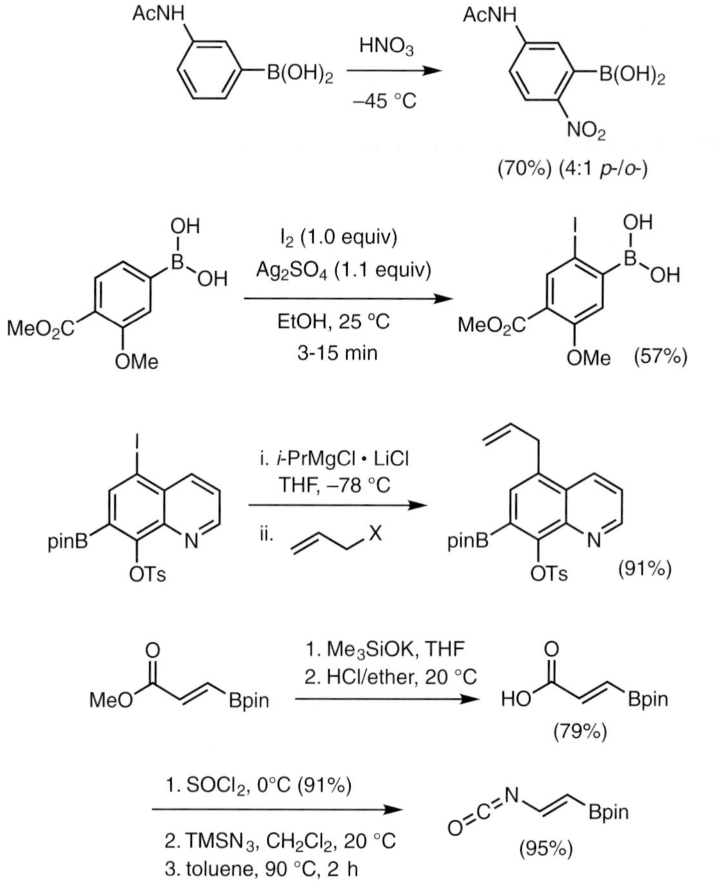

**Figure 1.31** Other chemoselective reactions compatible with boronic acid derivatives.

a copper catalyst that may insert into the B–C bond of arylboronic acids, but the addition of fluoride anion provides in situ protection [413]. Unless a special protecting group is employed, metalations of halides in the presence of a boronic acid or the corresponding pinacol esters are difficult due to the electrophilic properties of the boron atom [414]. Exceptionally, pinacol arylboronates containing an iodo substituent can undergo a successful iodine–magnesium exchange under conditions developed by Baron and Knochel, followed by electrophilic trapping (Equation 1.58) [415].

Schrock carbene formation is compatible with arylboronates [92], and radical additions to allyl or vinylboronates provide useful, functionalized alkylboronic esters [416]. Some alkenylboronates can be isomerized to allylboronates in high yields under Ru or Ir catalysis [417]. Pinacol alkenylboronates are robust enough to tolerate a number of transformations such as ester hydrolysis and a Curtius rearrangement (Equation 1.59, Figure 1.31) [418]. Various addition and cycloaddition chemistry of alkenyl- and alkenylboronic acid derivatives are possible, including radical additions, cyclopropanation, and [4 + 2] cycloadditions [258, 359]. Interestingly, pinacol aryl- and alkenylboronates containing a racemic secondary alcohol can be resolved using a lipase enzyme [419].

Alkylboronic esters can also tolerate a wide range of conditions, and problems, if any, are usually encountered in the purification steps rather than with the actual chemistry. The synthesis of 2-amino-3-boronopropionic acid, the boronic acid analogue of aspartic acid (**113**, Scheme 1.3), which included reactions such as carbethoxyester hydrolysis, a Curtius rearrangement, and hydrogenolysis, convincingly illustrates the range of possibilities [420]. Unlike the α-aminoalkylboronic acids, homologous (β-amino) compound **113** is stable and is thought to exist as an internal chelate or a chelated dimer in aqueous solution. Likewise, the lithium

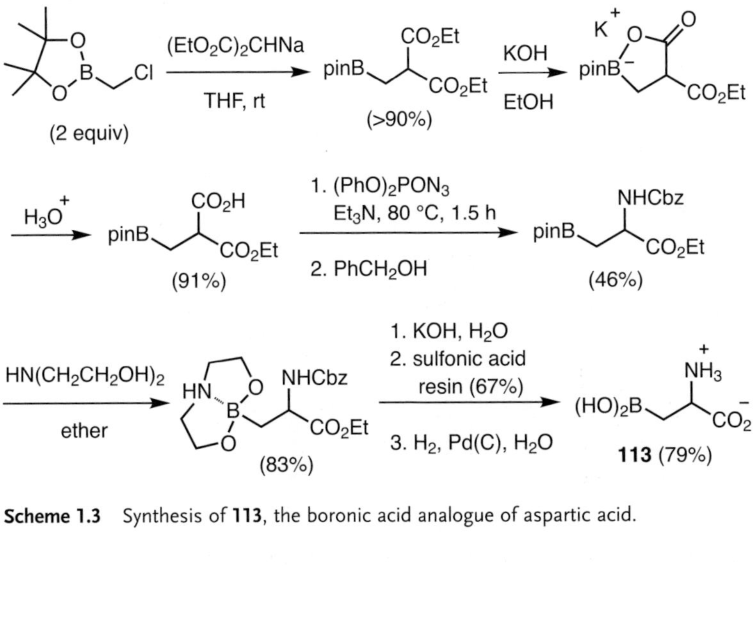

**Scheme 1.3** Synthesis of **113**, the boronic acid analogue of aspartic acid.

enolate of 3-oxo alkylboron pinacolates can be formed with LDA and exists in a Z chelated form [421].

### 1.3.8.6 Protection of Boronic Acids for Orthogonal Transformations

There are situations where free boronic acids and standard boronic esters such as pinacolates are not suitable to permit orthogonal reactivity. In situ protection of pinacol esters with alkoxide ligands can be helpful [418c], but permanent protection is a more general strategy. Toward this end, derivatives such as the large ester **42**, diethanolamine adducts **43**, trifluoroborate salts [194], N-methyldiaminoacetate (MIDA) **55** [156], and 1,8-diaminonaphthalene (dan) adducts **66** [171] have been developed (Figure 1.32). As shown in Equation 1.60 (Figure 1.33), the scope of compatible transformations can be further increased with the help of a bulky boronate ester to effectively protect the susceptible boron center in oxidations, reductions, and other reactions [120]. These boronates tolerate additions of organometallics onto aldehydes [422], where they can induce stereoselectivity and can also serve in the preparation of cyclopropylboronates [120c]. Boronates **42**, however, are difficult to cleave and are rather removed through a C–B bond transformation. Protection of the boronyl group as a diethanolamine ester allows a clean bromine/lithium exchange [423], which was used in the preparation of para- and meta-chlorosulfonyl arylboronic acids after trapping with sulfur dioxide (Equation 1.61, Figure 1.33) [423b]. Direct deprotonation of polyfluorinated arylboronic diethanolamine esters was reported [424]. Trifluoroborate salts are tolerant of a wide variety of transformations (see Section 1.2.3.8 and Chapter 11). Likewise, aryl and alkenyl MIDA boronates described in Section 1.2.3.3 can tolerate a wide complement of transformations and can even be carried out through multistep syntheses [425], although their main use is as masking groups in cross-coupling chemistry (Equation 1.62) [426]. Both MIDA boronates **55** [155, 426] and the 1,8-diaminonaphthalene adducts **66** [171, 427] have been employed successfully in iterative cross-coupling strategies to assemble oligoarenes (Scheme 1.4). A derivative of **66** may be used as an *ortho*-directing group for transition metal-catalyzed C–H activation/silylation [428].

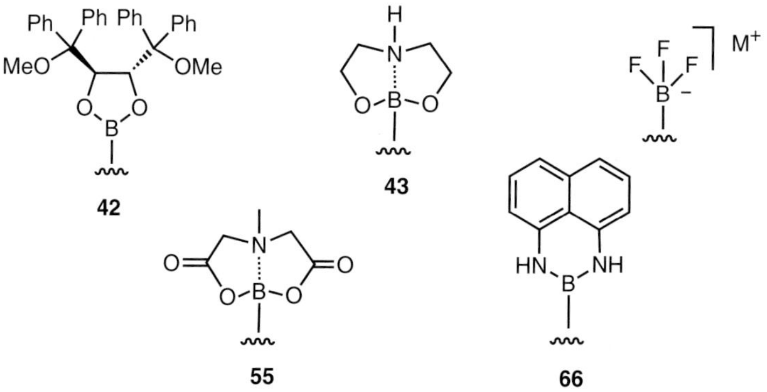

**Figure 1.32** Common protecting and masking groups for boronic acids.

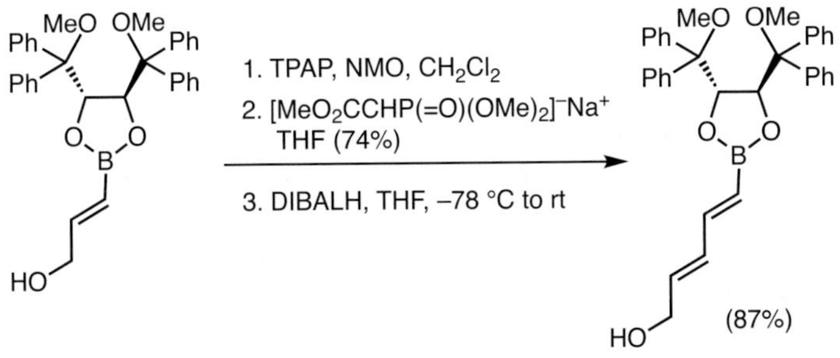


(1.60)

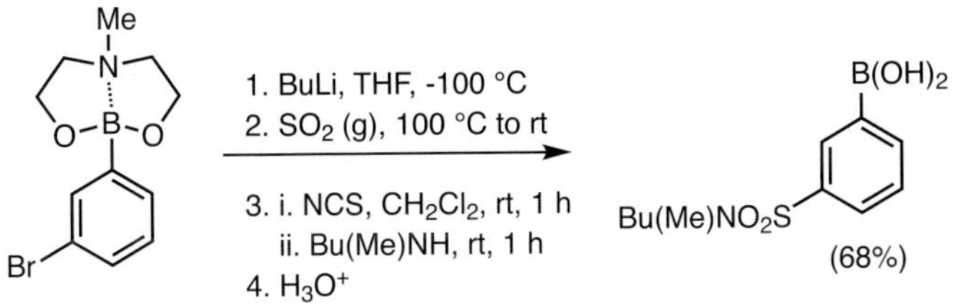


(1.61)

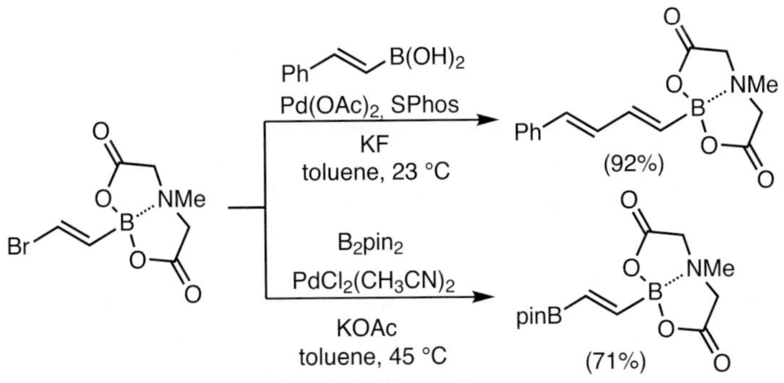


(1.62)

**Figure 1.33** Examples of selective transformations on protected boronic acids.

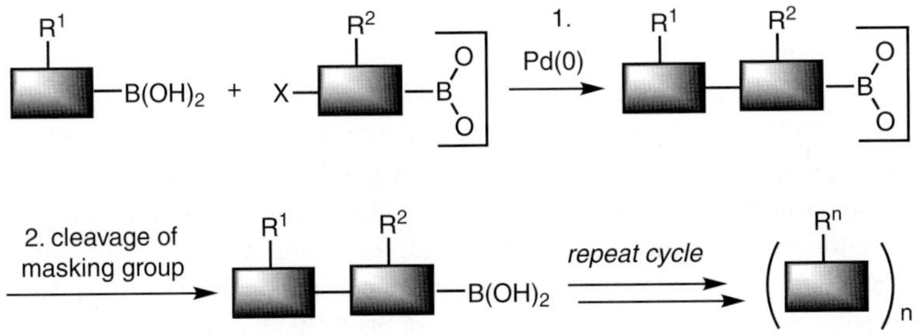


**Scheme 1.4** Iterative synthetic scheme for oligoarenes using masked boronic acids.

## 1.4
## Isolation and Characterization

As discussed in Section 1.2.2.2, the polar (and often amphiphilic) character of boronic acids tends to make their isolation and purification a difficult task. In some cases, nonpolar organic solvents may be used to precipitate small boronic acids dissolved in a polar organic solvent. At higher pH values where the hydroxyboronate species is predominant (Section 1.2.2.4.1), boronic acids may, however, be entirely miscible in water. For this reason, when extracting boronic acids from aqueous solutions, it is desirable to adjust the pH of the water phase to a neutral or slightly acidic level and to use a polar organic solvent for an efficient partition. The use of aqueous conditions is to be avoided for amphoteric boronic acids containing amino substituents, as they are soluble in water in the entire pH range (at pH > 8, the hydroxyboronate species is predominant, and at lower pH, the amine is protonated). The phase switch system with aqueous sorbitol described in Section 1.3.8 may be employed for isolating boronic acids or eliminating nonpolar impurities [392]. It is important to realize that commercial samples of boronic acids may contain various amounts of residual boric acid, which is silent in $^1$H NMR spectroscopy but can be detected by $^{11}$B NMR. Boric acid can usually be separated from boronic acids through a partition between water and chilled diethyl ether. In addition to these potential difficulties in isolating boronic acids, their tendency to form oligomeric anhydrides further complicates their characterization. To palliate these problems, boronic acids are often purified and characterized as esters. The following section provides a summary of useful methods and generalizations for the isolation and characterization of boronic acids and boronic esters.

### 1.4.1
**Recrystallization and Chromatography**

Most boronic acids can be recrystallized with ease. The choice of recrystallization solvent, however, greatly affects the relative proportions of free boronic acid and its corresponding anhydrides in the purified solid. Santucci and Gilman found that acids are usually obtained from aqueous solutions (i.e., water or aqueous ethanol), and anhydrides predominate when nonpolar recrystallization solvents like ethylene dichloride are employed [429]. Recrystallization in benzene gives some dehydration, but to a lesser extent. Several other solvents have been used for the recrystallization of arylboronic acids, including two-solvent systems. Most boronic acids are soluble in polar solvents like ether, methylene chloride, and ethyl acetate and are insoluble in pentane or hexanes. Much like carboxylic acids, most boronic acids interact strongly with silica gel. Depending on the degree of hydrophobicity of the boron substituent, chromatography and TLC on silica gel are possible despite the high retentivity of boronic acids. To this end, the eluent mix of 20–50% ethyl acetate/hexanes is generally suitable for most arylboronic acids, and those with additional polar groups may require methanol or acetic acid as a coeluent. Some electron-rich arylboronic acids tend to deboronate faster on silica gel; thus, prolonged exposure to silica from lengthy separations should be avoided. In such cases, filtration through a short plug of silica using acetone as coeluent [430] or the use of a polar eluent mixture made of $CH_2Cl_2$ and EtOAC was found suitable [409]. For example, a highly lipophilic trienylboronic acid was conveniently purified by silica gel chromatography [273].

## 1.4.2
### Solid Supports for Boronic Acid Immobilization and Purification

Recently, the increasing popularity of boronic acids as synthetic intermediates has motivated the development of solid supports and linkers to allow their immobilization and facilitate purification operations or derivatization (Figure 1.34). The appeal of these methods is particularly apparent in view of the difficulties often encountered in the isolation of pure boronic acids from both aqueous and organic solvent systems.

#### 1.4.2.1 Diethanolaminomethyl Polystyrene

Diol-based insoluble polystyrene resins that can form supported boronic esters are obvious choices for immobilizing boronic acids. Hall and coworkers reported the first example of solid support for boronic acids, the diethanolaminomethyl polystyrene resin (DEAM-PS, **114** in Figure 1.34), which is now commercially available [431, 432]. The immobilization of alkyl-, alkenyl-, and arylboronic acids with this resin is straightforward, consisting simply of mixing a slight excess of DEAM-PS, as a suspension, in an anhydrous solution containing the boronic acid [431a]. Tetrahydrofuran was found to be the solvent of choice as it dissolves most boronic acids. It is noteworthy that no azeotropic removal of the water released is needed, which comes as a benefit of the B–N coordination in the resulting adducts and of the highly hydrophobic nature of this polystyrene support. This simple procedure can be employed for purifying boronic acids (Equation 1.63, Figure 1.34) or for scavenging excess reagent from crude reaction mixtures [433], including amphoteric ones that would be otherwise difficult to isolate from aqueous solvent systems. Following

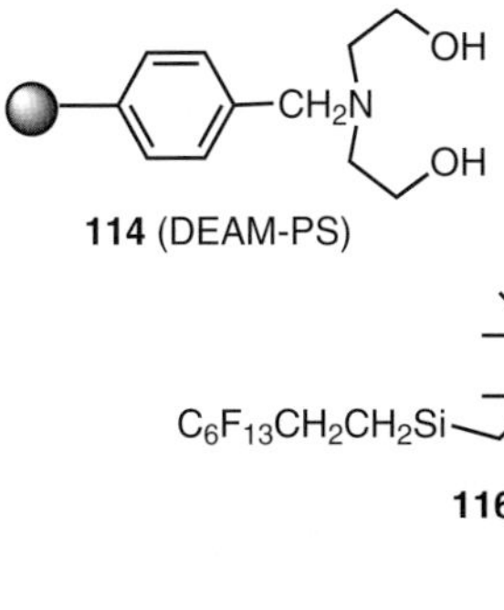

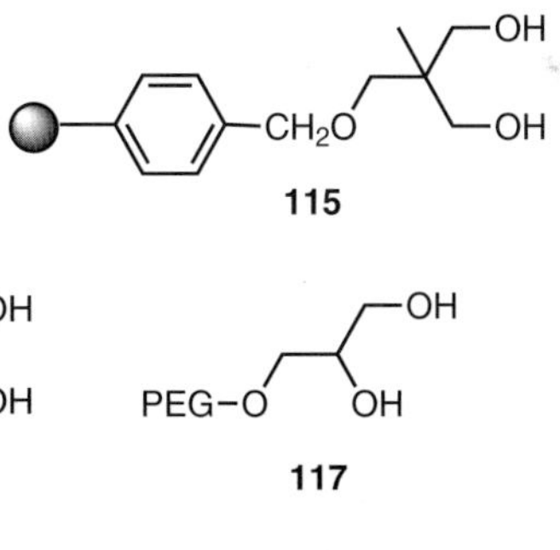

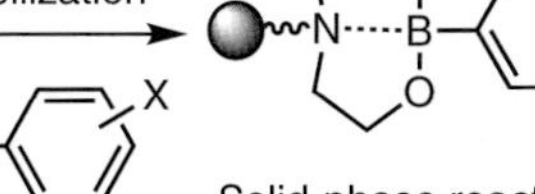

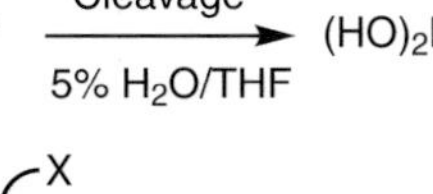

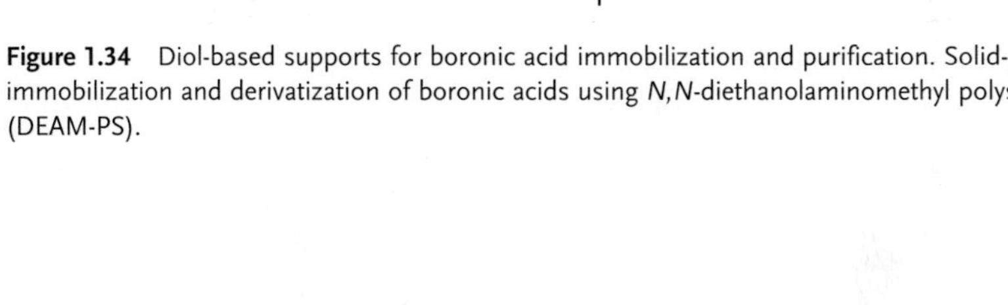

$$(1.63)$$

**Figure 1.34** Diol-based supports for boronic acid immobilization and purification. Solid-phase immobilization and derivatization of boronic acids using N,N-diethanolaminomethyl polystyrene (DEAM-PS).

resin washings, the desired boronic acid can be recovered upon treatment of the resin with a 5–10% solution of water in THF. A wide variety of arylboronic acids were immobilized with the DEAM-PS resin, and it has even been employed successfully in the derivatization of functionalized boronic acids [432]. Thus, amino-substituted arylboronic acids supported onto DEAM-PS were transformed into anilides and ureas, bromomethyl-substituted ones were reacted with amines, formyl-substituted ones were subjected to reductive amination with aldehydes, and carboxy-substituted phenylboronic acids were transformed into amides [432]. All these transformations afford new arylboronic acid derivatives in very high purity directly after cleavage from the resin. The DEAM-PS-supported boronic acids were also employed in the interesting concept of resin–resin transfer reactions (RRTR), whereby a phase transfer agent is used *in situ* to allow the transfer of one supported substrate to another resin-supported substrate. This convergent solid-phase synthetic strategy was applied to the Suzuki-Miyaura cross-coupling [434] and the borono-Mannich reactions [435].

#### 1.4.2.2 Other Solid-Supported Diol Resins

A macroporous polystyrene resin functionalized with a 1,3-diol unit, **115**, was described by Carboni *et al.* [436]. Although the immobilization and subsequent cleavage of boronic acids both require harsher conditions compared to DEAM-PS, this support was also proven useful in the derivatization of functionalized boronic acids, as well as in a number of elegant C–C bond forming/release procedures [437] and a traceless cleavage of arenes [438]. Analogous pinacol-like linkers were also described, although preattachment of the boronic acid prior to immobilization was required in these examples [439]. A ROMP gel diol was employed for the immobilization of allylboronates [440]. A catechol-functionalized polystyrene resin was also found to be effective in the immobilization and derivatization of functionalized arylboronic acids [441].

#### 1.4.2.3 Soluble Diol Approaches

Fluorous-phase purification methodologies using fluoroalkyl-tagged substrates combine the advantages of homogeneous reaction conditions of solution-phase reactions with the ease of purification of solid-phase methods. In this regard, pinacol-like and other diol-based polyfluoroalkyl linkers such as **116** were described [442]. The resulting fluorous boronates were employed in a variety of transformations and allowed a facile purification by simple partition between fluorous and organic solvents. A dendritic high-loading polyglycerol, **117**, was shown to be effective in immobilizing arylboronic acids and in facilitating the purification of biaryl products from homogeneous Suzuki cross-coupling reactions [443].

### 1.4.3
**Analytical and Spectroscopic Methods for Boronic Acid Derivatives**

#### 1.4.3.1 Melting Points, Combustion Analysis, and HPLC

The difficulty in measuring accurate and reproducible melting points for free boronic acids has long been recognized [444]. Rather than true melting points, these measurements are often more reflective of dehydration or decomposition points [212, 445]. The lack of reproducibility for a given boronic acid may originate from the water contents of the sample used, which affects the acid–anhydride transition. Moreover, as mentioned above, the water content also depends on the recrystallization solvent [429]. For these reasons, it is often more appropriate to report melting points of boronic acids as their diethanolamine esters (Section 1.2.3.2.1). Likewise, combustion analysis of free boronic acids may provide inaccurate results depending on the recrystallization method. Reverse-phase HPLC chromatography may be used for analyzing boronic acids and esters, albeit on-column hydrolysis can complicate the analysis of boronic esters. Fast methods suitable to arylboron pinacol esters have been reported [446].

### 1.4.3.2 Mass Spectrometry

One useful diagnostic detail in the mass spectrometric analysis of boronic acid derivatives is the observation of boron's isotopic pattern, which is constituted of $^{10}$B (20% distribution) and $^{11}$B (80% distribution). On the other hand, unless other functionalities help increase the sensitivity of a boronic acid-containing compound, it is often difficult to obtain intense signals with most ionization methods due to the low volatility of these compounds. This problem is exacerbated by the facile occurrence of gas-phase dehydration and anhydride (boroxine) formation in the ion source. Electrospray ionization in the negative mode tends to provide the best results with minimal fragmentation, the $[M-H]^-$ and $[2M-H_2O-H]^-$ fragments being most common using methanol, acetonitrile, water, or mixtures thereof as the most effective solvent systems. For amino-substituted boronic acid compounds, the ESI positive mode is usually effective, giving $[M + H]^+$ and $[M + Na]^+$ as common fragments. To minimize thermal reactions and improve volatility, cyclic boronates may be employed. These derivatives were even made on analytical scale [447]. The fragmentation patterns of various *para*-substituted arylboronic esters of 1,2-ethanediol were studied using electron impact ionization and several deboronative fragmentation pathways were observed [448]. The nature of the *para*-substituent was found to have a marked influence. In another study by GC-MS, *ortho*-substituents were found to interact strongly during fragmentation [447]. Boropeptides, a popular class of enzyme inhibitors (Section 1.6.5), and phenylboronic acid were characterized by positive-ion ammonia chemical ionization with different diols as benchtop derivatization agents [449].

### 1.4.3.3 Nuclear Magnetic Resonance Spectroscopy

Boron compounds, including boronic acid derivatives, can be conveniently analyzed by NMR spectroscopy [450]. Of the two isotopes, $^{11}$B is the most abundant (80%) and possesses properties that are more attractive toward NMR. Specifically, these attributes include its lower resonance frequency, spin state (3/2) and its quadrupole moment, a wide range of chemical shifts, and its relatively high magnetic receptivity (16% of $^1$H). Most boronic acids are soluble in dimethylsulfoxide (DMSO-d6), and it is a particularly effective NMR solvent ($-B(OH)_2$ resonance $\sim 8.3$ ppm). When analyzing boronic acids in nonhydroxylic solvents by NMR spectroscopy, it is often necessary to add a small amount of deuterated water (e.g., one or two drops) to the sample in order to break up the oligomeric anhydrides. Alternatively, analysis in anhydrous alcoholic solvents such as methanol will allow observation of the *in situ* formed methanolic ester. Observation of the $^{11}$B nucleus against a reference compound (e.g., $BF_3$) is straightforward with modern instruments, and can be especially revealing of the electronic characteristics [34] and coordination state of the boronate moiety. The boron resonance of free boronic acids and tricoordinate ester derivatives is generally detected in the 25–35 ppm range, and tetracoordinate derivative such as diethanolamine esters are detected at around 10 ppm [451]. In $^{13}$C analysis, carbons next to the boron atom tend to be broadened often beyond the limits of detection due to the quadrupolar relaxation of $^{11}$B. Consequently, with aromatic boronic acids, the signal from the quaternary carbon bearing the boron atom can be very difficult to observe over the background noise.

#### 1.4.3.4 Other Spectroscopic Methods

In spite of their limited structure determination capabilities, ultraviolet and infrared spectroscopies were determinant characterization techniques in the early days of boronic acid research [429]. Noted IR absorptions are the strong H-bonded OH stretch (3300–3200 cm$^{-1}$), and a very strong band attributed to B–O stretch (1380–1310 cm$^{-1}$). IR is particularly diagnostic of the presence of boronic anhydrides. Upon anhydride (boroxine) formation, the OH stretch disappears and a new strong absorption appears at 680–705 cm$^{-1}$ [75].

## 1.5
### Overview of the Reactions of Boronic Acid Derivatives

### 1.5.1
#### Metalation and Metal-Catalyzed Protodeboronation

In 1882, Michaelis and Becker described the preparation of phenylmercuric chloride (**118**) from the reaction of phenylboronic acid and aqueous mercuric chloride (Equation 1.64, Figure 1.35) [228b]. Benzylboronic acid was transformed to benzylmercuric chloride in the same manner, and both compounds were found to resist hydrolysis under the conditions of their preparation. Mechanistic studies later showed that this reaction proceeds through the hydroxyboronate ion [452]. Catechol and pinacol alkenylboronic esters were also found to be transmetalated into the corresponding organomercurial derivative with retention of configuration [453, 454]. One of the early observations on the reactivity of arylboronic acids was the realization that a number of metal ions (other than Hg(II)) can induce protodeboronation in water, presumably via the intermediacy of an arylmetal species (Equation 1.64). Thus, Ainley and Challenger found that hot solutions of phenylboronic acid with copper sulfate, cadmium bromide, or zinc chloride produce benzene [70]. As phenylboronic acid is stable to dilute hydrochloric acid, it was deduced that the deboronation occurred through the formation of transmetalated intermediates similar to **118** (Figure 1.35) and their reaction with water, and not from the possible release of acid by hydrolysis of the metal salt. Instead of giving benzene, cupric chloride and bromide were found to provide the respective phenyl chloride and bromide [70].

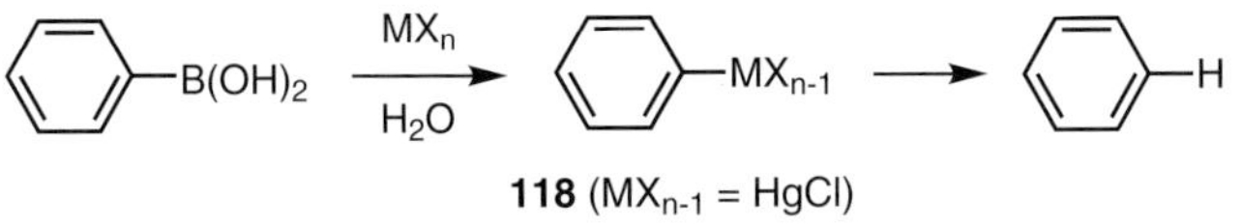

## 1.5 Overview of the Reactions of Boronic Acid Derivatives | 79

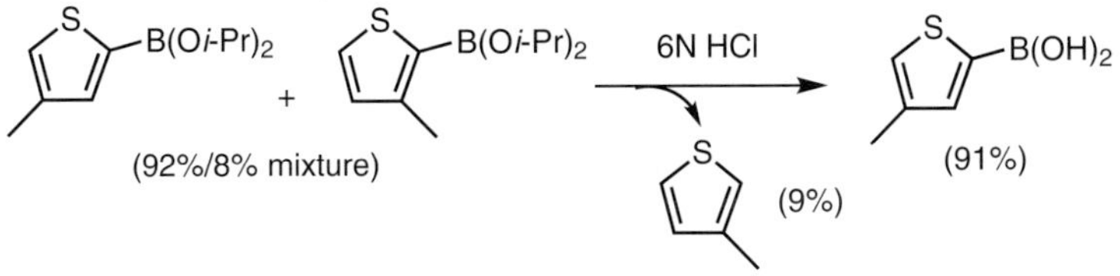

$$\text{Ph-B(OH)}_2 \xrightarrow[\text{H}_2\text{O}]{\text{MX}_n} \text{Ph-MX}_{n-1} \longrightarrow \text{Ph-H} \quad (1.64)$$

**118** (MX$_{n-1}$ = HgCl)

$$\text{(Thiophene-B(O}i\text{-Pr)}_2\text{ isomers)} \xrightarrow{\text{6N HCl}} \text{(91\%) + (9\%)} \quad (1.65)$$

(92%/8% mixture)

**Figure 1.35** Protodeboronation of boronic acids.

Halide salts of beryllium, magnesium, and calcium did not react with phenylboronic acid [70]. Arylboronic acids were transformed into arylthallium derivatives in a similar fashion [455], and alkylboronic acids were found to be unreactive under the same conditions [86]. Ammonical solutions of silver nitrate also induce protodeboronation of arylboronic acids with production of silver oxide [202]. Aliphatic boronic acids behave differently and rather tend to undergo a reductive coupling to give the dimeric alkane products [85]. Kuivila *et al.* studied the mechanism of metal ion catalysis in the aqueous protodeboronation of arylboronic acids [456]. Substituent effects and the influence of pH were investigated, and both base and cadmium catalysis pathways were evidenced for this reaction. The order of effectiveness of the different metal ions at effecting deboronation was established to be Cu(II) > Pb(II) > Ag(I) > Cd(II) > Zn(II) > Co(II) > Mg(II) > Ni(II). Boron–zinc exchange with boronic acids is a well-established synthetic process [457]. More recently, bismuth [458] and gold salts [459] were found to undergo a B–M transmetalation.

The silver nitrate-promoted protodeboronation method can be synthetically useful [438]. The regioselective protodeboronation of an isomeric mixture of heterocyclic boronic acids was employed as a separation strategy (Equation 1.65) [460]. On a synthetic chemistry standpoint, however, reaction of the metalated intermediates with electrophiles other than a proton is usually more attractive. Indeed, one of the most important recent developments in boronic acid chemistry strove from the discoveries that transition metals such as palladium(0), rhodium(I), and copper(I) can oxidatively insert into the B–C bond and undergo further chemistry with organic substrates. These processes are discussed in Sections 1.5.3 and 1.5.4 and several other chapters.

### 1.5.2
**Oxidative Replacement of Boron**

#### 1.5.2.1 Oxygenation
The treatment of arylboronic acids and esters with alkaline hydrogen peroxide to produce the corresponding phenols was first reported more than 75 years ago [70]. The oxidation of alkyl- and alkenylboronic acid derivatives leads to alkanols [42]

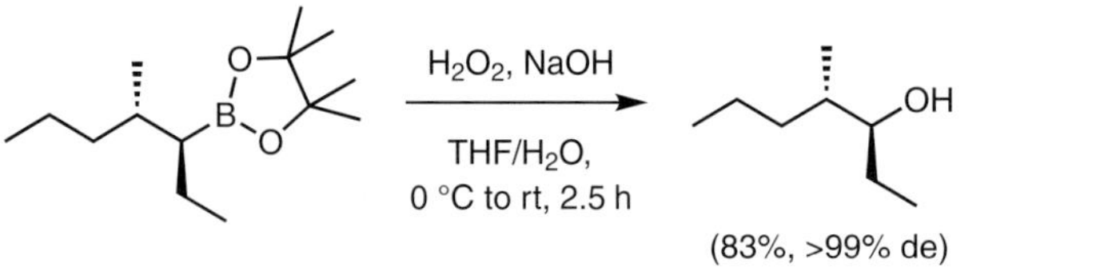

**Figure 1.36** Oxidation of boronic acids (esters).

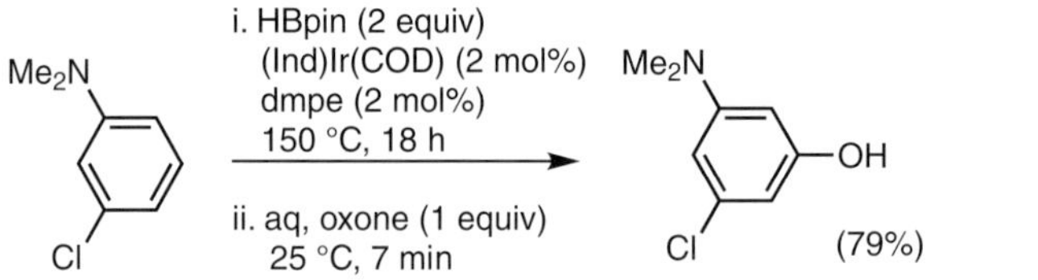

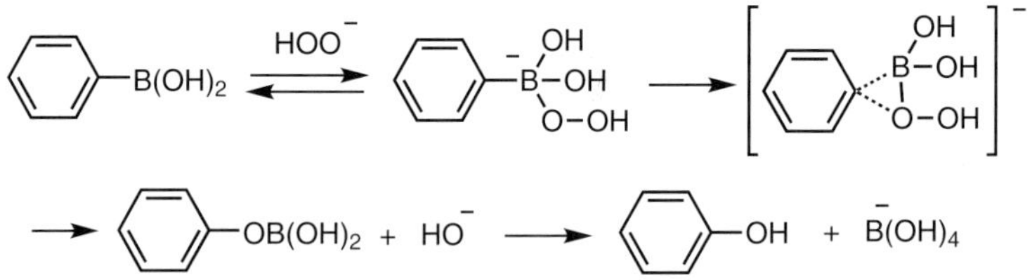

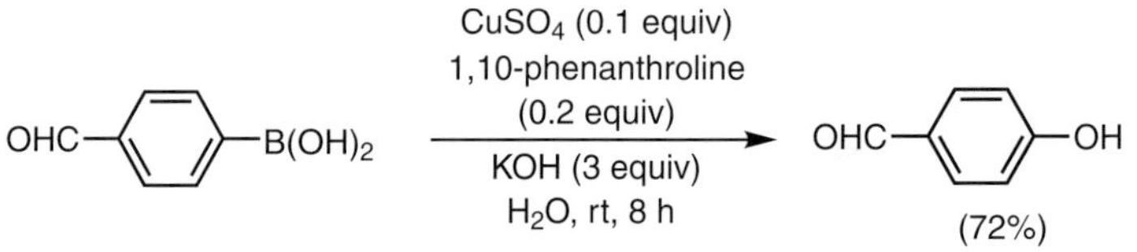

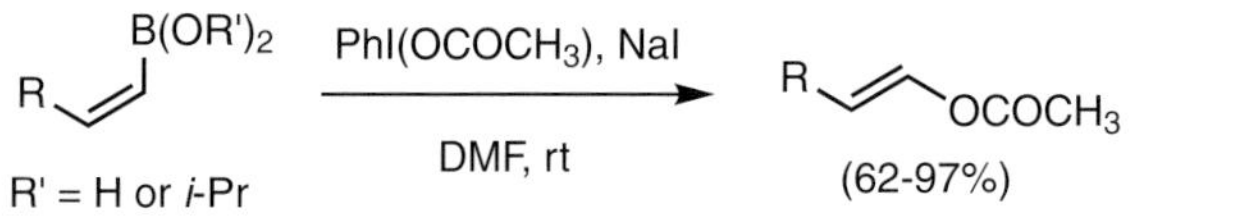

and aldehydes/ketones, respectively [95, 309a, 348, 399]. With chiral α-substituted alkylboronates, the reaction proceeds by retention of configuration (Equation 1.66, Figure 1.36) [137, 461]. In fact, the oxidation of boronic acids and esters is a synthetically useful process in the preparation of chiral aliphatic alcohols via asymmetric hydroboration of alkenes [369] or from the Matteson homologation chemistry [406]. On the other hand, the oxidation of arylboronic acids is usually not a popular and economical approach for preparing phenols. It was reported, however, that a one-pot C–H activation/borylation/oxidation sequence gives access to *meta*-substituted phenols that would be difficult to obtain by other means (Equation 1.67) [462]. The mechanism of the aqueous basic oxidation of phenylboronic acid was investigated by Kuivila [463]. The rate is first order each in boronic acid and hydroperoxide ion, which led the authors to propose the mechanism of Figure 1.36 (Equation 1.68). The transition state features a boron–to–oxygen migration of the ipso-carbon. Milder oxidants like anhydrous trimethylamine N-oxide [464], oxone [465], sodium perborate [466], and hydroxylamine [467] can also be employed for the oxidation of most types of boronic acid derivatives. It is noteworthy that perborate was found to give a cleaner oxidation of alkenylboronic acids into aldehydes compared to hydrogen peroxide [399]. Recently, mild room-temperature copper-catalyzed hydroxylations of arylboronic acids have appeared (Equation 1.69) [468]. Allylic boronic esters can be oxidized using nitrosobenzene with a SE' regioselectivity complementing the use of peroxide [469]. Interestingly, the combined use of diacetoxyiodobenzene and sodium iodide under anhydrous conditions transforms alkenylboronic acids and esters into enol acetates in a stereospecific manner (Equation 1.70) [470].

### 1.5.2.2 Amination and Amidation

Aryl azides can be accessed indirectly from arylboronic acids via *in situ* generated aryllead intermediates (Equation 1.71, Figure 1.37) [471]. A mild procedure for ipso-nitration of arylboronic acids was developed (Equation 1.72), and a mechanism was proposed [472]. Common methods and reagents for electrophilic amination, however, do not affect boronic acids and their esters. These processes require the intermediacy of more electrophilic boron substrates such as borinic acids or dichloroboranes. For example, enantiomerically pure propanediol boronates, which are accessible from the asymmetric hydroboration of alkenes with ipc$_2$BH followed by the acetaldehyde-promoted workup and transesterification, can be treated sequentially with MeLi and acetyl chloride. The resulting borinic ester is sufficiently electrophilic to react at room temperature with the amination reagent hydroxylamine-O-sulfonic acid with retention of stereochemistry to give primary amines in essentially 100% optical purity (Equation 1.73) [473]. The preparation of optically pure secondary amines from alkyl azides also requires the intermediacy of the highly electrophilic dichloroboranes (Equation 1.74) [191], which can be made from boronic esters and monoalkylboranes, as described in Section 1.2.3.6. Intramolecular variants of the reaction with alkyl azides provide access to pyrrolidines and piperidines [474]. A more contemporary amination of arylboronic acids affords primary anilines using a copper oxide catalyst and aqueous ammonia at ambient temperature (Equation 1.75) [475]. A copper-catalyzed coupling with nitrosoarenes gives diaryl amines [476] and a XeF$_2$-promoted reaction with nitriles gives anilides [477].

### 1.5.2.3 Halodeboronation

#### 1.5.2.3.1 Arylboronic Acids and Esters

As described above, cuprous chloride and bromide provided the corresponding ipso-substituted phenyl halides from benzeneboronic acid [70]. A modern stepwise one-pot version of these copper-promoted halogenations employs pinacol arylboronates made via transition metal-catalyzed

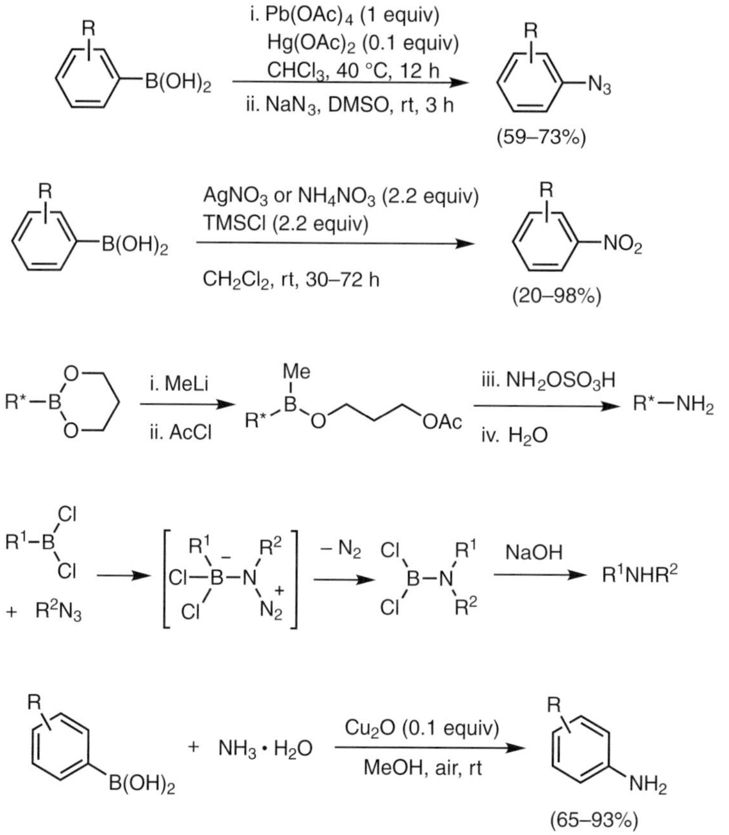

**Figure 1.37** Oxidative amination of boronic acid derivatives.

borylation [478]. Arylboronic acids are halodeboronated regioselectively by the action of aqueous chlorine, bromine, and aqueous iodine-containing potassium iodide [70]. Alkylboronic acids do not react under the same conditions [42]. The kinetics of bromonolysis in aqueous acetic acid has been studied by Kuivila and Easterbrook, who found that bases catalyze the reaction [479]. This observation and a Hammett plot of 10 arylboronic acids [480] are consistent with a proposed electrophilic ipso-substitution mechanism involving the usual weakening effect of the C–B bond through formation of a boronate anion (Equation 1.76, Figure 1.38). N-Bromo- and N-iodosuccinimides convert arylboronic acids to the corresponding aryl halides in good to excellent yields [481]. Most arylboronic acids react in refluxing acetonitrile, whereas the most activated ones such as 2-methoxyphenylboronic acid are iodinated at room temperature. Boronic esters provide significantly lower yields, and N-chlorosuccinimide is essentially unreactive even in the presence of bases. The use of 1,3-dibromo-5,5-dimethylhydantoin (DBDMH) under catalysis by sodium methoxide was shown to be an efficient bromodeboronation method for arylboronic acids when acetonitrile is used as the solvent (Equation 1.77, Figure 1.38) [482].

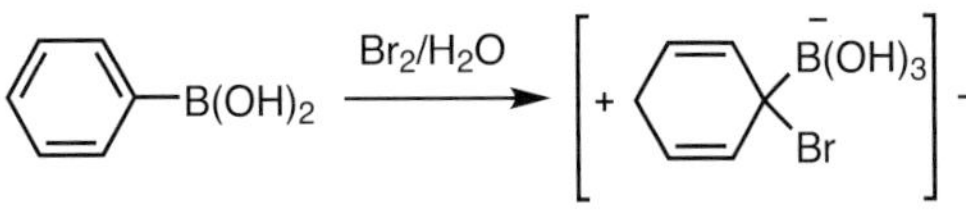

**Figure 1.38** Halodeboronation of arylboronic acids.

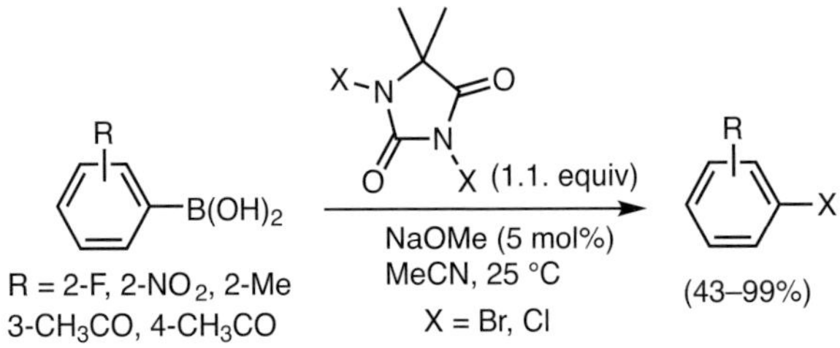

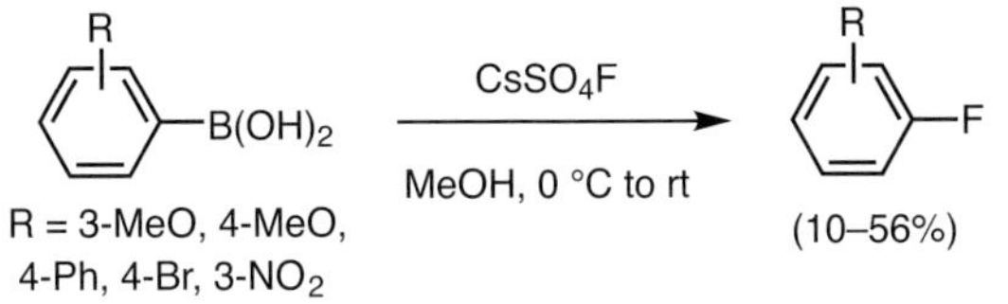

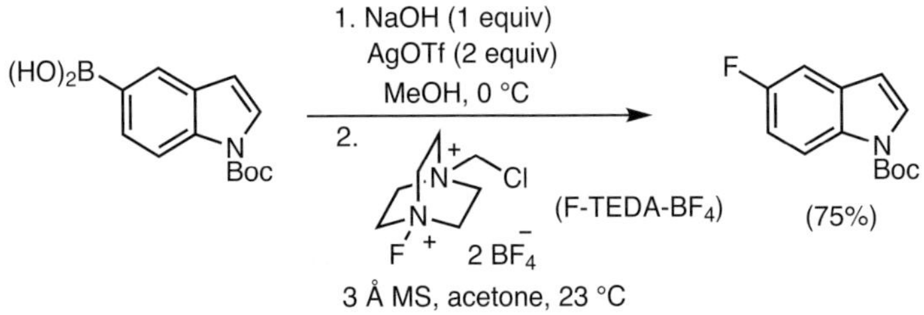

The corresponding reagent DCDMH leads to the isolation of arylchlorides. The combined use of chloramine-T and sodium iodide in aqueous THF affords aryliodides from 2,2-dimethylpropanediol boronates [483]. Aryltrifluoroborate salts are transformed into bromides by action of $n$-Bu$_4$NBr$_3$ under aqueous conditions [484]. Arylfluorides can be obtained in rather modest yield by treatment of arylboronic acids with cesium fluoroxysulfate (CsSO$_4$F) in methanol (Equation 1.78) [485]. Recently, however, interest in medical applications of positron emission tomography led to improved and more general fluorodeboronation procedures, including stepwise Pd(II)- [486] and Ag(I)-promoted methods (Equation 1.79) [487]. Aryl(phenyl)iodonium salts are formed by treatment of arylboronic acids with trifluoromethanesulfonic acid and diacetoxyiodobenzene in dichloromethane [488].

**1.5.2.3.2 Alkenylboronic Acids and Esters** The sequential treatment of alkenylboronic esters with bromine in ethereal anhydrous solvent and then with sodium hydroxide or alkoxides in a one-pot fashion provides the corresponding alkenyl bromides with inversion of olefin geometry (Equations 1.80 and 1.81, Figure 1.39) [489–491]. A reasonable mechanism to account for the inversion was

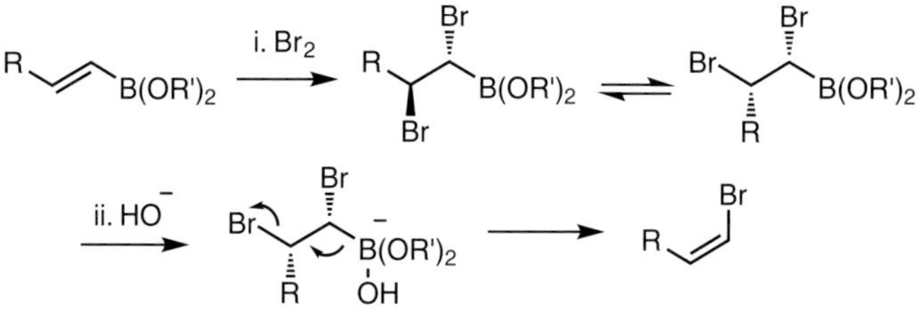

**Figure 1.39** Halodeboronation of alkenylboronic acids (esters).

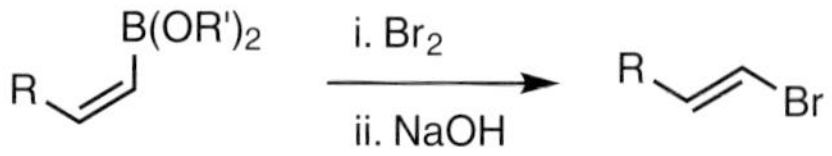

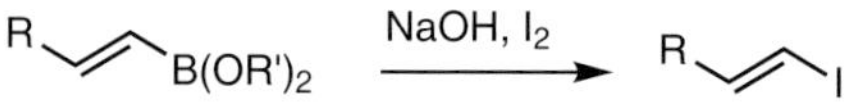

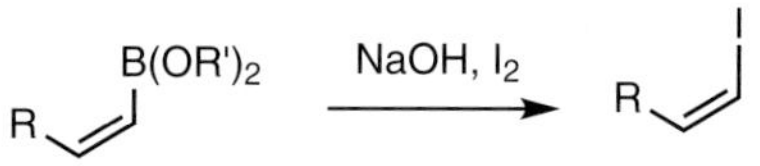

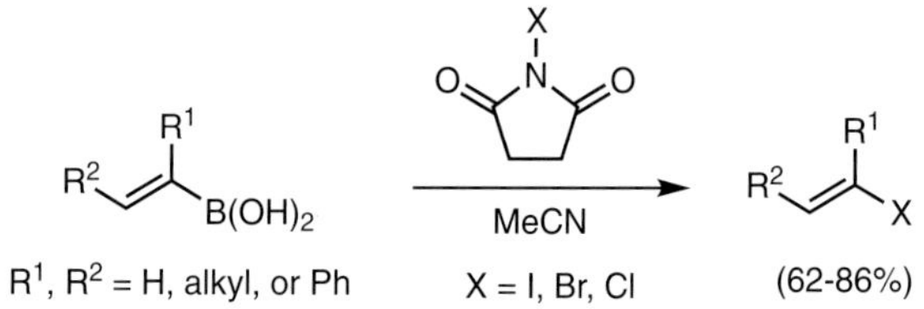

proposed based on the formation of a vicinal dibromide followed by a *trans*-bromodeboronation promoted by the addition of the base (Equation 1.80) [491]. The related iodinolysis process is complementary, giving alkenyl iodides with retention of olefin geometry (Equations 1.82 and 1.83) [492]. The procedure involves the simultaneous action of iodine and aqueous sodium hydroxide, and a tentative mechanism involving the *syn*-deboronation of an iodohydrin intermediate has been proposed to explain the stereochemistry of this reaction [491]. Like the bromination process, however, a sequential treatment of the alkenylboronic acid with iodine and then with sodium hydroxide generally provides the corresponding alkenyl iodides by inversion of geometry [491]. In both cases, boronic acids can be used directly with only 1 equiv of halogen, whereas boronic esters can be transformed effectively with at least 2 equiv of the requisite halogen. The use of ICl and sodium acetate was also demonstrated [493]. The combination of ICl and sodium methoxide as base was found to be more efficient in the case of hindered pinacol alkenylboronates, and both isomers can be obtained selectively from a single *E*-1-alkenylboronate depending on the order of addition [494]. Petasis and Zavialov reported a mild halogenation procedure for various types of alkenylboronic acids using halosuccinimides as reagents (Equation 1.84, Figure 1.39) [495]. The reactions proceed in acetonitrile at room temperature and provide high yields of alkenyl halide products with retention of olefin geometry. The chlorination variant with N-chlorosuccinimide requires the use of triethylamine as a base. The chlorination of alkenylboronic acids was also carried out with chlorine and occurs by inversion of olefin geometry [496].

### 1.5.3
### Carbon–Carbon Bond Forming Processes

#### 1.5.3.1 Transition Metal-Catalyzed Cross-Coupling with Carbon Halides and Surrogates (Suzuki–Miyaura Cross-Coupling)

The ability of boronic acids to undergo C–C bond formation in the presence of a stoichiometric quantity of palladium was recognized in 1975 [272]. A subsequent 1979 *Chemical Communications* paper by Miyaura and Suzuki reported findings generally regarded as the most important discovery in the recent history of boronic acid chemistry [497]. This paper described the palladium(0)-catalyzed coupling between alkenyl boranes or catecholates and aryl halides, in the presence of a base, providing arylated alkene products in high yields. Soon thereafter, a seminal paper on the synthesis of biaryls by coupling of phenylboronic acid with aryl bromides and halides was reported (Equation 1.85, Figure 1.40) [498]. Since then, significant improvements of this important synthetic methodology have been made through optimization of the different reaction parameters such as catalyst, ligands, base, solvent, and additives.

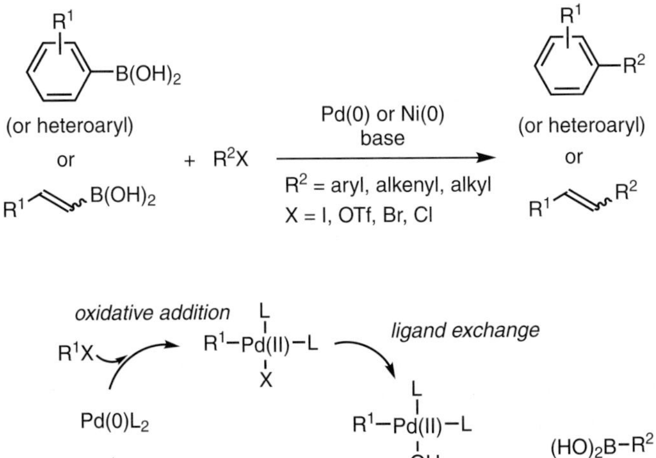

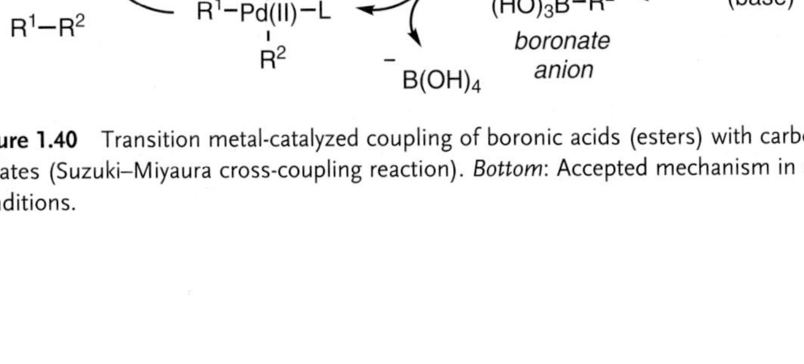

**Figure 1.40** Transition metal-catalyzed coupling of boronic acids (esters) with carbon halides/triflates (Suzuki–Miyaura cross-coupling reaction). *Bottom*: Accepted mechanism in aqueous conditions.

These advances have been reviewed regularly [499], including applications in natural product synthesis [499g]. All the contemporary aspects of the Suzuki–Miyaura cross-coupling reaction are covered in detail in Chapter 4; therefore, only a brief summary is provided in this section. The accepted mechanism for the aqueous basic variant involves oxidative addition of the halide substrate to give a Pd(II) intermediate, followed by a transmetalation, and a final reductive elimination that regenerates the Pd(0) catalyst (Figure 1.40) [500–502]. The two key catalytic intermediates have been observed by electrospray mass spectrometry [503], but ambiguities remain pertaining to the nature of the turnover-limiting step. Although the specific role and influence of the base remain unclear [504], it was suggested that the transmetalation is facilitated by a base-mediated formation of the tetracoordinate boronate anion [505], which is more electrophilic than the free boronic acid (Sections 1.5.1 and 1.5.2). A recent report, however, showed that when a weak base is used in aqueous solvents, transmetalation between a Pd hydroxy complex and trigonal boronic acid is possible [506]. A useful carbonylative variant has also been developed to access benzophenones [507], which can also be produced from the coupling of acid chlorides [508] or anhydrides [509]. Another variant allows the preparation of α,β-unsaturated carboxyesters from alkenylboronic esters [294]. In many of these reactions, a dreaded limitation with some *ortho*-substituted and electron-poor arylboronic acids is the possible occurrence of a competitive protolytic deboronation, which is exacerbated by the basic conditions and the use of a transition metal catalyst (Section 1.5.1). As a result, an excess of boronic acids is often needed, but a method employing a fluorescent dye was proposed as a way to monitor consumption of the boronic acid using a standard handheld UV lamp [510]. Methods to minimize this side reaction were developed, in particular the use of milder bases [511] like fluoride salts [512] and nonaqueous conditions [513]. Recently, a slow-release strategy using MIDA boronates was shown to allow effective coupling of α-heterocyclic and other sensitive boronic acids that are notorious for their tendency to protodeboronate [41]. Competitive homocoupling of the arylboronic acid can compete, but it can also be an attractive process for making symmetrical biaryls [514]. Despite these impediments, the venerable Suzuki–Miyaura cross-coupling reaction has become the most versatile method to synthesize a broad range of biaryl and heteroaryl compounds that find widespread uses as pharmaceutical drugs and materials. The reaction is particularly useful in combination with *ortho*-metalation approaches to generate the arylboronic acid substrate [515]. Alkenylboronic acids and esters, including vinylboronates [516], are also very useful substrates, in particular to access substituted olefins and dienyl moieties commonly encountered in several classes of bioactive natural products [351, 517]. To this end, Kishi and coworkers examined the influence of the base, and developed an optimal variant using thallium hydroxide [350] (Equation 1.86, Figure 1.41) [518]. The Suzuki–Miyaura cross-coupling can be applied to the use of allylic boronic acids [519] and alkylboronic acids [362, 520], including cyclopropylboronic acids [520d], and major recent advances were made with the use of alkyltrifluoroborate salts (Chapter 11) [521]. Hitherto known to be notorious for their tendency to undergo β-hydride elimination, alkyl bromides are now suitable as electrophiles under carefully optimized conditions that even allow couplings of secondary alkyl halides [522] and $Csp^3$–$Csp^3$ couplings with alkylboronic acids

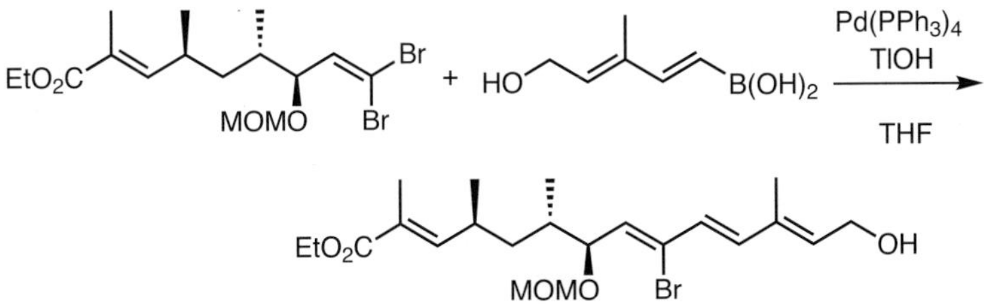

*1.5 Overview of the Reactions of Boronic Acid Derivatives* | 87

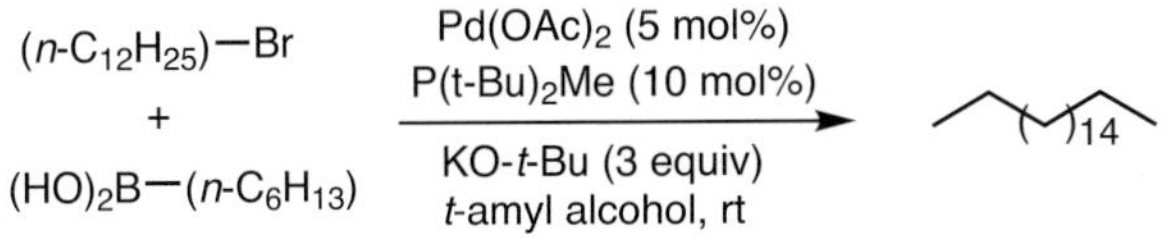

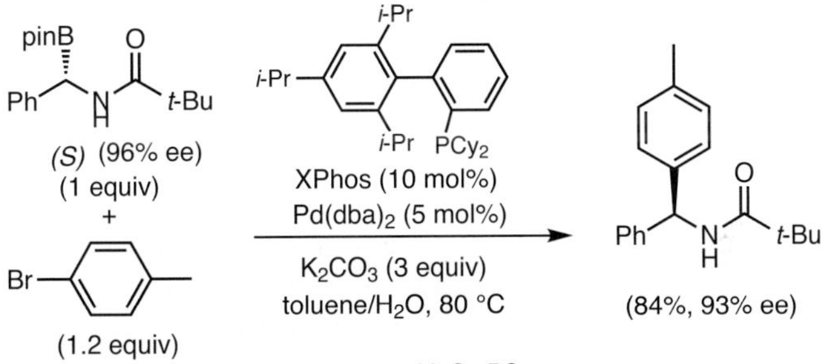

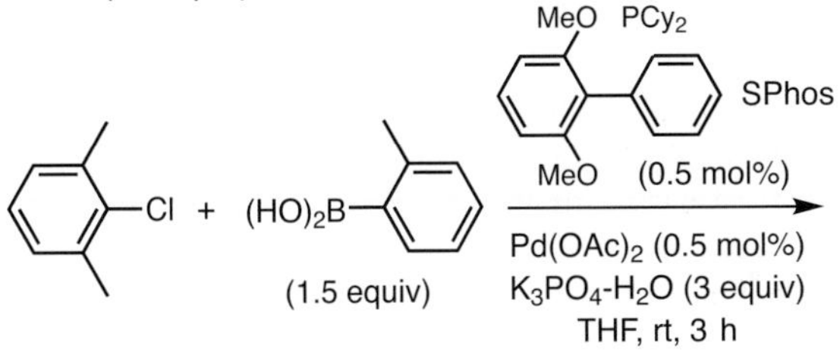

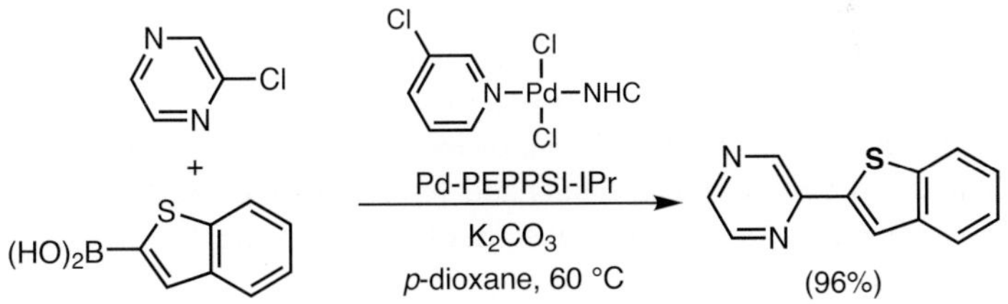

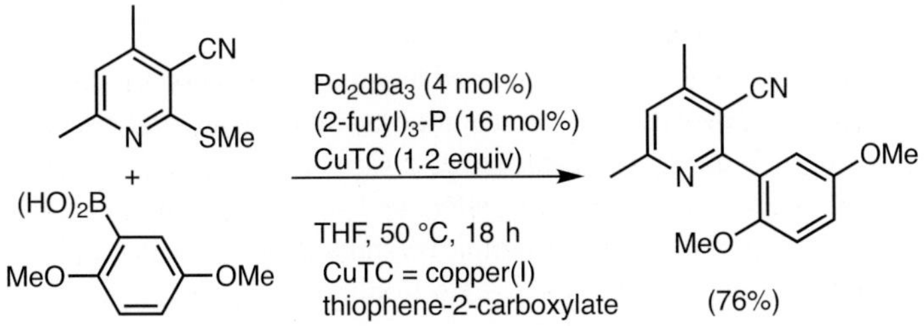

**Figure 1.41** Selected examples of Suzuki–Miyaura cross-coupling reactions.

(Equation 1.87) [523]. Methods for stereoselective cross-couplings of optically enriched alkylboronic acids have begun to emerge (Equation 1.88) [524]. The Suzuki reaction has also been applied very successfully in the fields of polymer chemistry [525], as well as solid-phase chemistry and combinatorial library synthesis [526]. It has been applied industrially [527], especially in medicinal chemistry, for example, in the production of the antihypertensive drug losartan [224]. As described in Section 1.3.8.6, the use of masking groups allows iterative cross-couplings for a controlled synthesis of oligoarenes [427] and polyenes, including naturally occurring ones [426].

In the past decade alone, several new and further improved catalysts and ligands have been developed for difficult substrates such as aryl chlorides, which are cheaper and more available than bromides [528]. Among other advances, new phosphine-based systems developed by Fu [529], Buchwald [530], Organ [531], and others [532] even allow room-temperature couplings with aryl and heteroaryl chlorides. For example, Buchwald and coworkers developed a universal palladium catalyst system based on SPhos, a rationally designed ligand with unprecedented stability and scope for couplings of hindered aryl chlorides at room temperature (Equation 1.89) [533]. Organ and coworkers developed Pd-PEPPSI, a class of very active and broadly applicable palladium complexes of N-heterocyclic carbenes (NHC) (Equation 1.90) [531]. These and other phosphine-free systems based on NHC ligands were shown to perform very well even with hindered boronic acids and electrophiles [534]. Other transition metals were found to catalyze the reaction, notably nickel [535], ruthenium [536], iron [537], and gold [538], albeit the range of suitable substrates tends to be more limited. Interestingly, advantageous ligand-free [539] and even "palladium-free" couplings have even been reported [540]. Other classes of substrates such as polyfluoroarenes [541], aryltosylates [542], arylammonium salts [543], arylcarbamates and carbonates [225, 544], aryl methyl ethers [545], allylic halides and esters [546], and allylic ethers [547] were recently uncovered to further expand the scope of this cross-coupling chemistry. Allylic alcohols can couple directly with alkyl-, alkenyl-, and arylboronic acids [548]. Moreover, arylsulfonium salts [549], thioesters [550], and thioethers [551] were shown to be suitable electrophilic substrates. For example, heteroaromatic thioethers couple to arylboronic acids under base-free conditions promoted by copper(I) thiophene-2-carboxylate (Equation 1.91) [552]. A more detailed description of the Liebeskind–Srogl cross-coupling [553] can be found in Chapter 7. Likewise, more details and recent advances in the Suzuki–Miyaura cross-coupling reaction are described in detail in Chapter 4.

#### 1.5.3.2 Transition Metal-Catalyzed Insertions, Cycloisomerizations, and C–H Functionalizations Based on Transmetalation of Boronic Acids

Numerous reaction processes have been reported based on exploiting the ability of boronic acids to transmetalate with Pd(II) and other transition metals, and a detailed overview would be beyond the scope of this chapter. Carbonylations and carboxylations of arylboronic esters to provide carboxyesters and acids are known (Equation 1.92, Figure 1.42) [554]. There are several examples of transition metal-catalyzed ring forming reactions employing boronic acids as electrophiles [555]. These processes are illustrated in a nice example by Murakami and coworkers of a palladium-catalyzed cyclization of 2-(alkynyl)aryl isocyanates terminated through a Pd(II) transmetalation/

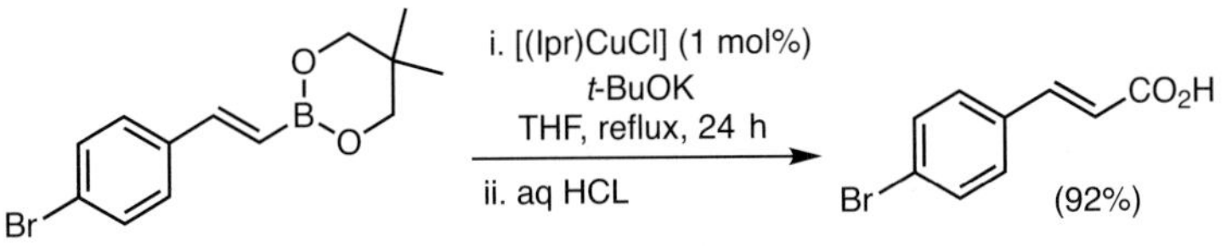

## 1.5 Overview of the Reactions of Boronic Acid Derivatives

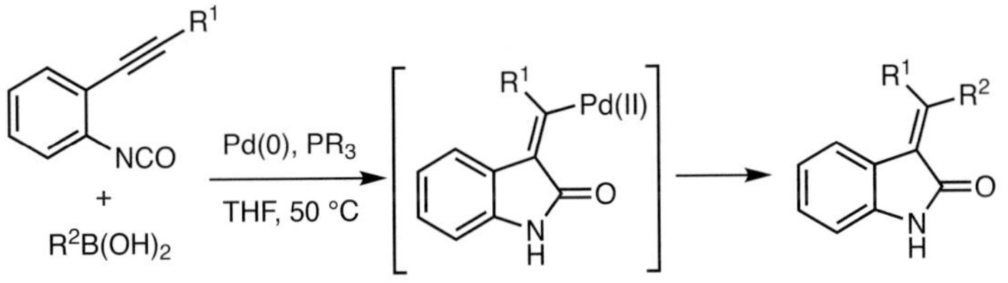

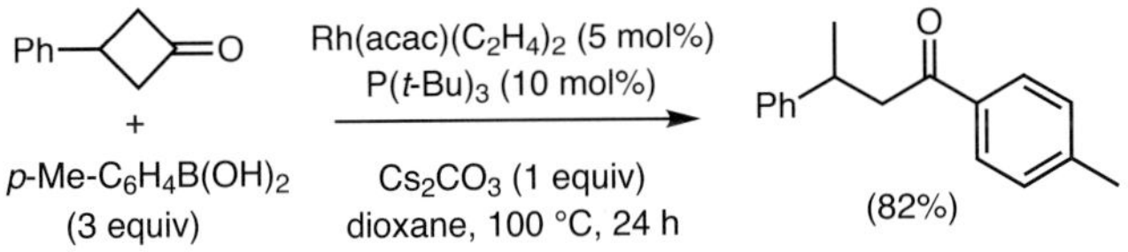

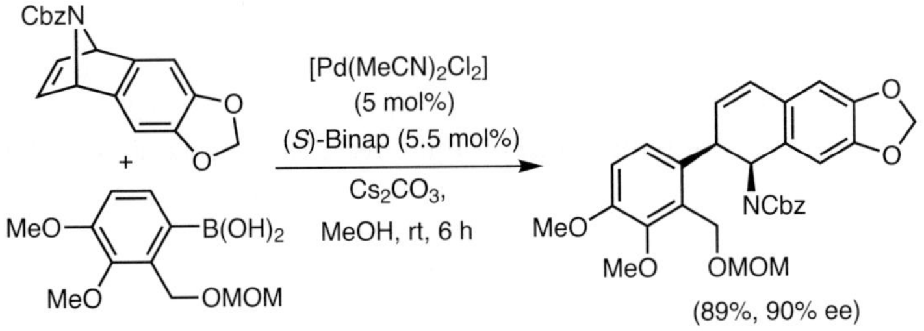

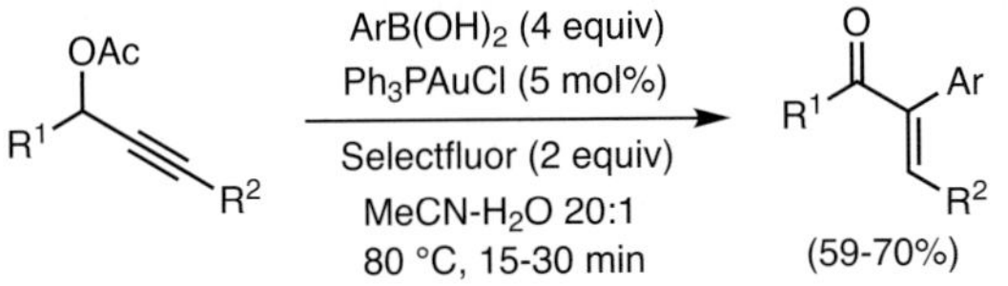

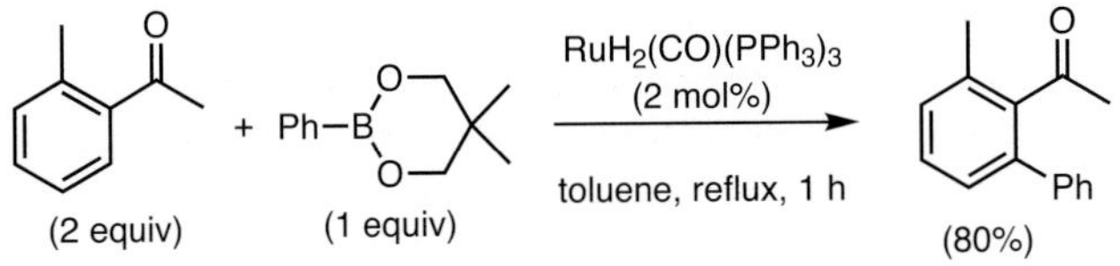

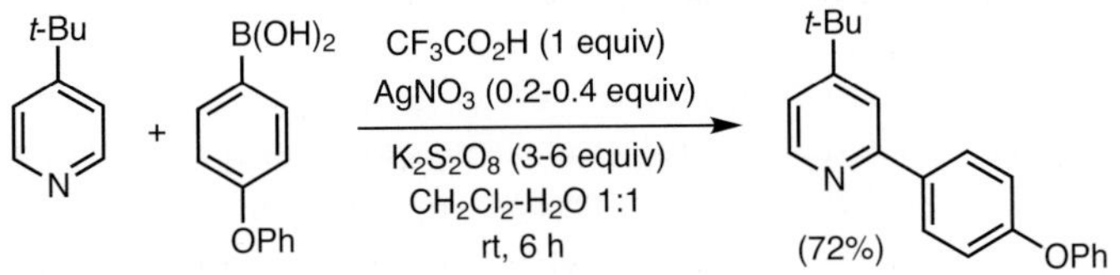

Figure 1.42 Other transition metal-catalyzed transformations of boronic acids (esters).

reductive elimination (Equation 1.93) [556]. Cyclobutanones undergo a C–C bond insertion/functionalization with arylboronic acids (Equation 1.94) [557]. Other recent examples include the use of diazoesters as substrates, affording α,β-diaryl acrylates as products [558], copper-catalyzed stereospecific couplings between arylboronic acids and allylic phosphates [559], and a useful aromatic trifluoromethylation reaction [560]. A bicyclic allylic carbamate was opened enantioselectively in a key Pd(II)-catalyzed step toward the synthesis (+)-homochelidonine (Equation 1.95) [561]. Gold-catalyzed oxidative couplings using boronic acids were reported (Equation 1.96) [562]. Patel and Jamison reported a nickel-catalyzed three-component reaction between alkynes, imines, and organoboron compounds such as alkenyl- and arylboronic acids [563]. The resulting allylic amines are obtained in high regioselectivity. A palladium-catalyzed three-component reaction between allenes, organic halides, and boronic acids was reported [564].

Recent interest in C–H activation/functionalization of arenes has motivated the use of boronic acids as partners. For example, arylboronic esters were used in a ruthenium-catalyzed *ortho*-arylation of aromatic ketones via C–H activation/functionalization (Equation 1.97) [565] or in a dealkoxylation/functionalization [566]. Several palladium-catalyzed variants using various directing groups [567], including functionalization of $sp^3$ centers [568], have been described recently using boronic acids to transmetalate with the Pd(II) intermediate of C–H activation. Silver-catalyzed α-arylation of pyridines (Equation 1.98) [569] and iron-mediated [570] direct arylations of unactivated arenes have also been reported.

#### 1.5.3.3 Heck-Type Coupling to Alkenes and Alkynes

A number of reports have highlighted the ability of boronic acids to undergo rhodium- [571], ruthenium- [572], iridium- [573], or palladium(II)- [574] catalyzed addition–dehydrogenation reactions (oxidative Heck reaction) on alkenes (Equation 1.99, Figure 1.43). The Pd(II)-catalyzed variant is particularly versatile, as demonstrated with the assembly of [2]rotaxanes [575]. An interesting C-glycosylation of glycals provides different isomeric alkenes dependent on the choice of oxidant (Equation 1.100) [576]. A copper-catalyzed "Sonogashira-like" variant affords an aerobic oxidative addition between terminal alkynes and arylboronic acids that produce internal alkynes [577]. A rhodium-catalyzed addition onto 1-arylethenyl acetates affords stilbene derivatives via *cine* substitution [578].

#### 1.5.3.4 Rhodium- and Other Transition Metal-Catalyzed Additions to Alkenes, Carbonyl Compounds, and Imine Derivatives

Another recent breakthrough in organoboron chemistry is the exciting discovery that rhodium(I) complexes catalyze the conjugate addition of boronic acids to carbonyl compounds [579] and a wide range of activated alkene substrates (Equations 1.101 and 1.102, Figure 1.44) [580]. The latter process, reviewed in Chapter 5, can even provide enantioselectivities over 99% with several classes of substrates [581]. Under certain conditions, diarylketones can be obtained in the Rh(I)-catalyzed addition of arylboronic acids to benzaldehydes [582]. A nickel–carbene catalyst was found effective directly from boronic esters under mild conditions [583]. Palladium

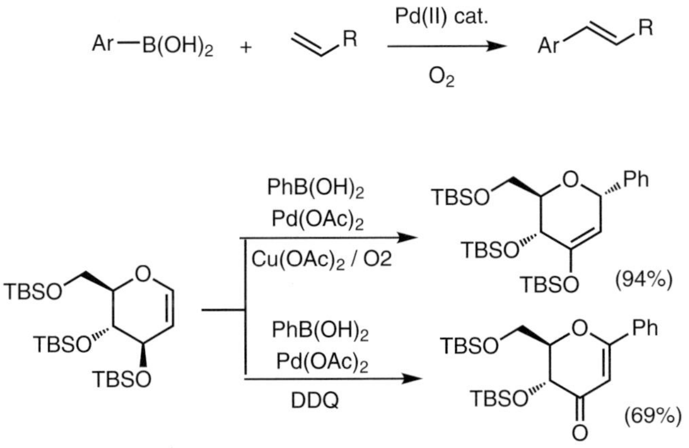

## 1.5 Overview of the Reactions of Boronic Acid Derivatives | 91

$$\text{Ar}-\text{B(OH)}_2 + \diagup\!\!\!\diagup\text{R} \xrightarrow[\text{O}_2]{\text{Pd(II) cat.}} \text{Ar}\diagup\!\!\!\diagup\text{R} \qquad (1.99)$$

(1.100)

**Figure 1.43** Heck-type reactions with boronic acids.

and nickel catalysts can also promote similar additions of boronic acids onto unactivated alkenes [584], alkynes (giving polysubstituted alkenes stereoselectively) [585], allenes [586], and 1,3-butadienes [587].

Selected recent examples of catalytic enantioselective additions of boronic acids and esters to aldehydes and ketones include rhodium- [588], copper- [589], and ruthenium-catalyzed methods (Equation 1.103) [590]. Rhodium-catalyzed additions to imine derivatives are possible [591]. For example, arylboroxines were shown to undergo a catalytic asymmetric addition to N-tosylarylimines [592]. This procedure

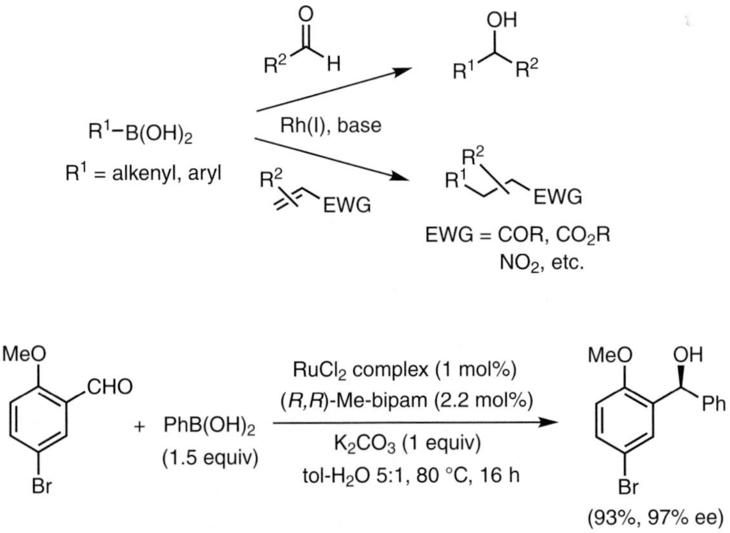

(1.101)

(1.102)

EWG = COR, CO$_2$R, NO$_2$, etc.

(1.103)

(93%, 97% ee)

**Figure 1.44** Rhodium- and ruthenium-catalyzed additions of boronic acids onto carbonyl compounds and activated alkenes.

and other asymmetric variants provide branched amines with high stereoselectivities (see Chapter 9).

#### 1.5.3.5 Diol-Catalyzed Additions of Boronic Esters to Unsaturated Carbonyl Compounds and Acetals

In addition to the above variant that makes use of transition metal catalysts, it was long known that strong Lewis acids can promote the conjugate addition of boronic esters to α,β-unsaturated carbonyl compounds [593]. More recently, it was shown that the B–C bond of alkynylboronic esters is labile enough to allow their uncatalyzed nucleophilic addition to enones, and a stoichiometric asymmetric procedure has been developed using binaphthyl alkynylboronates [594]. A catalytic variant employing chiral binaphthol catalysts was subsequently developed for both alkynyl- and alkenylboronates (Equation 1.104, Figure 1.45) [595]. The mechanism of these diol-catalyzed reactions has been debated to occur either through a simple transesterification (complete exchange to a more electrophilic arenediol ester) or via a mixed, Brønsted activated ester [596]. Using enals as substrates, secondary amine-catalyzed variants proceeding through iminium ion intermediates have been reported, albeit with limited scope and enantioselectivity [597]. Reagents other than allylic or propargylic boronates do not add spontaneously to carbonyl compounds. A recent report, however, describes enantioselective tartrate-derived diol-catalyzed additions of aryl- and alkenylboronates onto chromene acetals (Equation 1.105) [598].

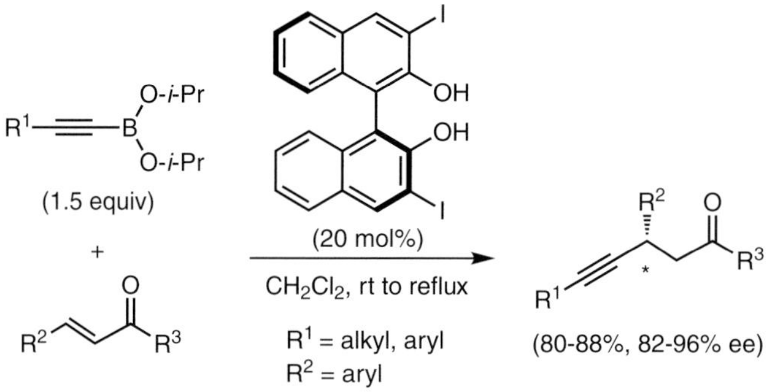


(1.104)

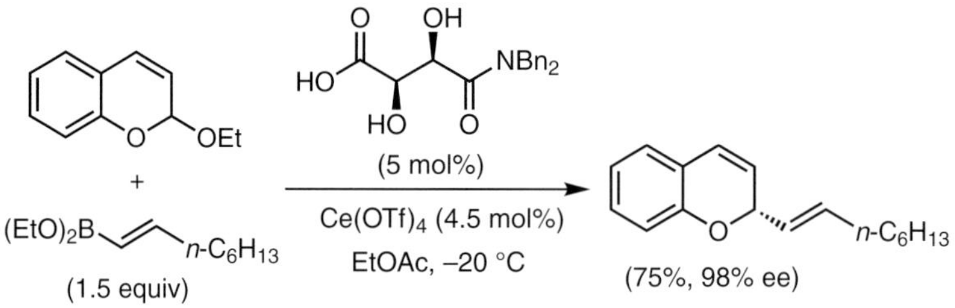


(1.105)

**Figure 1.45** Diol-catalyzed additions of boronic esters.

### 1.5.3.6 Allylation of Carbonyl Compounds and Imine Derivatives

The uncatalyzed addition of allylic boronates to aldehydes was first disclosed in 1974 [599]. This reaction has since found tremendous use in the stereoselective synthesis of acetate and propionate units found in numerous natural products (Equation 1.106, Figure 1.46) [600]. One of the most recent developments of this reaction is the discovery that additions of allylboronates to aldehydes can be catalyzed by Lewis [601] and Brønsted acids [602]. The dramatic rate acceleration observed in these variants allows a substantial decrease in the reaction temperature, which in turn leads to outstanding levels of diastereo- and enantioselectivity using chiral catalysts [603]. Since then, many newer catalytic procedures for additions of allylic boronates to carbonyl compounds and imine derivatives have been developed, as well as efficient methods for the preparation of functionalized reagents [600]. Many of these advances are described in Chapter 8.

### 1.5.3.7 Uncatalyzed Additions of Boronic Acids to Imines and Iminiums

In 1993, Petasis disclosed a novel Mannich-type multicomponent reaction between alkenylboronic acids, secondary amines, and paraformaldehyde [604]. Subsequently, a variant between α-ketoacids, amines, and boronic acids was developed, providing a novel synthetic route to α-amino acids (Equation 1.107, Figure 1.46) [605a]. The use of α-hydroxyaldehydes lends access to β-amino alcohols in high yields and stereoselectivity (Equation 1.108) [605b], and both alkenyl- and arylboronic acids can be employed. A catalytic asymmetric approach has been reported in 2008 [606]. The Petasis borono-Mannich reaction was reviewed recently [607], and its mechanism and applications are discussed with several examples in Chapter 9.

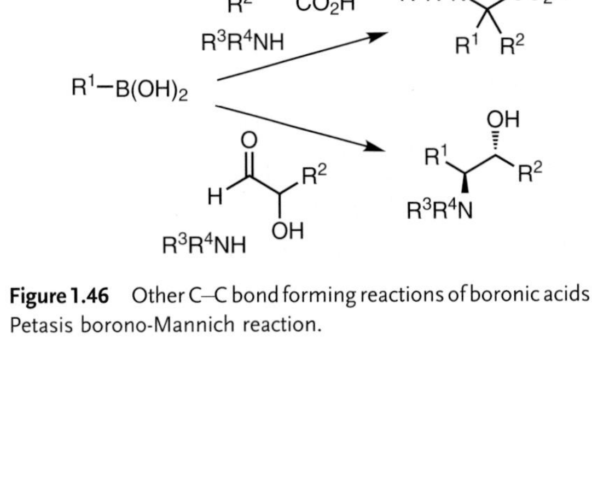

**Figure 1.46** Other C–C bond forming reactions of boronic acids (esters): carbonyl allylboration and Petasis borono-Mannich reaction.

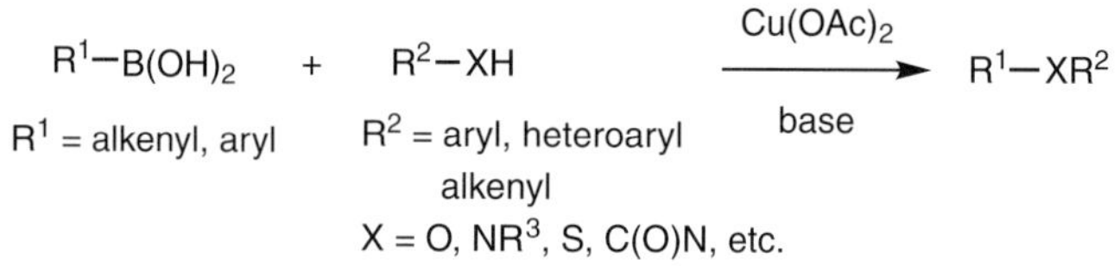

## 1 Structure, Properties, and Preparation of Boronic Acid Derivatives

$$R^1-B(OH)_2 + R^2-XH \xrightarrow[\text{base}]{Cu(OAc)_2} R^1-XR^2 \quad (1.109)$$

$R^1$ = alkenyl, aryl  
$R^2$ = aryl, heteroaryl, alkenyl  
$X$ = O, $NR^3$, S, C(O)N, etc.

[Reaction scheme: EtO-C(O)-CH(NHAc)-CH$_2$-(3,5-diiodo-4-hydroxyphenyl) + (HO)$_2$B-C$_6$H$_4$-OTBS (2 equiv) → with Cu(OAc)$_2$ (1 equiv), pyridine (5 equiv), Et$_3$N (5 equiv), CH$_2$Cl$_2$, 25 °C, 18 h → diaryl ether product (84%)]  (1.110)

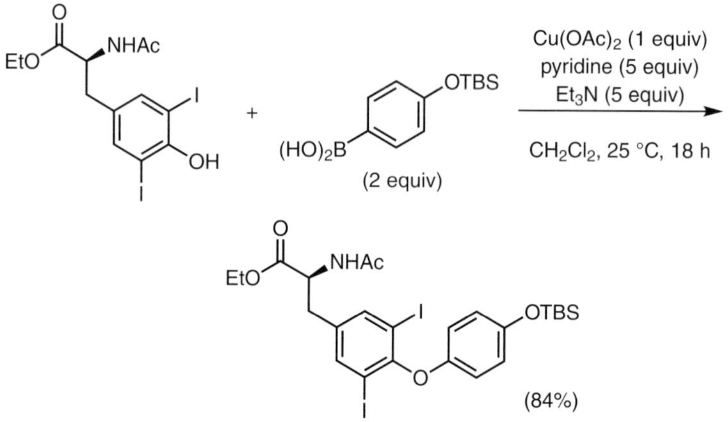

**Figure 1.47** Copper-catalyzed coupling of boronic acids with oxygen and nitrogen compounds.

### 1.5.4
### Carbon–Heteroatom Bond Forming Processes

#### 1.5.4.1 Copper-Catalyzed Coupling with Nucleophilic Oxygen and Nitrogen Compounds

In 1998, groups led by Chan, Evans, and Lam independently reported their observation that copper diacetate promotes the coupling of aryl and heteroaryl boronic acids to moderately acidic heteroatom-containing functionalities like phenols, thiols, amines, amides, and various heterocycles (Equation 1.109, Figure 1.47) [608–610]. The potential of this mild and general method was convincingly exemplified with the syntheses of the diaryl ether units of a thyroxine intermediate (Equation 1.110) [609] and the teicoplanin aglycon related to vancomycin [211]. This new reaction has since been extended to other classes of substrates, including applications in solid-phase synthesis [611]. A mechanism was suggested based on transmetalation of the boronic acid with Cu(OAc)$_2$, followed by ligand exchange with the nucleophilic substrate, and reductive elimination to give the coupling product [608]. These copper-catalyzed heterocoupling reactions of boronic acids constitute the main topic of Chapter 6.

### 1.5.5
### Other Reactions

1,3-Dicarbonyl compounds are arylated with arylboronic acids in the presence of lead tetraacetate and catalytic Hg(OAc)$_2$ under *in situ* conditions that promote a rapid boron-to-lead transmetalation (Equation 1.111, Figure 1.48) [612]. A more recent method for α-arylation and α-vinylation of carbonyl compounds consists in adding boroxines to α-diazocarbonyl compounds via palladium catalysis [613a] or thermal, base-promoted catalysis (Equation 1.112) [613b]. A similar, metal-free reductive

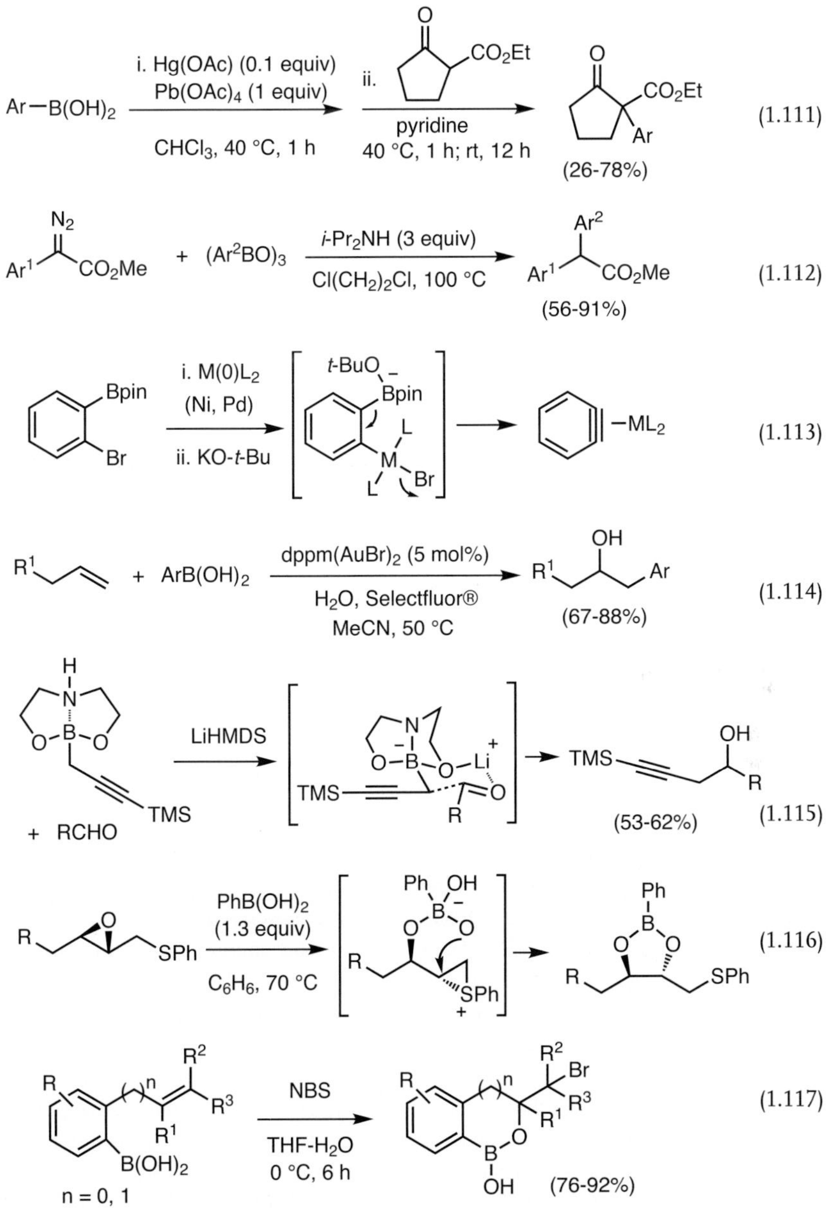

Figure 1.48  Selected examples of miscellaneous reactions of boronic acid derivatives.

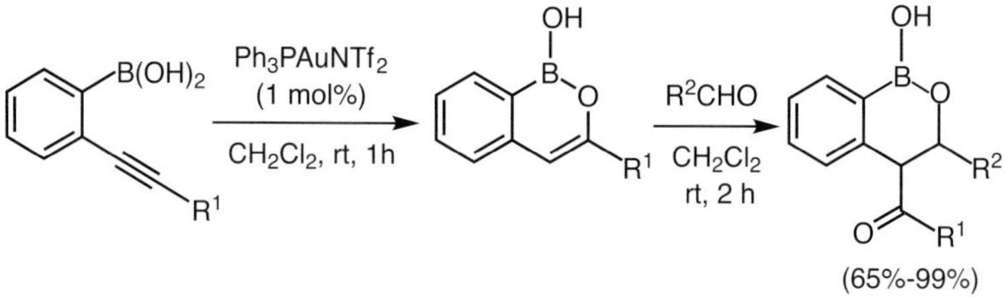

# 1 Structure, Properties, and Preparation of Boronic Acid Derivatives

(1.118)

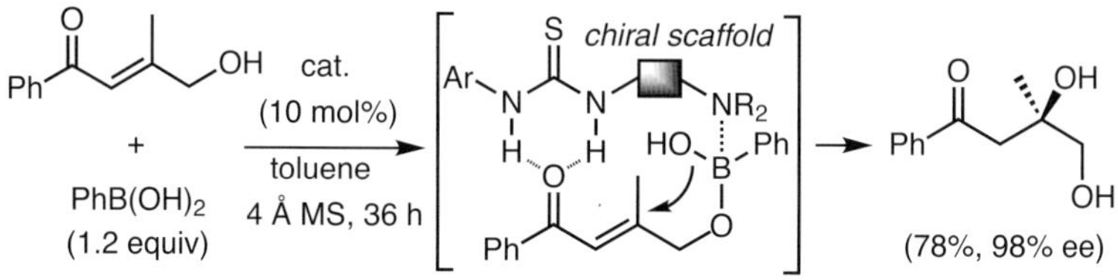

(1.119)

**Figure 1.48** (Continued)

coupling between tosylhydrazones and arylboronic acids was recently described [614]. Allylic carbonates [615] and even amines [616] provide cross-coupling products with boronic acids under nickel catalysis. The metalation of *ortho*-bromobenzeneboronic esters was shown to be an effective route to benzyne complexes of Group 10 metals (e.g., Ni, Pd) (Equation 1.113) [617]. Boronic acids have been employed in multicomponent reaction processes other than the Petasis reaction (Section 1.5.3.7). They were shown to react with diazocyclopentadiene and rhenium(I) tricarbonyl complex to give new monoalkylated cyclopentadienyl rhenium complexes [618]. Recently, fluoride-enabled cationic gold-catalyzed processes were reported that employ boronic acids as reagents. An interesting three-component oxidative alkoxy- and hydroxyarylation of alkenes was described (Equation 1.114) [619], purportedly via Au(III) activation of the alkene pi-bond, and a related hydration/functionalization of alkynes was reported [620]. A chemo- and regioselective Ru(II)-catalyzed cyclotrimerization involving alkynylboronates and two other alkynes can be turned into a four-component synthesis of polysubstituted arenes when combined with a one-pot Suzuki coupling [262a]. A stereoselective three-component reaction between zincated hydrazones, alkenylboronates, and electrophiles was described [621].

Diethylzinc can promote a B–Zn transmetalation of organoboronates followed by a Lewis acid-catalyzed asymmetric 1,2-addition to aldehydes or ketones [457]. Recently, it was found that diethanolamine propargyl boronates can be activated by a strong base and can undergo α-addition to aldehydes (Equation 1.115) [622]. This unique behavior contrasts with the traditional use of allylic and propargylic boronic esters, which add at the γ-carbon. Although nickel- [535] and iron- [537] catalyzed couplings are thought to involve radical intermediates, another uncommon role for boronic acids is their use as precursors of radicals. Exceptionally, radical cyclizations that are initiated by treatment of 2-arylboronic acids with manganese triacetate were recently reported [623].

Under favorable conditions, the hydroxyl group of boronic acids can serve as an internal nucleophile. For example, epoxy sulfides are opened stereoselectively by phenylboronic acid to afford diol products (Equation 1.116) [624]. A variant of this process makes use of a palladium catalyst [625]. Boronic acids have been employed as internal nucleophiles in a bromo-boronolactonization of olefins (Equation 1.117) [626]. Recently, Au(I) catalysis was applied to similar substrates, giving transient boron enolates that can be further reacted with aldehydes (Equation 1.118) [627]. Falck and coworkers developed an ingenious chiral amine-catalyzed, boronic acid-promoted oxy-Michael reaction (Equation 1.119) [628].

## 1.6
## Overview of Other Applications of Boronic Acid Derivatives

### 1.6.1
### Use as Reaction Promoters and Catalysts

By forming transient esters with alcohols, boronic acids have the capability to act as catalysts or templates for directed reactions [629]. In the early 1960s, Letsinger demonstrated that a bifunctional boronic acid, 8-quinolineboronic acid, accelerates the hydrolysis of certain chloroalkanols (Equation 1.120, Figure 1.49) [630], and that boronoarylbenzimidazole serves as catalyst for the etherification of chloroethanol [631]. Mechanisms involving covalent hemiester formation between the boronic acid in the catalyst and the alcohol substrate, combined with a basic or nucleophilic participation of the nitrogen, were invoked. Yamamoto and coworkers found that a number of electron-poor arylboronic acids, in particular 3,4,5-trifluorobenzeneboronic acid, catalyze the direct amidation reactions between carboxylic acids and amines [632]. Hall and coworkers recently identified improved catalysts such as *ortho*-iodobenzeneboronic acid, which functions at room temperature to give high yields of amides from aliphatic amines and acids (Equation 1.121) [633]. Arylboronic acids can also catalyze aldol reactions [634], various cycloadditions of $\alpha,\beta$-unsaturated caboxylic acids [635], Friedel–Crafts alkylation of benzylic alcohols (Equation 1.122) [636], and transpositions of allylic and propargylic alcohols [637]. Arylboronic acids can also catalyze the hydrolysis of salycylaldehyde imines [638] and affect the alkaline conversion of D-glucose into D-fructose [639]. Phenylboronic acid assists in the cyclodimerization of D-glucosamine into a pyrazine [640] and in the photocyclization of benzoin into 9,10-phenanthrenequinone [641].

Boronic acids can be employed to promote templating effects. Narasaka et al. demonstrated that phenylboronic acid can be employed to hold a diene and dienophile in such a way that the regiocontrol of a Diels–Alder reaction can even be inverted [642]. This templating strategy was elegantly exploited in the synthesis of a key intermediate in the total synthesis of taxol by Nicolaou *et al.* (Equation 1.123) [643]. By using a similar effect, phenols are *ortho*-alkylated with aldehydes through a proposed six-membered transition state where phenylboronic acid, used stoichiometrically, holds the two reactants in place (Equation 1.124) [644].

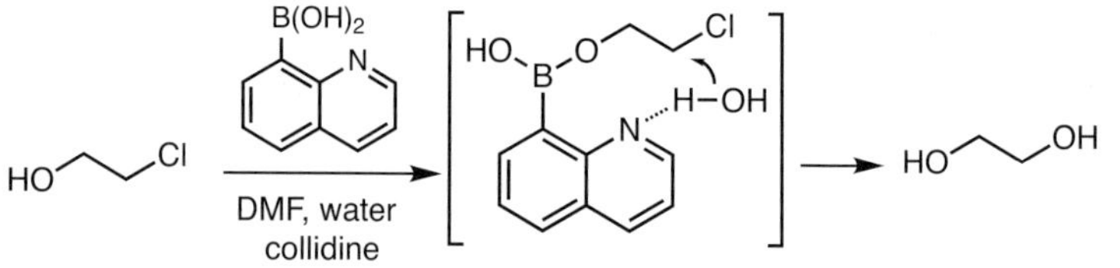


(1.120)

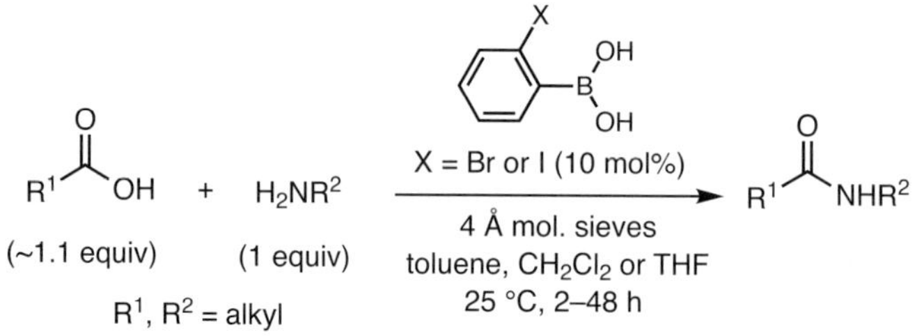


(1.121)

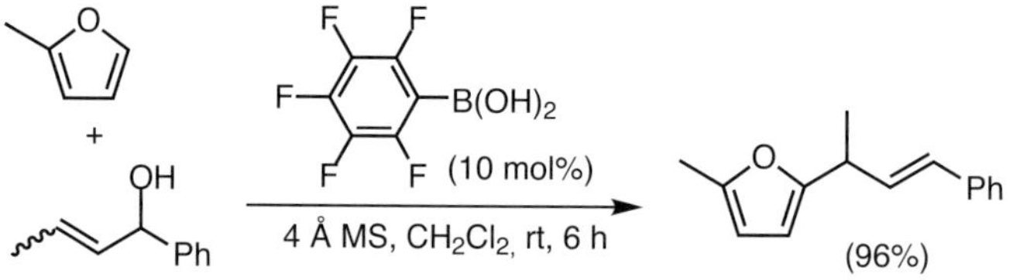


(1.122)

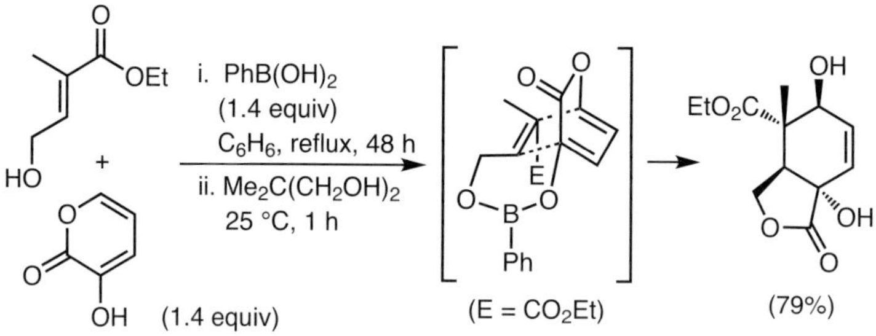


(1.123)

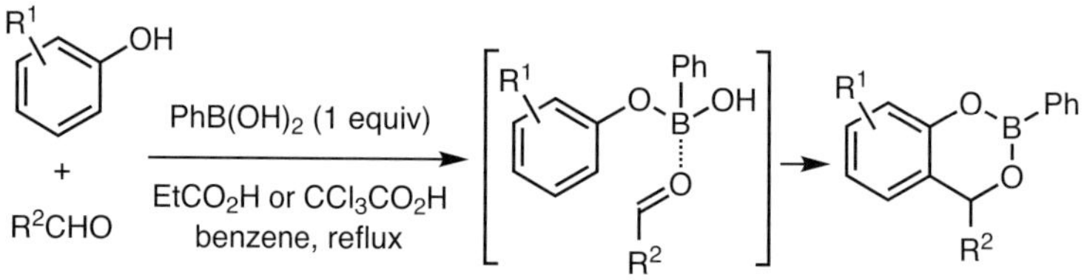


(1.124)

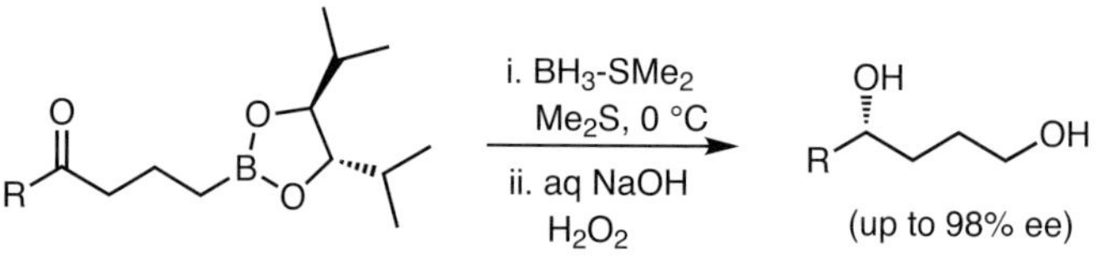


(1.125)

**Figure 1.49** Selected examples of applications of boronic acids (esters) as reaction promoters and catalysts.

Molander et al. have demonstrated the existence of neighboring group participation from a chiral boronate in the reduction of ketones (Equation 1.125) [645]. A highly ordered cyclic transition structure with boron–carbonyl coordination was invoked to explain the high level of remote stereoinduction. The reduction of imine derivatives was also performed with high selectivity [646].

Boronic acids and their derivatives are very popular as components of chiral Lewis acids and promoters for a variety of reaction processes [629]. Indeed, chiral acyloxyboranes and the oxazaborolidines (Section 1.2.3.6) and their protonated salts made a mark in organic synthesis [180–182]. A tartramide-derived dioxaborolane is a key chiral promoter in the asymmetric cyclopropanation of allylic alcohols [647]. More examples and details on the applications of boronic acid derivatives as reaction promoters and catalysts are provided in Chapter 12.

## 1.6.2
### Use as Protecting Groups for Diols and Diamines

The use of boronic acids to protect diol units in carbohydrate chemistry has been demonstrated several decades ago, in particular by the work of Ferrier [648] and Köster [649]. For example, whereas an excess of ethylboronic acid (as the boroxine) leads to a bisboronate furanose derivative of D-lyxose, equimolar amounts provided 2,3-O-ethylboranediyl-D-lyxofuranose (Equation 1.126, Figure 1.50) [650]. From the latter, a regioselective diacetylation reaction followed by treatment with HBr led to the desired α-D-lyxofuranosyl bromide in very high yield. An alternative method for the preparation of cyclic alkylboronic esters involves treatment of diols with lithium trialkylborohydrides [105]. Phenylboronic esters of carbohydrates have also been exploited in the regioselective sulfation of saccharides [651], and as a way to regioselectively alkylate diol units of pyranosides [652]. The reaction of phenylboronic acids with nucleosides and mononucleotides was described long ago [653]. The *ortho*-acetamidophenyl boronate group was employed to protect the vicinal 1,2-diol of adenosine [395]. It was found more resistant to hydrolysis than the corresponding phenylboronate, which was ascribed by the authors to the beneficial coordination effect of the *ortho*-substituent. Phenylboronic acid has also been used as a protecting group for 1,2- and 1,3-diol units of other natural products such as terpenes [654], macrolides [655], prostaglandins [656], quinic acid derivatives [657], anthracyclines [658], steroids [659], macrocyclic polyamines [660], and polyether antibiotics [661]. Typically, phenylboronates are made by a simple condensation with a diol, which can be eventually deprotected by exchange with another diol or by a destructive oxidation with hydrogen peroxide. For example, phenylboronic acid was employed to selectively protect the 1,3-diol unit of a triol (Equation 1.127, Figure 1.50) [661]. Oxidation of the remaining hydroxyl and oxidative deprotection of the phenylboronate led to a concomitant cyclization to give a pyran product. A high-yielding solid-state method for the protection of diols, polyols, and diamines with $PhB(OH)_2$ was described [662]. Phenylboronic acid was also employed as an *in situ* protective reagent in osmium tetraoxide-promoted dihydroxylation of alkenes [663]. In this variant, it serves as a water replacement for cleavage of the osmate intermediate while

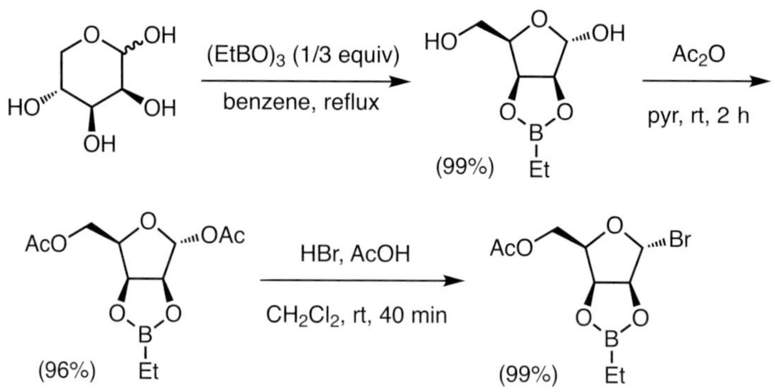

**Figure 1.50** Examples of the use of boronic acids for the protection of diol compounds.

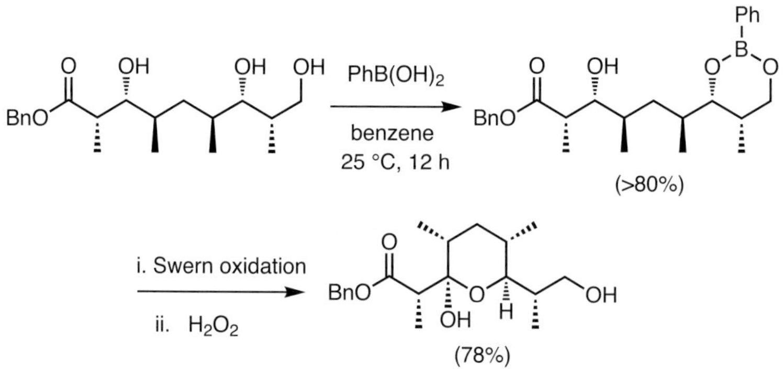

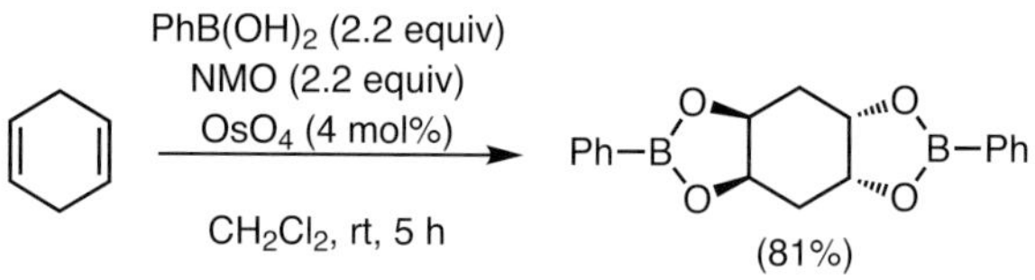

providing a nonpolar cyclic boronate derivative that is easier to extract in organic solvents compared to the free diol. Sharpless and coworkers applied this "boronate capture" procedure to the dihydroxylation of polyenes (Equation 1.128) and found several further advantages such as faster reaction times, minimization of overoxidation, and a marked effect on the diastereoselectivity of these multiple dihydroxylations [664].

## 1.6.3
### Use as Supports for Immobilization, Derivatization, Affinity Purification, Analysis of Diols, Sugars, and Glycosylated Proteins and Cells

The concept of immobilizing or enriching diol compounds with a boronic acid-conjugated support as a sort of heterogeneous protecting group strategy is the

## 1.6 Overview of Other Applications of Boronic Acid Derivatives | 101

antipode of the diol-based supports described in Section 1.4.2. Examples of such boronic acid matrices include polystyryl boronic acid resins (**119**) [665–667], the cellulose-derived support **120** [668], the methacrylic polymer **121** [669], and the polyacrylamide-supported nitroarylbenzene boronic acid **122** [670] (Figure 1.51). Recently, nanoparticles [671] and modified silica [672] have received significant attention. The applications of immobilized boronic acids have been reviewed and include the purification or analysis of carbohydrates and glycopeptides, diverse nucleic acid derivatives embedding rigid vicinal cis-diols, and catechols including L-DOPA, catechol estrogens, and catecholamines from urine [673, 674]. For instance, one of the most important biomedical uses of immobilized boronic acids is in the enrichment and quantification of glycosylated peptides and proteins [675], such as the level of glycosylated hemoglobin in red blood cells, which is an important indicator for the clinical analysis of diabetes. Boronic acid-functionalized composite nanoparticles were used to enrich glycoproteins from human colorectal cancer tissues to identify

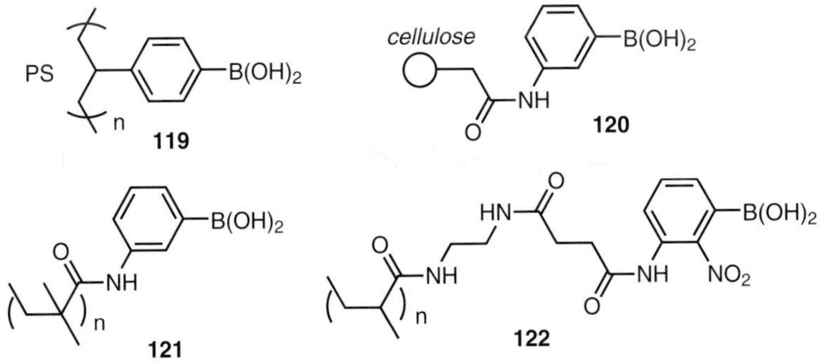

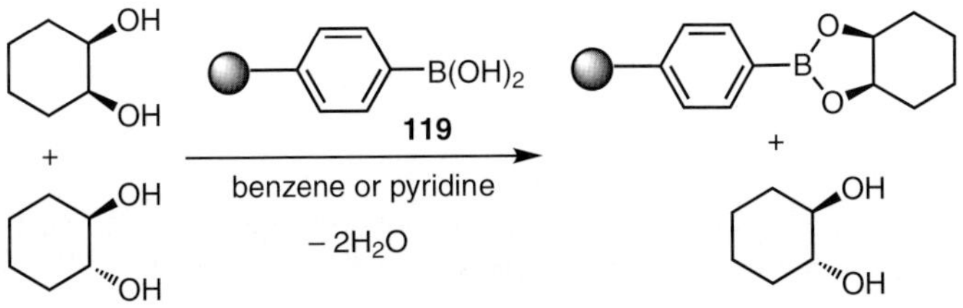

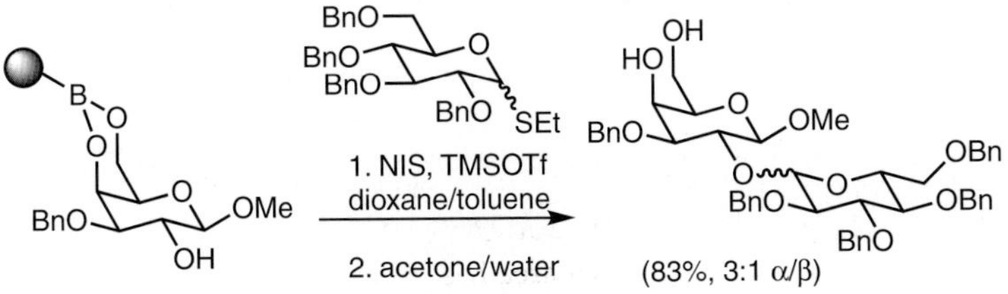

**Figure 1.51** Boronic acid supports for diol compounds.

N-glycosylation sites [671a] and to analyze diol-containing antibiotics in milk samples [671b]. In one other application, a water-soluble polyacrylamide copolymer was tested as a mitogen for lymphocytes [676]. Other supports have also been considered as components of sensing systems for glucose [677–679] and nucleotides such as AMP [680]. With hydrogels, the extent of carbohydrate binding can be correlated with swelling (change in volume) [679]. All of the above arylboronic acid supports demonstrate a selectivity profile similar to their homogeneous counterpart, and only *cis*-diols of a favorable coplanar geometry can be immobilized efficiently. For example, polystyryl boronic acid (**119**) was put to use in the fractionation of carbohydrates and in the separation of isomeric diols [665, 681]. In agreement with the stereochemical arguments discussed in previous sections, of the *cis*- and *trans*-1,2-cyclohexenadiol isomeric mixtures, only the former bound to resin **119**, thereby allowing an effective separation of the two isomers (Equation 1.129, Figure 1.51) [681]. The boronic acid-substituted methacrylic polymer **121** was employed to separate ribonucleosides and deoxyribonucleoside mixtures [669]. The selectivity profile of support **120** in the binding of various nucleic acid and sugar derivatives was studied. Not surprisingly, the heterogeneous boronate formation process in a chromatography column was found to be more efficient at a higher pH, with diols of favorable geometry, and also dependent on the ionic strength and the nature of the cations in the eluent [668]. A Wulff-type (cf. Section 1.2.2.4.1) amino-boronic acid-functionalized copolymeric monolith, however, was claimed to bind diol-containing biomolecules as neutral pH [682]. Polyacrylamide support **122** was employed in the purification of transfer ribonucleic acids [670]. Due to the low p$K_a$ (about 7) of its electron-poor boronic acid unit, the immobilization process was performed efficiently at neutral pH, and recovery of the tRNA from the column occurred at pH 4.5. In hopes of further increasing affinity and selectivity in carbohydrate binding, the technique of molecular imprinting polymerization was tested with boronic acid-containing monomers [66a, 683, 684].

Fréchet also demonstrated the utility of resin **119** in the selective immobilization and transformation of carbohydrate derivatives [666a, 685]. Inspired by this work, Boons and coworkers used the same resin as a reusable linker system for the solid-phase synthesis of oligosaccharides (Equation 1.130, Figure 1.51) [686]. In exciting recent applications, boronic acid-functionalized surfaces were employed in the preparation of microarrays of carbohydrates [687], Fc-fused lectins [688], and the electrochemically addressable immobilization of cells [689].

## 1.6.4
### Use as Receptors and Sensors for Carbohydrates and Other Small Molecules

The ability of boronic acids to form esters reversibly with *cis*-diols (Section 1.2.3.2.3) has been a central theme in the intensive area of sensor and receptor development for oligosaccharides. This very active research area has been reviewed regularly [674, 690], including the previous edition of this monograph. These molecules can be used for a variety of applications such as derivatizing agents for the chromatographic detection of carbohydrates, and in particular in the important social health issue of blood glucose monitoring for diabetes patients. A two-component system based on boronic acid-appended viologen dyes is making significant progress toward this application [691]. Progress has also been made in the development of selective receptors for complex oligosaccharides [692] and glycoproteins [693]. Some of these most recent advances in the field of carbohydrate sensing and recognition with boronic acids are reviewed in Chapter 13.

Mixed receptors containing boronic acids and charged functionalities were also developed for the recognition of sugar acids [694] and even for heparin [394], a polysulfated saccharide. Boronic acid sensors can also target catechols like L-DOPA and dopamine [695], catecholamines in urea [696], polyphenols in green tea [697], α-hydroxycarboxylic acids [698], and receptors selective for tartrate were reported [699].

### 1.6.5
### Use as Antimicrobial Agents and Enzyme Inhibitors

Although there has been significant activity in the area of boron therapeutics over the past decade [700]. Michaelis and Becker first noted the toxicity of phenylboronic acid against microorganisms and its relative harmlessness against higher animals more than a century ago [228]. The antimicrobial properties of simple arylboronic acid derivatives have been further examined in the 1930s [202]. Interestingly, the activity of arylboronic acids in plants has been investigated thoroughly, and several of them were found to promote root growth [60, 62]. Several boronic acids and their benzodiaza- and benzodioxaborole derivatives were evaluated as sterilants of houseflies [61]. A number of boronic acids and esters display potent antifungal activity [701]. For instance, the diazaborine family, exemplified by the thienodiazaborine **123** (Figure 1.52), has long been known to possess potent activity against a wide range of Gram-negative bacteria [702] by targeting the NAD(P)H-dependent enoyl acyl carrier protein reductase [703]. This enzyme is involved in the last reductive step of fatty acid synthase in bacteria, and the structure of the inhibitory complex with diazaborines in the presence of the nucleotide cofactor was elucidated by X-ray crystallography [704]. Interestingly, the bisubstrate complex shows a covalent bond between boron, in a tetracoordinate geometry, and the 2′-hydroxyl of the nicotinamide ribose. In addition to their potential in the fight against microbial resistance in *Mycobacterium tuberculosis* and other strains, diazaborine compounds may find other medicinal applications as estrogen mimics [705]. A prostaglandin mimetic where a boronyl group replaces the carboxylate, **124**, was found to be moderately active [706]. Other boronic acid compounds have been identified as inhibitors of β-lactamase [707], histone deacetylase [708], tubulin polymerization [709], and carboxypeptidase [710], among others. The cyclic hemiboronic ester 4-fluorobenzoxaborole, **125**, was found to inhibit the terminal nucleotide in the editing site of the tRNA-isoleucyl synthetase complex, **126**, by forming a hydrogen-bonded boronate

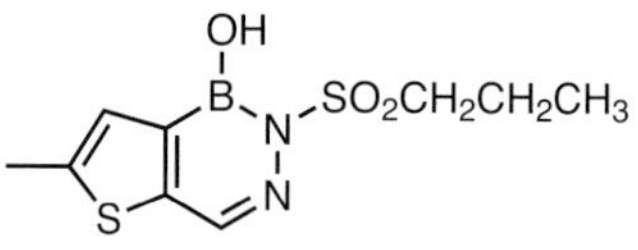

**Figure 1.52** Examples of biologically active boronic acids. *Note*: Compound **128** is the dipeptidyl boronic acid antineoplastic drug bortezomib, a proteasome inhibitor.

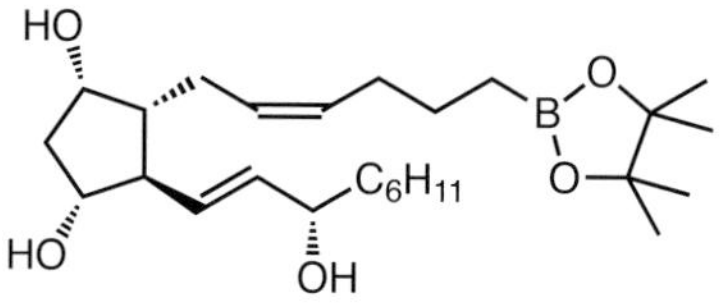

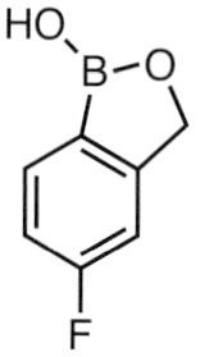

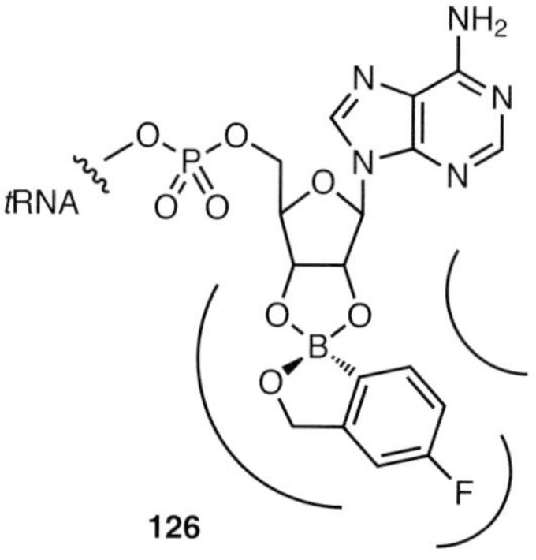

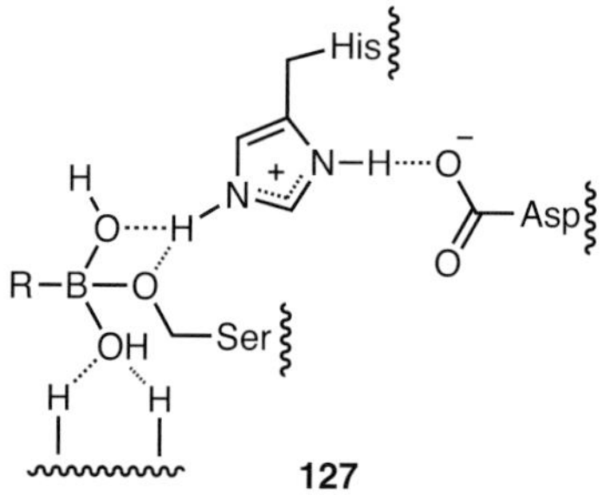

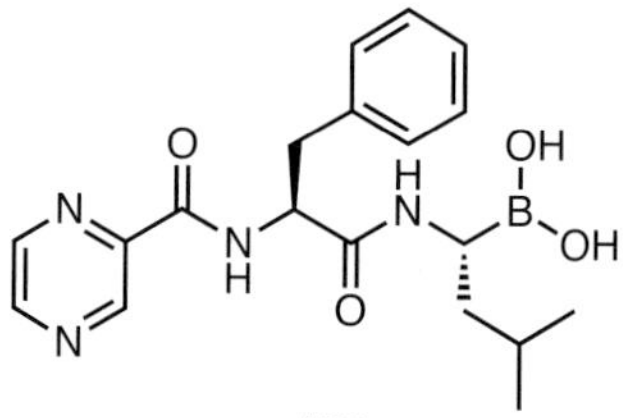

(Figure 1.52) [711]. This antifungal compound is currently entering phase 3 clinical studies for the treatment of onychomysis. Related compounds with this new boron pharmacophore have been recently reported [712]. A multivalent boroxole-functionalized polymer shows potential as a vaginal microbicide targeting the pg120 HIV viral envelope, with minimal cytotoxicity to human cells [713].

Boronic acids have long been known to inhibit hydrolytic enzymes such as serine proteases [700], and the efficiency of a sepharose-based arylboronic acid sorbent in the chromatographic purification of this class of enzymes has been demonstrated [714]. In the development of boronic acid-based enzyme inhibitors as pharmaceutical drugs, target specificity within a wide family is crucial in order to avoid side effects. The development of the α-aminoalkylboronic acid analogues of α-amino acids was key in the recent development of potent peptidylboronic acid analogues with improved specificity. The usual mechanism of inhibition is believed to be the formation of a tetracoordinate boronate complex (**127**, Figure 1.52) with the side chain hydroxyl nucleophile of the active serine residue, thus mimicking the tetrahedral intermediate for amidolysis [715]. Other modes of inhibition have been identified, however, involving formation of covalent adducts with the serine or histine residues of the active site [716, 717]. The validity of this concept was confirmed with the commercialization of the peptidylboronic acid antineoplastic drug bortezomib, Velcade™ (**128**), for the treatment of relapsed and refractory multiple myeloma [718, 719]. Bortezomib is the first boronic acid drug on the market. This discovery and other recent efforts in the medicinal chemistry of boronic acids are described in Chapter 13.

### 1.6.6
**Use in Neutron Capture Therapy for Cancer**

Several boronic acids such as 4-boronophenylalanine (**68**, Figure 1.17) have been evaluated as boron carriers for their potential use in a form of therapy for malignant brain tumors and other locally advanced cancers (head, neck) based on the technology of soft neutron capture [720]. Although this technology is making steady progress, it is still in experimental stage, giving promising outcomes in a few reported cases [721]. The selective delivery of sufficient concentrations of boron to the tumor site is a major issue for success [722].

### 1.6.7
**Use in Transmembrane Transport**

As first demonstrated with monosaccharides by Shinbo et al., the ability of boronic acids to complex diols can be exploited in the study of molecular transport across lipophilic membranes [723]. Compounds that possess such carrier properties have potential applications in drug delivery. For example, Mohler and Czarnik demonstrated the ability of a cholanyl 3-pyridiniumboronic acid derivative (**129**, Figure 1.53) to transport ribonucleosides across a dichloroethane liquid membrane [724]. Other examples of boronic acid-based systems include a three-component amino acid transport system [725], the catecholamine transporter **130** [726], and various carriers for monosaccharides such as fructose [59]. In fact, one of the most important potential applications of boronic acid carriers is in the area of development of selective fructose-permeable liquid membranes [727]. D-Fructose is the sweetest and most valuable of all common natural sweeteners. Its current production as a "high-fructose corn syrup," enriched from crudes containing other sugars, is an energy-intensive industrial process involving the evaporation of large quantities of water. The use of membrane-based technology could be highly advantageous due to its potential amenability to a continuous automated process. Detailed reviews on the use of boronic acids in membrane transport last appeared in 2004 [728]. Recent advances include the use of lipophilic Wulff-type 2-(aminomethyl) phenylboronic acid [729] and the use of boronic acid **131** (Figure 1.53) for target-selective, controllable vesicle membrane fusion as demonstrated with inositol triggering [730].

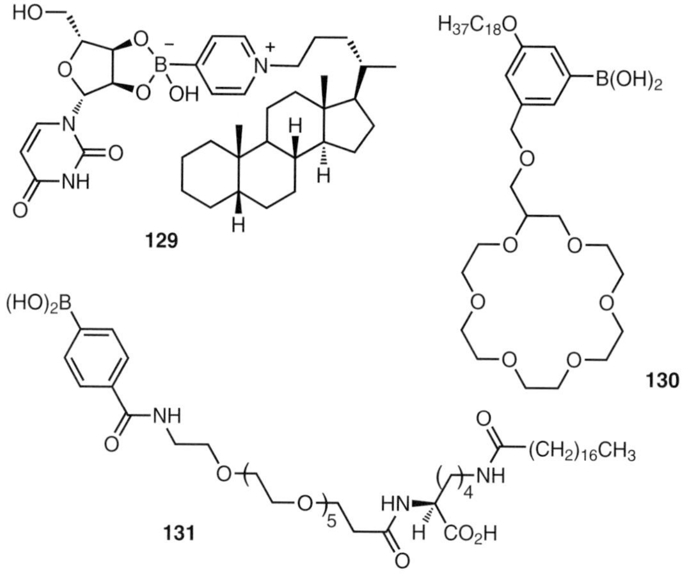

**Figure 1.53** Examples of boronic acid-based transporters.

## 1.6.8
### Use in Bioconjugation and Labeling of Proteins and Cell Surface

Proteins and enzymes can be linked covalently to 3-aminophenyl boronic acid, and the resulting conjugates were shown to bind to small cis-diol molecules and glycated hemoglobin [731]. Studies both in solution and using gel chromatography confirmed the low affinity of the boronate interaction. To address this problem, a conjugation method was developed based on the relatively stronger salicylhydroxamic acid–boronate interaction [177, 732]. As demonstrated on a diboronic acid–alkaline phosphatase conjugate **132** (Figure 1.54), higher affinity over a wider range of pH can be achieved by taking advantage of polyvalent interactions with the complexing sepharose support. An elegant and more contemporary approach to the fluorescent labeling of proteins was disclosed whereby an optimal tetraserine (tetraol) peptide sequence expressed on a protein terminal recognizes a diboronic acid dye with submicromolar affinity (Chapter 13) [733]. An alternative approach to covalent labeling employs an iodoaryl-modified mutant protein that can undergo Suzuki–Miyaura cross-coupling with a boronic acid "label" under physiological conditions [734].

A benzophenone boronic acid, **133** (Figure 1.54), was employed for probing altered specificity of chemically modified mutant subtilisin enzymes by photoaffinity labeling [735]. As discussed in Section 1.6.3, boronic acid supports can be employed to purify glycohemoglobin. A related soluble and colored arylboronic acid was reported for the quantification of these proteins [736]. More than three decades ago, a dansyl-labeled arylboronic acid (**134**) was reported to bind to the cell wall of the bacteria *Bacillus subtilis* presumably via boronate ester formation with the sugar

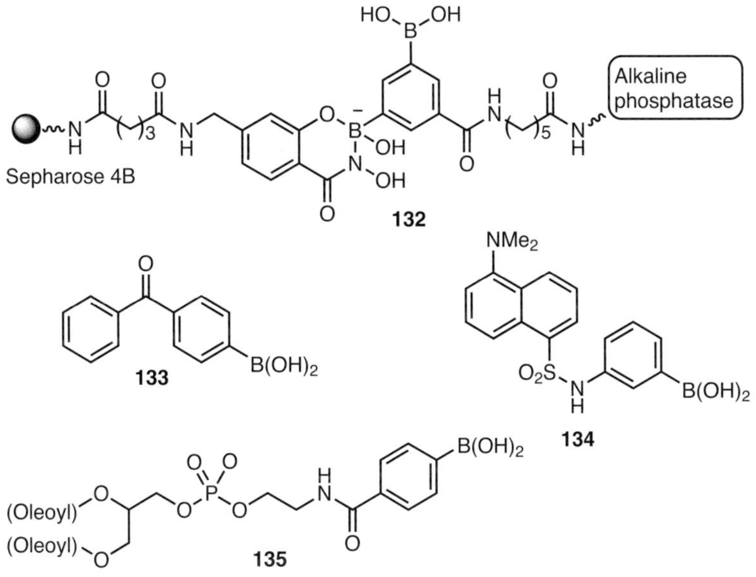

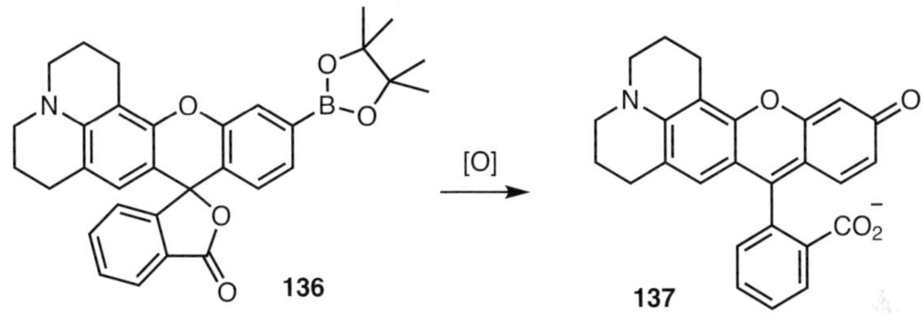

Figure 1.54 Boronic acid compounds used in protein labeling and conjugation, and as probes in chemical biology.

coating [737]. In the same study, a diboronic acid was found to agglutinate erythrocytes. Smith and coworkers designed liposomes containing a phospholipid bearing an arylboronic acid (e.g., **135**), and demonstrated the binding of these liposomes to erythrocytes presumably through interaction with the glycocalyx [738]. A specific diboronic acid sensor was shown to bind to tumor cells overexpressing the fucosylated sialyl Lewis X trisaccharide (Chapter 13) [739].

### 1.6.9
### Use in Chemical Biology

There has been increasing interest in the use of boronic acids as probes to study cell biology and as components of synthetic proteins. For example, Chang and coworkers developed several fluorescent polyarylboronic acid dyes of various emission colors as *in vivo* indicators of reactive oxygen species such as hydrogen peroxide [740, 741].

For example, peroxyoorange 1 (**136**) turns to its quinone-phenolic form **137** upon exposure to peroxide, as demonstrated in macrophages (Equation 1.131, Figure 1.54) [741]. Catalytic antibodies with amide hydrolase activity were generated using a boronic acid hapten based on the concept of protease inhibition described in Section 1.6.5 [742]. More recently, unnatural amino acid mutagenesis was utilized to site-selectively insert 4-boronophenylalanine as a "genetically encoded chemical warhead," and the resulting proteins were shown to bind to an acyclic aglycon [743]. Hoeg-Jensen *et al.* designed a new concept for peptide or protein protraction by soluble reversible self-assembly based on boronic acid–diol interactions. This concept was demonstrated with hexameric insulin, which could be disassembled and released by addition of sorbitol or glucose [744]. More details on the applications of boronic acids in chemical biology can be found in Chapter 13.

### 1.6.10
### Use in Materials Science and Self-Assembly

One of the major new directions in the application of boronic acids is their use as building blocks in the design and preparation of new materials [745] and in self-assembly [746] (e.g., Figure 1.14). For example, the formation of rigid oligomeric boronic anhydrides led to the preparation of a new class of covalent organic frameworks such as the crystalline porous solid COF-1 (**138**, Figure 1.55) [100]. COF-1 demonstrates high surface area and is thermally stable up to 500 °C. Many other bonding modes of boronic acids can be exploited in the design of new materials, such as hydrogen bonding dimerization (as in Figure 1.3b), boronic ester formation, Lewis base coordination, and even mixed bonding modes involving reversible covalent interactions with appended aldehydes, amines, and amino and other functionalities.

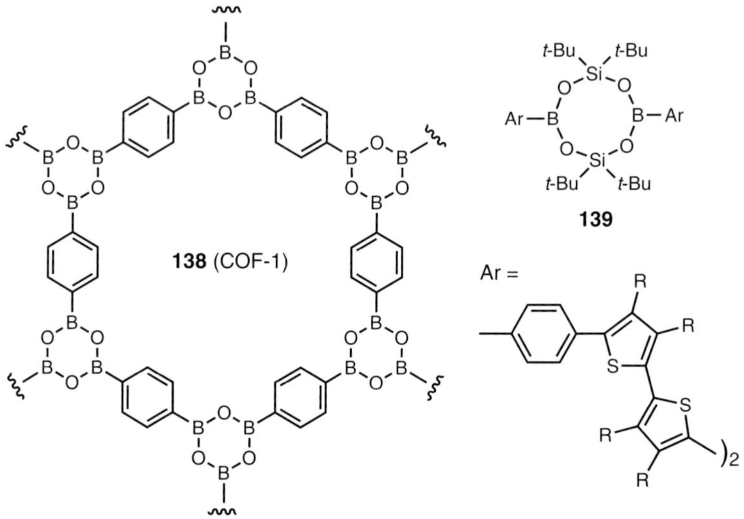

**Figure 1.55** Examples of functional materials based on boronic acid components.

For example, mixed boronic siloxanes of type **139** were developed as conjugated polymer sensors [747]. Existing polyol materials such as lignin may be modified with boronic acids [748]. More applications of boronic acids in materials science are described in detail in Chapter 14.

## References

1. (a) Frankland, E. and Duppa, B.F. (1860) *Justus Liebigs Ann. Chem.*, **115**, 319; (b) Frankland, E. and Duppa, B. (1860) *Proc. R. Soc. Lond.*, **10**, 568; (c) Frankland, E. (1862) *J. Chem. Soc.*, **15**, 363.
2. Brown, H.C. (1975) *Organic Synthesis via Boranes*, John Wiley & Sons, Inc., New York.
3. Matteson, D.S. (1995) *Stereodirected Synthesis with Organoboranes*, Springer, Berlin.
4. Lappert, M.F. (1956) *Chem. Rev.*, **56**, 959–1064.
5. Pelter, A., Smith, K., and Brown, H.C. (1988) *Borane Reagents*, Academic Press, New York.
6. Mikhailov, B.M. and Bubnov, Y.N. (1984) *Organoboron Compounds in Organic Synthesis*, Harwood Academics, Glasgow.
7. Vaultier, M. and Carboni, B. (1995) Chapter 9, in *Comprehensive Organometallic Chemistry II*, vol. **11** (eds E.V. Abel and G. Wilkinson), Pergamon Press, Oxford, pp. 191–276.
8. Kaufmann, D.E. and Matteson, D.S. (eds) (2003) *Science of Synthesis: Vol. 6: Boron Compounds*, Thieme, Stuttgart, Germany.
9. Rettig, S.J. and Trotter, T. (1977) *Can. J. Chem.*, **55**, 3071–3075.
10. Cambridge Crystallographic Database Compound (CCDC) number 222652 (www.ccdc.cam.ac.uk).
11. Soundararajan, S., Duesler, E.N., and Hageman, J.H. (1993) *Acta Crystallogr.*, **C49**, 690–693.
12. Parry, P.R., Wang, C., Batsanov, A.S., Bryce, M.R., and Tarbit, B. (2002) *J. Org. Chem.*, **67**, 7541–7543 (CCDC numbers: bromide, 184781 and 184782; chloride, 184783).
13. Fournier, J.-H., Maris, T., Wuest, J.D., Guo, W., and Galoppini, E. (2003) *J. Am. Chem. Soc.*, **125**, 1002–1006.
14. Sana, M., Leroy, G., and Wilante, C. (1991) *Organometallics*, **10**, 264–270.
15. Ho, O.C., Soundararajan, R., Lu, J., Matteson, D.S., Wang, Z., Chen, X., Wei, M., and Willett, R.D. (1995) *Organometallics*, **14**, 2855–2860.
16. Zhdankin, V.V., Persichini, P.J., III, Zhang, L., Fix, S., and Kiprof, P. (1999) *Tetrahedron Lett.*, **40**, 6705–6708.
17. Adamczyk-Wozniak, A., Cyranski, M.K., Zubrowska, A., and Sporzynski, A. (2009) *J. Organomet. Chem.*, **694**, 3533–3541.
18. Rettig, S.J. and Trotter, J. (1975) *Can. J. Chem.*, **53**, 1393–1401.
19. Csuk, R., Müler, N., and Sterk, H. (1985) *Z. Naturforsch. B*, **40**, 987–989.
20. Mancilla, T., Contreras, R., and Wrackmeyer, B. (1986) *J. Organomet. Chem.*, **307**, 1–6.
21. Bien, J.T., Eschner, M.J., and Smith, B.D. (1995) *J. Org. Chem.*, **60**, 4525–4529.
22. Matteson, D.S., Michnick, T.J., Willett, R.D., and Patterson, C.D. (1989) *Organometallics*, **8**, 726–729.
23. Cammidge, A.N., Goddard, V.H.M., Gopee, H., Harrison, N.L., Hughes, D.L., Schubert, C.J., Sutton, B.M., Watts, G.C., and Whitehead, A.J. (2006) *Org. Lett.*, **8**, 4071–4074.
24. Kaiser, P.F., White, J.M., and Hutton, C.A. (2008) *J. Am. Chem. Soc.*, **130**, 16450–16451.
25. Matteson, D.S. (1995) Chapter 1, in *Stereodirected Synthesis with Organoboranes*, Springer, Berlin, pp. 1–20.
26. (a) Yamashita, M., Yamamoto, Y., Akiba, K.-y., and Nagase, S. (2000) *Angew. Chem., Int. Ed.*, **39**, 4055–4058; (b) Yamashita, M., Yamamoto, Y., Akiba, K.-y., Hashizume, D., Iwasaki, F., Takagi, N., and Nagase, S. (2005) *J. Am. Chem. Soc.*, **127**, 4354–4371.

27 Yamashita, M., Kamura, K., Yamamoto, Y., and Akiba, K.-y. (2002) *Chem. Eur. J.*, **8**, 2976–2979.

28 Yamamoto, Y. and Moritani, I. (1975) *J. Org. Chem.*, **40**, 3434–3437.

29 Matteson, D.S. (1970) *Progress in Boron Chemistry*, vol. **3** (eds R.J. Brotherton and H. Steinberg), Pergamon, New York, pp. 117–176.

30 Good, C.D. and Ritter, D.M. (1962) *J. Am. Chem. Soc.*, **84**, 1162–1166.

31 Hall, L.W., Odom, J.D., and Ellis, P.D. (1975) *J. Am. Chem. Soc.*, **97**, 4527–4531.

32 Holliday, A.K., Reade, W., Johnstone, R.A.W., and Neville, A.F. (1971) *J. Chem. Soc., Chem. Commun.*, 51–52.

33 Bachler, V. and Metzler-Nolte, N. (1998) *Eur. J. Inorg. Chem.*, 733–744.

34 Beachell, H.C. and Beistel, D.W. (1964) *Inorg. Chem.*, **3**, 1028–1032.

35 Matteson, D.S. and Mah, R.W.H. (1963) *J. Org. Chem.*, **28**, 2171–2174.

36 Matteson, D.S. (1962) *J. Org. Chem.*, **27**, 4293–4300.

37 (a) Matteson, D.S. and Peacock, K. (1960) *J. Am. Chem. Soc.*, **82**, 5759–5760; (b) Matteson, D.S. and Waldbillig, J.O. (1963) *J. Org. Chem.*, **28**, 366–369.

38 Matteson, D.S. and Talbot, M.L. (1967) *J. Am. Chem. Soc.*, **89**, 1123–1126.

39 Evans, D.A., Golob, A.M., Mandel, N.S., and Mandel, G.S. (1978) *J. Am. Chem. Soc.*, **100**, 8170–8174.

40 Singleton, D.A. (1992) *J. Am. Chem. Soc.*, **114**, 6563–6564.

41 Knapp, D.M., Gillis, E.P., and Burke, M.D. (2009) *J. Am. Chem. Soc.*, **131**, 6961–6963.

42 Snyder, H.R., Kuck, J.A., and Johnson, J.R. (1938) *J. Am. Chem. Soc.*, **60**, 105–111.

43 Johnson, J.R. and Van Campen, M.G., Jr. (1938) *J. Am. Chem. Soc.*, **60**, 121–124.

44 Soloway, A.H., Whitman, B., and Messer, J.R. (1960) *J. Pharmacol. Exp. Ther.*, **129**, 310–314.

45 Benderdour, M., Bui-Van, T., Dicko, A. et al. (1998) *J. Trace Elem. Med. Biol.*, **12**, 2–7.

46 Soloway, A.H., Whitman, B., and Messer, J.R. (1962) *J. Med. Pharm. Chem.*, **7**, 640.

47 Soloway, A.H. (1958) *Science*, **128**, 1572.

48 Matteson, D.S., Soloway, A.H., Tomlinson, D.W., Campbell, J.D., and Nixon, G.A. (1964) *J. Med. Chem.*, **7**, 640–643.

49 Weir, R.J., Jr., and Fisher, R.S. (1972) *Toxicol. Appl. Pharmacol.*, **23**, 351.

50 Linden, C.H., Hall, A.H., and Kulig, K.W. (1986) *J. Toxicol. Clin. Toxicol.*, **24**, 269–279.

51 Restuccio, A., Mortensen, M.E., and Kelley, M.T. (1992) *Am. J. Emerg. Med.*, **10**, 545–547.

52 Lorand, J.P. and Edwards, J.O. (1959) *J. Org. Chem.*, **24**, 769–774.

53 Branch, G.E.K., Yabroff, D.L., and Bettmann, B. (1934) *J. Am. Chem. Soc.*, **56**, 937–941.

54 Yabroff, D.L., Branch, G.E.K., and Bettmann, B. (1934) *J. Am. Chem. Soc.*, **56**, 1850–1857.

55 Bettman, B., Branch, G.E.K., and Yabroff, D.L. (1934) *J. Am. Chem. Soc.*, **56**, 1865–1870.

56 Soundararajan, S., Badawi, M., Kohlrust, C.M., and Hageman, J.H. (1989) *Anal. Biochem.*, **178**, 125–134.

57 Yan, J., Springsteen, G., Deeter, S., and Wang, B.H. (2004) *Tetrahedron*, **60**, 11205–11209.

58 Babcock, L. and Pizer, R. (1980) *Inorg. Chem.*, **19**, 56–61.

59 Westmark, P.R., Gardiner, S.J., and Smith, B.D. (1996) *J. Am. Chem. Soc.*, **118**, 11093–11100.

60 Torssell, K., McLendon, J.H., and Somers, G.F. (1958) *Acta Chem. Scand.*, **12**, 1373–1385.

61 Settepani, J.A., Stokes, J.B., and Borkovek, A.B. (1970) *J. Med. Chem.*, **13**, 128–131.

62 Torssell, T. (1964) *Progress in Boron Chemistry*, vol. **1** (eds H. Steinberg and A.L. McCloskey), Pergamon, New York, pp. 369–415.

63 Mulla, H.R., Agard, N.J., and Basu, A. (2004) *Bioorg. Med. Chem. Lett.*, **14**, 25–27.

64 Fisher, F.C. and Havinga, E. (1974) *Recl. Trav. Chim. Pays Bas*, **93**, 21–24.

65 Yang, W., Yan, J., Springsteen, G., Deeter, S., and Wang, B. (2003) *Bioorg. Med. Chem. Lett.*, **13**, 1019–1022.

**66** (a) Wulff, G. (1982) *Pure Appl. Chem.*, **54**, 2093–2102; (b) Wiskur, S.L., Lavigne, J.J., Ait-Haddou, H., Lynch, V., Chiu, Y.H., Canary, J.W., and Anslyn, E.W. (2001) *Org. Lett.*, **3**, 1311–1314; (c) Juillard, J. and Geuguen, N. (1967) *C.R. Acad. Sci (Paris)*, **C264**, 259–261.

**67** McDaniel, D.H. and Brown, H.C. (1955) *J. Am. Chem. Soc.*, **77**, 3756–3763.

**68** Singhal, R.P., Ramamurthy, B., Govindraj, N., and Sarwar, Y. (1991) *J. Chromatogr.*, **543**, 17–38.

**69** (a) Ni, W.J., Kaur, G., Springsteen, G., Franzen, S., and Wang, B. (2004) *Bioorg. Chem.*, **32**, 571–581; (b) Zhu, L., Shabbir, S.H., Gray, M., Lynch, V.M., Sorey, S., and Anslyn, E.V. (2006) *J. Am. Chem. Soc.*, **128**, 1222–1232; (c) Collins, B.E., Sorey, S., Hargrove, A.E., Shabbir, S.H., Lynch, V.M., and Anslyn, E.V. (2009) *J. Org. Chem.*, **74**, 4055–4060.

**70** Ainley, A.D. and Challenger, F. (1930) *J. Chem. Soc.*, 2171–2180.

**71** Dewar, M.J.S. and Jones, R. (1967) *J. Am. Chem. Soc.*, **89**, 2408–2410.

**72** Dewar, M.J.S. (1964) *Progress in Boron Chemistry*, vol. 1 (eds H. Steinberg and A.L. McCloskey), Pergamon, New York, pp. 235–263.

**73** Carboni, B. and Monnier, L. (1999) *Tetrahedron*, **55**, 1197–1248.

**74** Yabroff, D.L. and Branch, G.E.K. (1933) *J. Am. Chem. Soc.*, **55**, 1663–1665.

**75** Snyder, H.R., Konecky, M.S., and Lennarz, W.J. (1958) *J. Am. Chem. Soc.*, **80**, 3611–3615.

**76** Reetz, M.T., Niemeyer, C.M., and Harms, K. (1991) *Angew. Chem., Int. Ed. Engl.*, **30**, 1472–1474.

**77** Reetz, M.T., Niemeyer, C.M., Hermes, M., and Goddard, R. (1992) *Angew. Chem., Int. Ed. Engl.*, **31**, 1017–1019.

**78** Reetz, M.T., Huff, J., and Goddard, R. (1994) *Tetrahedron Lett.*, **35**, 2521–2524.

**79** Nozaki, K., Yoshida, M., and Takaya, H. (1994) *Angew. Chem., Int. Ed. Engl.*, **33**, 2452–2454.

**80** Nozaki, K., Tsutsumi, T., and Takaya, H. (1995) *J. Org. Chem.*, **60**, 6668–6669.

**81** Sakurai, H., Iwasawa, N., and Narasaka, K. (1996) *Bull. Chem. Soc. Jpn.*, **69**, 2585–2594.

**82** Katz, H.E. (1989) *J. Org. Chem.*, **54**, 2179–2183.

**83** Lemieux, V., Spantulescu, M.D., Baldridge, K.K., and Branda, N.R. (2008) *Angew. Chem. Int. Ed.*, **47**, 5034–5037.

**84** Hawkins, V.F., Wilkinson, M.C., and Whiting M. (2008) *Org. Proc. Res. Devel.*, **12**, 1265–1268.

**85** Johnson, J.R., Van Campen, M.G., Jr., and Grummit, O. (1938) *J. Am. Chem. Soc.*, **60**, 111–115.

**86** Challenger, F. and Richards, O.V. (1934) *J. Chem. Soc.*, 405–411.

**87** Snyder, H.R. and Wyman, F.W. (1948) *J. Am. Chem. Soc.*, **70**, 234–237.

**88** Kuivila, H.G. and Nahabedian, K.V. (1961) *J. Am. Chem. Soc.*, **83**, 2159–2163.

**89** Soloway, A.H. (1959) *J. Am. Chem. Soc.*, **81**, 3017–3019.

**90** Dick, G.R., Knapp, D.M., Gillis, E.P., and Burke, M.D. (2010) *Org. Lett.*, **12**, 2314–2317.

**91** (a) Kuivila, H.G. and Nahabedian, K.V. (1961) *J. Am. Chem. Soc.*, **83**, 2164–2166; (b) Nahabedian, K.V. and Kuivila, H.G. (1961) *J. Am. Chem. Soc.*, **83**, 2167–2174.

**92** McKiernan, G.J. and Hartley, R.C. (2003) *Org. Lett.*, **5**, 4389–4392.

**93** Mothana, S., Chahal, N., Vanneste, S., and Hall, D.G. (2007) *J. Comb. Chem.*, **9**, 193–196.

**94** Beckett, M.A., Gilmore, R.J., and Idrees, K. (1993) *J. Organomet. Chem.*, **455**, 47–49.

**95** (a) Brown, H.C., Basavaiah, D., and Kulkarni, S.U. (1982) *J. Org. Chem.*, **47**, 3808–3810; (b) Brown, H.C., Basavaiah, D., Kulkarni, S.U., Lee, H.D., Negishi, E.-i., and Katz, J.J. (1986) *J. Org. Chem.*, **51**, 5270–5276.

**96** Korich, A.L. and Iovine, P.M. (2010) *Dalton Trans.*, **39**, 1423–1431.

**97** Schleyer, P.V., Jiao, H.J., Hommes, N.J.R.V., Malkin, V.G., and Malkina, O.L. (1997) *J. Am. Chem. Soc.*, **119**, 12669–12770.

**98** Brock, C.P., Minton, R.P., and Niedenzu, K. (1987) *Acta Crystallogr.*, **C43**, 1775–1779.

**99** Morgan, A.B., Jurs, J.L., and Tour, J.M. (2000) *J. Appl. Polym. Sci.*, **76**, 1257–1268.

100 Côté, A.P., Benin, A.I., Ockwig, N.W., O'Keefe, M., Matzger, A.J., and Yaghi, O.M. (2005) *Science*, **310**, 1166–1170.
101 Li, Y., Ding, J., Day, M., Tao, Y., Lu, J., and D'iorio, M. (2003) *Chem. Mater.*, **15**, 4936–4943.
102 Tokunaga, Y., Ueno, H., Shimomura, Y., and Seo, T. (2002) *Heterocycles*, **57**, 787–790.
103 Farfan, N., Höpfl, H., Barba, V., Ochoa, M.E., Santillan, R., Gómez, E., and Gutiérrez, A. (1999) *J. Organomet. Chem.*, **581**, 70–81, and references cited therein.
104 Schnürch, M., Holzweber, M., Mihovilovic, M.D., and Stanetty, P. (2007) *Green Chem.*, **9**, 139–145.
105 Garlaschelli, L., Mellerio, G., and Vidari, G. (1989) *Tetrahedron Lett.*, **30**, 597–600.
106 Dahlhoff, W.V. and Köster, R. (1982) *Heterocycles*, **18**, 421–449.
107 Kuivila, H.G., Keough, A.H., and Soboczenski, E.J. (1954) *J. Org. Chem.*, **8**, 780–783.
108 Sugihara, J.M. and Bowman, C.M. (1958) *J. Am. Chem. Soc.*, **80**, 2443–2446.
109 Meiland, M., Heinze, T., Guenther, W., and Liebert, T. (2009) *Tetrahedron Lett.*, **50**, 469–472.
110 Ahn, Y.H. and Chang, Y.-T. (2004) *J. Comb. Chem.*, **6**, 293–296.
111 (a) Roy, C.D. and Brown, H.C. (2007) *J. Organomet. Chem.*, **692**, 784–790; (b) Roy, C.D. and Brown, H.C. (2007) *Monatsh. Chem.*, **138**, 879–887.
112 (a) Letsinger, R.L. and Skoog, I. (1955) *J. Am. Chem. Soc.*, **77**, 2491–2494; (b) Musgrave, O.C. and Park, T.O. (1955) *Chem. Ind. (London)*, **48**, 1552.
113 Weidman, H. and Zimmerman, H.K., Jr. (1958) *Justus Liebigs Ann. Chem.*, **619**, 28–35.
114 (a) Haruta, R., Ishiguro, M., Ikeda, N., and Yamamoto, Y. (1982) *J. Am. Chem. Soc.*, **104**, 7667–7669; (b) Roush, W.R., Walts, A.G., and Hoong, L.K. (1985) *J. Am. Chem. Soc.*, **107**, 8186–8190.
115 Ketuly, K.A. and Hadi, A.H.A. (2010) *Molecules*, **15**, 2347–2356.
116 Ditrich, K., Bube, T., Stürmer, R., and Hoffmann, R.W. (1986) *Angew. Chem., Int. Ed. Engl.*, **25**, 1028–1030.
117 Matteson, D.S. and Kandil, A.A. (1986) *Tetrahedron Lett.*, **27**, 3831–3834.
118 Ray, R. and Matteson, D.S. (1980) *Tetrahedron Lett.*, **21**, 449–450.
119 Herold, T., Schrott, U., and Hoffmann, R.W. (1981) *Chem. Ber.*, **111**, 359–374.
120 (a) Luithle, J.E.A. and Pietruszka, J. (1999) *J. Org. Chem.*, **64**, 8287–8297; (b) Luithle, J.E.A. and Pietruszka, J. (2000) *J. Org. Chem.*, **65**, 9194–9200; (c) Pietruszka, J. and Solduga, G. (2009) *Eur. J. Org. Chem.*, 5998–6008.
121 Matteson, D.S. and Ray, R. (1980) *J. Am. Chem. Soc.*, **102**, 7590–7591.
122 Matteson, D.S., Sadhu, K.M., and Lienhard, G.E. (1981) *J. Am. Chem. Soc.*, **103**, 5241–5242.
123 Matteson, D.S., Ray, R., Rocks, R.R., and Tsai, D.J.S. (1983) *Organometallics*, **2**, 1536–1543.
124 Matteson, D.S. and Sadhu, K.M. (1984) *Organometallics*, **3**, 614–618.
125 Martichonok, V. and Jones, J.B. (1996) *J. Am. Chem. Soc.*, **118**, 950–958.
126 Brown, H.C. and Rangaishenvi, M.V. (1988) *J. Organomet. Chem.*, **358**, 15–30.
127 Matteson, D.S. and Kim, G.Y. (2002) *Org. Lett.*, **4**, 2153–2155.
128 Inglis, S.R., Woon, E.C.Y., Thompson, A.L., and Schofield, C.J. (2010) *J. Org. Chem.*, **75**, 468–471.
129 Yuen, A.K.L. and Hutton, C.A. (2005) *Tetrahedron Lett.*, **46**, 7899–7903.
130 Pennington, T.E., Cynantya, K.B., and Hutton, C.A. (2004) *Tetrahedron Lett.*, **45**, 6657–6660.
131 Wityak, J., Earl, R.A., Abelman, M.M., Bethel, Y.B., Fisher, B.N., Kauffman, G.S., Kettner, C.A., Ma, P., McMillan, J.L., Mersinger, L.J., Pesti, J., Pierce, M.E., Rankin, F.W., Chorvat, R.J., and Confalone, P.N. (1995) *J. Org. Chem.*, **60**, 3717–3722.
132 Coutts, S.J., Adams, J., Krolikowski, D., and Snow, R.J. (1994) *Tetrahedron Lett.*, **35**, 5109–5112.
133 Yan, J., Jin, S., and Wang, B. (2005) *Tetrahedron Lett.*, **46**, 8503–8505.
134 Bowie, R.A. and Musgrave, O.C. (1963) *J. Chem. Soc., Chem. Commun.*, 3945–3949.
135 Matteson, D.S. and Man, H.W. (1996) *J. Org. Chem.*, **61**, 6047–6051.

136 Matteson, D.S., Soundararajan, R., Ho, O.C., and Gatzweiler, W. (1996) *Organometallics*, **15**, 152–163.
137 Tripathy, P.B. and Matteson, D.S. (1990) *Synthesis*, 200–206.
138 Kettner, C.A. and Shenvi, A.B. (1984) *J. Biol. Chem.*, **259**, 15106–15114.
139 Wulff, G., Lauer, M., and Böhnke, H. (1984) *Angew. Chem., Int. Ed. Engl.*, **23**, 741–742.
140 Edwards, J.O., Morrison, G.C., Ross, V., and Schultz, J.W. (1955) *J. Am. Chem. Soc.*, **77**, 266–268.
141 Springsteen, G. and Wang, B. (2002) *Tetrahedron*, **58**, 5291–5300.
142 Barker, S.A., Chopra, A.K., Hatt, B.W., and Somers, P.J. (1973) *Carbohydr. Res.*, **26**, 33–40.
143 Pizer, R. and Tihal, C. (1992) *Inorg. Chem.*, **31**, 3243–3247.
144 Iwatsuki, S., Nakajima, S., Inamo, M., Takagi, H.D., and Ishihara, K. (2007) *Inorg. Chem.*, **46**, 354–356.
145 Iwasawa, N. and Takahagi, H. (2007) *J. Am. Chem. Soc.*, **129**, 7754–7755.
146 Bielecki, M., Eggert, H., and Norrild, J.C. (1999) *J. Chem. Soc., Perkin Trans. 2*, 449–455.
147 Draffin, S.P., Duggan, P.J., and Fallon, G.D. (2004) *Acta Crystallogr. E*, **E60**, o1520–o1522.
148 Draffin, S.P., Duggan, P.J., Fallon, G.D., and Tyndall, E.M. (2005) *Acta Crystallogr. E*, **E61**, o1733–o1735.
149 Nagai, Y., Kobayashi, K., Toi, H., and Aoyama, Y. (1993) *Bull. Chem. Soc. Jpn.*, **66**, 2965–2971.
150 Stones, D., Manku, S., Lu, L., and Hall, D.G. (2004) *Chem. Eur. J.*, **10**, 92–100.
151 (a) Dowlut, M. and Hall, D.G. (2006) *J. Am. Chem. Soc.*, **128**, 4226–4227; (b) Bérubé, M., Dowlut, M., and Hall, D.G. (2008) *J. Org. Chem.*, **73**, 6471–6479.
152 (a) Springsteen, G. and Wang, B. (2001) *Chem. Commun.*, 1608–1609; (b) Arimori, S. and James, T.D. (2002) *Tetrahedron Lett.*, **43**, 507–509.
153 Bosch, L.I., Fyles, T.M., and James, T.D. (2004) *Tetrahedron*, **60**, 11175–11190.
154 Bromba, C., Carrie, P., Chui, J.K.W., and Fyles, T.M. (2009) *Supramol. Chem.*, **21**, 81–88.
155 Gillis, E.P. and Burke, M.D. (2007) *J. Am. Chem. Soc.*, **129**, 6716–6717.
156 Gillis, E.P. and Burke, M.D. (2009) *Aldrichimica Acta*, **42**, 17–27.
157 Woods, W.G. and Strong, P.L. (1966) *J. Am. Chem. Soc.*, **88**, 4667–4671.
158 Brown, H.C. and Gupta, S.K. (1971) *J. Am. Chem. Soc.*, **93**, 1816–1818.
159 Tucker, C.E., Davidson, J., and Knochel, P. (1992) *J. Org. Chem.*, **57**, 3482–3485.
160 Panek, J.S. and Xu, F. (1992) *J. Org. Chem.*, **57**, 5288–5290.
161 Thaisrivongs, S. and Wuest, J.D. (1977) *J. Org. Chem.*, **42**, 3243–3246.
162 (a) Furuta, K., Miwa, Y., Iwanaga, K., and Yamamoto, H. (1988) *J. Am. Chem. Soc.*, **110**, 6254–6255; (b) Furuta, K., Maruyama, T., and Yamamoto, H. (1991) *J. Am. Chem. Soc.*, **113**, 1041–1042.
163 (a) Takasu, M. and Yamamoto, H. (1990) *Synlett.*, 194–196; (b) Sartor, D., Saffrich, J., and Helmchen, G. (1990) *Synlett.*, 197–198; (c) Kiyooka, S.-i., Kaneko, Y., Komura, M., Matsuo, H., and Nakano, M. (1991) *J. Org. Chem.*, **56**, 2276–2278.
164 (a) Marder, T.B. and Norman, N.C. (1998) *Top. Catal.*, **5**, 63–73; (b) Ishiyama, T. and Miyaura, N. (2000) *J. Organomet. Chem.*, **611**, 392–402.
165 Iwadate, N. and Suginome, M. (2010) *J. Am. Chem. Soc.*, **132**, 2548–2549.
166 Ishiyama, T., Murata, M., Ahiko, T.-a., and Miyaura, N. (2000) *Org. Synth.*, **77**, 176–182.
167 Anastasi, N.R., Waltz, K.M., Weerakoon, W.L., and Hartwig, J.F. (2003) *Organometallics*, **22**, 365–369.
168 Letsinger, R.L. and Hamilton, S.B. (1958) *J. Am. Chem. Soc.*, **80**, 5411–5413.
169 Dewar, M.J.S., Kubba, V.P., and Pettit, R. (1958) *J. Chem. Soc.*, 3076–3079.
170 Hawkins, R.T. and Snyder, H.R. (1960) *J. Am. Chem. Soc.*, **82**, 3863–3866.
171 Noguchi, H., Hojo, K., and Suginome, M. (2007) *J. Am. Chem. Soc.*, **129**, 758–759.
172 Yanagisawa, A., Inanami, H., and Yamamoto, H. (1998) *Chem. Commun.*, 1573–1574.
173 Mohler, L.K. and Czarnik, A.W. (1993) *J. Am. Chem. Soc.*, **115**, 7037–7038.
174 Shull, B.K., Spielvogel, D.E., Gopalaswamy, R., Sankar, S., Boyle, P.D.,

Head, G., and Devito, K. (2000) *J. Chem. Soc., Perkin Trans. 2*, 557–561.

175 (a) Chissick, S.S., Dewar, M.J.S., and Maitlis, P.M. (1959) *J. Am. Chem. Soc.*, **81**, 6329–6330; (b) Chissick, S.S., Dewar, M.J.S., and Maitlis, P.M. (1961) *J. Am. Chem. Soc.*, **83**, 2708–2711.

176 Pailer, M. and Fenzl, W. (1961) *Monatsh. Chem.*, **92**, 1294–1299.

177 Stolowitz, M.L., Ahlem, C., Hughes, K.A., Kaiser, R.J., Kesicki, E.A., Li, G., Lund, K.P., Torkelson, S.M., and Wiley, J.P. (2001) *Bioconjug. Chem.*, **12**, 229–239.

178 (a) Corey, E.J. and Loh, T.-P. (1991) *J. Am. Chem. Soc.*, **113**, 8966–8967; (b) Corey, E.J., Cywin, C.L., and Roper, T.D. (1992) *Tetrahedron Lett.*, **33**, 6907–6010; (c) Ishihara, K., Kondo, S., and Yamamoto, H. (2000) *J. Org. Chem.*, **65**, 9125–9128; (d) Kinugasa, M., Harada, T., and Oku, A. (1997) *J. Am. Chem. Soc.*, **119**, 9067–9068.

179 Vedejs, E., Chapman, R.W., Lin, S., Müller, M., and Powell, D.R. (2000) *J. Am. Chem. Soc.*, **122**, 3047–3052.

180 Corey, E.J. (2002) *Angew. Chem., Int. Ed.*, **41**, 1650–1667.

181 (a) Corey, E.J. and Helal, C.J. (1998) *Angew. Chem., Int. Ed.*, **37**, 1986–2012; (b)Cho, B.T. (2006) *Tetrahedron*, **62**, 7621–7643.

182 Corey, E.J. (2009) *Angew. Chem., Int. Ed.*, **48**, 2100–2117.

183 Groziak, M.P., Ganguly, A.D., and Robinson, P.D. (1994) *J. Am. Chem. Soc.*, **116**, 7597–7605.

184 Snyder, H.R., Reedy, A.J., and Lennarz, W.J. (1958) *J. Am. Chem. Soc.*, **80**, 835–838.

185 Lennarz, W.J. and Snyder, H.R. (1960) *J. Am. Chem. Soc.*, **82**, 2172–2175.

186 (a) Soloway, A.H. (1960) *J. Am. Chem. Soc.*, **82**, 2442–2444; (b) Dewar, M.J.S. and Dougherty, R.C. (1964) *J. Am. Chem. Soc.*, **86**, 433–436; (c) Tschampel, P. and Snyder, H.R. (1964) *J. Org. Chem.*, **29**, 2168–2172; (d) Matteson, D.S., Biernbaum, M.S., Bechtold, R.A., Campbell, J.D., and Wilcsek, R.J. (1978) *J. Org. Chem.*, **43**, 950–954; (e) Boldyreva, O., Dorokhov, V.A., and Mikhailov, B.M. (1985) *Izv. Akad. Nauk SSSR, Ser. Khim.*, 428–430;

(f) Hughes, M.P. and Smith, B.D. (1997) *J. Org. Chem.*, **62**, 4492–4499; (g) Zhuo, J.-C., Soloway, A.H., Beeson, J.C., Ji, W., Barnum, B.A., Rong, F.-G., Tjarks, W., Jordan, G.T., IV, Liu, J., and Shore, S.G. (1999) *J. Org. Chem.*, **64**, 9566–9574; (h) Ruman, T., Dlugopolska, K., Kusnierz, A., Jurkiewicz, A., Les, A., and Rode, W. (2009) *Bioorg. Chem.*, **37**, 65–69; (i) Ruman, T., Dlugopolska, K., Kusnierz, A., and Rode, W. (2009) *Bioorg. Chem.*, **37**, 180–184; (j) Ruman, T., Dlugopolska, K., and Rode, W. (2010) *Bioorg. Chem.*, **38**, 33–36.

187 Robinson, P.D., Groziak, M.P., and Yi, L. (1996) *Acta Crystallogr.*, **C52**, 2826–2830.

188 Matteson, D.S. and Cheng, T.-C. (1968) *J. Org. Chem.*, **33**, 3055–3060.

189 Brown, H.C., Salunkhe, A.M., and Argade, A.B. (1992) *Organometallics*, **11**, 3094–3097.

190 Brown, H.C., Cole, T.E., Bakshi, R.K., Srebnik, M., and Singaram, B. (1986) *Organometallics*, **5**, 2303–2307.

191 Brown, H.C., Salunkhe, A.M., and Singaram, B. (1991) *J. Org. Chem.*, **56**, 1170–1175.

192 Brown, H.C., Cole, T.E., and Singaram, B. (1984) *Organometallics*, **3**, 774–777.

193 (a) Vedejs, E., Chapman, R.W., Fields, S.C., Lin, S., and Schrimpf, M.R. (1995) *J. Org. Chem.*, **60**, 3020–3027; (b) Vedejs, E., Fields, S.C., Hayashi, R., Hitchcock, S.R., Powell, D.R., and Schrimpf, M.R. (1999) *J. Am. Chem. Soc.*, **121**, 2460–2470.

194 (a) Darses, S. and Genêt, J.-P. (2003) *Eur. J. Org. Chem.*, 4313–4327; (b) Molander, G.A. and Ellis, N. (2007) *Acc. Chem. Res.*, **40**, 275–286.

195 Ting, R., Harwig, C., auf der Keller, U., McCormick, S., Austin, P., Overall, C.M., Adam, M.J., Ruth, T.J., and Perrin, D.M. (2008) *J. Am. Chem. Soc.*, **130**, 12045–12055.

196 Kim, B.J. and Matteson, D.S. (2004) *Angew. Chem., Int. Ed.*, **43**, 3056–3058.

197 Molander, G.A. and Ribagorda, M. (2003) *J. Am. Chem. Soc.*, **125**, 11148–11149.

198 Yamamoto, Y., Takizawa, M., Yu, X.-Q., and Miyaura, N. (2008) *Angew. Chem., Int. Ed.*, **47**, 928–931.

199 Gilman, H. and Moore, L.O. (1958) *J. Am. Chem. Soc.*, **80**, 3609–3611.
200 (a) Khotinsky, E. and Melamed, M. (1909) *Ber. Dtsch. Chem. Ges.*, **54**, 2784; (b) Khotinsky, E. and Melamed, M. (1909) *Ber. Dtsch. Chem. Ges.*, **42**, 3090.
201 Krause, E. and Nitsche, R. (1921) *Ber. Dtsch. Chem. Ges.*, **55**, 1261–1265.
202 Seaman, W. and Johnson, J.R. (1931) *J. Am. Chem. Soc.*, **53**, 711–723.
203 Bean, F.R. and Johnson, J.R. (1932) *J. Am. Chem. Soc.*, **54**, 4415–4425.
204 Cladingboel, D.E. (2000) *Org. Process Res. Dev.*, **4**, 153–155.
205 Zenk, R. and Partzsch, S. (2003) *Chim. Oggi*, **21**, 70–73.
206 Jendralla, H., Wagner, A., Mollath, M., and Wunner, J. (1995) *Liebigs Ann.*, 1253–1257.
207 Brown, H.C. and Cole, T.E. (1983) *Organometallics*, **2**, 1316–1319.
208 Brown, H.C., Srebnik, M., and Cole, T.E. (1986) *Organometallics*, **5**, 2300–2303.
209 Li, W., Nelson, D.P., Jensen, M.S., Hoerrner, R.S., Cai, D., Larsen, R.D., and Reider, P.J. (2002) *J. Org. Chem.*, **67**, 5394–5397.
210 Das, S., Alexeev, V.L., Sharma, A.C., Geib, S.J., and Asher, S.A. (2003) *Tetrahedron Lett.*, **44**, 7719–7722.
211 Evans, D.A., Katz, J.L., Peterson, G.S., and Hintermann, T. (2001) *J. Am. Chem. Soc.*, **123**, 12411–12413.
212 Gilman, H., Santucci, L., Swayampati, D.R., and Ranck, R.D. (1957) *J. Am. Chem. Soc.*, **79**, 3077–3081.
213 Knochel, P., Dohle, W., Gommermann, N., Kneisel, F.F., Kopp, F., Korn, T., Sapountzis, I., and Vu, V.A. (2003) *Angew. Chem., Int. Ed.*, **42**, 4302–4320.
214 (a) Collibee, S.E. and Yu, J. (2005) *Tetrahedron Lett.*, **46**, 4453–4455; (b) Demory, E., Blandin, V., Einhorn, J., and Chavant, P. Y. (2011) *Org. Process Res. Dev.* **15**, 710–716.
215 Wong, K.-T., Chien, Y.Y., Liao, Y.-L., Lin, C.-C., Chou, M.Y., and Leung, M.-K. (2002) *J. Org. Chem.*, **67**, 1041–1044.
216 Garg, N.K., Sarpong, R., and Stoltz, B.M. (2002) *J. Am. Chem. Soc.*, **124**, 13179–13184.
217 Chavant, P.Y. and Vaultier, M. (1993) *J. Organomet. Chem.*, **455**, 37–46.
218 Marr, G., Moore, R.E., and Rockett, B.W. (1967) *J. Organomet. Chem.*, **7**, P11–P13.
219 Hawkins, R.T. and Stroup, D.B. (1969) *J. Org. Chem.*, **34**, 1173–1174.
220 Lauer, M. and Wulff, G. (1983) *J. Organomet. Chem.*, **256**, 1–9.
221 Sharp, M.J. and Snieckus, V. (1985) *Tetrahedron Lett.*, **49**, 5997–6000.
222 Sharp, M.J., Cheng, W., and Snieckus, V. (1987) *Tetrahedron Lett.*, **28**, 5093–5096.
223 Alo, B.I., Kandil, A., Patil, P.A., Sharp, M.J., Siddiqui, M.A., and Snieckus, V. (1991) *J. Org. Chem.*, **96**, 3763–3768.
224 Larsen, R.D., King, A.O., Chen, C.Y., Corley, E.G., Foster, B.S., Roberts, F.E., Yang, C., Lieberman, D.R., Reamer, R.A., Tschaen, D.M., Verhoeven, T.R., Reider, P.J., Lo, Y.S., Rossano, L.T., Brookes, A.S., Meloni, D., Moore, J.R., and Arnett, J.F. (1994) *J. Org. Chem.*, **59**, 6391–6394.
225 Antoft-Finch, A., Blackburn, T., and Snieckus, V. (2009) *J. Am. Chem. Soc.*, **131**, 17750–17752.
226 Caron, S. and Hawkins, J.M. (1998) *J. Org. Chem.*, **63**, 2054–2055.
227 Kristensen, J., Lysén, M., Vedso, P., and Begtrup, M. (2001) *Org. Lett.*, **3**, 1435–1437.
228 (a) Michaelis, A. and Becker, P. (1880) *Ber. Dtsch. Chem. Ges.*, **13**, 58; (b) Michaelis, A. and Becker, P. (1882) *Ber. Dtsch. Chem. Ges.*, **15**, 180–185.
229 Breuer, S.W., Thorpe, F.G., and Podestá, J.C. (1974) *Tetrahedron Lett.*, **42**, 3719–3720.
230 Haubold, W., Herdtle, J., Gollinger, W., and Einholz, W. (1986) *J. Organomet. Chem.*, **315**, 1–8.
231 Zhao, Z. and Snieckus, V. (2005) *Org. Lett.*, **7**, 2523–2526.
232 Ishiyama, T., Murata, M., and Miyaura, N. (1995) *J. Org. Chem.*, **60**, 7508–7510.
233 (a) Murata, M., Watanabe, S., and Masuda, Y. (1997) *J. Org. Chem.*, **62**, 6458–6459; (b) Murata, M., Oyama, T., Watanabe, S., and Masuda, Y. (2000) *J. Org. Chem.*, **65**, 164–168; (c) Baudoin, O., Guénard, D., and Guéritte, F. (2000) *J. Org. Chem.*, **65**, 9268–9271.

234 Baudoin, O., Décor, A., Cesario, M., and Guéritte, F. (2003) *Synlett*, 2009–2012.
235 Song, Y.-L. and Morin, C. (2001) *Synlett*, 266–268.
236 Yoburn, J.-C. and Van Vranken, D.L. (2003) *Org. Lett.*, **5**, 2817–2820.
237 Lin, S. and Danishefsky, S.J. (2001) *Angew. Chem., Int. Ed.*, **40**, 1967–1970.
238 (a) Fang, H., Kaur, G., Yan, J., and Wang, B. (2005) *Tetrahedron Lett.*, **46**, 1671–1674; (b) Diemer, V., Chaumeil, H., Defoin, A., and Carré, C. (2010) *Tetrahedron*, **66**, 918–929.
239 Larson, K.K. and Sarpong, R. (2009) *J. Am. Chem. Soc.*, **131**, 13244–13245.
240 Nakamura, H., Fujiawara, M., and Yamamoto, Y. (1998) *J. Org. Chem.*, **63**, 7529–7530.
241 Molander, G.A., Trice, S.L.J., and Dreher, S.D. (2010) *J. Am. Chem. Soc.*, **132**, 17701–17703.
242 Ishiyama, T., Ishida, K., and Miyaura, N. (2001) *Tetrahedron*, **57**, 9813–9816.
243 Billingsley, K.L. and Buchwald, S.L. (2008) *J. Org. Chem.*, **73**, 5589–5591.
244 Fürstner, A. and Seidel, G. (2002) *Org. Lett.*, **4**, 541–543.
245 Ma, Y., Song, C., Jiang, W., Xue, G., Cannon, J.F., Wang, X., and Andrus, M.B. (2003) *Org. Lett.*, **5**, 4635–4638.
246 Wilson, D.A., Wilson, C.J., Rosen, B.M., and Percec, V. (2008) *Org. Lett.*, **10**, 4879–4882.
247 Wilson, D.A., Wilson, C.J., Moldoveanu, C., Resmerita, A.-M., Corcoran, P., Hoang, L.M., Rosen, B.M., and Percec, V. (2010) *J. Am. Chem. Soc.*, **132**, 1800–1801.
248 (a) Zhu, W. and Ma, D. (2006) *Org. Lett.*, **8**, 261–263; (b) Kleeberg, C., Dang, L., Lin, Z., and Marder, T.B. (2009) *Angew. Chem., Int. Ed.*, **48**, 5350–5354.
249 Mkhalid, I.A.I., Barnard, J.H., Marder, T.B., Murphy, J.M., and Hartwig, J.F. (2010) *Chem. Rev.*, **110**, 890–931.
250 (a) Waltz, K.M. and Hartwig, J.F. (1997) *Science*, **277**, 211–213; (b) Chen, H. and Hartwig, J.F. (1999) *Angew. Chem., Int. Ed.*, **38**, 3391–3393.
251 (a) Iverson, C.N. and Smith, M.R., III (1999) *J. Am. Chem. Soc.*, **121**, 7696–7697; (b) Cho, J.-Y., Iverson, C.N., and Smith, M.R., III (2000) *J. Am. Chem. Soc.*, **122**, 12868–12869.
252 Chen, H., Schlecht, S., Temple, T.C., and Hartwig, J.F. (2000) *Science*, **287**, 1995–1997.
253 Ishiyama, T., Takagi, J., Ishida, K., Miyaura, N., Anastasi, N.R., and Hartwig, J.F. (2002) *J. Am. Chem. Soc.*, **124**, 390–391.
254 Shimada, S., Batsanov, A.S., Howard, J.A.K., and Marder, T.B. (2001) *Angew. Chem., Int. Ed.*, **40**, 2168–2171.
255 Chotana, G.A., Rak, M.A., and Smith, M.R., III (2005) *J. Am. Chem. Soc.*, **127**, 10539–10544.
256 Hurst, T.E., Macklin, T.K., Becker, M., Hartmann, E., Kugel, W., Parisienne-La Salle, J.C., Batsanov, A.S., Marder, T.B., and Snieckus, V. (2010) *Chem. Eur. J.*, **16**, 8155–8161.
257 Kawamorita, S., Ohmiya, H., Hara, K., Fukuoka, A., and Sawamura, M. (2009) *J. Am. Chem. Soc.*, **131**, 5058–5059.
258 Hilt, G. and Bolze, P. (2005) *Synthesis*, 2091–2115.
259 Moore, J.E., York, M., and Harrity, J.P.A. (2005) *Synlett*, 860–862.
260 Delaney, P.M., Browne, D.L., Adams, H., Plant, A., and Harrity, J.P.A. (2008) *Tetrahedron*, **64**, 866–873.
261 Davies, M.W., Johnson, C.N., and Harrity, J.P.A. (2001) *J. Org. Chem.*, **66**, 3525–3532.
262 (a) Yamamoto, Y., Ishii, J.-i., Nishiyama, H., and Itoh, K. (2004) *J. Am. Chem. Soc.*, **126**, 3712–3713; (b) Yamamoto, Y., Ishii, J.-i., Nishiyama, H., and Itoh, K. (2005) *Tetrahedron*, **61**, 11501–11510.
263 Auvinet, A.-L., Harrity, J.P.A., and Hilt, G. (2010) *J. Org. Chem.*, **75**, 3893–3896.
264 Mo, F.Y., Jiang, Y.B., Qiu, D., Zhang, Y., and Wang, J.B. (2010) *Angew. Chem., Int. Ed.*, **49**, 1846–1849.
265 Nielsen, D.R. and McEwen, W.E. (1957) *J. Am. Chem. Soc.*, **79**, 3081–3084.
266 Kaufman, D. (1987) *Chem. Ber.*, **120**, 901–905.
267 IJpeij, E.G., Beijer, F.H., Arts, H.J., Newton, C., de Vries, J.G., and Gruter, G.J.-M. (2002) *J. Org. Chem.*, **67**, 169–176.
268 Schilling, B., Kaiser, V., and Kaufmann, E.D. (1998) *Eur. J. Org. Chem.*, 701–709.

269 Tyrrell, E. and Brookes, P. (2004) *Synthesis*, 469–483.
270 Fuller, A.A., Hester, H.R., Salo, E.V., and Stevens, E.P. (2003) *Tetrahedron Lett.*, **44**, 2935–2938.
271 (a) Helm, M.D., Moore, J.E., Plant, A., and Harrity, J.P.A. (2005) *Angew. Chem., Int. Ed.*, **44**, 3889–3892; (b) Delaney, P.M., Huang, J., Macdonald, S.J.F., and Harrity, J.P.A. (2008) *Org. Lett.*, **10**, 781–783; (c) Huang, J., Macdonald, S.J.F., Cooper, A.W.J., Fisher, G., and Harrity, J.P.A. (2009) *Tetrahedron Lett.*, **50**, 5539–5541; (d) Browne, D.L., Vivat, J.F., Plant, A., Gomez-Bengoa, E., and Harrity, J.P.A. (2009) *J. Am. Chem. Soc.*, **131**, 7762–7769.
272 Dieck, H.A. and Heck, R.F. (1975) *J. Org. Chem.*, **40**, 1083–1090.
273 Brown, H.C. and Bhat, N.G. (1988) *Tetrahedron Lett.*, **29**, 21–24.
274 Uenishi, J., Matsui, K., and Wada, A. (2003) *Tetrahedron Lett.*, **44**, 3093–3096.
275 Matteson, D.S. (1960) *J. Am. Chem. Soc.*, **82**, 4228–4233.
276 Tivola, P.B., Deagostino, A., Prandi, C., and Venturello, P. (2002) *Org. Lett.*, **4**, 1275–1277.
277 Passafaro, M.S. and Keay, B.A. (1996) *Tetrahedron Lett.*, **37**, 429–432.
278 Rauniyar, V., Zhai, H.M., and Hall, D.G. (2008) *Synth. Commun.*, **38**, 3984–3995.
279 Gao, F. and Hoveyda, A.H. (2010) *J. Am. Chem. Soc.*, **132**, 10961–10963.
280 (a) Farinola, G.M., Fiandanese, V., Mazzone, L., and Naso, F. (1995) *J. Chem. Soc., Chem. Commun.*, 2523–2524; (b) Babudri, F., Farinola, G.M., Fiandanesse, V., Mazzone, L., and Naso, F. (1998) *Tetrahedron*, **54**, 1085–1094.
281 Mikhail, I. and Kaufmann, D. (1990) *J. Organomet. Chem.*, **398**, 53–57.
282 Itami, K., Kamei, T., and Yoshida, J.-i. (2003) *J. Am. Chem. Soc.*, **125**, 14670–14671.
283 Cole, T.E., Quintanilla, R., and Rodewald, S. (1991) *Organometallics*, **10**, 3777–3781.
284 Takahashi, K., Takagi, J., Ishiyama, T., and Miyaura, N. (2000) *Chem. Lett.*, 126–127.
285 Takagi, J., Takahashi, K., Ishiyama, T., and Miyaura, N. (2002) *J. Am. Chem. Soc.*, **124**, 8001–8006.
286 Ishiyama, T., Takagi, J., Kamon, A., and Miyaura, N. (2003) *J. Organomet. Chem.*, **687**, 284–290.
287 Murata, M., Oyama, T., Watanabe, S., and Masuda, Y. (2000) *Synthesis*, 778–780.
288 Brown, H.C. and Subba Rao, B.C. (1956) *J. Am. Chem. Soc.*, **78**, 5694–5695.
289 Brown, H.C. (1962) *Hydroboration*, Benjamin, Reading MA.
290 Brown, H.C. and Zweifel, G. (1961) *J. Am. Chem. Soc.*, **83**, 3834–3840.
291 Zweifel, G. and Brown, H.C. (1963) *J. Am. Chem. Soc.*, **85**, 2066–2072.
292 Zweifel, G., Clark, G.M., and Polston, N.L. (1971) *J. Am. Chem. Soc.*, **93**, 3395–3399.
293 Brown, H.C., Scouten, C.G., and Liotta, R. (1979) *J. Am. Chem. Soc.*, **101**, 96–99.
294 Miyaura, N. and Suzuki, A. (1981) *Chem. Lett.*, 879–882.
295 Hoffman, R.W. and Dresely, S. (1988) *Synthesis*, 103–106.
296 Brown, H.C., Mandal, A.K., and Kulkarni, S.U. (1977) *J. Org. Chem.*, **42**, 1392–1398.
297 Martinez-Fresneda, P. and Vaultier, M. (1989) *Tetrahedron Lett.*, **30**, 2929–2932.
298 Kamabuchi, A., Moriya, T., Miyaura, N., and Suzuki, A. (1993) *Synth. Commun.*, **23**, 2851–2859.
299 Gravel, M., Touré, B.B., and Hall, D.G. (2004) *Org. Prep. Proced. Int.*, **36**, 573–580.
300 Kalinin, A.V., Scherer, S., and Snieckus, V. (2003) *Angew. Chem., Int. Ed.*, **42**, 3399–3404.
301 (a) Brown, H.C. and Gupta, S.K. (1972) *J. Am. Chem. Soc.*, **94**, 4370–4371; (b) Brown, H.C. and Gupta, S.K. (1975) *J. Am. Chem. Soc.*, **97**, 5249–5255; (c) Lane, C.F. and Kabalka, G.W. (1976) *Tetrahedron*, **32**, 981–990.
302 Arase, A., Hoshi, M., Mijin, A., and Nishi, K. (1995) *Synth. Commun.*, **25**, 1957–1962.
303 Scheidt, K.A., Tasaka, A., Bannister, T.D., Wendt, M.D., and Roush, W.R. (1999) *Angew. Chem., Int. Ed.*, **38**, 1652–1655.
304 Brown, H.C. and Campbell, J.B. (1980) *J. Org. Chem.*, **45**, 389–395.

305 Hassner, A. and Soderquist, J.A. (1977) *J. Organomet. Chem.*, **131**, C1–C4.
306 Soundararajan, R. and Matteson, D.S. (1990) *J. Org. Chem.*, **55**, 2274–2275.
307 Josyula, K.V.B., Gao, P., and Hewitt, C. (2003) *Tetrahedron Lett.*, **44**, 7789–7792.
308 Brown, H.C. and Imai, T. (1984) *Organometallics*, **3**, 1392–1395.
309 (a) Brown, H.C., Basavaiah, D., and Kulkarni, S.U. (1982) *J. Org. Chem.*, **47**, 3808–3810; (b) Brown, H.C., Imai, T., and Bhat, N.G. (1986) *J. Org. Chem.*, **51**, 5277–5282.
310 Hata, T., Kitagawa, H., Masai, H., Kurahashi, T., Shimizu, M., and Hiyama, T. (2001) *Angew. Chem., Int. Ed.*, **40**, 790–792.
311 Männig, D. and Nöth, H. (1985) *Angew. Chem., Int. Ed. Engl.*, **24**, 878–879.
312 (a) Burgess, K. and Ohlmeyer, M.J. (1991) *Chem. Rev.*, **91**, 1179–1191; (b) Beletskaya, I. and Pelter, A. (1997) *Tetrahedron*, **53**, 4957–5026.
313 He, X. and Hartwig, J.F. (1996) *J. Am. Chem. Soc.*, **118**, 1696–1702.
314 Pereira, S. and Srebnik, M. (1995) *Organometallics*, **14**, 3127–3128.
315 Wang, Y.D., Kimball, G., Prashad, A.S., and Wang, Y. (2005) *Tetrahedron Lett.*, **46**, 8777–8780.
316 Praveen Ganesh, N., d'Hondt, S., and Chavant, P.Y. (2007) *J. Org. Chem.*, **72**, 4510–4514.
317 Pereira, S. and Srebnik, M. (1996) *Tetrahedron Lett.*, **37**, 3283–3286.
318 Gridnev, I.D., Miyaura, N., and Suzuki, A. (1993) *Organometallics*, **12**, 589–592.
319 Lipshutz, B.H. and Bosković, Z.V. (2008) *Angew. Chem., Int. Ed.*, **47**, 10183–10186.
320 Satoh, M., Nomoto, Y., Miyaura, N., and Suzuki, A. (1989) *Tetrahedron Lett.*, **30**, 3789–3792.
321 Yamamoto, Y., Fujikawa, R., Yamada, A., and Miyaura, N. (1999) *Chem. Lett.*, 1069–1070.
322 Ohmura, T., Yamamoto, Y., and Miyaura, N. (2000) *J. Am. Chem. Soc.*, **122**, 4990–4991.
323 Gao, D. and O'Doherty, G.A. (2010) *Org. Lett.*, **12**, 3752–3755.
324 Renaud, J. and Ouellet, S.G. (1998) *J. Am. Chem. Soc.*, **120**, 7995–7996.
325 Connon, S.J. and Blechert, S. (2003) *Angew. Chem., Int. Ed.*, **42**, 1900–1923.
326 Blackwell, H.E., O'Leary, D.J., Chatterjee, A.K., Washenfelder, R.A., Bussmann, D.A., and Grubbs, R.H. (2000) *J. Am. Chem. Soc.*, **122**, 58–71.
327 Morrill, C. and Grubbs, R.H. (2003) *J. Org. Chem.*, **68**, 6031–6034.
328 Njardarson, J.T., Biswas, K., and Danishefsky, S.J. (2002) *Chem. Commun.*, 2759–2761.
329 (a) Renaud, J., Graf, C.-D., and Oberer, L. (2000) *Angew. Chem., Int. Ed.*, **39**, 3101–3104. (b) Micalizio, G.C. and Schreiber, S.L. (2002) *Angew. Chem., Int. Ed.*, **41**, 3272–3276.
330 Kim, M. and Lee, D. (2005) *Org. Lett.*, **7**, 1865–1868.
331 Coapes, B.R., Souza, F.E.S., Thomas, R.L., Hall, J.J., and Marder, T.B. (2003) *Chem. Commun.*, 614–615.
332 (a) Ishiyama, T., Matsuda, N., Miyaura, N., and Suzuki, A. (1993) *J. Am. Chem. Soc.*, **115**, 11018–11019; (b) Ishiyama, T., Matsuda, N., Murata, M., Ozawa, F., Suzuki, A., and Miyaura, N. (1996) *Organometallics*, **15**, 713–720.
333 (a) Iverson, C.N. and Smith, M.R. (1996) *Organometallics*, **15**, 5155–5165; (b) Lesley, M.J.G., Nguyen, P., Taylor, N.J., Marder, T.B., Scott, A.J., Clegg, W., and Norman, N.C. (1996) *Organometallics*, **15**, 5137–5154; (c) Anderson, K.M., Lesley, M.J.G., Norman, N.C., Orpen, A.G., and Starbuck, J. (1999) *New J. Chem.*, **23**, 1053–1055.
334 (a) Ishiyama, T., Yamamoto, M., and Miyaura, N. (1996) *Chem. Commun.*, 2073–2074; (b) Ishiyama, T., Kitano, T., and Miyaura, N. (1998) *Tetrahedron Lett.*, **39**, 2357–2360.
335 Takahashi, K., Ishiyama, T., and Miyaura, N. (2001) *J. Organomet. Chem.*, **625**, 47–53.
336 Lee, J.-E., Kwon, J., and Yun, J. (2008) *Chem. Commun.*, 733–734.
337 Suginome, M., Ohmura, T., Miyake, Y., Mitani, S., Ito, Y., and Murakami, M. (2003) *J. Am. Chem. Soc.*, **125**, 11174–11175.
338 Suginome, M., Yamamoto, A., and Murakami, M. (2003) *J. Am. Chem. Soc.*, **125**, 6358–6359.

339 Mannathan, S., Jeganmohan, M., and Cheng, C.-H. (2009) *Angew. Chem., Int. Ed.*, **48**, 2192–2195.
340 (a) Olsson, V.J. and Szabó, K.J. (2007) *Angew. Chem., Int. Ed.*, **46**, 6891–6893; (b) Olsson, V.J. and Szabó, K.J. (2009) *J. Org. Chem.*, **74**, 7715–7723.
341 Srebnik, M., Bhat, N.G., and Brown, H.C. (1988) *Tetrahedron Lett.*, **29**, 2635–2638.
342 Deloux, L. and Srebnik, M. (1994) *J. Org. Chem.*, **59**, 6871–6873.
343 Molander, G.A. and Ellis, N.M. (2008) *J. Org. Chem.*, **73**, 6841–6844.
344 (a) Hansen, E.C. and Lee, D. (2005) *J. Am. Chem. Soc.*, **127**, 3252–3253; (b) Suginome, M., Shirakura, M., and Yamamoto, A. (2006) *J. Am. Chem. Soc.*, **128**, 14438–14439; (c) Nishihara, Y., Miyasaka, M., Okamoto, M., Takahashi, H., Inoue, E., Tanemura, K., and Takagi, K. (2007) *J. Am. Chem. Soc.*, **129**, 12634–12635; (d) Geny, A., Leboeuf, D., Rouquié, G., Vollhardt, K.P.C., Malacria, M., Gandon, V., and Aubert, C. (2007) *Chem. Eur. J.*, **13**, 5408–5424.
345 Tonogaki, K., Itami, K., and Yoshida, J.-i. (2006) *J. Am. Chem. Soc.*, **128**, 1464–1465.
346 Shimizu, M., Kitagawa, H., Kurahashi, T., and Hiyama, T. (2001) *Angew. Chem., Int. Ed.*, **40**, 4283–4286.
347 Matteson, D.S. and Majumdar, D. (1983) *Organometallics*, **2**, 230–236.
348 Matteson, D.S., Moody, R.J., and Jesthi, P.K. (1975) *J. Am. Chem. Soc.*, **97**, 5608–5609.
349 Matteson, D.S. and Moody, R.J. (1982) *Organometallics*, **1**, 20–28.
350 Uenishi, J.-i., Beau, J.-M., Armstrong, R.W., and Kishi, Y. (1987) *J. Am. Chem. Soc.*, **109**, 4756–4758.
351 Evans, D.A., Ng, H.P., and Rieger, D.L. (1993) *J. Am. Chem. Soc.*, **115**, 11446–11459.
352 Endo, K., Hirokami, M., and Shibata, T. (2010) *J. Org. Chem.*, **75**, 3469–3472.
353 Takai, K., Shinomiya, N., Kaihara, H., Yoshida, N., and Moriwake, T. (1995) *Synlett*, 963–964.
354 Raheem, I.T., Goodman, S.N., and Jacobsen, E.N. (2004) *J. Am. Chem. Soc.*, **126**, 706–707.
355 Lightfoot, A.P., Maw, G., Thirsk, C., Twiddle, S.J.R., and Whiting, A. (2003) *Tetrahedron Lett.*, **44**, 7645–7648.
356 Heinrich, M.R., Sharp, L.A., and Zard, S.Z. (2005) *Chem. Commun.*, 3977–3079.
357 Satoh, Y., Serizawa, H., Miyaura, N., Hara, S., and Suzuki, A. (1988) *Tetrahedron Lett.*, **29**, 1811–1814.
358 Daini, M., Yamamoto, A., and Suginome, M. (2008) *J. Am. Chem. Soc.*, **130**, 2918–2919.
359 Carboni, B. (2005) Chapter 1, in *Boronic Acids*, 1st edn (ed. D.G. Hall), Wiley-VCH Verlag GmbH, Weinheim.
360 Matteson, D.S. and Peacock, K. (1963) *J. Org. Chem.*, **28**, 369–371.
361 Brown, H.C., Bhat, N.G., and Srebnik, M. (1988) *Tetrahedron Lett.*, **29**, 2631–2634.
362 Sato, M., Miyaura, N., and Suzuki, A. (1989) *Chem. Lett.*, 1405–1408.
363 (a) Rathke, M.W., Chao, E., and Wu, G. (1976) *J. Organomet. Chem.*, **122**, 145–149; (b) Wuts, P.G.M. and Thompson, P.A. (1982) *J. Organomet. Chem.*, **234**, 137–141; (c) Sadhu, K.M. and Matteson, D.S. (1985) *Organometallics*, **4**, 1687–1689.
364 Michnick, T.J. and Matteson, D.S. (1991) *Synlett*, 631–632.
365 (a) Matteson, D.S. and Majumdar, D. (1979) *J. Organomet. Chem.*, **170**, 259–264; (b) Phillion, D.P., Neubauer, R., and Andrew, S.S. (1986) *J. Org. Chem.*, **51**, 1610–1612.
366 Pintaric, C., Olivero, S., Gimbert, Y., Chavant, P.-Y., and Dunach, E. (2010) *J. Am. Chem. Soc.*, **132**, 11825–11827.
367 Brown, H.C. and Singaram, B. (1988) *Acc. Chem. Res.*, **21**, 287–293.
368 Brown, H.C. and Singaram, B. (1984) *J. Am. Chem. Soc.*, **106**, 1797–1800.
369 (a) Crudden, C.M. and Edwards, D. (2003) *Eur. J. Org. Chem.*, 4695–4712; (b) Carroll, A.-M., O'Sullivan, T.P., and Guiry, P.J. (2005) *Adv. Synth. Catal.*, **347**, 609–631.
370 (a) Smith, S.M. and Takacs, J.M. (2010) *J. Am. Chem. Soc.*, **132**, 1740–1741; (b) Smith, S.M. and Takacs, J.M. (2010) *Org. Lett.*, **12**, 4612–4615; Carroll, A.-M., O'Sullivan, T.P., and Guiry, P.J. (2005) *Adv. Synth. Catal.*, **347**, 609–631.

371 Lee, Y. and Hoveyda, A.H. (2009) *J. Am. Chem. Soc.*, **131**, 3160–3161.
372 Rubina, M., Rubin, M., and Gevorgyan, V. (2003) *J. Am. Chem. Soc.*, **125**, 7198–7199.
373 Hohn, E., Palecek, J., and Pietruszka, J. (2009) *Eur. J. Org. Chem.*, 3765–3782.
374 Hupe, E., Marek, I., and Knochel, P. (2002) *Org. Lett.*, **4**, 2861–2863.
375 (a) Ueda, M., Satoh, A., and Miyaura, N. (2002) *J. Organomet. Chem.*, **642**, 145–147; (b) Morgan, J.B. and Morken, J.P. (2004) *J. Am. Chem. Soc.*, **126**, 15338–15339; (c) Moran, M.J. and Morken, J.P. (2006) *Org. Lett.*, **8**, 2413–2415; (d) Paptchikhine, A., Cheruku, P., Engman, M., and Andersson, P.G. (2009) *Chem. Commun.*, 5996–5998.
376 Laitar, D.S., Tsui, E.Y., and Sadighi, J.P. (2006) *J. Am. Chem. Soc.*, **128**, 11036–11037.
377 (a) Burks, H.E. and Morken, J.P. (2007) *Chem. Commun.*, 4717–4718; (b) Kliman, L., Mlynarski, S.N., and Morken, J.P. (2009) *J. Am. Chem. Soc.*, **131**, 13210–13211.
378 Lee, Y., Jang, H., and Hoveyda, A.H. (2009) *J. Am. Chem. Soc.*, **131**, 18234–18235.
379 Sasaki, Y., Zhong, C.M., Sawamura, M., and Itoh, H. (2010) *J. Am. Chem. Soc.*, **132**, 1226–1227.
380 Schiffner, J.A., Müther, K., and Oestreich, M. (2010) *Angew. Chem., Int. Ed.*, **49**, 1194–1196.
381 (a) Mun, S., Lee, J.-E., and Yun, J. (2006) *Org. Lett.*, **8**, 4887–4889; (b) Lee, J.-E. and Yun, J. (2008) *Angew. Chem., Int. Ed.*, **47**, 145–147; (c) Sim, H.-S., Feng, X., and Yun, J. (2009) *Chem. Eur. J.*, **15**, 1939–1943; (d) Chea, H., Sim, H., and Yun, J. (2009) *Adv. Synth. Catal.*, **351**, 855–858; (e) Lillo, V., Prieto, A., Bonet, A., Díaz-Requejo, M.M., Ramírez, J., Pérez, P.J., and Fernández, E. (2009) *Organometallics*, **28**, 659–662; (f) Fleming, W.J., Müller-Bunz, H., Lillo, V., Fernández, E., and Guiry, P.J. (2009) *Org. Biomol. Chem.*, **7**, 2520–2524; (g) Lillo, V., Geier, M.J., Westcott, S.A., and Fernández, E. (2009) *Org. Biomol. Chem.*, **7**, 4674–4676; (h) Lee, Y. and Hoveyda, A.H. (2009) *J. Am. Chem. Soc.*, **131**, 3160–3161; (i) Chen, I.-H., Yin, L., Itano, W., Kanai, M., and Shibasaki, M. (2009) *J. Am. Chem. Soc.*, **131**, 11664–11665; (j) Shiomi, T., Adachi, T., Toribatake, K., Zhou, L., and Nishiyama, H. (2009) *Chem. Commun.*, 5987–5989.
382 (a) Lee, K.-S., Zhugralin, A.R., and Hoveyda, A.H. (2009) *J. Am. Chem. Soc.*, **131**, 7253–7255; (b) Bonet, A., Gulyás, H., and Fernández, E. (2010) *Angew. Chem., Int. Ed.*, **49**, 5130–5134.
383 Lee, J.C.H. and Hall, D.G. (2010) *J. Am. Chem. Soc.*, **132**, 5544–5545.
384 (a) Kennedy, J.W.J. and Hall, D.G. (2005) Chapter 6, in *Boronic Acids*, 1st edn (ed. D.G. Hall), Wiley-VCH Verlag GmbH, Weinheim; (b) Hall, D.G. (2008) *Pure Appl. Chem.*, **80**, 913–927.
385 (a) Carosi, L. and Hall, D.G. (2007) *Angew. Chem., Int. Ed.*, **27**, 5913–5915; (b) Lessard, S., Peng, F., and Hall, D.G. (2009) *J. Am. Chem. Soc.*, **131**, 9612–9613; (c) Guzman-Martinez, A. and Hoveyda, A.H. (2010) *J. Am. Chem. Soc.*, **132**, 10634–10637.
386 (a) Olsson, V.J., Sebelius, S., Selander, N., and Szabo, K.J. (2006) *J. Am. Chem. Soc.*, **128**, 4588–4589; (b) Selander, N., Kipke, A., Sebelius, S., and Szabo, K.J. (2007) *J. Am. Chem. Soc.*, **129**, 13723–13731. (c) Dutheuil, G., Selander, N., Szabo, K.J., Aggarwal, V.K. (2008) *Synthesis*, 2293–2297.
387 (a) Wu, J.Y., Moreau, B., and Ritter, T. (2009) *J. Am. Chem. Soc.*, **131**, 12915–12916; (b) Ely, R.J. and Morken, J.P. (2010) *J. Am. Chem. Soc.*, **132**, 2534–2535.
388 (a) Morin, C. (1994) *Tetrahedron*, **50**, 12521–12569; (b) Srebnik, M. (2003) *Tetrahedron*, **59**, 579–593.
389 Laplante, C. and Hall, D.G. (2001) *Org. Lett.*, **3**, 1487–1490.
390 Matteson, D.S. and Liedtke, J.D. (1965) *J. Am. Chem. Soc.*, **87**, 1526–1531.
391 Matteson, D.S. (1999) *J. Organomet. Chem.*, **581**, 51–65.
392 Mothana, S., Grassot, J.-M., and Hall, D.G. (2010) *Angew. Chem., Int. Ed.*, **49**, 2883–2887.

393 Michaelis, A. (1901) *Justus Liebigs Ann. Chem.*, **315**, 19.
394 Zhong, Z. and Anslyn, E.V. (2002) *J. Am. Chem. Soc.*, **124**, 9014–9015.
395 Cai, S.X. and Keana, J.F.W. (1991) *Bioconjug. Chem.*, **2**, 317–322.
396 (a) Matteson, D.S. and Arne, K.H. (1978) *J. Am. Chem. Soc.*, **100**, 1325–1326; (b) Matteson, D.S. and Arne, K.H. (1982) *Organometallics*, **1**, 280–288.
397 Matteson, D.S. and Majumdar, D. (1980) *J. Chem. Soc., Chem. Commun.*, 39–40.
398 Jagannathan, S., Forsyth, T.P., and Kettner, C.A. (2001) *J. Org. Chem.*, **66**, 6375–6380.
399 Matteson, D.S. and Moody, R.J. (1980) *J. Org. Chem.*, **45**, 1091–1095.
400 Matteson, D.S. (1975) *Synthesis*, 147–158.
401 Endo, K., Hirokami, M., and Shibata, T. (2009) *Synlett*, 1131–1135.
402 Matteson, D.S. and Wilson, J.W. (1985) *Organometallics*, **4**, 1690–1692.
403 (a) Knochel, P. (1990) *J. Am. Chem. Soc.*, **112**, 7431–7433; (b) Sakai, M., Saito, S., Kanai, G., Suzuki, A., and Miyaura, N. (1996) *Tetrahedron Lett.*, **52**, 915–924.
404 Waas, J.R., Sidduri, A., and Knochel, P. (1992) *Tetrahedron Lett.*, **33**, 3717–3720.
405 Brown, H.C. and Imai, T. (1983) *J. Am. Chem. Soc.*, **105**, 6285–6289.
406 (a) Matteson, D.S. (1988) *Tetrahedron*, **54**, 10555–10607; (b) Matteson, D.S. (2005) Chapter 8, in *Boronic Acids*, 1st edn (ed. D.G. Hall), Wiley-VCH Verlag GmbH, Weinheim.
407 Wityak, J., Earl, R.A., Abelman, M.M., Bethel, Y.B., Fisher, B.N., Kauffman, G.K., Kettner, C.A., Ma, P., McMillan, J.L., Mersinger, L.J., Pesti, J., Pierce, M.E., Rankin, F.W., Chorvat, R.J., and Confalone, P.N. (1995) *J. Org. Chem.*, **60**, 3717–3722.
408 Thomas, S.P., French, R.M., Jheengut, V., and Aggarwal, V.K. (2009) *Chem. Rec.*, **9**, 24–39.
409 Ni, W., Fang, H., Springsteen, G., and Wang, B. (2004) *J. Org. Chem.*, **69**, 1999–2007.
410 Kuivila, H.G., Benjamin, L.E., Murphy, C.J., Price, A.D., and Polevy, J.H. (1962) *J. Org. Chem.*, **27**, 825–829.
411 Al-Zoubi, R.M. and Hall, D.G. (2010) *Org. Lett.*, **12**, 2480–2483.
412 Qiu, D., Mo, F., Zheng, Z., Zhang, Y., and Wang, J. (2010) *Org. Lett.*, **12**, 5474–5477.
413 Dai, C., Li, M. and Wang, B. (2009) *Chem. Commun.*, 5251–5253.
414 Gilman, H., Swayampati, D.R., and Ranck, R.O. (1958) *J. Am. Chem. Soc.*, **80**, 1355–1357.
415 Baron, O. and Knochel, P. (2005) *Angew. Chem., Int. Ed.*, **44**, 3133–3135.
416 Lopes-Ruiz, H. and Zard, S.Z. (2001) *Chem. Commun.*, 2618–2619.
417 Moriya, T., Suzuki, A., and Miyaura, N. (1995) *Tetrahedron Lett.*, **36**, 1887–1888.
418 (a) Rasset-Deloge, C., Martinez-Fresneda, P., and Vaultier, M. (1992) *Bull. Soc. Chim. Fr.*, **129**, 285–290; (b) Jehanno, E. and Vaultier, M. (1995) *Tetrahedron Lett.*, **36**, 4439–4442. (c) Châtaigner, I., Lebreton, J., Zammatio, F., and Villiéras, J. (1997) *Tetrahedron Lett.* **38**, 3719–3722.
419 Andrade, L.H. and Barcellos, T. (2009) *Org. Lett.*, **11**, 3052–3055.
420 Kinder, D.H. and Ames, M.M. (1987) *J. Org. Chem.*, **52**, 2452–2454.
421 Curtis, A.D.M. and Whiting, A. (1991) *Tetrahedron Lett.*, **32**, 1507–1510.
422 Chan, K.-F. and Wong, H.N.C. (2001) *Org. Lett.*, **3**, 3991–3994.
423 (a) Yamamoto, Y., Seko, T., and Nemoto, H. (1989) *J. Org. Chem.*, **54**, 4734–4736; (b) Vedsø, P., Olesen, P.H., and Hoeg-Jensen, T. (2004) *Synlett*, 892–894; (c) Dabrowski, M., Kurach, P., Lulinski, S., and Serwatowski, J. (2007) *Appl. Organomet. Chem.*, **21**, 234–238.
424 Durka, K., Kurach, P., Lulinski, S., and Serwatowski, J. (2009) *Eur. J. Org. Chem.*, 4325–4332.
425 Gillis, E.P. and Burke, M.D. (2008) *J. Am. Chem. Soc.*, **130**, 14084–14085.
426 (a) Lee, S.J., Gray, K.G., Paek, J.S., and Burke, M.D. (2008) *J. Am. Chem. Soc.*, **130**, 466–468; (b) Lee, S.J., Anderson, T.M., and Burke, M.D. (2010) *Angew. Chem., Int. Ed.*, **49**, 8860–8863.
427 Noguchi, H., Shioda, T., Chou, C.-M., and Suginome, M. (2008) *Org. Lett.*, **10**, 377–380.
428 Ihara, H. and Suginome, M. (2009) *J. Am. Chem. Soc.*, **131**, 7502–7503.
429 Santucci, L. and Gilman, H. (1958) *J. Am. Chem. Soc.*, **80**, 193–196.

430 Cao, H., McGill, T., and Heagy, M.D. (2004) *J. Org. Chem.*, **69**, 2959–2966.

431 (a) Hall, D.G., Tailor, J., and Gravel, M. (1999) *Angew. Chem., Int. Ed.*, **38**, 3064–3067; (b) Arimori, S., Hartley, J.H., Bell, M.L., Oh, C.S., and James, T.D. (2000) *Tetrahedron Lett.*, **41**, 10291–10294.

432 Gravel, M., Thompson, K.A., Zak, M., Bérubé, C., and Hall, D.G. (2002) *J. Org. Chem.*, **67**, 3–15.

433 Tsukamoto, H., Suzuki, T., Sato, M., and Kondo, Y. (2007) *Tetrahedron Lett.*, **48**, 8438–8441.

434 Gravel, M., Bérubé, C., and Hall, D.G. (2000) *J. Comb. Chem.*, **2**, 228–231.

435 Thompson, K.A. and Hall, D.G. (2000) *Chem. Commun.*, 2379–2380.

436 Carboni, B., Pourbaix, C., Carreaux, F., Deleuze, H., and Maillard, B. (1999) *Tetrahedron Lett.*, **40**, 7979–7983.

437 Pourbaix, C., Carreaux, F., and Carboni, B. (2001) *Org. Lett.*, **3**, 803–805.

438 Pourbaix, C., Carreaux, F., Carboni, B., and Deleuze, H. (2000) *Chem. Commun.*, 1275–1276.

439 (a) Li, W. and Burgess, K. (1999) *Tetrahedron Lett.*, **40**, 6527–6530; (b) Dunsdon, R.M., Greening, J.R., Jones, P.S., Jordan, S., and Wilson, F.X. (2000) *Bioorg. Med. Chem. Lett.*, **10**, 1577–1579.

440 Arnauld, T., Barrett, A.G.M., and Seifried, R. (2001) *Tetrahedron Lett.*, **42**, 7889–7901.

441 Wang, W.Q., Gao, X.M., Springsteen, G., and Wang, B. (2002) *Tetrahedron Lett.*, **43**, 6339–6342.

442 (a) Chen, D., Qing, F.-L., and Huang, Y. (2002) *Org. Lett.*, **4**, 1003–1006; (b) Huang, Y., Chen, D., and Qing, F.L. (2003) *Tetrahedron*, **59**, 7879–7886.

443 Hebel, A. and Haag, R. (2002) *J. Org. Chem.*, **67**, 9452–9455.

444 Dennis, L.M. and Shelton, R.S. (1930) *J. Am. Chem. Soc.*, **52**, 3128–3132.

445 Gilman, H., Santucci, L., Swayampati, D.R., and Ranck, R.O. (1957) *J. Am. Chem. Soc.*, **79**, 2898–2901.

446 Duran, D., Wu, N., Mao, B., and Xu, J. (2006) *J. Liq. Chromatogr. Relat. Technol.*, **29**, 661–672.

447 Longstaff, C. and Rose, M.E. (1982) *Org. Mass Spectrom.*, **17**, 508–518.

448 Kaminski, J.J. and Lyle, R.E. (1978) *Org. Mass Spectrom.*, **13**, 425–428.

449 Haas, M.J., Blom, K.F., and Schwarz, C.H., III (1999) *Anal. Chem.*, **71**, 1574–1578.

450 Hermánek, S. (1992) *Chem. Rev.*, **92**, 325–362.

451 Nöth, H. and Wrackmeyer, B. (1978) *Nuclear Magnetic Resonance Spectroscopy of Boron Compounds*, NMR Basic Principles and Progress Series 14 (eds P. Diehl, E. Fluck, and R. Kosfeld), Springer, Berlin.

452 Matteson, D.S. and Krämer, E. (1968) *J. Am. Chem. Soc.*, **90**, 7261–7267.

453 Larock, R.C., Gupta, S.K., and Brown, H.C. (1972) *J. Am. Chem. Soc.*, **94**, 4371–4373.

454 Rozema, M.J., Rajagopal, D., Tucker, C.E., and Knochel, P. (1992) *J. Organomet. Chem.*, **438**, 11–27.

455 Challenger, F. and Parker, B. (1931) *J. Chem. Soc.*, 1462–1467.

456 Kuivila, H.G., Reuwer, J.F., and Mangravite, J.A. (1964) *J. Am. Chem. Soc.*, **86**, 2666–2670.

457 (a) Bolm, C. and Rudolph, J. (2003) *J. Am. Chem. Soc.*, **125**, 14850–14851; (b) Jimeno, C., Sayalero, S., Fjermestad, T., Colet, G., Maseras, F., and Pericàs, M.A. (2008) *Angew. Chem., Int. Ed.*, **47**, 1098–1101; (c) Fandrick, D.R., Fandrick, K.R., Reeves, J.T., Tan, Z., Johnson, C.S., Lee, H., Song, J.J., Yee, N.K., and Senanayake, C.H. (2010) *Org. Lett.*, **12**, 88–91; (d) Fandrick, K.R., Fandrick, D.R., Gao, J.J., Reeves, J.T., Tan, Z.L., Li, W.J., Song, J.H.J., Lu, B., Yee, N.K., and Senanayake, C.H. (2010) *Org. Lett.*, **12**, 3748–3751.

458 Matano, Y., Begum, S.A., Miyamatsu, T., and Suzuki, H. (1998) *Organometallics*, **17**, 4332–4334.

459 (a) Partyka, D.V., Zeller, M., Hunter, A.D., and Gray, T.G. (2006) *Angew. Chem., Int. Ed.*, **45**, 8188–8191; (b) Seidel, G., Lehmann, C.W., and Fürstner, A. (2010) *Angew. Chem., Int. Ed.*, **49**, 8466–8470.

460 Klingensmith, L.M., Bio, M.M., and Moniz, G.A. (2007) *Tetrahedron Lett.*, **48**, 8242–8245.

461 Brown, H.C. and Zweifel, G. (1961) *J. Am. Chem. Soc.*, **83**, 2544–2551.

462 Maleczka, R.E., Jr., Shi, F., Holmes, D., and Smith, M.R., III (2003) *J. Am. Chem. Soc.*, **125**, 7792–7793.

463 (a) Kuivila, H.G. (1954) *J. Am. Chem. Soc.*, **76**, 870–874; (b) Kuivila, H.G. and Armour, A.G. (1957) *J. Am. Chem. Soc.*, **79**, 5659–5662.

464 Kabalka, G.W. and Hedgecock, H.C., Jr. (1975) *J. Org. Chem.*, **40**, 1776–1779.

465 Webb, K.S. and Levy, D. (1995) *Tetrahedron Lett.*, **36**, 5117–5118.

466 (a) Matteson, D.S. and Ray, R. (1980) *J. Am. Chem. Soc.*, **102**, 7591–7593; (b) Fontani, P., Carboni, B., Vaultier, M., and Maas, G. (1991) *Synthesis*, 605–609.

467 Kianmehr, E., Yahyaee, M., and Tabatabai, K. (2007) *Tetrahedron Lett.*, **48**, 2713–2715.

468 (a) Prakash, G.K.S., Chacko, S., Panja, C., Thomas, T.E., Gurung, L., Rasul, G., Mathew, T., and Olah, G.A. (2009) *Adv. Synth. Catal.*, **351**, 1567–1574; (b) Xu, J., Wang, X., Shao, C., Su, D., Cheng, G., and Hu, Y. (2010) *Org. Lett.*, **12**, 1964–1967.

469 Kyne, R.E., Ryan, M.C., Kliman, L.T., and Morken, J.P. (2010) *Org. Lett.*, **12**, 3796–3799.

470 Murata, M., Satoh, K., Watanabe, S., and Masuda, Y. (1998) *J. Chem. Soc., Perkin Trans. 1*, 1465–1466.

471 Huber, M.-L. and Pinhey, J.T. (1990) *J. Chem. Soc., Perkin Trans. 1*, 721–722.

472 (a) Stefan, S., Jurgen, S., Prakash, G.K.S., Petasis, N.A., and Olah, G.A. (2000) *Synlett*, 1845–1847; (b) Prakash, G.K.S., Panja, C., Mathew, T., Surampudi, V., Petasis, N.A., and Olah, G.A. (2004) *Org. Lett.*, **6**, 2205–2207.

473 Brown, H.C., Kim, K.-W., Cole, T.E., and Singaram, B. (1986) *J. Am. Chem. Soc.*, **108**, 6761–6764.

474 (a) Jego, J.-M., Carboni, B., Vaultier, M., and Carrié, R. (1989) *J. Chem. Soc., Chem. Commun.*, 142–143; (b) Jego, J.-M., Carboni, B., and Vaultier, M. (1992) *Bull. Soc. Chim. Fr.*, **129**, 554–565.

475 Rao, H., Fu, H., Jiang, Y., and Zhao, Y. (2009) *Angew. Chem., Int. Ed.*, **48**, 1114–1116; Jiang, Z.Q., Wu, Z.Q., Wang, L.X., Wu, D., and Zhou, X.G. (2010) *Can. J. Chem.*, **88**, 964–968.

476 Yu, Y., Srogl, J., and Liebeskind, L.S. (2004) *Org. Lett.*, **6**, 2631–2634.

477 Prakash, G.K.S., Moran, M.D., Mathew, T., and Olah, G.A. (2009) *J. Fluor. Chem.*, **130**, 806–809.

478 Murphy, J.C., Liao, X., and Hartwig, J.F. (2007) *J. Am. Chem. Soc.*, **129**, 15434–15435.

479 Kuivila, H.G. and Easterbrook, E.K. (1951) *J. Am. Chem. Soc.*, **73**, 4629–4632.

480 Kuivila, H.G. and Hendrickson, A.R. (1952) *J. Am. Chem. Soc.*, **74**, 5068–5070.

481 Thiebes, C., Prakash, G.K.S., Petasis, N.A., and Olah, G.A. (1998) *Synlett*, 141–142.

482 Szumigala, R.H., Jr., Devine, P.N., Gauthier, D.R., Jr., and Volante, R.P. (2004) *J. Org. Chem.*, **69**, 566–569.

483 Thompson, A.L.S., Kabalka, G.W., Akula, M.R., and Huffman, J.W. (2005) *Synthesis*, 547–550.

484 Yao, M.-L., Reddy, M.S., Yong, L., Walfish, I., Blevins, D.W., and Kabalka, G.W. (2010) *Org. Lett.*, **12**, 700–703.

485 (a) Diorazio, L.J., Widdowson, D.A., and Clough, J.M. (1992) *Tetrahedron*, **48**, 8073–8088; (b) Clough, J.M., Diorazio, L.J., and Widdowson, D.A. (1990) *Synlett*, 761–762.

486 Furuya, T., Kaiser, H.M., and Ritter, T. (2008) *Angew. Chem., Int. Ed.*, **47**, 5993–5996.

487 Furuya, T. and Ritter, T. (2009) *Org. Lett.*, **11**, 2860–2863.

488 Carroll, M.A., Pike, V.W., and Widdowson, D.A. (2000) *Tetrahedron Lett.*, **41**, 5393–5396.

489 Brown, H.C., Hamaoka, T., and Ravindran, N. (1973) *J. Am. Chem. Soc.*, **95**, 6456–6457.

490 Brown, H.C., Bhat, N.G., and Rajagopalan, S. (1986) *Synthesis*, 480–482.

491 Brown, H.C., Subrahmanyam, C., Hamaoka, T., Ravindran, N., Bowman, D.H., Misumi, S., Unni, M.K., Somayaji, V., and Bhat, N.G. (1989) *J. Org. Chem.*, **54**, 6068–6075.

492 (a) Brown, H.C., Hamaoka, T., and Ravindran, N. (1973) *J. Am. Chem. Soc.*, **95**, 5786–5788; (b) Brown, H.C. and Somayaji, V. (1984) *Synthesis*, 919–920; (c) Brown, H.C., Hamaoka, T.,

Ravindran, N., Subrahmanyam, C., Somayaji, V., and Bhat, N.G. (1989) *J. Org. Chem.*, **54**, 6075–6079.

493 Kabalka, G.W., Gooch, E.E., and Hsu, H.C. (1981) *Synth. Commun.*, **11**, 247–251.

494 (a) Stewart, S.K. and Whiting, A. (1995) *Tetrahedron Lett.*, **36**, 3929–3932; (b) Henaff, N., Stewart, S. K., Whiting, A. (1997) *Tetrahedron Lett.* **38**, 4525–4526.

495 Petasis, N.A. and Zavialov, I.A. (1996) *Tetrahedron Lett.*, **37**, 567–570.

496 Kunda, S.A., Smith, T.L., Hylarides, M.D., and Kabalka, G.W. (1985) *Tetrahedron Lett.*, **26**, 279–280.

497 Miyaura, N. and Suzuki, A. (1979) *J. Chem. Soc., Chem. Commun.*, 866–867.

498 Miyaura, N., Yanagi, T., and Suzuki, A. (1981) *Synth. Commun.*, **11**, 513–519.

499 (a) Miyaura, N. and Suzuki, A. (1995) *Chem. Rev.*, **95**, 2457–2483; (b) Suzuki, A. (1999) *J. Organomet. Chem.*, **576**, 147–168; (c) Miyaura, N. (2002) *Top. Curr. Chem.*, **219**, 11–59; (d) Kotha, S., Lahiri, K., and Kashinath, D. (2002) *Tetrahedron*, **58**, 9633–9695; (e) Hassan, J., Sévignon, M., Gozzi, C., Schulz, E., and Lemaire, M. (2002) *Chem. Rev.*, **102**, 1359–1469; (f) Suzuki, A. (2004) *Proc. Jpn Acad., Ser. B*, **80**, 359–371; (g) Nicolaou, K.C., Bulger, P.G., and Sarlah, D. (2005) *Angew. Chem., Int. Ed.*, **44**, 4442–4489.

500 Miyaura, N., Yamada, K., Suginome, H., and Suzuki, A. (1985) *J. Am. Chem. Soc.*, **107**, 972–980.

501 Smith, G.B., Dezeny, G.C., Hughes, D.L., King, A.O., and Verhoeven, T.R. (1994) *J. Org. Chem.*, **59**, 8151–8156.

502 Moreno-Manas, M., Pérez, M., and Pleixats, R. (1996) *J. Org. Chem.*, **61**, 2346–2351.

503 Aliprantis, A.O. and Canary, J.W. (1994) *J. Am. Chem. Soc.*, **116**, 6985–6986.

504 Miyaura, N. (2002) *J. Organomet. Chem.*, **653**, 54–57.

505 (a) Martin, A.R. and Yang, Y. (1993) *Acta. Chem. Scand.*, **47**, 221–230; (b) Matos, K. and Soderquist, J.A. (1998) *J. Org. Chem.*, **63**, 461–470; (c) Braga, A.A.C., Morgon, N.H., Ujaque, G., Lledos, A., and Maseras, F. (2005) *J. Am. Chem. Soc.*, **127**, 9288–9307; (d) Braga, A.A.C., Morgon, N.H., Ujaque, G., Lledos, A., and Maseras, F. (2006) *J. Organomet. Chem.*, **691**, 4459–4466.

506 Carrow, B.P. and Hartwig, J.F. (2011) *J. Am. Chem. Soc.*, **133**, 2116–2119.

507 (a) Ishiyama, T., Kizaki, H., Miyaura, N., and Suzuki, A. (1993) *Tetrahedron Lett.*, **34**, 7595–7598; (b) Ishiyama, T., Kizaki, H., Hayashi, T., Suzuki, A., and Miyaura, N. (1998) *J. Org. Chem.*, **63**, 4726–4731.

508 (a) Bumagin, N.A. and Korolev, D.N. (1999) *Tetrahedron Lett.*, **40**, 3057–3060; (b) Haddach, M. and McCarthy, J.R. (1999) *Tetrahedron Lett.*, **40**, 3109–3112.

509 Goossen, L.J. and Ghosh, K. (2001) *Angew. Chem., Int. Ed.*, **40**, 3458–3460.

510 Barder, T.E. and Buchwald, S.L. (2007) *Org. Lett.*, **9**, 137–139.

511 (a) Gronowitz, S., Hörnfeldt, A.-B., and Yang, Y.-H. (1986) *Chem. Scr.*, **26**, 383–386; (b) Chaumeil, H., Signorella, S., and Le Drian, C. (2000) *Tetrahedron*, **56**, 9655–9662.

512 Wright, S.W., Hageman, D.L., and McClure, L.D. (1994) *J. Org. Chem.*, **59**, 6095–6097.

513 Watanabe, T., Miyaura, N., and Suzuki, A. (1992) *Synlett*, 207–209.

514 Yoshida, H., Yamaryo, Y., Ohshita, J., and Kunai, A. (2003) *Tetrahedron Lett.*, **44**, 1541–1544.

515 Cai, X. and Snieckus, V. (2004) *Org. Lett.*, **6**, 2293–2295.

516 (a) Kerins, F. and O'Shea, D.F. (2002) *J. Org. Chem.*, **67**, 4968–4971; (b) Lightfoot, A.P., Twiddle, S.J.R., and Whiting, A. (2005) *Synlett*, 529–531.

517 (a) Miyaura, N., Suginome, H., and Suzuki, A. (1983) *Tetrahedron*, **39**, 3271–3277; (b) Miyaura, N. and Suzuki, A. (1990) *Org. Synth.*, **68**, 130–136.

518 Roush, W.R., Reilly, M.L., Koyama, K., and Brown, B.B. (1997) *J. Org. Chem.*, **62**, 8708–8721.

519 Sebelius, S., Olsson, V.J., Wallner, O.A., and Szabo, K.J. (2006) *J. Am. Chem. Soc.*, **128**, 8150–8151.

520 (a) Chemler, S.R., Trauner, D., and Danishefsky, S.J. (2001) *Angew. Chem., Int. Ed.*, **40**, 4544–4568, and references

cited therein; (b) Zou, G., Reddy, Y.K., and Falck, J.R. (2001) *Tetrahedron Lett.*, **42**, 7213–7215; (c) Kataoka, N., Shelby, Q., Stambuli, J.P., and Hartwig, J.F. (2002) *J. Org. Chem.*, **67**, 5553–5566; (d) Zhou, S.-M., Deng, M.-Z., Xia, L.-J., and Tang, M.-H. (1998) *Angew. Chem., Int. Ed.*, **37**, 2845–2847.

521 Doucet, H. (2008) *Eur. J. Org. Chem.*, 2013–2030.

522 Gonzalez-Bobes, F. and Fu, G.C. (2006) *J. Am. Chem. Soc.*, **128**, 5360–5361.

523 Kirchhoff, J.H., Netherton, M.R., Hills, I.D., and Fu, G.C. (2002) *J. Am. Chem. Soc.*, **124**, 13662–13663.

524 (a) Crudden, C.M., Glasspoole, B.W., and Lata, C.J. (2009) *Chem. Commun.*, 6704–6716; (b) Imao, D., Glasspoole, B.W., Laberge, V.S., and Crudden, C.M. (2009) *J. Am. Chem. Soc.*, **131**, 5024–5025; (c) He, A. and Falck, J.R. (2010) *J. Am. Chem. Soc.*, **132**, 2524–2525; (d) Ohmura, T., Awano, T., and Suginome, M. (2010) *J. Am. Chem. Soc.*, **132**, 13191–13193; (e) Sandrock, D.L., Jean-Gérard, L., Chen, C.Y., and Dreher, S.D. (2010) *J. Am. Chem. Soc.*, **132**, 17108–17110.

525 Sakamoto, J., Rehahn, M., Wegner, G., and Schluter, A.D. (2009) *Macromol. Rapid Commun.*, **30**, 653–687.

526 (a) Franzen, R. (2000) *Can. J. Chem.*, **78**, 957–962; (b) Brase, S., Kirchhoff, J.H., and Kobberling, J. (2003) *Tetrahedron*, **59**, 885–939.

527 (a) Urawa, Y., Naka, H., Miyazawa, M., Souda, S., and Ogura, K. (2002) *J. Organomet. Chem.*, **653**, 269–278; (b) Yasuda, N. (2002) *J. Organomet. Chem.*, **653**, 279–287.

528 Littke, A.F. and Fu, G.C. (2002) *Angew. Chem., Int. Ed.*, **41**, 4176–4211.

529 (a) Littke, A.F. and Fu, G.C. (1998) *Angew. Chem., Int. Ed.*, **37**, 3387–3388; (b) Littke, A.F., Dai, C., and Fu, G.C. (2000) *J. Am. Chem. Soc.*, **122**, 4020–4028; (c) Review: Fu, G.C. (2008) *Acc. Chem. Res.*, **41**, 1555–1564.

530 (a) Old, D.W., Wolfe, J.P., and Buchwald, S.L. (1998) *J. Am. Chem. Soc.*, **120**, 9722–9723; (b) Wolfe, J.P., Singer, R.A., Yang, B.H., and Buchwald, S.L. (1999) *J. Am. Chem. Soc.*, **121**, 9550–9561; (c) Yin, J., Rainka, M.P., Zhang, X.-X., and Buchwald, S.L. (2002) *J. Am. Chem. Soc.*, **124**, 1162–1163; (d) Billingsley, K. and Buchwald, S.L. (2007) *J. Am. Chem. Soc.*, **129**, 3358–3366; (e) Martin, R. and Buchwald, S.L. (2008) *Acc. Chem. Res.*, **41**, 1461–1473.

531 O'Brien, C.J., Kantchev, E.A.B., Valente, C., Hadei, N., Chass, G.A., Lough, A., Hopkinson, A.C., and Organ, M.E. (2006) *Chem. Eur. J.*, **12**, 4743–4778.

532 (a) Gstottmayr, C.W.K., Bohn, V.P.M., Herdtweck, E., Grosche, M., and Hermann, W.A. (2002) *Angew. Chem., Int. Ed.*, **41**, 1363–1365; (b) Stambuli, J.P., Kuwano, R., and Hartwig, J.F. (2002) *Angew. Chem., Int. Ed.*, **41**, 4746–4748.

533 Walker, S.D., Barder, T.E., Martinelli, J.R., and Buchwald, S.L. (2004) *Angew. Chem., Int. Ed.*, **43**, 1871–1876.

534 Navarro, O., Kelly, R.A., III, and Nolan, S.P. (2003) *J. Am. Chem. Soc.*, **125**, 16194–16195.

535 Percec, V., Bae, J.-Y., and Hill, D.H. (1995) *J. Org. Chem.*, **60**, 1060–1065.

536 Na, Y., Park, S., Han, S.B., Han, H., Ko, S., and Chang, S. (2004) *J. Am. Chem. Soc.*, **126**, 250–258.

537 (a) Guo, Y., Young, D.J., and Hot, T.S.A. (2008) *Tetrahedron Lett.*, **49**, 5620–5621; (b) Hatakeyama, T., Hashimoto, T., Kondo, Y., Fujiwara, Y., Seike, H., Takaya, H., Tamada, Y., Ono, T., and Nakamura, M. (2010) *J. Am. Chem. Soc.*, **132**, 10674–10676.

538 (a) González-Arellano, C., Corma, A., Iglesias, M., and Sánchez, F. (2006) *J. Catal.*, **238**, 497–501; (b) Mankad, N.P. and Toste, F.D. (2010) *J. Am. Chem. Soc.*, **132**, 12859–12861.

539 Sakurai, H., Tsukuda, T., and Hirao, T. (2002) *J. Org. Chem.*, **67**, 2721–2722, and references cited therein.

540 Leadbeater, N.E. and Marco, M. (2003) *J. Org. Chem.*, **68**, 5660–5667. A reassessment of this procedure revealed that minute levels of Pd contaminants in the inorganic base employed are responsible for the generation of biaryl products: Arvela, R. K., Leadbeater, N. E., Sangi, M. S.,

Williams, V.A., Granados, P., and Singer, R. D. (2005) *J. Org. Chem.* **70**, 161–168.

541 Schaub, T., Backes, M., and Radius, U. (2006) *J. Am. Chem. Soc.*, **128**, 15964–15965.

542 (a) Nguyen, H.N., Huang, X., and Buchwald, S.L. (2003) *J. Am. Chem. Soc.*, **125**, 11818–11819; (b) Percec, V., Golding, G.M., Smidrkal, J., and Weichold, O. (2004) *J. Org. Chem.*, **69**, 3447–3452; (c) Ang, Z.-Y. and Hu, Q.-S. (2004) *J. Am. Chem. Soc.*, **126**, 3058–3059.

543 Blakey, S.B. and MacMillan, D.W.C. (2003) *J. Am. Chem. Soc.*, **125**, 6046–6047.

544 (a) Quasdorf, K.W., Tian, X., and Garg, N.K. (2008) *J. Am. Chem. Soc.*, **130**, 14422–14423; (b) Guan, B.-T., Wang, Y., Li, B.-J., Yu, D.-G., and Shi, Z.-J. (2008) *J. Am. Chem. Soc.*, **130**, 14468–14470; (c) Quasdorf, K.W., Riener, M., Petrova, K.V., and Garg, N.K. (2009) *J. Am. Chem. Soc.*, **131**, 17748–17749.

545 Tobisu, M., Shimasaki, T., and Chatani, N. (2008) *Angew. Chem., Int. Ed.*, **47**, 4866–4869.

546 (a) Ohmiya, H., Makida, Y., Li, D., Tanabe, M., and Sawamura, M. (2010) *J. Am. Chem. Soc.*, **132**, 879–889; (b) Pigge, F.C. (2010) *Synthesis*, 1745–1762.

547 Nishikata, T. and Lipshutz, B.H. (2009) *J. Am. Chem. Soc.*, **131**, 12103–12105.

548 (a) Tsukamoto, H., Sato, M., and Kondo, Y. (2004) *Chem. Commun.*, 1200–1201; (b) Kayaki, Y., Koda, T., and Ikariya, T. (2004) *Eur. J. Org. Chem.*, 4989–4993; (c) Yoshida, M., Gotou, T., and Ihara, M. (2004) *Tetrahedron Lett.*, **45**, 5573–5575.

549 Srogl, J., Allred, G.D., and Liebeskind, L.S. (1997) *J. Am. Chem. Soc.*, **119**, 12376–12377.

550 (a) Savarin, C., Srogl, J., and Liebeskind, L.S. (2000) *Org. Lett.*, **2**, 3229–3231; (b) Liebeskind, L.S. and Srogl, J. (2000) *J. Am. Chem. Soc.*, **122**, 11260–11261.

551 (a) Savarin, C., Srogl, J., and Liebeskind, L.S. (2001) *Org. Lett.*, **3**, 91–93; (b) Liebeskind, L.S. and Srogl, J. (2002) *Org. Lett.*, **4**, 979–981; (c) Kusturin, C.L. and Liebeskind, L.S. (2002) *Org. Lett.*, **4**, 983–985.

552 Kusturin, C., Liebeskind, L.S., Rahman, H., Sample, K., Schweitzer, B., Srogl, J., and Neumann, W.L. (2003) *Org. Lett.*, **5**, 4349–4352.

553 Prokopcova, H. and Kappe, C.O. (2009) *Angew. Chem., Int. Ed.*, **48**, 2276–2286.

554 (a) Ukai, K., Aoki, M., Takaya, J., and Iwasawa, N. (2006) *J. Am. Chem. Soc.*, **128**, 8706–8707; (b) Ohishi, T., Nishiura, M., and Hou, Z.M. (2008) *Angew. Chem., Int. Ed.*, **47**, 5792–5795.

555 Miura, T. and Murakami, M. (2007) *Chem. Commun.*, 217–224.

556 Miura, T., Toyashima, T., and Murakami, M. (2008) *Org. Lett.*, **10**, 4887–4889.

557 Matsuda, T., Makino, M., and Murakami, M. (2004) *Org. Lett.*, **6**, 1257–1259.

558 Tsoi, Y.-T., Zhou, Z., Chan, A.S.C., and Yu, W.-Y. (2010) *Org. Lett.*, **12**, 4506–4509.

559 Ohmiya, H., Yokokawa, N., and Sawamura, M. (2010) *Org. Lett.*, **12**, 438–440.

560 Chu, L. and Qing, F.-L. (2010) *Org. Lett.*, **12**, 5060–5063.

561 McManus, H.A., Fleming, M.J., and Lautens, M. (2007) *Angew. Chem., Int. Ed.*, **46**, 433–436.

562 (a) Zhang, G., Peng, Y., Cui, L., and Zhang, L. (2009) *Angew. Chem., Int. Ed.*, **48**, 3112–3115; (b) Zhang, G., Cui, L., Wang, Y., and Zhang, L. (2010) *J. Am. Chem. Soc.*, **132**, 1474–1475.

563 Patel, S.J. and Jamison, T.F. (2003) *Angew. Chem., Int. Ed.*, **42**, 1364–1367.

564 Huang, T.-H., Chang, H.-M., Wu, M.-Y., and Cheng, C.-H. (2002) *J. Org. Chem.*, **67**, 99–105.

565 Kakiuchi, F., Kan, S., Igi, K., Chatani, N., and Murai, S. (2003) *J. Am. Chem. Soc.*, **125**, 1698–1699.

566 Kakiuchi, F., Usui, M., Ueno, S., Chatani, N., and Murai, S. (2004) *J. Am. Chem. Soc.*, **126**, 2706–2707.

567 (a) Delcamp, J.H., and White, M.C. (2006) *J. Am. Chem. Soc.*, **128**, 15076–15077; (b) Giri, R., Mangel, N., Li, J.-J., Wang, D.-H., Breazzano, S.P., Saunders, L.B., and Yu, J.-Q. (2007) *J. Am. Chem. Soc.*, **129**, 3510–3511; (c) Sun, C.-L., Liu, N., Li, B.-J., Yu, D.-G., Wang, Y., and Shi, Z.-J. (2010) *Org. Lett.*, **12**, 184–187.

568 Chen, X., Goodhue, C.E., and Yu, J.-Q. (2006) *J. Am. Chem. Soc.*, **128**, 12634–12635.

569 Seiple, I.B., Su, S., Rodriguez, R.A., Gianatassio, R., Fujiwara, Y., Sobel, A.L., and Baran, P.S. (2010) *J. Am. Chem. Soc.*, **132**, 13194–13196.

570 Wen, J., Zhang, J., Chen, S.-Y., Li, J., and Yu, X.-Q. (2008) *Angew. Chem., Int. Ed.*, **47**, 8897–8900.

571 Zou, G., Wang, Z., Zhu, J., and Tang, J. (2003) *Chem. Commun.*, 2438–2439.

572 Farrington, E.J., Brown, J.M., Barnard, C.F.J., and Rowsell, E. (2002) *Angew. Chem., Int. Ed.*, **41**, 169–171.

573 Koike, T., Du, X., Sanada, T., Danda, Y., and Mori, A. (2003) *Angew. Chem., Int. Ed.*, **42**, 89–92.

574 (a) Cho, C.S. and Uemura, S. (1994) *J. Organomet. Chem.*, **465**, 85–92; (b) Du, X., Suguro, M., Hirabayashi, K., Mori, A., Nishikata, T., Hagiwara, N., Kawata, K., Okeda, T., Wang, H.F., Fugami, K., and Kosugi, M. (2001) *Org. Lett.*, **3**, 3313–3316; (c) Jung, Y.C., Mishra, R.K., Yoon, C.H., and Jung, K.W. (2003) *Org. Lett.*, **5**, 2231–2234; (d) Lindh, J., Enquist, P.-A., Pilotti, A., Nilsson, P., and Larhead, M. (2007) *J. Org. Chem.*, **72**, 7957–7962; (e) Ruan, J., Li, X., Saidi, O., and Xiao, J. (2008) *J. Am. Chem. Soc.*, **130**, 2424–2425; (f) Delcamp, J.H., Brucks, A.P., and White, M.C. (2008) *J. Am. Chem. Soc.*, **130**, 11270–11271; (g) Su, Y. and Jiao, N. (2009) *Org. Lett.*, **11**, 2980–2983; (h) Werner, E.W. and Sigman, M.S. (2010) *J. Am. Chem. Soc.*, **132**, 13981–13983.

575 Crowley, J.D., Hanni, K.D., Lee, A.-L., and Leigh, D.A. (2007) *J. Am. Chem. Soc.*, **129**, 12092–12093.

576 Xiong, D.-C., Zhang, L.-H., and Xe, X.-S. (2009) *Org. Lett.*, **11**, 1709–1712.

577 Rao, H., Fu, H., Jiang, Y., and Zhao, Y. (2010) *Adv. Synth. Catal.*, **352**, 458–462.

578 Yu, J.-Y., Shimizu, R., and Kuwano, R. (2010) *Angew. Chem., Int. Ed.*, **49**, 6396–6399.

579 Sakai, M., Ueda, M., and Miyaura, N. (1998) *Angew. Chem., Int. Ed.*, **37**, 3279–3281.

580 Sakai, M., Hayashi, H., and Miyaura, N. (1997) *Organometallics*, **16**, 4229–4231.

581 Hayashi, T. and Yamasaki, K. (2003) *Chem. Rev.*, **103**, 2829–2844.

582 Qin, C., Chen, J., Wu, H., Cheng, J., Zhang, Q., Zuo, B., Su, W., and Ding, J. (2008) *Tetrahedron Lett.*, **49**, 1884–1888.

583 Bouffard, J. and Itami, K. (2009) *Org. Lett.*, **11**, 4410–4413.

584 Iwai, Y., Gligorich, K.M., and Sigman, M.S. (2008) *Angew. Chem., Int. Ed.*, **47**, 3219–3222.

585 (a) Oh, C.H., Jung, H.H., Kim, K.S., and Kim, N. (2003) *Angew. Chem., Int. Ed.*, **42**, 805–808; (b) Zou, G., Zhu, J., and Tang, J. (2003) *Tetrahedron Lett.*, **44**, 8709–9711; (c) Gupta, A.K., Kim, K.S., and Oh, C.H. (2005) *Synlett*, 457–460.;

586 (a) Ma, S., Jiao, N., and Ye, L. (2003) *Chem. Eur. J.*, **9**, 6049–6056; (b) Oh, C.H., Ahn, T.W., and Reddy, V.R. (2003) *Chem. Commun.*, 2622–2623.

587 Shirakawa, E., Takahashi, G., Tsuchimoto, T., and Kawakami, Y. (2002) *Chem. Commun.*, 2210–2211.

588 Morikawa, S., Michigami, K., and Amii, H. (2010) *Org. Lett.*, **12**, 2520–2523.

589 Tomita, D., Kanai, M., and Shibasaki, M. (2006) *Chem. Asian J.*, **1**, 161–166.

590 Yamamoto, Y., Kurihara, K., and Miyaura, N. (2009) *Angew. Chem., Int. Ed.*, **48**, 4414–4416.

591 Ueda, M., Saito, A., and Miyaura, N. (2000) *Synlett*, 1637–1639.

592 (a) Kuriyama, M., Soeta, T., Hao, X., Chen, Q., and Tomioka, K. (2004) *J. Am. Chem. Soc.*, **126**, 8128–8129; (b) Beenan, M.A., Weix, D.J., and Ellman, J.A. (2006) *J. Am. Chem. Soc.*, **128**, 6304–6305.

593 Takada, E., Hara, S., and Suzuki, A. (1993) *Tetrahedron Lett.*, **34**, 7067–7070.

594 Chong, J.M., Shen, L., and Taylor, N.J. (2000) *J. Am. Chem. Soc.*, **122**, 1822–1823.

595 (a) Wu, T.R. and Chong, J.M. (2005) *J. Am. Chem. Soc.*, **127**, 3244–3245; (b) Wu, T.R. and Chong, J.M. (2007) *J. Am. Chem. Soc.*, **129**, 3244–3245.

596 Pellegrinet, S.C. and Goodman, J.M. (2006) *J. Am. Chem. Soc.*, **128**, 3116–3117.

597 (a) Lee, S. and MacMillan, D.W.C. (2007) *J. Am. Chem. Soc.*, **129**, 15438–15439; (b) Kim, S.-G. (2008) *Tetrahedron Lett.*, **49**,

6148–6151; (c) Choi, K.-S. and Kim, S.-G. (2010) *Tetrahedron Lett.*, **51**, 5203–5206.

598 Moquist, P.N., Kodama, T., and Schaus, S.E. (2010) *Angew. Chem., Int. Ed.*, **49**, 7096–7100.

599 Blais, J., L'Honoré, A., Soulié, J., and Cadiot, P. (1974) *J. Organomet. Chem.*, **78**, 323–337.

600 (a) Hall, D.G. (2007) *Synlett*, 1644–1655; (b) Lachance, H. and Hall, D.G. (2009) *Organic Reactions*, vol. **73** (ed. S.E. Denmark), John Wiley & Sons, Inc., New York, pp. 1–624.

601 (a) Kennedy, J.W.J. and Hall, D.G. (2002) *J. Am. Chem. Soc.*, **124**, 11586–11587; (b) Ishiyama, T., Ahiko, T.-a., and Miyaura, N. (2002) *J. Am. Chem. Soc.*, **124**, 12414–12415; (c) Rauniyar, V. and Hall, D.G. (2004) *J. Am. Chem. Soc.*, **126**, 4518–4519.

602 Yu, S.H., Ferguson, M.J., McDonald, R., and Hall, D.G. (2005) *J. Am. Chem. Soc.*, **127**, 12808–12809.

603 (a) Rauniyar, V. and Hall, D.G. (2007) *Angew. Chem., Int. Ed.*, **45**, 2426–2428; (b) Rauniyar, V., Zhai, H.M., and Hall, D.G. (2008) *J. Am. Chem. Soc.*, **130**, 8481–8490; (c) Rauniyar, V. and Hall, D.G. (2009) *J. Org. Chem.*, **74**, 4236–4241; (d) Jain, P. and Antilla, J. C. (2010) *J. Am. Chem. Soc.* **132**, 11884–11886.

604 Petasis, N.A. and Akritopoulou, I. (1993) *Tetrahedron Lett.*, **34**, 583–586.

605 (a) Petasis, N.A. and Zavialov, I.A. (1997) *J. Am. Chem. Soc.*, **119**, 445–446; (b) Petasis, N.A. and Zavialov, I.A. (1998) *J. Am. Chem. Soc.*, **120**, 11798–11799.

606 Lou, S. and Schaus, S.E. (2008) *J. Am. Chem. Soc.*, **130**, 6922–6923.

607 Candeias, N.R., Montalbano, F., Cal, P.M.S.D., and Gois, P.M.P. (2010) *Chem. Rev.*, **110**, 6169–6193.

608 Chan, D.M.T., Monaco, K.L., Wang, R.-P., and Winters, M.P. (1998) *Tetrahedron Lett.*, **39**, 2933–2936.

609 Evans, D.A., Katz, J.L., and West, T.R. (1998) *Tetrahedron Lett.*, **39**, 2937–2940.

610 Lam, P.Y.S., Clark, C.G., Saubern, S., Adams, J., Winters, M.P., Chan, D.M.T., and Combs, A. (1998) *Tetrahedron Lett.*, **39**, 2941–2944.

611 (a) Combs, A.P., Saubern, S., Rafalski, M., and Lam, P.Y.S. (1999) *Tetrahedron Lett.*, **40**, 1623–1626; (b) Combs, A.P., Tadesse, S., Rafalski, M., Haque, T.S., and Lam, P.Y.S. (2002) *J. Comb. Chem.*, **4**, 179–182.

612 Morgan, J. and Pinhey, J.T. (1990) *J. Chem. Soc., Perkin Trans. 1*, 715–722.

613 (a) Peng, C., Wang, Y., and Wang, J. (2008) *J. Am. Chem. Soc.*, **130**, 1566–1567; (b) Peng, C., Zhang, W., Yan, G., and Wang, J. (2009) *Org. Lett.*, **11**, 1667–30.

614 Barluenga, J., Tomás-Gamasa, M., Aznar, F., and Valdés, C. (2009) *Nat. Chem.*, **1**, 494–499.

615 Mizojiri, R. and Kobayashi, Y. (1995) *J. Chem. Soc., Perkin Trans. 1*, 2073–2075.

616 Trost, B.M. and Spagnol, M.D. (1995) *J. Chem. Soc., Perkin Trans. 1*, 2083–2096.

617 Retbøll, M., Edwards, A.J., Rae, A.D., Willis, A.C., Bennett, M.A., and Wenger, E. (2002) *J. Am. Chem. Soc.*, **124**, 8348–8360.

618 Minutolo, F. and Katzenellenbogen, J.A. (1998) *J. Am. Chem. Soc.*, **120**, 13264–13625.

619 Melhado, A.D., Brenzovich, W.E., Jr., Lackner, A.D., and Toste, F.D. (2010) *J. Am. Chem. Soc.*, **132**, 8885–8887.

620 Wang, W., Jasinski, J., Hammond, G.B., and Xu, B. (2010) *Angew. Chem., Int. Ed.*, **49**, 7247–7252.

621 Nakamura, M., Hatakeyama, T., Hara, K., Fukudome, H., and Nakamura, E. (2004) *J. Am. Chem. Soc.* **126**, 14344–14345.

622 Fandrick, D.R., Reeves, J.T., Tan, Z., Lee, H., Song, J.J., Yee, N.K., and Senanayake, C.H. (2009) *Org. Lett.*, **11**, 5458–5461.

623 Dickschat, A. and Studer, A. (2010) *Org. Lett.*, **12**, 3972–3974.

624 Hayakawa, H., Okada, N., and Miyashita, M. (1999) *Tetrahedron Lett.*, **40**, 3191–3194.

625 Hirai, A., Yu, X.-Q., Tonooka, T., and Miyashita, M. (2003) *Chem. Commun.*, 2482–2483.

626 Falck, J.R., Bondlela, M., Venkataraman, S.K., and Srinivas, D. (2001) *J. Org. Chem.*, **66**, 7148–7150.

627 Körner, C., Starkov, F., and Sheppard, T.D. (2010) *J. Am. Chem. Soc.*, **132**, 5968–5969.

628 Li, D.R., Murugan, A., and Falck, J.R. (2008) *J. Am. Chem. Soc.*, **130**, 46–48.
629 (a) Ishihara, K. and Yamamoto, H. (1999) *Eur. J. Org. Chem.*, 527–538; (b) Georgiou, I., Ilyashenko, G., and Whiting, A. (2009) *Acc. Chem. Res.*, **42**, 756–768.
630 (a) Letsinger, R.L., Dandegaonker, S., Vullo, W.J., and Morrison, J.D. (1963) *J. Am. Chem. Soc.*, **85**, 2223–2227; (b) Letsinger, R.L. and Morrison, J.D. (1963) *J. Am. Chem. Soc.*, **85**, 2227–2229.
631 Letsinger, R.L. and MacLean, D.B. (1963) *J. Am. Chem. Soc.*, **85**, 2230–2236.
632 Ishihara, K., Ohara, S., and Yamamoto, H. (1996) *J. Org. Chem.*, **61**, 4196–4197.
633 Al-Zoubi, R.M., Marion, O., and Hall, D.G. (2008) *Angew. Chem., Int. Ed.*, **47**, 2876–2879.
634 Aelvoet, K., Batsanov, A.S., Blatch, A.J., Grosjean, C., Patrick, L.G.F., Smethurst, C.A., and Whiting, A. (2008) *Angew. Chem., Int. Ed.*, **47**, 768–770.
635 (a) Zheng, H. and Hall, D.G. (2010) *Chem. Eur. J.*, **16**, 5454–5460; (b) Zheng, H. and Hall, D.G. (2010) *Tetrahedron Lett.*, **51**, 3561–3564.
636 McCubbin, J.A., Hosseini, H., and Krokhin, O.V. (2010) *J. Org. Chem.*, **75**, 959–962.
637 Zheng, H., Lejkowski, M., and Hall, D.G. (2011) *Chemical Science*, **2**, 1305–1310.
638 Rao, G. and Philipp, M. (1991) *J. Org. Chem.*, **56**, 505–512.
639 Barker, S.A., Hatt, B.W., and Somers, P.J. (1973) *Carbohydr. Res.*, **26**, 41–53.
640 Rohovec, J., Kotek, J., Peters, J.A., and Maschmeyer, T. (2001) *Eur. J. Org. Chem.*, 3899–3901.
641 de Vries, J.G. and Hubbard, S.A. (1988) *J. Chem. Soc., Chem. Commun.*, 1172–1173.
642 Narasaka, K., Shimada, G., Osoda, K., and Iwasawa, N. (1991) *Synthesis*, 1171–1172.
643 Nicolaou, K.C., Liu, J.J., Yang, Z., Ueno, H., Sorensen, E.J., Claiborne, C.F., Guy, R.K., Hwang, C.-K., Nakada, M., and Nantermet, P.G. (1995) *J. Am. Chem. Soc.*, **117**, 634–644.
644 (a) Nagata, W., Okada, K., and Aoki, T. (1979) *Synthesis*, 365–368; (b) Murphy, W.S., Tuladhar, S.M., and Duffy, B. (1992) *J. Chem. Soc., Perkin Trans. 1*, 605–609; (c) Pettigrew, J.D., Cadieux, J.A., So, S.S.S., and Wilson, P.D. (2005) *Org. Lett.*, **7**, 467–470; (d) Zheng, H. and Hall, D.G. (2010) *Tetrahedron Lett.*, **51**, 4256–4259.
645 (a) Molander, G.A., Bobbitt, K.L., and Murry, C.K. (1992) *J. Am. Chem. Soc.*, **114**, 2759–2760; (b) Molander, G.A. and Bobbitt, K.L. (1993) *J. Am. Chem. Soc.*, **115**, 7517–7518.
646 Sailes, H.E., Watts, J.P., and Whiting, A. (2000) *J. Chem. Soc., Perkin Trans. 1*, 3362–3374.
647 Charette, A.B., Juteau, H., Lebel, H., and Molinaro, C. (1998) *J. Am. Chem. Soc.*, **120**, 11943–11952.
648 (a) Ferrier, R.J., Prasad, D., and Rudowski, A. (1965) *J. Chem. Soc.*, 858–863; (b) Ferrier, R.J. (1978) *Adv. Carbohydr. Chem. Biochem.*, **35**, 31–80.
649 (a) Dahlhoff, W.V. and Köster, R. (1982) *Heterocycles*, **18**, 421–449; (b) Dahlhoff, W.V., Fenzl, W., and Köster, R. (1990) *Liebigs Ann. Chem.*, 807–810.
650 Dahlhoff, W.V. and Köster, R. (1980) *Synthesis*, **1980**, 936–937.
651 Langston, S., Bernet, B., and Vasella, A. (1994) *Helv. Chim. Acta*, **77**, 2341–2353.
652 (a) Oshima, K., Kitazono, E.-i., and Aoyama, Y. (1997) *Tetrahedron Lett.*, **38**, 5001–5004; (b) Oshima, K. and Aoyama, Y. (1999) *J. Am. Chem. Soc.*, **121**, 2315–2316; (c) Kaji, E., Nishino, T., Ishige, K., Ohya, Y., and Shirai, Y. (2010) *Tetrahedron Lett.*, **51**, 1570–1573.
653 Yurkevich, A.M., Kolodkina, I.I., Varshavskaya, L.S., Borodulina-Shvetz, V.I., Rudakova, I.P., and Preobrazhenski, N.A. (1969) *Tetrahedron*, **25**, 477–484.
654 McMurry, J.E. and Erion, M.D. (1985) *J. Am. Chem. Soc.*, **107**, 2712–2720.
655 Perun, T.J., Martin, J.R., and Egan, R.S. (1974) *J. Org. Chem.*, **39**, 1490–1493.
656 Liljebris, C., Nilsson, B.M., Resul, B., and Hacksell, U. (1996) *J. Org. Chem.*, **61**, 4028–4034.
657 Flores-Parra, A., Paredes-Tepox, C., Joseph-Nathan, P., and Contreras, R. (1990) *Tetrahedron*, **46**, 4137–4148.

658 Bertounesque, E., Florent, J.-C., and Monneret, C. (1991) *Synthesis*, 270–272.
659 Hungerford, N.L., McKinney, A.R., Stenhouse, A.M., and McLeod, M.D. (2006) *Org. Biomol. Chem.*, **4**, 3951–3959.
660 (a) Ishihara, K., Kuroki, Y., Hanaki, N., Ohara, S., and Yamamoto, H. (1996) *J. Am. Chem. Soc.*, **118**, 1569–1570; (b) Kuroki, Y., Ishihara, K., Hanaki, N., Ohara, S., and Yamamoto, H. (1998) *Bull. Chem. Soc. Jpn.*, **71**, 1221–1230.
661 (a) Evans, D.A. and Polniaszek, R.P. (1986) *Tetrahedron Lett.*, **27**, 5683–5686; (b) Evans, D.A., Polniaszek, R.P., DeVries, K.M., Guinn, D.E., and Mathre, D.J. (1991) *J. Am. Chem. Soc.*, **113**, 7613–7630.
662 Kaupp, G., Naimi-Jamal, M.R., and Stepanenko, V. (2003) *Chem. Eur. J.*, **9**, 4156–4160.
663 Iwasawa, N., Kato, T., and Narasaka, K. (1988) *Chem. Lett.*, 1721–1724.
664 (a) Gypser, A., Michel, D., Nirschl, D.S., and Sharpless, K.B. (1998) *J. Org. Chem.*, **63**, 7322–7327; (b) Hovelmann, C. H. and Muniz, K. (2005) *Chem. Eur. J.* **11**, 3951–3958.
665 Barker, S.A., Hatt, B.W., Somers, P.J., and Woodbury, R.R. (1973) *Carbohydr. Res.*, **26**, 55–64.
666 (a) Seymour, E. and Fréchet, J.M.J. (1976) *Tetrahedron Lett.*, **15**, 1149–1152; (b) Farrall, M.J. and Fréchet, J.M.J. (1976) *J. Org. Chem.*, **41**, 3877–3882.
667 Bullen, N.P., Hodge, P., and Thorpe, F.G. (1981) *J. Chem. Soc., Perkin 1*, 1863–1867.
668 (a) Weith, H.L., Wiebers, J.L., and Gilham, P.T. (1970) *Biochemistry*, **9**, 4396–4401; (b) Rosenberg, M., Wiebers, J.L., and Gilham, P.T. (1972) *Biochemistry*, **11**, 3623–3628.
669 (a) Schott, H. (1972) *Angew. Chem., Int. Ed. Engl.*, **11**, 824–825; (b) Schott, H., Rudloff, E., Schmidt, P., Roychoudhury, R., and Kössel, H. (1973) *Biochemistry*, **12**, 932–938.
670 Johnson, B.J.B. (1981) *Biochemistry*, **20**, 6103–6108.
671 (a) Zhang, L., Xu, Y., Yao, H., Xie, L., Yao, J., Lu, H., Yang, P., Dou, P., Liang, L., He, J., Liu, Z., and Chen, H.-Y. (2009) *J. Chromatogr. A*, **1216**, 7558–7563; (b) Frasconi, M., Tel-Vered, R., Riskin, M., and Willner, I. (2010) *Anal. Chem.*, **82**, 2512–2519.
672 (a) Li, X., Pennington, J., Stobaugh, J.F., and Schöneich, C. (2008) *Anal. Biochem.*, **372**, 227–236; (b) Xu, Y., Wu, Z., Zhang, L., Lu, H., Yang, P., Webley, P.A., and Zhao, D. (2009) *Anal. Chem.*, **81**, 503–508.
673 (a) Mazzeo, J.R. and Krull, I.S. (1989) *Biochromatography*, **4**, 124–130; (b) Singhal, R.P. and DeSilva, S.S.M. (1992) *Adv. Chromatogr.*, **31**, 293–335; (c) Liu, X.-C. and Scouten, W.H. (2005) Chapter 8, in *Handbook of Affinity Chromatography*, 2nd edn (ed. D.S. Hage), CRC Press, pp. 215–226;(d) Liu, X.-C. (2006) *Chin. J. Chromatogr.*, **24**, 73–80.
674 Nishiyabu, R., Kubo, Y., James, T.D., and Fossey, J.S. (2010) *Chem. Commun.*, 1106–1123.
675 (a) Mallia, A.K., Hermanson, G.T., Krohn, R.I., Fujimoto, E.K., and Smith, P.K. (1981) *Anal. Lett.*, **14** (B8), 649–661; (b) Middle, F.A., Bannister, A., Bellingham, A.J., and Dean, P.D.G. (1983) *Biochem. J.*, **209**, 771–779; (c) Zhang, Q., Schepmoes, A.A., Brock, J.W.C., Wu, S., Moore, R.J., Purvine, S.O., Baynes, J.W., Smith, R.D., and Metz, T.O. (2008) *Anal. Chem.*, **80**, 9822–9829.
676 Miyazaki, H., Kikuchi, A., Koyama, Y., Okano, T., Sakurai, Y., and Kataoka, K. (1993) *Biochem. Biophys. Res. Commun.*, **195**, 829–836.
677 Kataoka, K., Miyazaki, H., Okano, T., and Sakurai, Y. (1994) *Macromolecules*, **27**, 1061–1062.
678 Kikuchi, A., Suzuki, K., Okabayashi, O., Hoshino, H., Kataoka, K., Sakurai, Y., and Okano, T. (1996) *Anal. Chem.*, **68**, 823–828.
679 Gabai, R., Sallacan, N., Chegel, V., Bourenko, T., Katz, E., and Willner, I. (2001) *J. Phys. Chem. B*, **105**, 8196–8202.
680 Kanekiyo, Y., Sano, M., Iguchi, R., and Shinkai, S. (2000) *J. Pol. Sci. A*, **38**, 1302–1310.
681 Seymour, E. and Fréchet, J.M.J. (1976) *Tetrahedron Lett.*, **17**, 3669–3672.
682 Ren, L., Liu, Z., Liu, Y., Dou, P., and Chen, H.-Y. (2009) *Angew. Chem., Int. Ed.*, **48**, 6704–6707.

683 Glad, M., Norrlöw, O., Sellergren, B., Siegbahn, N., and Mosbach, K. (1985) *J. Chromatogr.*, **347**, 11–23.

684 Wulff, G. and Schauhoff, S. (1991) *J. Org. Chem.*, **56**, 395–400.

685 Fréchet, J.M.J., Nuyens, L.J., and Seymour, E. (1979) *J. Am. Chem. Soc.*, **101**, 432–436.

686 (a) Belogi, G., Zhu, T., and Boons, G.-J. (2000) *Tetrahedron Lett.*, **41**, 6965–6968; (b) Belogi, G., Zhu, T., and Boons, G.-J. (2000) *Tetrahedron Lett.*, **41**, 6969–6972.

687 Hsiao, H.-Y., Chen, M.-L., Wu, H.-T., Huang, L.D., Chien, W.-T., Yu, C.-C., Jan, F.-D., Sahabuddin, S., Chang, T.-C., and Lin, C.-C. (2010) *Chem. Commun.*, **47**, 1187–1189.

688 Chen, M.L., Adak, A.K., Yeh, N.C., Yang, W.B., Chuang, Y.J., Wong, C.H., Hwang, K.C., Hwu, J.R.R., Hsieh, S.L., and Lin, C.C. (2008) *Angew. Chem., Int. Ed.*, **47**, 8627–8630.

689 Polsky, R., Harper, J.C., Wheeler, D.R., Arango, D.C., and Brozik, S.M. (2008) *Angew. Chem., Int. Ed.*, **47**, 2631–2634.

690 (a) James, T.D., Sandanayake, K.R.A.S., and Shinkai, S. (1996) *Angew. Chem., Int. Ed.*, **35**, 1910–1922; (b) James, T.D. and Shinkai, S. (2002) *Topp. Curr. Chem.*, **218**, 159–200; (c) Wang, W., Gao, X., and Wang, B. (2002) *Curr. Org. Chem.*, **6**, 1285–1317; (d) Cao, H.S. and Heagy, M.D. (2004) *J. Fluoresc.*, **14**, 569–584; (e) Yan, J., Fang, H., and Wang, B.H. (2005) *Med. Res. Rev.*, **25**, 490–520; (f) James, T.D. (2007) *Top. Curr. Chem.*, **277**, 107–152; (g) Galbraith, E. and James, T.D. (2010) *Chem. Commun.*, **39**, 3831–3842; (h) Jin, S., Cheng, Y.F., Reid, S., Li, M.Y., and Wang, B.H. (2010) *Med. Res. Rev.*, **30**, 171–257.

691 Schiller, A., Wessling, R.A., and Singaram, B. (2007) *Angew. Chem., Int. Ed.*, **46**, 6457–6459.

692 Pal, A., Bérubé, M., and Hall, D.G. (2010) *Angew. Chem., Int. Ed.*, **49**, 1492–1495.

693 Li, M., Liu, N., Huang, Z., Du, L., Altier, C., Fang, H., and Wang, B. (2008) *J. Am. Chem. Soc.*, **130**, 12636–12638.

694 (a) Zhao, J., Fyles, T.M., and James, T.D. (2004) *Angew. Chem., Int. Ed.*, **43**, 3461–3464; (b) Levonis, S.M., Kiefel, M.J., and Houston, T.A. (2009) *Chem. Commun.*, 2278–2280; (c) Matsumoto, A., Sato, N., Kataoka, K., and Miyahara, Y. (2009) *J. Am. Chem. Soc.*, **131**, 12022–12023.

695 (a) Coskun, A. and Akkaya, E.U. (2004) *Org. Lett.*, **6**, 3107–3109; (b) Secor, K.E. and Glass, T.E. (2004) *Org. Lett.*, **6**, 3727–3730.

696 Maue, M. and Schrader, T. (2005) *Angew. Chem., Int. Ed.*, **44**, 2265–2270.

697 Hagihara, S., Tanaka, H., and Matile, S. (2008) *J. Am. Chem. Soc.*, **130**, 5656–5657.

698 (a) Gray, C.W., Jr., and Houston, T.A. (2002) *J. Org. Chem.*, **67**, 5426–5428; (b) Houston, T.A., Levonis, S.M., and Kiefel, M.J. (2007) *Aust. J. Chem.*, **60**, 811–815.

699 (a) Lavigne, J.J. and Anslyn, E.V. (1999) *Angew. Chem., Int. Ed.*, **38**, 3666–3669; (b) Nguyen, B.T., Wiskur, S.L., and Anslyn, E.V. (2004) *Org. Lett.*, **6**, 2499–2501.

700 (a) Groziak, M. (2001) *Am. J. Ther.*, **8**, 321–328; (b) Yang, W., Gao, X., and Wang, B. (2003) *Med. Res. Rev.*, **23**, 346–368;(c) Yang, W., Gao, X., and Wang, B.H. (2005) Chapter 13, in *Boronic Acids*, 1st edn (ed. D.G. Hall), Wiley-VCH Verlag GmbH, Weinheim.

701 (a) Vogels, C.M., Nikolcheva, L.G., Spinney, H.A., Norman, D.W., Baerlocher, M.O., Baerlocher, F.J., and Westcott, S.A. (2001) *Can. J. Chem.*, **79**, 1115–1123; (b) King, A.S., Nikolcheva, L.G., Graves, C.R., Kaminski, A., Vogels, C.M., Hudson, R.H.E., Ireland, R.J., Duffy, S.J., and Westcott, S.A. (2002) *Can. J. Chem.*, **80**, 1217–1222; (c) Irving, A.M., Vogels, C.M., Nikolcheva, L.G., Edwards, J.P., He, X.-F., Hamilton, M.G., Baerlocher, M.O., Baerlocher, F.J., Decken, A., and Westcott, S.A. (2003) *New J. Chem.*, **27**, 1419–1424; (d) Jabbour, A., Steinberg, D., Dembitsky, V.M., Moussaieff, A., Zaks, B., and Srebnik, M. (2004) *J. Med. Chem*, **47**, 2409–2410.

702 (a) Gronowitz, S., Dalgren, T., Namtvedt, J., Roos, C., Sjöberg, B., and Forsgren, U. (1971) *Acta Pharm. Suec.*, **8**,

377–390; (b) Högenauer, G. and Woisetschläger, M. (1981) *Nature*, **293**, 662–664.

703 Baldock, C., de Boer, G.-J., Rafferty, J.B., Stuitje, A.R., and Rice, D.W. (1998) *Biochem. Pharmacol.*, **55**, 1541–1549.

704 Baldock, C., Rafferty, J.B., Sedelnikova, S.E., Baker, P.J., Stuitje, A.R., Slabas, A.R., Hawkes, T.R., and Rice, D.W. (1996) *Science*, **274**, 2107–2110.

705 Robinson, P.D. and Groziak, M.P. (1999) *Acta. Crystallogr.*, **C55**, 1701–1704.

706 Feng, Z. and Hellberg, M. (2000) *Tetrahedron Lett.*, **41**, 5813–5814.

707 Tondi, D., Calo, S., Shoichet, B.K., and Costi, M.P. (2010) *Bioorg. Med. Chem. Lett.*, **20**, 3416–3419.

708 Suzuki, N., Suzuki, T., Ota, Y., Nakano, T., Kurihara, M., Okuda, H., Yamori, T., Tsumoto, H., Nakagawa, H., and Miyata, N. (2009) *J. Med. Chem.*, **52**, 2909–2922.

709 Nakamura, H., Kuroda, H., Saito, H., Suzuki, R., Yamori, T., Maruyama, K., and Haga, T. (2006) *ChemMedChem*, **1**, 729–740.

710 Inglis, S.R., Zervosen, A., Woon, E.C.Y., Gerards, T., Teller, N., Fischer, D.S., Luxen, A., and Schofield, C.J. (2009) *J. Med. Chem.*, **52**, 6097–6106.

711 Rock, F.L. et al. (2007) *Science*, **316**, 1759–1761.

712 (a) Akama, T., Baker, S.J., Zhang, Y.K., Hernandez, V., Zhou, H.C., Sanders, V., Freund, Y., Kimura, R., Maples, K.R., and Plattner, J.J. (2009) *Bioorg. Med. Chem. Lett.*, **19**, 2129–2132; (b) Ding, D.Z. et al. (2010) *ACS Med. Chem. Lett.*, **1**, 165–169.

713 Jay, J.I., Lai, B.E., Myszka, D.G., Mahalingam, A., Laugheinrich, K., Katz, D.F., and Kiser, P.F. (2010) *Mol. Pharm.*, **7**, 116–129.

714 (a) Akparov, V.K. and Stepanov, V.M. (1978) *J. Chromatogr.*, **155**, 329–336; (b) Cartwright, S. and Waley, S. (1984) *Biochem. J.*, **221**, 505–512.

715 Tsilikounas, E., Kettner, C.A., and Bachovkin, W.W. (1992) *Biochemistry*, **31**, 12839–12846.

716 Tsilikounas, E., Kettner, C.A., and Bachovkin, W.W. (1993) *Biochemistry*, **32**, 12651–12655.

717 Stoll, V.S., Eger, B.T., Hynes, R.C., Martichonok, V., Jones, J.B., and Pai, E.F. (1998) *Biochemistry*, **37**, 451–462.

718 Adams, J.A. (1998) *Bioorg. Med. Chem. Lett.*, **8**, 333–338.

719 Paramore, A. and Frantz, S. (2003) *Nat. Rev. Drug Discov.*, **2**, 611–612.

720 (a) Soloway, A.H., Tjarks, W., Barnum, B.A., Rong, F.-G., Barth, R.F., Codogni, I.M., and Wilson, J.G. (1998) *Chem. Rev.*, **98**, 1515–1562; (b) Barth, R.F., Coderre, J.A., Vicente, M.G.H., and Blue, T.E. (2005) *Clin. Cancer Res.*, **11**, 3987–4002; (c) Yamamoto, T., Nakai, K., and Matsumura, A. (2008) *Cancer Lett.*, **262**, 143–152; (d) Barth, R.F. (2009) *Appl. Radiat. Isot.*, **67**, S3–S6.

721 Kankaanranta, L., Seppala, T., Koivunoro, H., Saarilahti, K., Atula, T., Collan, J., Salli, E., Kortesniemi, M., Uusi-Simola, J., Makitle, A., Seppanen, M., Minn, H., Kotiluoto, P., Auterinen, I., Savolainen, S., Kouri, M., and Joensuu, H. (2007) *Int. J. Radiat. Oncol. Biol. Phys.*, **69**, 475–482.

722 Yanagie, H., Ogata, A., Sugiyama, H. et al. (2008) *Expert Opin. Drug Deliv.*, **5**, 427–443.

723 Shinbo, T., Nishimura, K., Yamaguchi, T., and Sugiura, M. (1986) *J. Chem. Soc., Chem. Commun.*, 349–351.

724 Mohler, L.K. and Czarnik, A.W. (1993) *J. Am. Chem. Soc.*, **115**, 2998–2999.

725 Reetz, M.T., Huff, J., Rudolph, J., Töllner, K., Deege, A., and Goddard, R. (1994) *J. Am. Chem. Soc.*, **116**, 11588–11589.

726 Paugam, M.-F., Bien, J.T., Smith, B.D., Chrisstoffels, L.A.J., de Jong, F., and Reinhoudt, D.N. (1996) *J. Am. Chem. Soc.*, **118**, 9820–9825.

727 Duggan, P.J. (2004) *Aust. J. Chem.*, **57**, 291–299.

728 Smith, B.D. and Gardiner, S.J. (1999) *Advances in Supramolecular Chemistry*, vol. 5 (ed. G.W. Gokel), JAI Press, pp. 157–202.

729 Duggan, P.J., Houston, T.A., Kiefel, M.J., Levonis, S.M., Smith, B.D., and Szydzik, M.L. (2008) *Tetrahedron*, **64**, 7122–7126.

730 Kashiwada, A., Tsuboi, M., and Matsuda, K. (2009) *Chem. Commun.*, 695–697.

731 Frantzen, F., Grimsrud, K., Heggli, D.-E., and Sundrehagen, E. (1995) *J. Chromatogr. B*, **670**, 37–45.

732 Wiley, J.P., Hughes, K.A., Kaiser, R.J., Kesicki, E.A., Lund, K.P., and Stolowitz, M.L. (2001) *Bioconjug. Chem.*, **12**, 240–250.

733 Halo, T.L., Appelbaum, J., Hobert, E.H., Balkin, D.M., and Schepartz, A. (2009) *J. Am. Chem. Soc.*, **131**, 438–439.

734 Chalker, J.M., Wood, C.S., and Davis, B.G. (2009) *J. Am. Chem. Soc.*, **131**, 16346–16347.

735 DeSantis, G., Paech, C., and Jones, J.B. (2000) *Bioorg. Med. Chem.*, **8**, 563–570.

736 Frantzen, F., Grimsrud, K., Heggli, D.-E., and Sundrehagen, E. (1997) *Clin. Chim. Acta*, **263**, 207–224.

737 Burnett, T.J., Peebles, H.C., and Hageman, J.H. (1980) *Biochem. Biophys. Res. Commun.*, **96**, 157–162.

738 Vandenburg, Y.R., Zhang, Z.-Y., Fishkind, D.J., and Smith, B.D. (2000) *Chem. Commun.*, 149–150.

739 (a) Yang, W., Gao, S., Gao, X., Karnati, V.V.R., Ni, W., Wang, B., Hooks, W.B., Carson, J., and Weston, B. (2002) *Bioorg. Med. Chem. Lett.*, **12**, 2175–2177; (b) Yang, W., Fan, H., Gao, X., Gao, S., Karnati, V.V.R., Ni, W., Hooks, W.B., Carson, J., Weston, B., and Wang, W. (2004) *Chem. Biol.*, **11**, 439–448.

740 Miller, E.W., Albers, A.E., Pralle, A., Isacoff, E.Y., and Chang, C.J. (2005) *J. Am. Chem. Soc.*, **127**, 16652–16659.

741 Dickinson, B.C., Huynh, C., and Chang, C.J. (2010) *J. Am. Chem. Soc.*, **132**, 5906–5915.

742 Gao, C., Lavey, B.J., Lo, C.-H.L., Datta, A., Wenworth, P., Jr., and Janda, K.D. (1998) *J. Am. Chem. Soc.*, **120**, 2212–2217.

743 (a) Brustad, E., Bushey, M.L., Lee, J.W., Groff, D., Liu, W., and Schultz, P.G. (2008) *Angew. Chem., Int. Ed.*, **47**, 8220–8223; (b) Liu, C.C., Mack, A.V., Brustad, E.M., Mills, J.H., Groff, D., Smider, V.V., and Schultz, P.G. (2009) *J. Am. Chem. Soc.*, **131**, 9616–9617.

744 Hoeg-Jensen, T., Havelund, S., Nielsen, P.K., and Markussen, J. (2005) *J. Am. Chem. Soc.*, **127**, 6158–6159.

745 Fujita, N., Shinkai, S., and James, T.D. (2008) *Chem. Asian J.*, **3**, 1076–1091.

746 Nishiyabu, R., Kubo, Y., James, T.D., and Fossey, J.S. (2010) *Chem. Commun.*, 1124–1150.

747 (a) Liu, W., Pink, M., and Lee, D. (2009) *J. Am. Chem. Soc.*, **131**, 8703–8707; (b) Liu, W., Huang, W., Pink, M., and Lee, D. (2010) *J. Am. Chem. Soc.*, **132**, 11844–11846.

748 Korich, A.L., Clarke, K.M., Wallace, D., and Iovine, P.M. (2009) *Macromolecules*, **42**, 5906–5908.

# 2
# Metal-Catalyzed Borylation of C−H and C−Halogen Bonds of Alkanes, Alkenes, and Arenes for the Synthesis of Boronic Esters

*Tatsuo Ishiyama and Norio Miyaura*

## 2.1
### Introduction

Since the first isolation of a boronic acid by Frankland in 1860, transmetalation between $B(OR)_3$ and RLi or RMgX has been a method of choice for the synthesis of organoboronic acids [1–5]. Although this traditional protocol is now very common for large-scale preparation of boronic acids and esters, catalyzed reactions are an attractive strategy for the synthesis of boronic acids without using air- and water-sensitive lithium and magnesium reagents. Thus, there have been extensive studies on metal-catalyzed addition reactions of pinacolborane (1, HBpin) [3, 5–7], bis (pinacolato)diboron (2, $B_2pin_2$) [3, 5, 8], 9-RS-9-BBN [9], $R_3SiB(OR)_2$ [3, 5, 10], $X_2B$-CN [11], and $R_3SnB(NR_2)_2$ [3, 5, 12] for monoborylation or diborylation of alkenes and alkynes. Coupling reaction of B−B or B−H compounds with aryl, vinyl, allyl, and benzyl halides or triflates using palladium catalysts is a new entry for achieving a direct borylation of organic electrophiles [3, 5, 7, 8f, 13]. Because of the availability of various electrophiles and mild reaction conditions, this protocol has allowed convenient access to organoboronic esters that have a variety of functional groups. An extension of this methodology to aliphatic or aromatic C−H borylation is also of significant value for the preparation of organoboron compounds from economical alkanes, alkenes, and arenes [4, 8f, 14]. Among all these extensive studies on metal-catalyzed borylation of organic compounds, this chapter focuses on coupling reactions of H−B or B−B compounds with C−H and C−halogen bonds. Metal-catalyzed addition reactions of these boron compounds to alkenes and alkynes have been reviewed elsewhere [5, 8].

The reaction between Pt(0) complexes and $B_2pin_2$ or $B_2cat_2$ produces a single crystal of *cis*-Pt(Bpin)$_2$(L)$_2$ (3) consisting of a distorted square planar coordination geometry for the Pt atom containing two *cis*-boryl and phosphine ligands [15–17], which allows insertion of alkenes and alkynes into the Pt−B bond (Scheme 2.1). This reaction is a key process for achieving catalyzed additions of B−B [3, 5, 8], B−Si [3, 5, 10], and B−Sn [3, 5, 12] compounds to alkenes and alkynes with Pd(0), Pt (0), or Rh(I) complexes. On the other hand, oxidative addition of HBcat or HBpin to

*Boronic Acids: Preparation and Applications in Organic Synthesis, Medicine and Materials*, Second Edition.
Edited by Dennis G. Hall.
© 2011 Wiley-VCH Verlag GmbH & Co. KGaA. Published 2011 by Wiley-VCH Verlag GmbH & Co. KGaA.

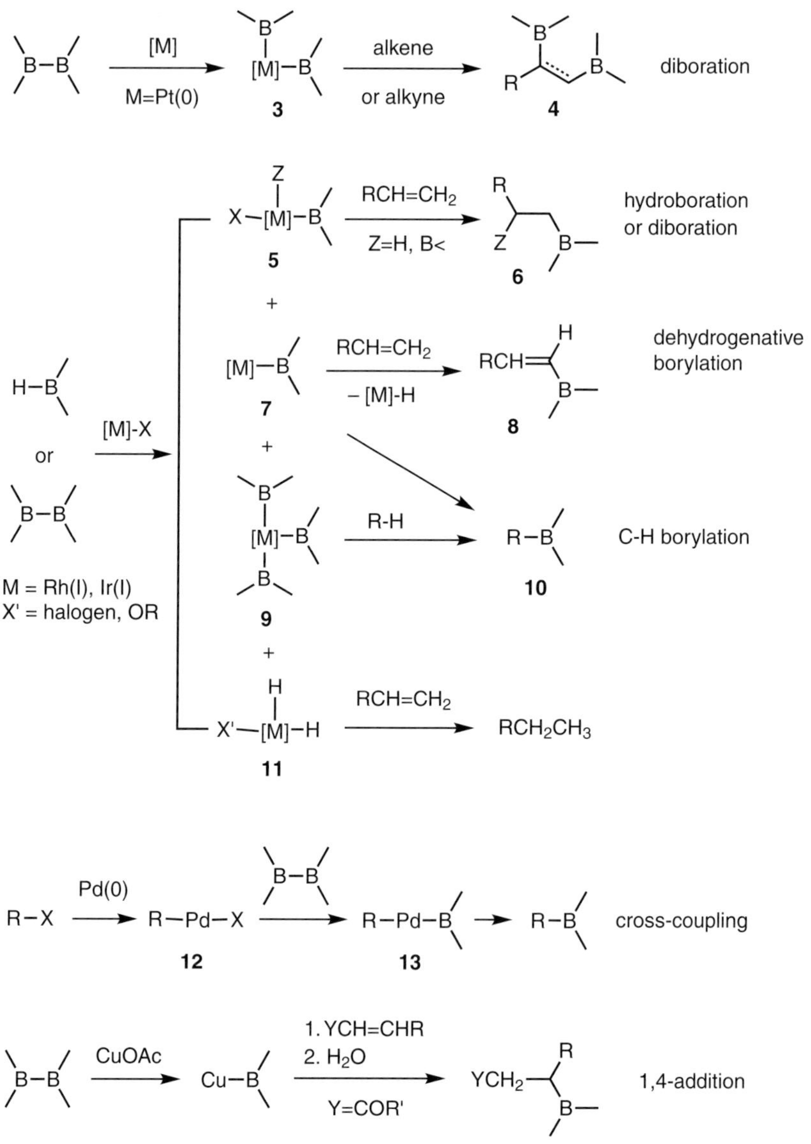

**Scheme 2.1** Borylation via coupling reactions of H–B and B–B compounds.

rhodium(I) or iridium(I) complexes ([M]-halogen) yields a metal hydride species (**5**, Z = H), which is an active species for catalyzed hydroboration of alkenes and alkynes [5–7, 18]. Further reaction of HBpin, HBcat, B$_2$cat$_2$, or B$_2$pin$_2$ provides a monoboryl complex (**7**, Z = H) [19], a bisboryl complex (**7**, Z = B<) [20], or a trisboryl complex (**9**) [19, 21, 22], which undergoes dehydrogenative borylation of alkenes (**8**) [23–25] or C–H borylation of alkanes, alkenes, and arenes (**10**) [4, 8f, 14]. Another convenient access to such boryl-transition metal complexes (**13**) is transmetalation between R-Pd-X (X = halogen, OR) and B—B compounds, which has been used for catalytic borylation of organic halides and triflates with $B_2pin_2$ [26, 27]. Transmetalation between $B_2pin_2$ and CuOAc provides a borylcopper(I) species (CuBpin), which undergoes 1,4-addition to enones and borylation of allylic acetates [28].

Thus, borylmetal species originating from transmetalation or oxidative addition of H—B or B—B compounds play a dominant role in catalytic C—B bond forming reactions. The synthesis, characterization, bonding, and reactivity of these catalytically important species have been reviewed [8]. There have also been extensive theoretical studies on the metal–boron bond and its role in catalytic cycles [29, 30].

## 2.2
## Borylation of Halides and Triflates via Coupling of H—B and B—B Compounds

In 1995, $B_2pin$ (**2**) was first introduced as a boron nucleophile for palladium-catalyzed borylation of organic halides and triflates [26, 27]. HBpin (**1**) is an economical alternative reported Murata *et al.* in 1997 for borylation of haloarenes [31]. These discoveries were followed by new design of H—B and B—B compounds allowing the synthesis of boronates sensitive to hydrolysis with water or boronates stable to silica gel required in chromatographic isolation, for example, **14** [32–34], **15** [35], **16** [36], **17** [37], **18** [38], **19** [39], and **20** [40] (Figure 2.1).

The high air and water stability of pinacol esters are convenient for isolation and handling; however, they are very resistant to hydrolysis under both acidic and basic conditions. Pinacol esters obtained by coupling can be converted into boronic acids when transesterification with diethanolamine is followed by hydrolysis with dilute

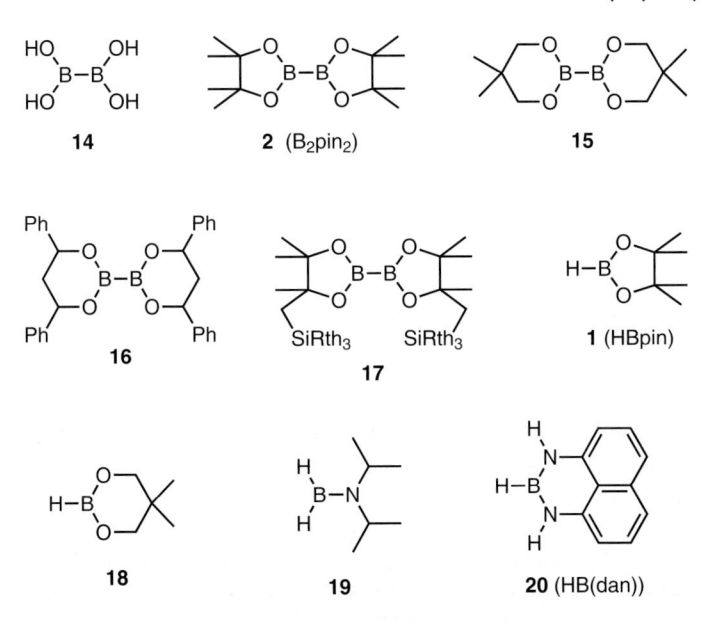

**Figure 2.1** H—B and B—B reagents for metal-catalyzed borylation.

aqueous acid or oxidative cleavage of the liberated diol with sodium periodate (A and B in Equation 2.1) [41]. Pinacol esters are also deprotected by destruction of the diol with boron tribromide or by conversion to the corresponding trifluoroborate with potassium hydrogen fluoride and subsequent hydroxylation with trimethylsilyl chloride and water (C and D in Equation 2.1) [42]. On the other hand, boronic esters obtained from **15, 18,** and **20** can be easily converted to boronic acids via hydrolysis with water (Equation 2.2). Benzyl diol ester (**16**) has been introduced for use in deprotection by catalyzed hydrogenolysis [36]. Fluorous bis(pinacolato) diboron (**17**, Rth = $n$-$C_6F_{13}CH_2CH_2$) was designed for carrying out borylation in $PhCF_3$ [37]. Diboronic acid (**14**), which achieves excellent results for allylic substrates, is an ideal compound for this purpose, but the scope for other substrates remains to be explored [32–34].

$$R-B(O)_2\text{pin} \xrightarrow[\substack{A: (1) HN(CH_2CH_2OH)_2, (2) HCl \\ B: NH_4OAc, NaIO_4 \\ C: (1) BBr_3, (2) MeOH \\ D: (1) KHF_2, (2) TMS\text{-}Cl, H_2O}]{} RB(OH)_2 \quad (2.1)$$

$$R-B(O)_2\text{(benzyl diol)} \xrightarrow[\text{2. } CF_3CO_2H]{\text{1. KOH}} RB(OH)_2 \quad (2.2)$$

### 2.2.1
### Borylation of Aryl Halides and Triflates

Representative conditions reported for the borylation of aryl halides or triflates with B–B compounds (**2, 15**) are summarized in Table 2.1. The phenyl groups of $PPh_3$ often participate in the coupling when palladium–triphenylphosphine complexes are employed for electron-rich halides. Thus, the use of $PdCl_2(dppf)$ (entry 1) [26] or Pd(OAc)$_2$ (entry 2) [43] is recommended for selective coupling. Highly donating $PCy_3$ [44], dicyclohexyl(biaryl)phosphines (**22** and **23**, entry 3) [45], or N-heterocyclic carbenes (NHCs) [46] are recommended for chloroarenes and electron-rich aryl bromides or triflates (entry 4). The reaction can be further accelerated by irradiation with microwaves [47]. Diborons do not transmetalate to Pd-halogen bonds under neutral conditions, but Ar-Pd-X (X = halogen, OTf) intermediate generated by oxidative addition undergoes smooth transmetalation with $B_2pin_2$ in the presence of a base. KOAc is recognized to be the best base for representative halides and triflates. The reaction is much faster when $K_2CO_3$ and $K_3PO_4$ are used, but these bases will increase the competitive formation of homocoupling products of haloarenes. This homocoupling reaction using 0.5 equiv of $B_2pin_2$ provides a method for the synthesis of symmetric biaryls and π-conjugated poly(phenylene)s from haloarenes [48]. On the other hand, such bases are not required for diazonium salts, which generate a cationic palladium(II) species (ArPd$^+$) via oxidative addition to palladium (0) complex (entries 5 and 6) [49, 50].

## 2.2 Borylation of Halides and Triflates via Coupling of H—B and B—B Compounds | 139

**Table 2.1** Borylation of aryl halides and triflates with diborons.

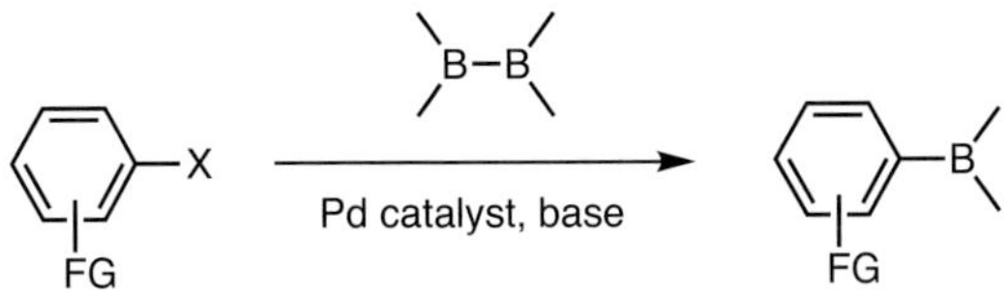

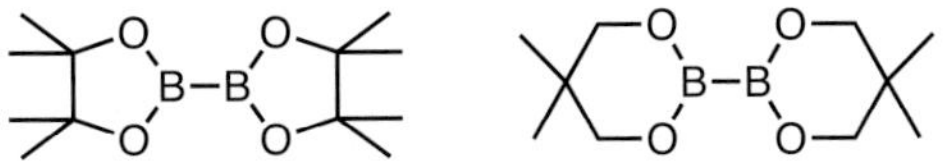

| Entry | Substrate | Catalyst, base, solvent, temperature |
|---|---|---|
| 1 | X = I, Br | PdCl$_2$(dppf), KOAc, DMSO or DMF, 80 °C |
| 2 | X = I, Br | Pd(OAc)$_2$, KOAc, DMF, 80 °C |
| 3 | X = Cl | Pd(dba)$_2$-2PCy$_3$, **22** or **23**, KOAc, dioxane, 80 °C |
| 4 | X = Cl | Pd(OAc)$_2$-2NHC, KOAc, THF, reflux |
| 5 | X = N$_2$BF$_4$ | PdCl$_2$(dppf), MeOH, rt |
| 6 | X = N$_2$BF$_4$ | Pd(OAc)$_2$-2NHC, THF, rt |

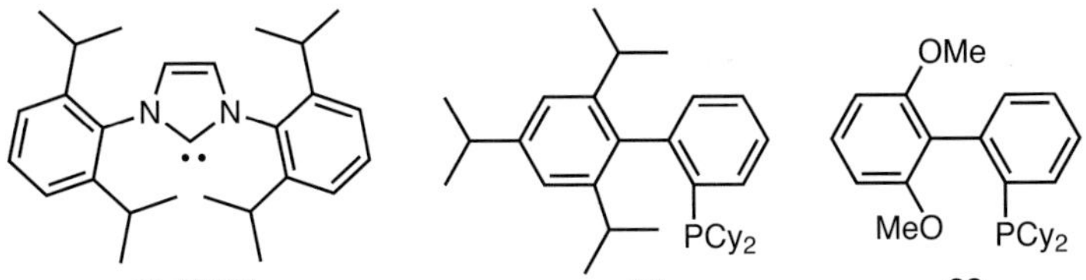

**21 (NHC)**   **22**   **23**

The reaction offers a convenient access to functionalized arylboronic esters in high yields (Figure 2.2). The reaction works well for aryl halides and triflates possessing nitrogen atoms, even for substrates having strongly coordinating pyridyl rings (**29** [51], **30** [52], **33** [53]). The reaction is also suitable for haloarenes possessing aldehyde, ketone, ester, and amide carbonyl groups (**24** [54], **25** [55], **26** [56], **27** [57], **28** [58]), bromoarenes having an iron carbonyl group (**32** [59]), and haloarenes supported on polymer resins such as **34** [60]. Precursors for the synthesis of π-conjugated poly(phenylene)s such as **31** were synthesized by double borylation of dihalides [61]. The reaction has been applied extensively to cross-coupling reactions of arylboronates in the synthesis of natural products and drug candidates [62], functional molecules and materials [63], and oligoarenes and polyarenes [64].

HBpin is an economical alternative for borylation of aryl halides or triflates (Table 2.2) [31]. It is very interesting that reducible functional groups such as ketone and ester carbonyl groups in haloarenes remain completely intact at 80 °C. However, the reaction is generally prone to produce dehalogenation by-products. This side reaction giving ArH increases in the case of electron-deficient haloarenes having

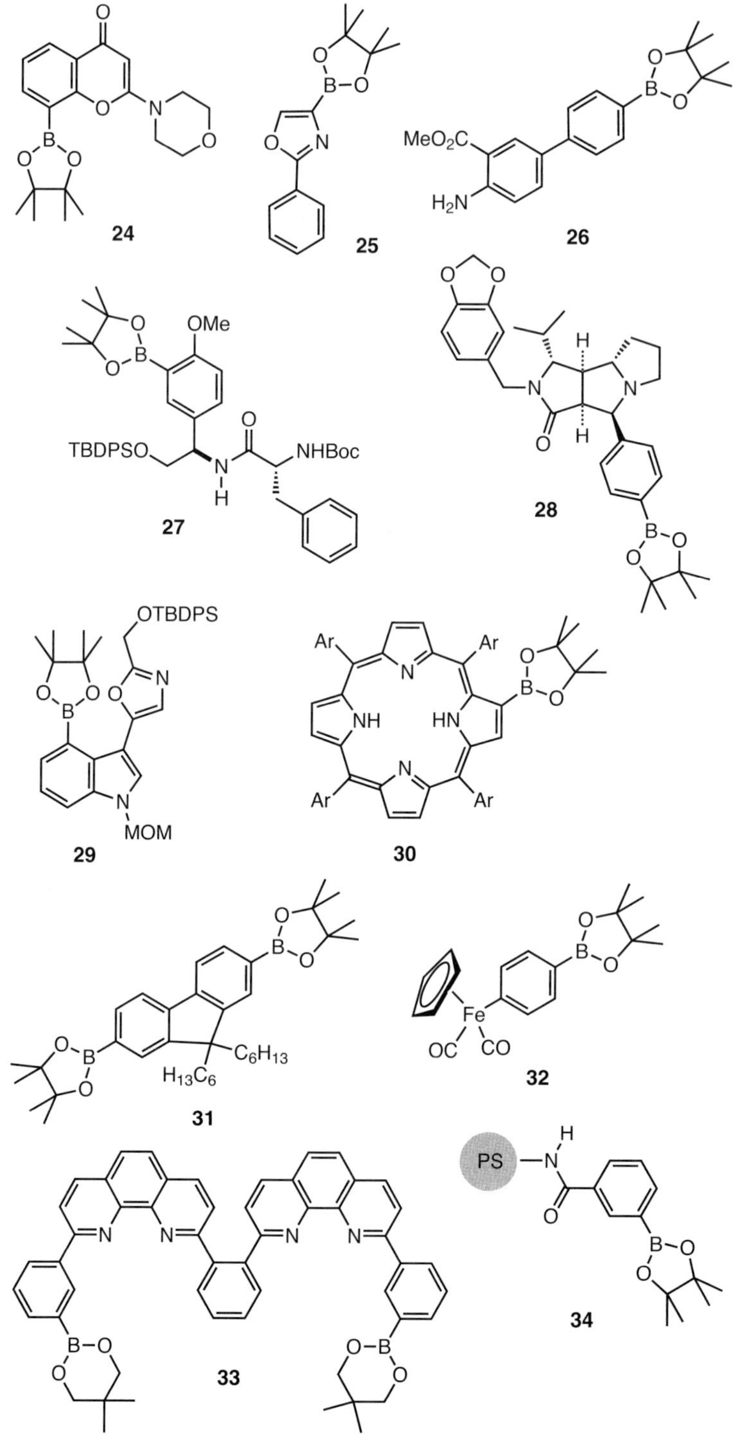

**Figure 2.2** Selected arylboronates synthesized using $B_2pin_2$.

## 2.2 Borylation of Halides and Triflates via Coupling of H—B and B—B Compounds

**Table 2.2** Borylation of aryl halides and triflates by B—H compounds.

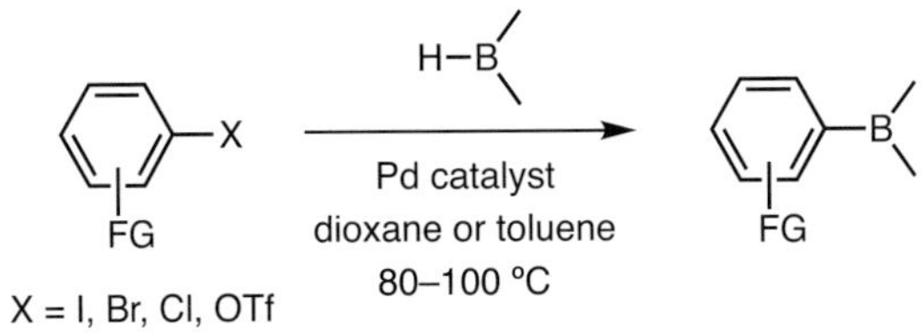

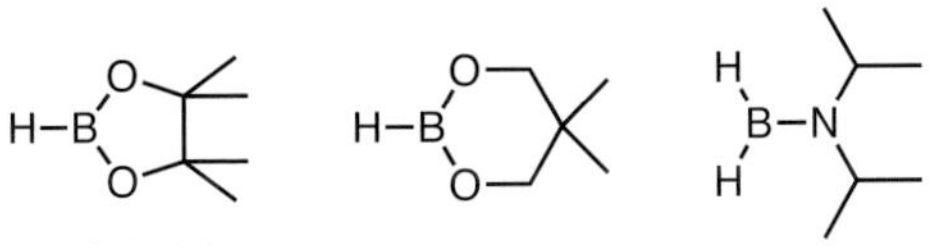

Reaction scheme: Ar-X (X = I, Br, Cl, OTf) with H-B, Pd catalyst, dioxane or toluene, 80–100 °C, giving Ar-B with FG.

Boranes: **1** (HBpin), **18** (H-B with 1,3-dioxo ring), **19** (H-B-N(iPr)₂ with two H on B)

| Entry | Borane | Substrate | Catalyst, solvent, base, temperature |
|---|---|---|---|
| 1 | HBpin | X = I, Br, OTf | PdCl$_2$(dppf), dioxane or toluene, Et$_3$N, 80–100 °C |
| 2 | HBpin | X = Br, Cl | Pd(dba)$_2$-D-t-BPF or t-Bu-DPEphos, dioxane, Et$_3$N, 80–120 °C |
| 3 | HBpin | X = I, Br, Cl | Pd(OAc)$_2$-23 or 35, dioxane, Et$_3$N, 80 °C |
| 4 | HBpin | X = Br | Pd(P(t-Bu)$_3$)$_2$, THF, Et$_3$N, 40 °C |
| 5 | HBpin | X = I, Br | PdCl$_2$(36)$_2$, Et$_3$N, dioxane, reflux |
| 6 | HBpin | X = I | CuI, THF, NaH, rt |
| 7 | HBpin or 18 | X = I, Br | NiCl$_2$(dppp), dppp, toluene, Et$_3$N, 100 °C |
| 8 | 19 | X = I, Br | PdCl$_2$(PPh$_3$)$_2$, Et$_3$N, dioxane, 80 °C |

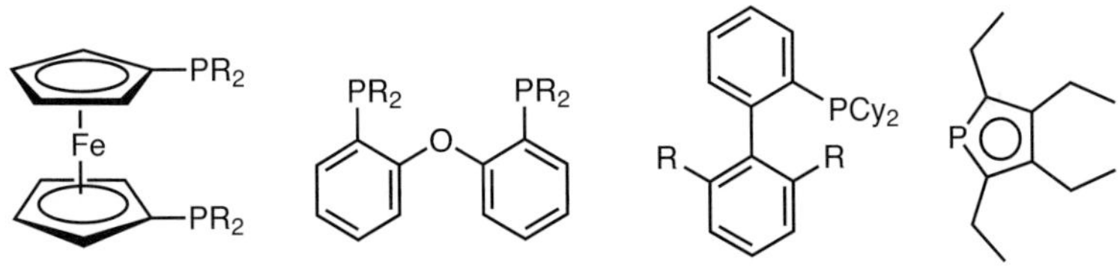

Ligands:
dppf (R = Ph)
D-t-BPF (R = t-Bu)
DPEphos (R = Ph)
t-Bu-DPEphos (R = t-Bu)
**35** (R = H)
**23** (R = OMe)
**36**

electron-withdrawing groups and decreases in the case of electron-rich halides. The presence of Et$_3$N plays a key role in not only preventing the production of ArH but also facilitating the B—C bond formation. The mechanism is not known, but displacement of Pd(II)-X with a weakly nucleophilic boryl anion (Et$_3$NH$^+$ Bpin$^-$) or σ-bond metathesis between H-Pd-Bpin and ArX has been proposed for the process leading to an Ar-Pd-Bpin intermediate [31]. PdCl$_2$(dppf) was originally recommended as a catalyst (entry 1), but recent studies have shown that bulky and donating phosphines such as D-t-BPF and t-Bu-DPEphos (entry 2) [65], **23** and **35** (entry 3) [66, 67], t-Bu$_3$P (entry 4) [68], and **36** (entry 5) [69] achieve high yields and high selectivities even for electron-deficient haloarenes. Nickel catalysts have advantages not only in low costs but also in high efficiency for electron-rich haloarenes or

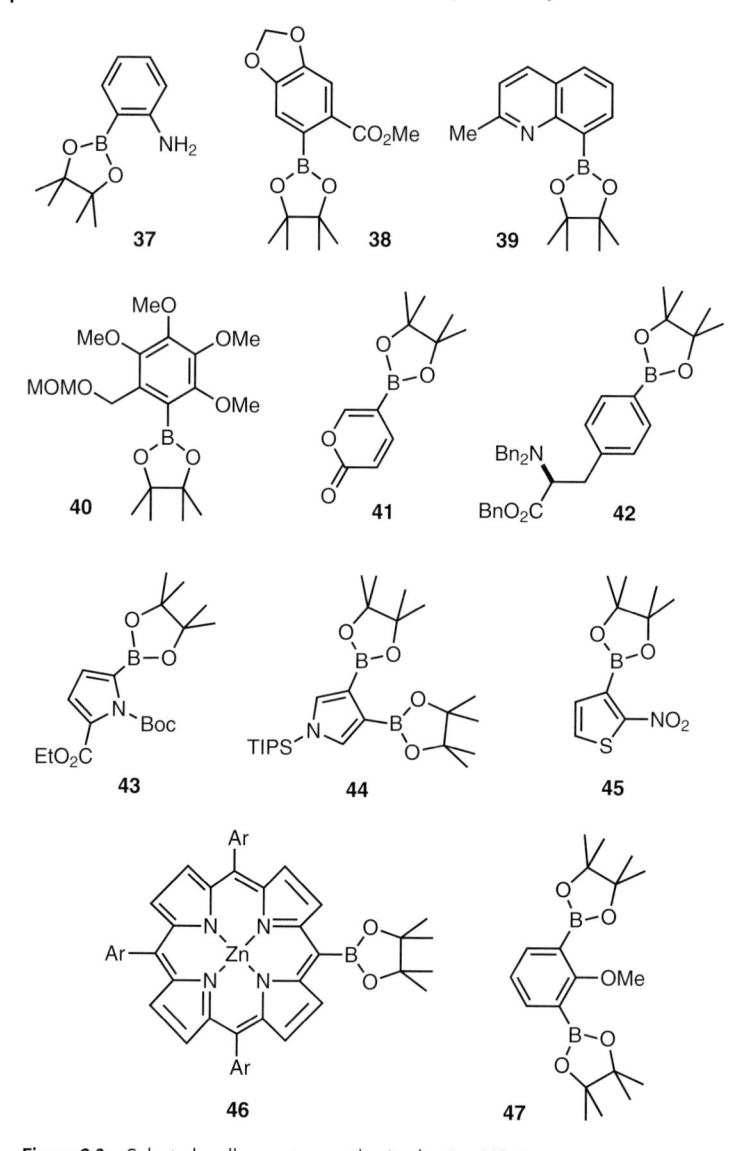

**Figure 2.3** Selected arylboronates synthesized using HBpin.

chloroarenes (entry 7) [38]. CuI is a new entry recommended for borylation of iodoarenes at room temperature (entry 6) [70]. (Diisopropylamino)borane (**19**) can be the best reagent for the synthesis of arylboronic acids via hydrolysis with water (entry 8) [39]. Only one hydrogen atom of **19** participates in coupling with haloarenes.

Representative arylboronates synthesized using HBpin are shown in Figure 2.3. The scope and limitation are similar to those of $B_2pin_2$ shown in Figure 2.2, although the reaction results in good yields for electron-rich halides rather than for electron-deficient halides. The reaction tolerates carbonyl groups including ketones, esters, and amides such as **38** [71], **41** [72], **42** [70–73], nitrogen atoms in aryl amines (**37**) [74], pyridine rings (**39**) [31], pyrrole rings (**43** [75], and **44** [76]), porphyrin rings (**46**) [77], and nitro groups (**45**) [68]. The reaction also works well for coupling at sterically hindered *ortho*-positions of haloarenes (**40** [78] and **47** [79]). Other selected examples are arylboronates used in the synthesis of bioactive compounds [80] and functional molecules or materials including oligo- and polyarenes [81]. In most cases, borylation has been directly followed by cross-coupling for the preparation of biaryls or polyaryls.

## 2.2.2
### Alkenyl Halides and Triflates

The borylation of 1-alkenyl halides and triflates with $B_2pin_2$ proceeds with complete retention of stereochemistry and is compatible with various functional groups [82, 83]. The reaction requires a much stronger base than that of haloarenes to avoid the formation of by-products originating from Heck coupling or homocoupling of haloalkenes. Powdered PhOK suspended in dioxane or toluene is recommended for unactivated/unconjugated bromides or triflates (Table 2.3, entry 1) [82, 83]. On the other hand, $K_2CO_3$ suspended in dioxane is recommended for base-sensitive electron-deficient substrates possessing a carbonyl group at $R^1$ or $R^2$ (entry 2) [84]. The conditions for borylation of haloarenes also work for iodoarenes (entry 3) [85]. Although HBpin suffers from low yields and positional and stereochemical isomerization of double bonds, it results in high yields for cycloalkenyl halides and triflates (entry 4) [86]. Pinacol alkenylboronates thus obtained can be isolated by chromatography on silica gel or they can be directly used for cross-coupling reactions without isolation of the boron intermediates.

**Table 2.3** Borylation of alkenyl halides and triflates.

X = I, Br, Cl, OTf

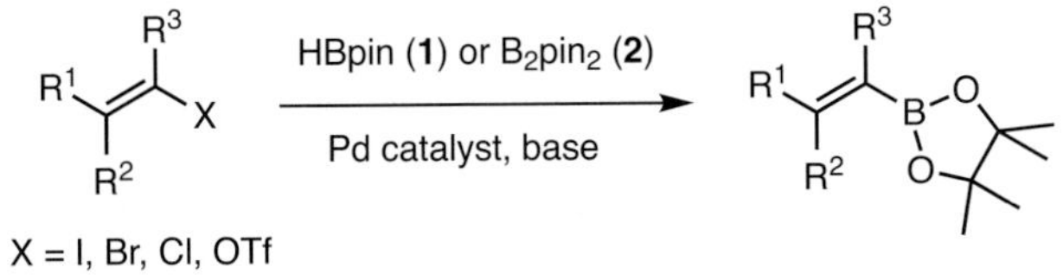

| Entry | Reagent | Catalyst | Base, solvent, temperature |
|---|---|---|---|
| 1 | $B_2pin_2$ | $PdCl_2(PPh_3)_2$/2PPh$_3$ | PhOK, toluene or dioxane, 50–80 °C |
| 2 | $B_2pin_2$ | $PdCl_2(PPh_3)_2$/2PPh$_3$ | $K_2CO_3$, dioxane, 80 °C |
| 3 | $B_2pin_2$ | $PdCl_2(dppf)$ | AcOK, DMSO, 80 °C |
| 4 | HBpin | $PdCl_2(dppf)$/4AsPh$_3$ | Et$_3$N, dioxane, 80 °C |

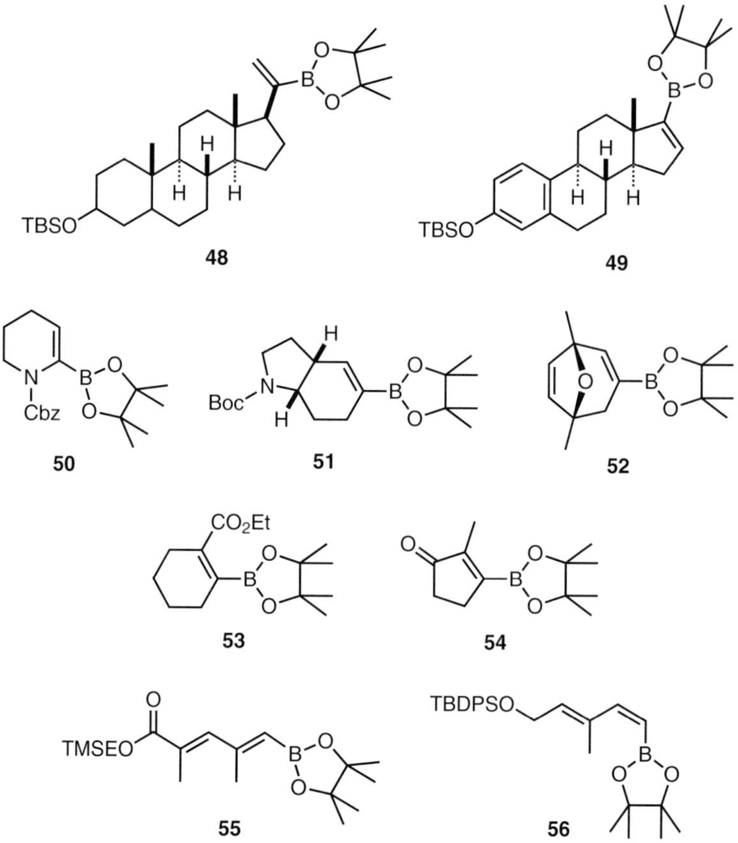

**Figure 2.4** Selected 1-alkenylboronates.

Syntheses of 1-alkenylboronates via coupling with $B_2pin_2$ or HBpin are shown in Figure 2.4. Boronates that are not available by conventional hydroboration are selected. Vinyl triflates and nonaflates afford 1-alkenylboronates such as **48** [82b], **49** [82b], **51** [87], and **52** [88] in high yields under conditions shown in entry 1 in Table 2.3. For the preparation of **50** [89], **53** [84], and **54** [84], $K_2CO_3$ was used as the basis for borylation of triflates with $B_2pin_2$ (entry 2, Table 2.3). 1,3-Dienyl iodides are borylated with retention of stereochemistry under conditions analogous to those of haloarenes using $PdCl_2(dppf)$ and KOAc in DMSO (**55** [85a], **56** [85b]).

Allylboron compounds can be selectively obtained by the coupling of alkenyl triflates. When a mixture of HBpin, alkenyl triflates, a $Pd(OAc)_2$-TANIAPHOS catalyst, and $PhNMe_2$ in dioxane was stirred at 25 °C and then aldehydes were added, the corresponding homoallyl alcohols were obtained via allylboration in good yields with high enantioselectivities (Equation 2.3) [86b].

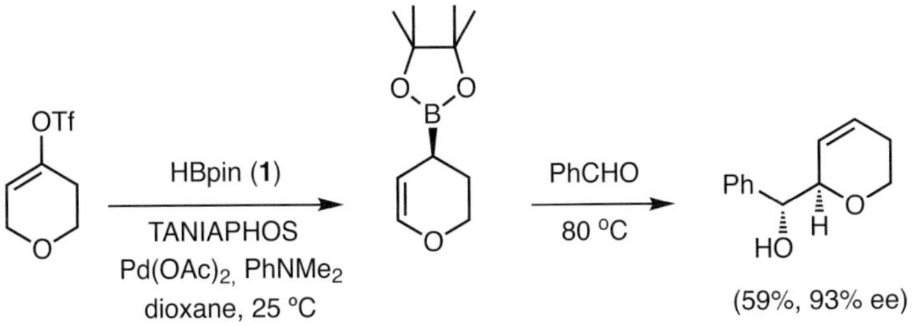

$$\text{(2.3)}$$

[Scheme: OTf-dihydropyran + HBpin (1), TANIAPHOS, Pd(OAc)$_2$, PhNMe$_2$, dioxane, 25 °C → Bpin-dihydropyran + PhCHO, 80 °C → Ph-CH(OH)-dihydropyran product (59%, 93% ee)]

## 2.2.3
### Allylic Halides, Allylic Acetates, and Allylic Alcohols

Palladium(0)-catalyzed coupling reaction of B$_2$pin$_2$ with allylic acetates [90], allylic chlorides (Ishiyama, T., Momota, S., and Miyaura, N., unpublished results), or alcohols [91] provides allylboronates (**58**, **61**) (Table 2.4). Bases are not necessary for acetates and alcohols, but allylic chlorides react in the presence of KOAc. The order of reactivity for oxidative addition to palladium(0) complexes is Cl > OAc ≫ OH. Thus, palladium-catalyzed reactions of allylic alcohols are generally very slow even at a high temperature; however, it is interesting that palladacycle (**59**) smoothly catalyzes the reaction at 50 °C in the presence of an acid cocatalyst (TsOH) [91]. The coupling occurs at the less hindered terminal carbon to produce thermally stable (*E*)-allylboronates for both **57** and **60**, thus suggesting B–C bond formation through a *syn*-π-allylpalladium intermediate generated by oxidative addition.

Diboronic acid (**14**, [B(OH)$_2$]$_2$) was first introduced for use in borylation of allylic alcohols (Table 2.5) [32]. The products are isolated as air- and water-stable crystalline potassium trifluoroborates (**64**) via treatment with KHF$_2$. The regioselectivity and stereoselectivity giving terminal (*E*)-allylboronates are the same as those shown in Table 2.4. The borylation occurs with complete inversion of configuration for both **66** and **68** presumably via oxidative addition to palladium(0) complex with inversion and B–C bond formation with retention of the configuration.

Borylation of vinylcyclopropanes possessing two electron-withdrawing groups such as **70** with B$_2$(OH)$_4$ gives ring-opened coupling products (**71**) (Equation 2.4) [33, 34]. The boron atom selectively couples at the terminal carbon.

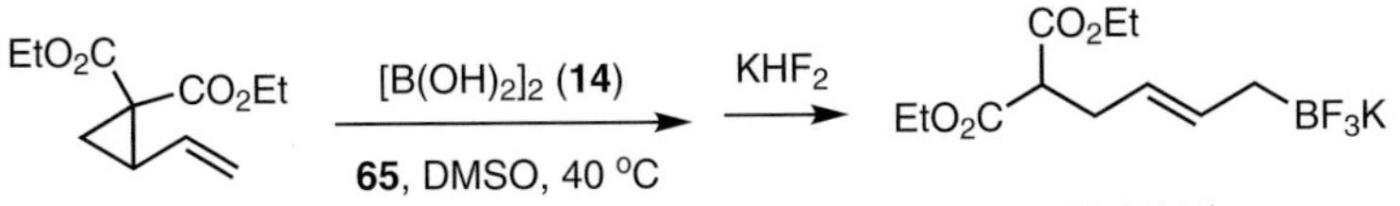

$$\text{(2.4)}$$

[Scheme: vinylcyclopropane **70** (EtO$_2$C, CO$_2$Et substituents) + [B(OH)$_2$]$_2$ (**14**), KHF$_2$, **65**, DMSO, 40 °C → **71** (82%): EtO$_2$C-C(CO$_2$Et)=CH-CH$_2$-CH$_2$-BF$_3$K]

A variety of 5-5, 6-5, and 7-5 *cis*-fused exomethylene cyclopentanols such as **74** are synthesized from β-ketoesters or diketones via a tandem cross-coupling/intramolecular allylboration reaction (Equation 2.5) [92].

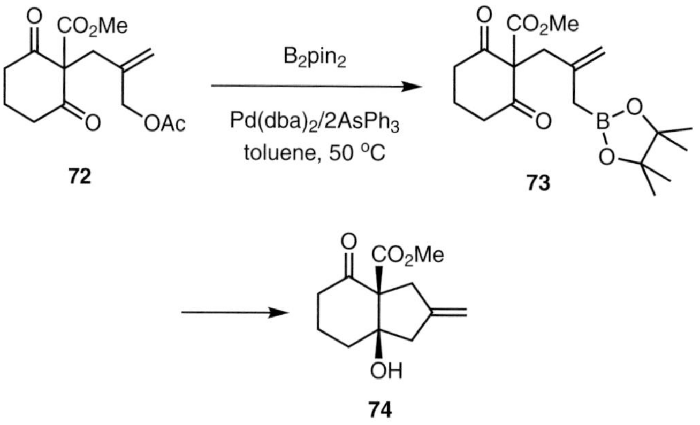

$$\text{72} \xrightarrow[\text{toluene, 50 °C}]{\text{B}_2\text{pin}_2 \atop \text{Pd(dba)}_2/2\text{AsPh}_3} \text{73} \quad (2.5)$$

$$\longrightarrow \text{74}$$

**Table 2.4** Borylation of allylic chlorides, acetates, and alcohols.

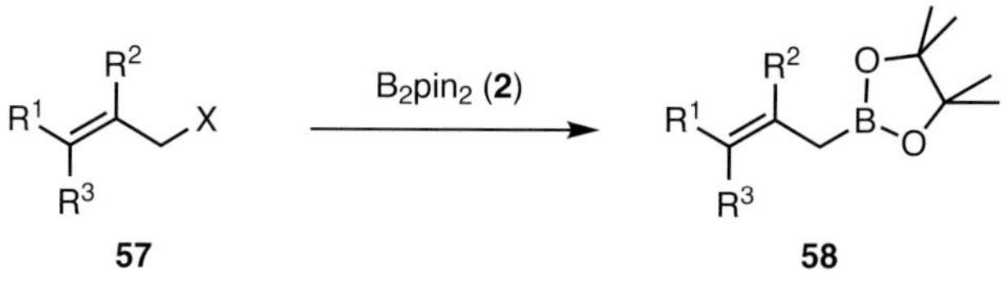

57 → 58

X = Cl    Pd(dba)$_2$-2AsPh$_3$, AcOK, toluene, 50 °C
X = OAc   Pd(dba)$_2$, DMSO, 50 °C
X = OH    59, TsOH, DMSO-MeOH, 50 °C

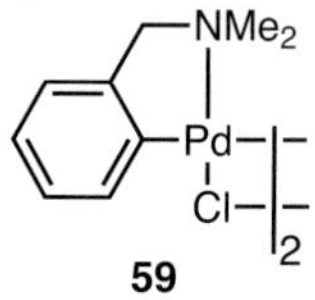

| Entry | R$^1$ | R$^2$ | R$^3$ | X   | Time (h) | Yield (%) |
|-------|-------|-------|-------|-----|----------|-----------|
| 1     | Ph    | H     | H     | Cl  | 10       | 70        |
| 2     | Ph    | H     | H     | OAc | 26       | 73        |
| 3     | Ph    | H     | H     | OH  | 16       | 81–96     |
| 4     | H     | Ph    | H     | Cl  | 5        | 71        |
| 5     | H     | Ph    | H     | OAc | 16       | 89        |
| 6     | Me    | H     | Me    | Cl  | 5        | 78        |

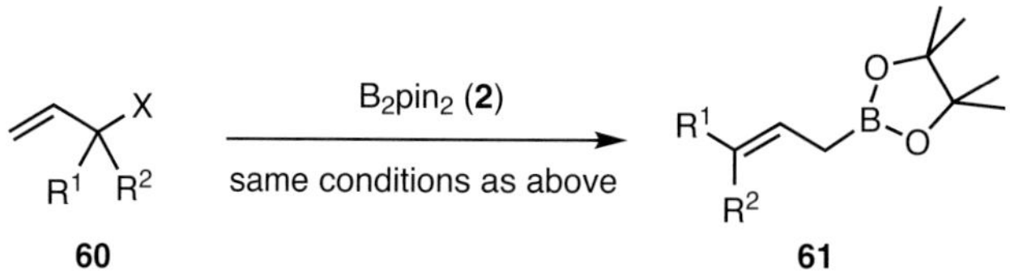

60 → 61, B$_2$pin$_2$ (2), same conditions as above

| Entry | R$^1$ | R$^2$ | X   | Time (h) | Yield (%) |
|-------|-------|-------|-----|----------|-----------|
| 7     | Ph    | H     | Cl  | 10       | 64        |
| 8     | Ph    | H     | OAc | 16       | 83        |
| 9     | Ph    | H     | OH  | 16       | 76–96     |
| 10    | Me    | Me    | OAc | 16       | 83        |
| 11    | Me    | Me    | OH  | 16       | 81        |

## 2.2 Borylation of Halides and Triflates via Coupling of H—B and B—B Compounds

**Table 2.5** Borylation of allylic alcohols with $B_2(OH)_4$.

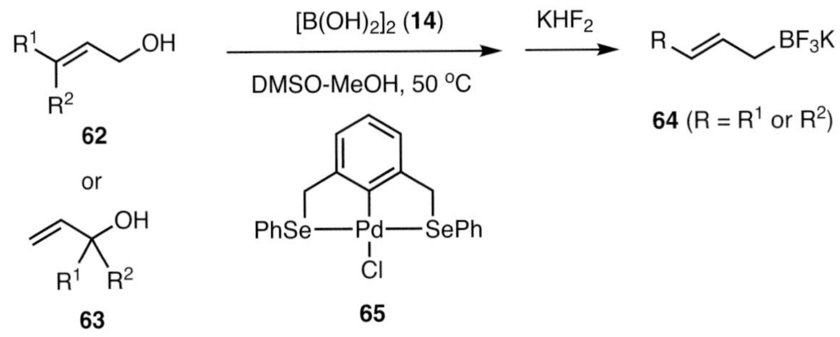

| Entry | 62 or 63 | $R^1$ | $R^2$ | Yield (%) |
|---|---|---|---|---|
| 1 | 62 | Ph | H | 92 |
| 2 | 62 | $C_3H_7$ | H | 94 |
| 3 | 62 | H | $CH_2OBn$ | 87 |
| 4 | 62 | H | $CH_2OH$ | 74 |
| 5 | 63 | Ph | H | 86 |
| 6 | 63 | Me | Me | 98 |

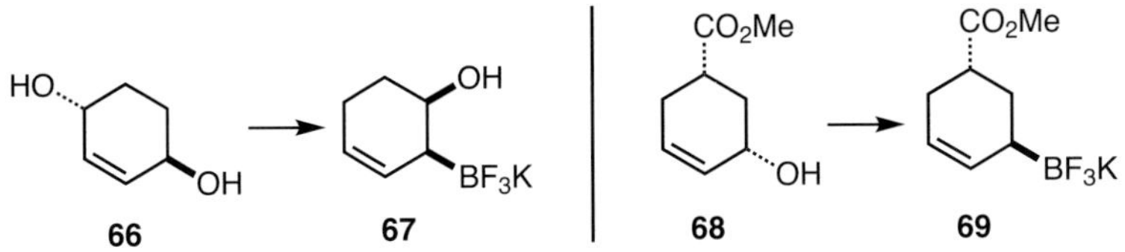

A method for preparation of a borylcopper species (**75**) from diboron and CuOAc was developed [28]. This reagent couples with allylic acetates to give allylboronates such as **76** (Equation 2.6) [93, 94]. The reaction can be carried out catalytically when carbonates (**77**) are used as substrates (Equation 2.7) [95]. Displacement occurs with inversion of the configuration and allylic rearrangement.

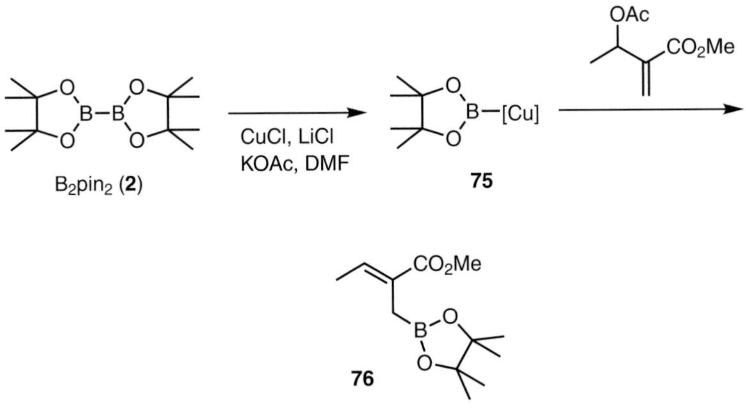

(2.6)

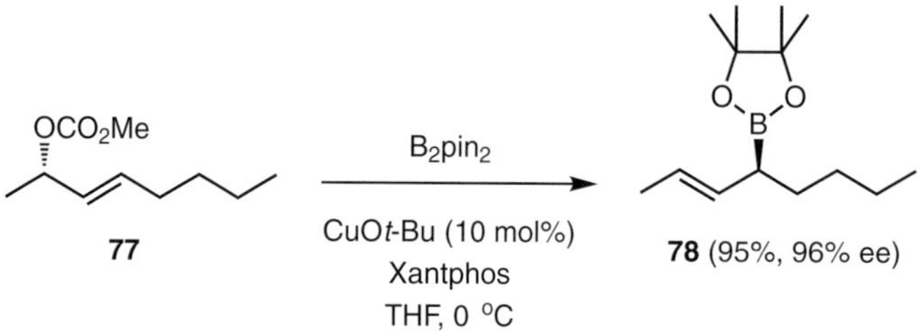

$$\underset{77}{\text{OCO}_2\text{Me}\text{-CH=CH-CH}_2\text{CH}_2\text{CH}_3} \xrightarrow[\substack{\text{CuO}t\text{-Bu (10 mol\%)} \\ \text{Xantphos} \\ \text{THF, 0 °C}}]{\text{B}_2\text{pin}_2} \underset{78\ (95\%,\ 96\%\ ee)}{\text{Bpin-CH-CH=CH-CH}_2\text{CH}_2\text{CH}_3} \quad (2.7)$$

### 2.2.4
### Benzylic Halides

Benzyl chlorides and bromides are borylated with $B_2pin_2$, a palladium catalyst and a base (Equation 2.8) [96, 97].

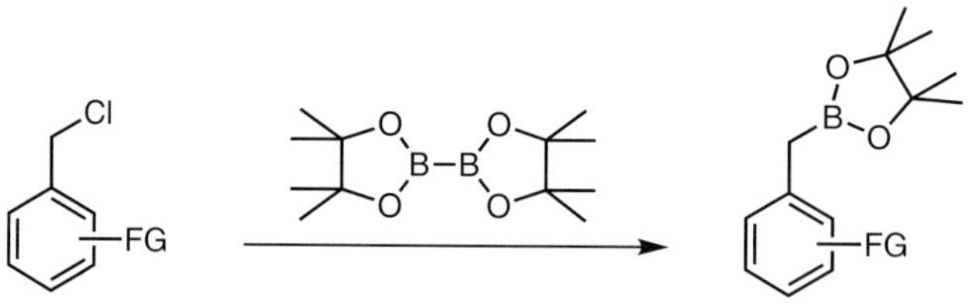

$$\underset{79}{\text{ArCH}_2\text{Cl-FG}} \xrightarrow{B_2pin_2} \underset{80}{\text{ArCH}_2\text{Bpin-FG}} \quad (2.8)$$

A: $Pd(dba)_2$-$2P(4\text{-MeOC}_6H_4)_3$, KOAc, toluene, rt
B: $PdCl_2(dppf)$, $Cs_2CO_3$, DMF, 80 °C

## 2.3
## Borylation via C—H Activation

### 2.3.1
### Aliphatic C—H Bonds

Because of the availability of various hydrocarbons, direct borylation of alkane, alkene, and arene C—H bonds via C—H activation is of significant value for the preparation of organoboron compounds from economical hydrocarbons [4, 8f, 14]. Some mechanistic key steps in putative catalytic cycles have been established by Hartwig and coworkers by stoichiometric borylation of alkanes and arenes with (boryl)metal complexes [98]. These discoveries were followed by the development of catalytic processes (Table 2.6). The catalytic borylation of alkanes was first demonstrated by photochemical borylation of alkanes with $Cp^*Re(CO)_3$ [99]. The rhodium complex $Cp^*Rh(\eta^4\text{-}C_6Me_6)$ (81), which easily generates a highly unsaturated $Cp^*Rh(I)$ species, was then introduced as a highly active catalyst for borylation of nonactivated alkanes and arenes under thermal conditions [100]. Under these conditions, $B_2pin_2$ afforded almost 2 equiv of borylalkanes. $[Cp^*RuCl_2]_2$ is also an

## 2.3 Borylation via C–H Activation | 149

**Table 2.6** C–H borylation of alkanes.

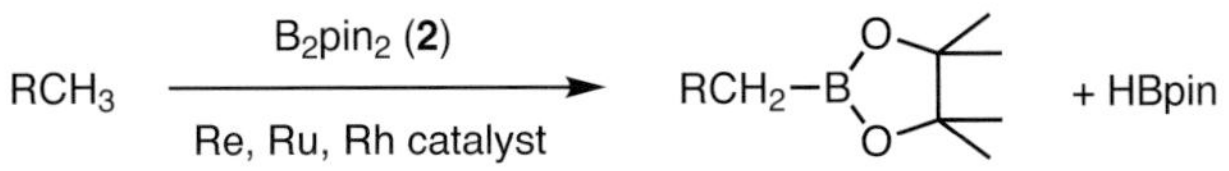

$$RCH_3 \xrightarrow[\text{Re, Ru, Rh catalyst}]{B_2pin_2 \text{ (2)}} RCH_2\text{–Bpin} + HBpin$$

| Entry | Catalyst, temperature |
|---|---|
| 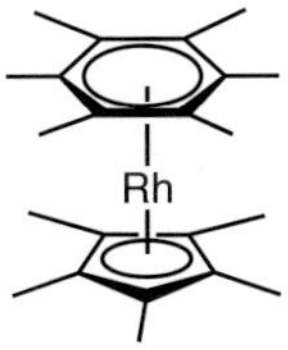 | **81** (Cp*Rh(C$_6$Me$_6$)) |
| 1 | [Cp*Re(CO)$_3$], $h\nu$, 25 °C |
| 2 | [Cp*RuCl$_2$]$_2$, 150 °C |
| 3 | **81**, 150 °C |

efficient catalyst for borylation of alkanes (entry 2) [101]. In all catalysts shown in entries 1–3, alkanes specifically react at the less hindered terminal carbon, presumably by a process proceeding through isomerization of a *sec*-alkylmetal intermediate to an *n*-alkyl isomer before reductive elimination of a C–B bond.

Figure 2.5 shows the electronic effect on regioselectivity [100b]. The reaction occurs preferentially at the methyl group closer to the heteroatom, and the effect of more electronegative oxygen in the ethers is larger than that of nitrogen in amines

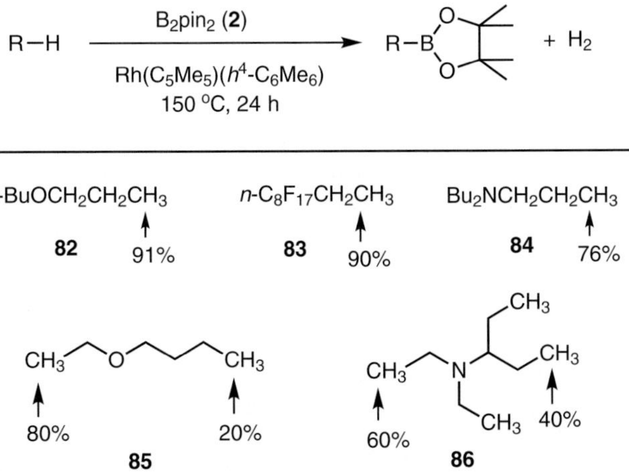

$$R\text{–H} \xrightarrow[\substack{Rh(C_5Me_5)(\eta^4\text{-}C_6Me_6) \\ 150\,°C,\,24\,h}]{B_2pin_2 \text{ (2)}} R\text{–Bpin} + H_2$$

*t*-BuOCH$_2$CH$_2$CH$_3$    *n*-C$_8$F$_{17}$CH$_2$CH$_3$    Bu$_2$NCH$_2$CH$_2$CH$_3$
**82** 91%    **83** 90%    **84** 76%

CH$_3$–O–...–CH$_3$ (**85**): 80%, 20%

CH$_3$–N(CH$_3$)–CH(CH$_3$)–CH$_3$ (**86**): 60%, 40%

**Figure 2.5** Regioselectivity in alkane C–H borylation.

(85 versus 86). A competitive reaction between (perfluoro-*n*-octyl)ethane (83) and octane gives a 94:6 ratio of products that favored borylation of the fluoroalkanes. Thus, there is an accelerating role of inductive effect of heteroatoms on reaction rates and coupling positions.

Rhodium-catalyzed borylation with $B_2pin_2$ at 150 °C provides functionalized polyolefins (Equation 2.9) [102]. One methyl group per main chain can be hydroxylated by the borylation–oxidation sequence (89).

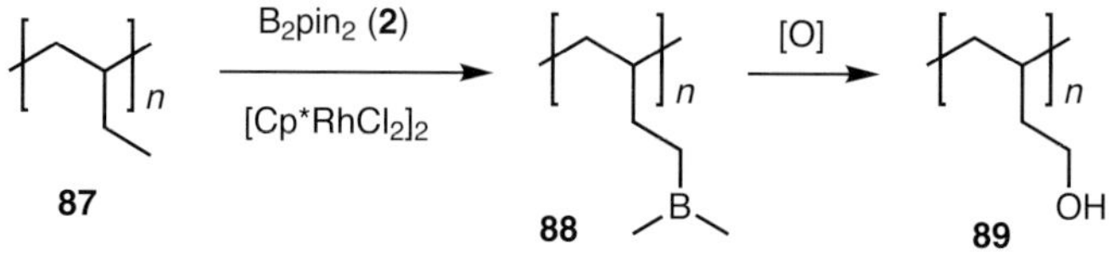

$$\text{87} \xrightarrow[\text{[Cp*RhCl}_2]_2]{B_2pin_2 \text{ (2)}} \text{88} \xrightarrow{[O]} \text{89} \tag{2.9}$$

Allylic C−H borylation can be catalyzed by iridium complexes. When a mixture of $B_2pin_2$, DBU, and a catalytic amount of [IrCl(cod)]$_2$ in neat cyclohexene was stirred at 70 °C and then aldehydes were added, the corresponding homoallyl alcohols were obtained via allylboration in good yields (Equation 2.10) [103].

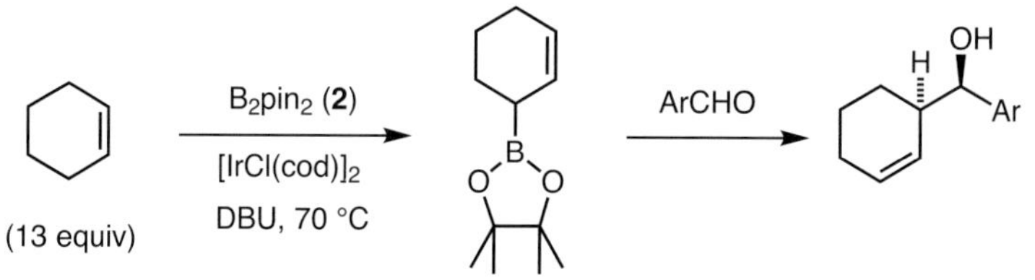

$$\text{(13 equiv)} \xrightarrow[\text{DBU, 70 °C}]{B_2pin_2 \text{ (2)} \atop [IrCl(cod)]_2} \xrightarrow{ArCHO} \tag{2.10}$$

Borylation of arenes possessing an alkyl substituent often provides benzylboronate along with an arylboronate, but no selective rhodium or iridium catalysts are available for borylation of the benzylic C−H bond preferentially over the aromatic C−H bond [100d,104]. Pd/C is the only catalyst now available for selective benzylic C−H borylation of alkylbenzenes by $B_2pin_2$ or HBpin (Equation 2.11) [105]. Toluene, xylenes, and mesitylene are all viable substrates; however, the reaction can be retarded by heteroatoms such as MeO and F substituents on the arene. Ethylbenzene results in a 3:1 mixture of pinacol 1-phenylethylboron and 2-phenylethylboron derivatives.

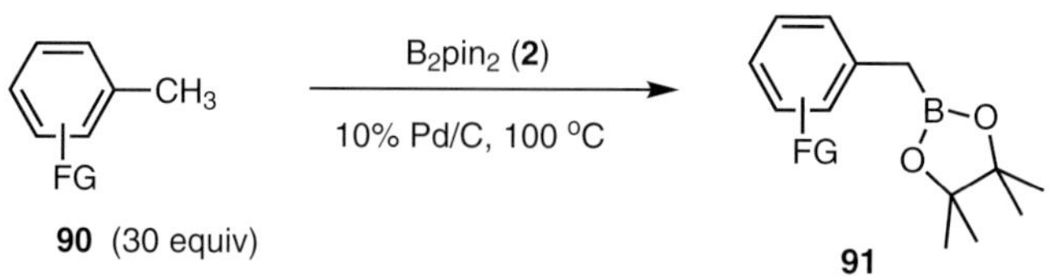

$$\text{90 (30 equiv)} \xrightarrow[\text{10\% Pd/C, 100 °C}]{B_2pin_2 \text{ (2)}} \text{91} \tag{2.11}$$

FG = H (74%), FG = 2-Me (77%), FG = 3-Me (79%), FG = 4-Me (72%)
FG = 3,5-Me$_2$ (64%), FG = 4-F (13%), FG = 4-MeO (13%)

**Table 2.7** Dehydrogenative coupling for the preparation of alkenylboron compounds.

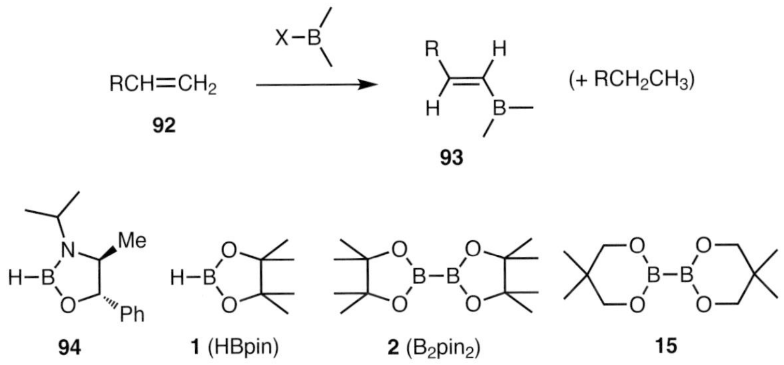

$$\text{RCH}=\text{CH}_2 \quad \xrightarrow{\text{X-B}\diagup} \quad \underset{\mathbf{93}}{\overset{R}{\underset{H}{>}}=\overset{H}{\underset{B-}{<}}} \quad (+ \text{RCH}_2\text{CH}_3)$$

**92**

**94**     **1 (HBpin)**     **2 (B$_2$pin$_2$)**     **15**

| Alkene | Reagent | Catalyst, solvent | Yield (%) |
| --- | --- | --- | --- |
| 4-MeOPhCH=CH$_2$ | 94 | [RhCl(alkene)$_2$]$_2$, toluene | 98 |
| 4-ClPhCH=CH$_2$ | 1 | [RhCl(alkene)$_2$]$_2$, toluene | 99 |
| 4-MeOPhCH=CH$_2$ | 2 | RhCl(CO)(PPh$_3$)$_2$, benzene-CH$_3$CN | 93 |
| 4-MeOPhCH=CH$_2$ | 14 | RhCl(CO)(PPh$_3$)$_2$, benzene-CH$_3$CN | 90 |
| PhCH=CH$_2$ | 1 | [RhCl(cod)]$_2$, toluene | 81 |
| 2-MeO$_2$CPhCH=CH$_2$ | 1 | RhCl(cod)]$_2$, toluene | 80 |
| Ph(Me)C=CH$_2$ | 2 | RhCl(CO)(PPh$_3$)$_2$, benzene-CH$_3$CN | 90 |
| Ph$_2$C=CH$_2$ | 2 | RhCl(CO)(PPh$_3$)$_2$, benzene-CH$_3$CN | 99 |

## 2.3.2
## Alkenyl C–H Bonds

There are two procedures for the borylation of vinylic C–H bonds, dehydrogenative coupling and the C–H activation method discussed in Scheme 2.1. Dehydrogenative coupling is attractive as a method for the synthesis of (*E*)-1-alkenylboronates **93** from alkenes (Table 2.7) [23–25]. The phosphine-free rhodium(I) catalyst is recommended for oxazaborolidine **94** [25] and HBpin [23, 24]; however, Rh(CO)(PPh$_3$)$_2$ has recently been recognized to be an efficient catalyst for B$_2$pin$_2$ and **15** [24c]. This catalyst exceptionally achieves high selectivities for borylation of exomethylene cycloalkanes. Mechanistically, the insertion of RCH=CH$_2$ **92** into [M]-B< (**7**) giving RCH$_2$C([M])(B<)H$_2$ can be followed by β-hydride elimination to yield **93** [25]. Styrene derivatives are recognized to be the best substrates for this purpose and the reaction often requires 2 equiv of vinylarene because H$_2$ thus generated by β-elimination hydrogenates a proportion of the alkene substrate.

The C–H activation procedure was reported for the borylation of vinylic ethers. A simple extension of the C–H activation protocol to typical alkenes results in a complex mixture of boron compounds since interactions between rhodium(I) or iridium(I) complexes with HBpin or B$_2$pin$_2$ yield several species that promote

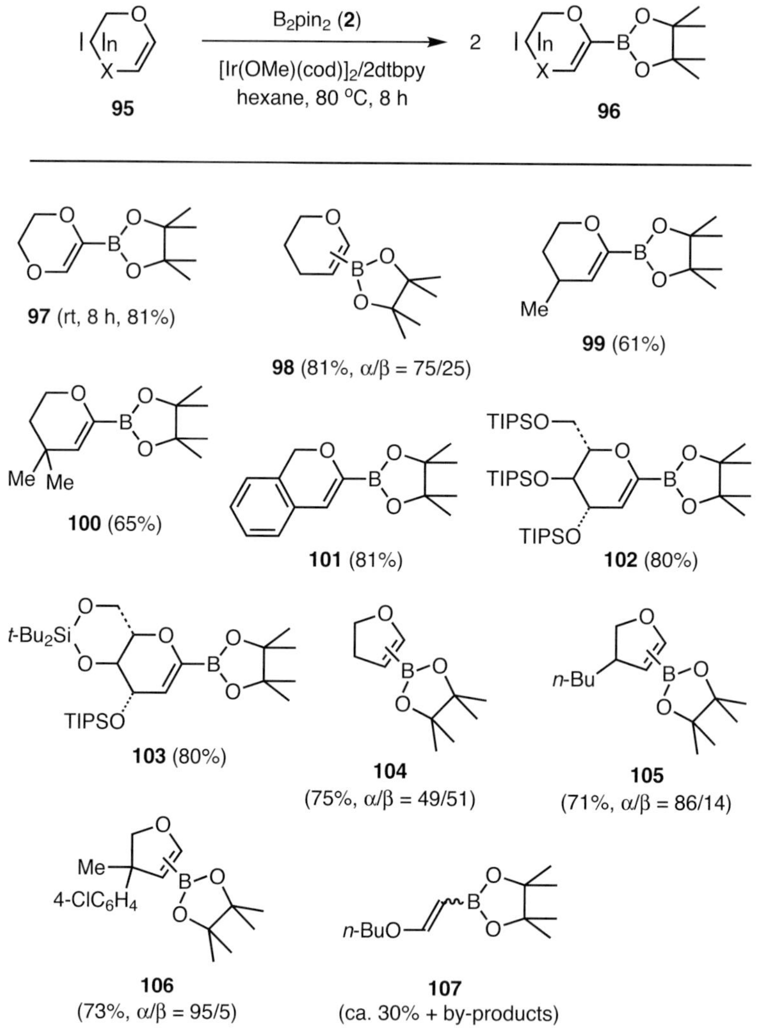

**Figure 2.6** Vinylic C—H borylation via C—H activation.

catalyzed hydroboration (**5**, Z = H), diboration (**5**, Z = B<), dehydrogenative coupling (**7**), and C—H borylation (**9**) (Scheme 2.1). Thus, difficulty in achieving a selective C—H coupling for alkenes is attributable to the high susceptibility of alkenes for insertion into [M](H)(X)(B<) (**5**) or [M](B<) (**7**) in advance of formation of [M](B<)$_3$ (**9**). However, cyclic vinyl ethers (**95**), which are slow for such an insertion process, allow selective borylation at the sp$^2$ C—H bond (Figure 2.6) [106]. Two boron atoms in B$_2$pin$_2$ participate in the coupling, thus allowing the formation of 2 equiv of **96** from 1 M of B$_2$pin$_2$. The coupling selectively occurs at the α-carbon of six-membered ethers possessing a substituent at the γ-position, although unsubstituted dihydropyrans (**98**) and dihydrofurans (**104–106**) are prone to giving a mixture of α- and β-coupling products. The role of an oxygen atom is attributable to its high electronegativity that accelerates oxidative addition of a neighboring C–H bond. Indeed, the dioxene precursor **97** reacts at room temperature, whereas dihydropyrans and dihydrofurans are borylated at 80 °C. The donating property of oxygen can also be critical for slowing down the insertion of alkene into other metal–boron species such as **5** and **7**.

C-Glucals are an important class of compounds due to the frequent occurrence of these fragments in pharmaceuticals. A key skeleton (**110**) of vimeomycinone B2 methyl ester was synthesized in 83% yield when the preparation of **109** was followed by cross-coupling (Equation 2.12 [106b]).

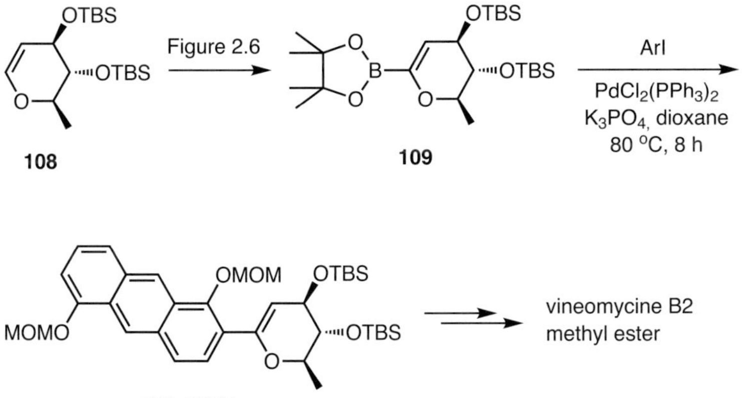

### 2.3.3
### Aromatic C–H Bonds

Under conditions similar to those used for aliphatic and vinylic C–H borylation, rhenium(I), rhodium(I), and iridium(I) complexes catalyze the borylation of arenes with HBpin or $B_2pin_2$. The rhenium complex $Cp^*Re(CO)_3$ catalyzes the reaction under irradiation of light (Table 2.8, entry 1) [99]. $Cp^*Rh(\eta^4\text{-}C_6Me_6)$ (**81**) is an excellent catalyst, giving 92% yield after 2.5 h with 5 mol% catalyst loading and 82% yield with 0.5 mol% loading at 150 °C for benzene (entry 2) [107–109]. The high efficiency of iridium catalysts has been studied extensively by Smith III and coworkers. It was found that [Ir(ind)(cod)] is inefficient, but addition of a small and electron-donating phosphine such as $PMe_3$ or dmpe substantially increases the catalyst activity and turnover number and suppresses side reactions at the benzylic C–H bond or aliphatic C–halogen bonds (entry 3) [110]. A class of iridium(I) complexes possessing a dtbpy ligand exhibits excellent activity and selectivity for aromatic C–H borylation with $B_2pin_2$ [22, 111, 112] or HBpin [113] (entries 4 and 5). The reaction was first demonstrated at 80–100 °C by Ir–Cl complexes, but it smoothly occurs at room temperature when the catalyst is prepared from $[Ir(OMe)(COD)]_2$ and dtbpy [112]. Other ligands reported for iridium catalysts are hydrotris(pyrazolyl)borate (Tp, entry 6) [114], DIPA (entry 7) [115], and N-heterocyclic carbene

**Table 2.8** Conditions for aromatic C–H borylation.

**2** ($B_2pin_2$) **1** (HBpin) **20** (HB(dan))

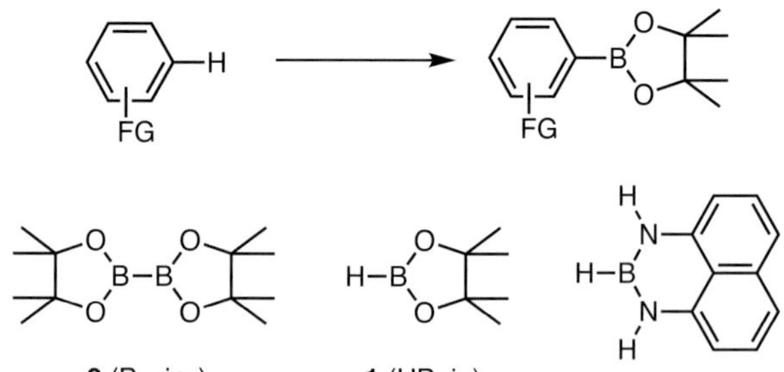

| Entry | Reagent | Catalyst, solvent, temperature |
|---|---|---|
| 1 | $B_2pin_2$ | $Cp^*Re(CO)_3$, *hν* neat, 25 °C |
| 2 | HBpin | $Cp^*Rh(C_6Me_6)$, *c*-$C_6H_{12}$,150 °C |
| 3 | HBpin | Ir(ind)(cod)-dmpe, neat, 150 °C |
| 4 | $B_2pin_2$ | $[Ir(OMe)(cod)]_2$-2dtbpy, hexane, 25 °C |
| 5 | HBpin | $[Ir(OMe)(cod)]_2$-2dtbpy, hexane, 25–80 °C |
| 6 | HBpin | Ir(Tp)(cod), neat, 120 °C |
| 7 | $B_2pin_2$ | $[IrCl(cod)]_2$-2DIPA, neat, 80 °C |
| 8 | HBpin | $Ir(CF_3CO_2)(NHC)_2(cod)$, neat, 40 °C |
| 9 | $Et_3SiBpin$ | $[Ir(OMe)(cod)]_2$-2dtbpy, neat, 80 °C |
| 10 | 20 | $[Ir(OMe)(cod)]_2$-dppe, neat, 80 °C |
| 11 | $B_2pin_2$ or HBpin | $[IrCl(cod)]_2$-BPDCA, neat, 80 °C |

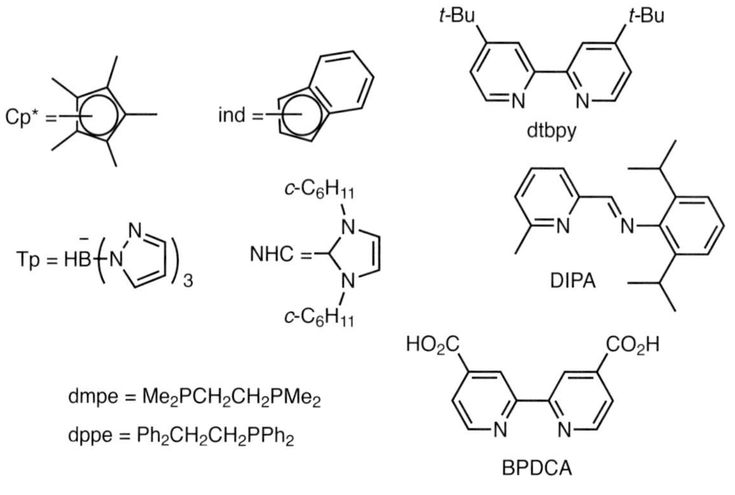

dmpe = $Me_2PCH_2CH_2PMe_2$

dppe = $Ph_2CH_2CH_2PPh_2$

(entry 8) [116]. The use of NHC ligand achieves 100% conversions at 40 °C for benzene, toluene, iodobenzene, and o-xylene. Silylborane Et$_3$SiBpin is a new entry used for borylation of arene C–H bonds and benzylic C–H bonds (entry 9) [100d]. The new reagent HB(dan) (**20**) affords arylboronates that are stable to silica gel for their isolation and easily hydrolyzed to boronic acids via treatment with aqueous acid (entry 10) [40]. A complex between [IrCl(cod)]$_2$ and 2,2′-dipyridinedicarboxylic acid (DPDCA) provides a recyclable catalyst that can be reused more than 10 times (entry 11) [117].

The reaction affords variously functionalized arylboronates directly from arenes (Figure 2.7). The reaction is suitable for arenes possessing I, Br, Cl, CO$_2$Me, CN, CF$_3$, and alkyl groups having a benzylic C–H bond. The electronic effect of substituents is rather small; thus, the coupling positions can be mainly controlled by the steric effect of substituents. The proportion of coupling products at the *ortho*-carbon is negligible because of the high sensitivity of the catalyst to steric hindrance, and monosubstituted arenes (**111**) result in a mixture of *meta*- and *para*-products in statistical ratios (about 2 : 1) [107–113]. Both 1,2- and 1,4-disubstituted arenes bearing identical substituents yield single isomers (**112**) [107–113]. 1,3-Disubstituted arenes are borylated at the common *meta*-position; therefore, isomerically pure products are obtained even with arenes containing two distinct substituents, for example, **113** [107–113], **115** [118], and **117** [119]. Borylation of naphthalene, pyrene, and perylene provides bis(boryl) or tetra(boryl) derivatives (**118** and **119**) as intermediates for synthesis of π-conjugate materials via cross-coupling [120]. On the other hand, sterically small F and CN groups are prone to direct *ortho*-borylation (**114, 116**) [114, 121]. Borylation of a cyclopentadienyl ring of ferrocene [122] and that of a five-membered ring of azulene [123] have also been studied.

Under conditions analogous to those used for typical arenes, heteroarenes are borylated with B$_2$pin$_2$ or HBpin (Figure 2.8). Thiophene, pyrrole, and furan derivatives are selectively borylated at the α-carbon of a heteroatom (**120** [124], **121** [124], **122** [124], **123** [111], **126** [111,112b], and **127** [111,112b]), whereas coupling occurs at the β-carbon when two α-carbons have substituents (**125** [124]) or blocked by steric hindrance of an *N*-substituent (**124**) [111]. Reactions of pyrrole, thiophene, and furan, which have two active C–H bonds, result in a mixture of monoborylation and diborylation products. Monoborylation is predominant when 10 equiv of a substrate is used toward B$_2$pin$_2$. On the other hand, 2,5-diboryl compounds are formed selectively when equimolar amounts of heteroarenes and B$_2$pin$_2$ are reacted (**126** and **127**) [112b]. Although unsubstituted pyridines display strong resistance, chloro and methyl groups at the α-carbon effectively block coordination to the catalyst. Thus, 2-chloropyridine yields a mixture of 4- and 5-borylpyridines (29/71) at room temperature, 2,6-disubstituted derivatives give 4-borylpyridines (**129**), and 2,3-disubstituted pyridine produces 5-borylpyridine (**132**) [125]. Thus, the reaction proceeds at the β- or γ-carbon of pyridine derivatives, whereas α-borylation has been reported exceptionally for bipyridines (**131**) [126]. These reactions of heteroaryl rings are significantly faster than those of typical arenes. Thus, benzo-fused furan and thiophene and indoles are selectively borylated at the five-membered

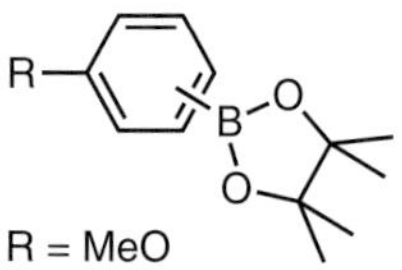

R = MeO
R = Me
R = CF₃
**111** (o/m/p = 0/60/30, entry 2)

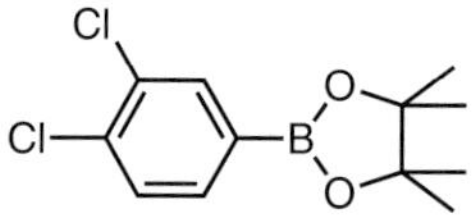

**112** (entry 4)

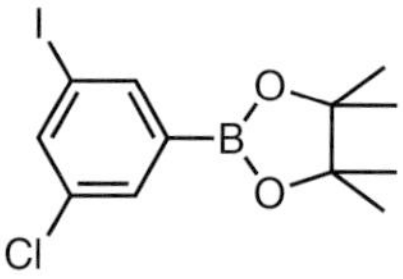

**113** (entry 4)

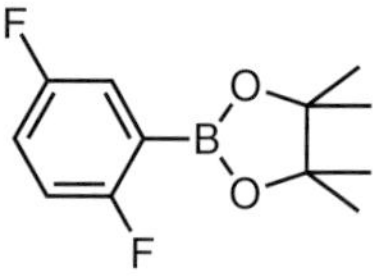

**114** (entry 6)

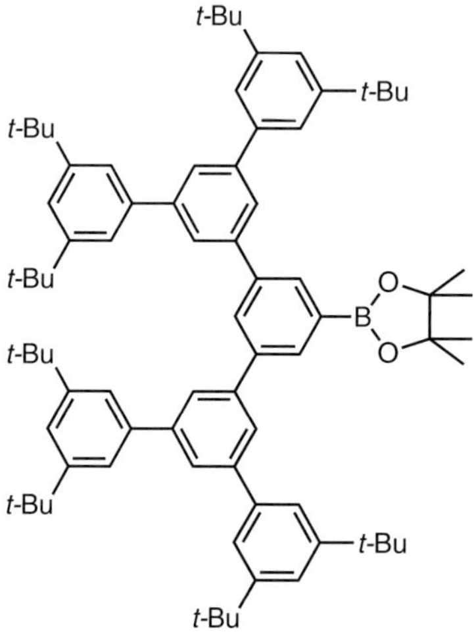

**115** (80 °C, entry 4)

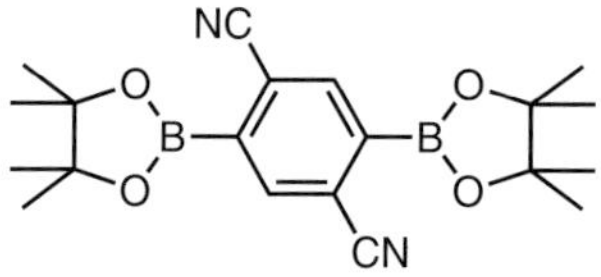

**116** (entry 5)

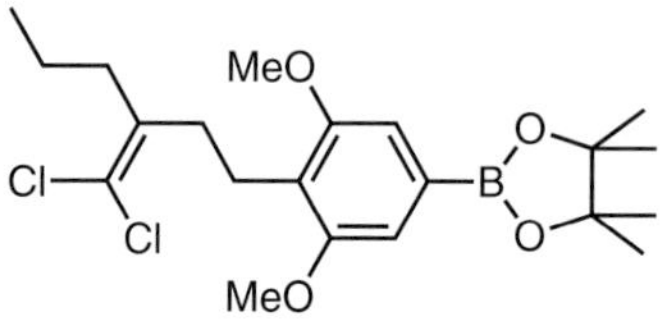

**117** (microwave, entry 4)

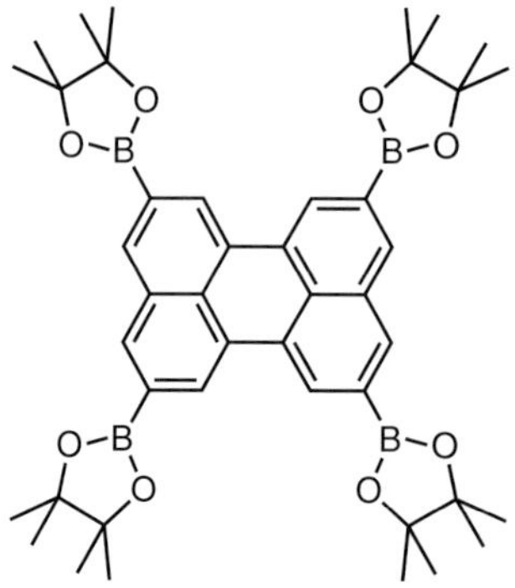

**118** (80 °C, entry 4)

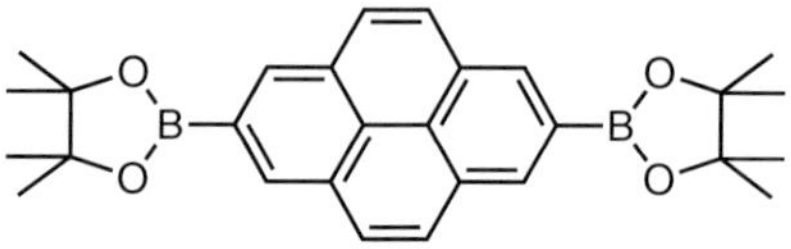

**119** (80 °C, entry 4)

**Figure 2.7** Selected arylboronates synthesized using conditions of Table 2.8.

ring [112a]. However, it is interesting that indoles possessing an α-substituent are selectively borylated at the 7-position, presumably due to N-chelation to iridium (**128**) [127]. Borylation of porphyrins takes place at the C2 carbon (**133**), in contrast to the orientation in electrophilic bromination occurring at the C3 carbon [128].

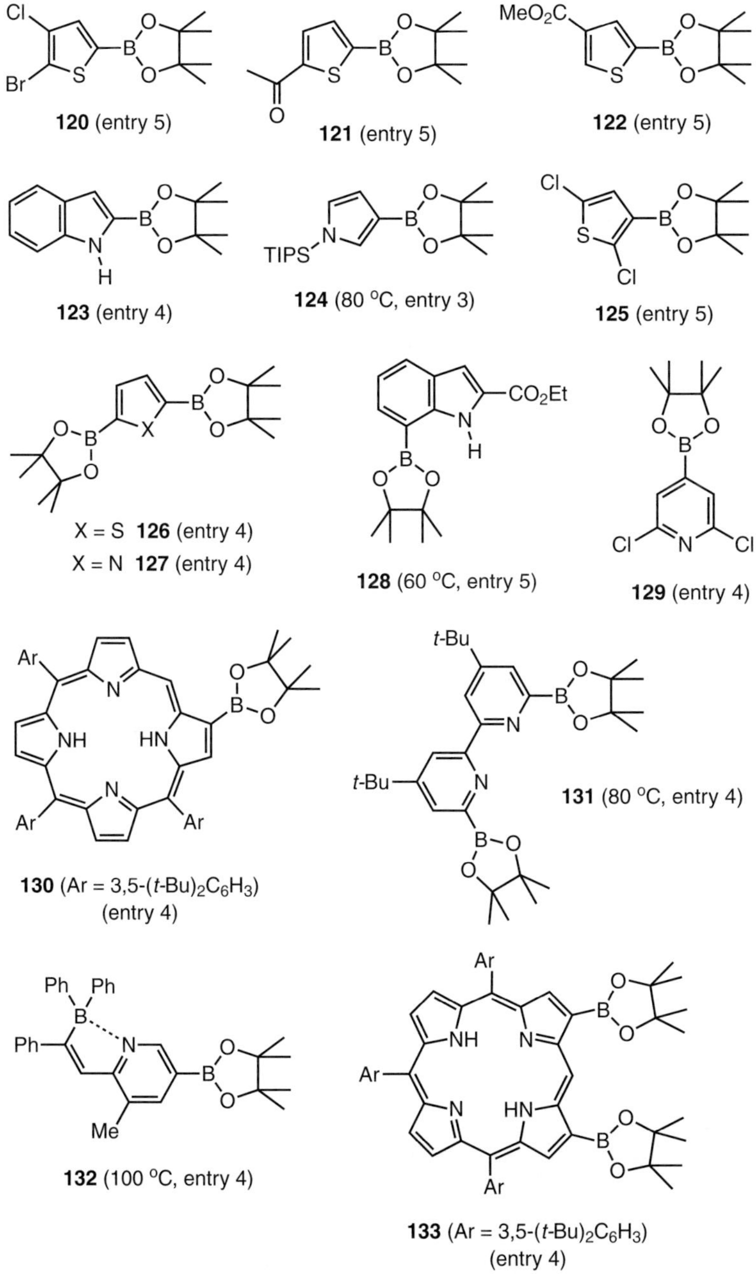

**120** (entry 5)

**121** (entry 5)

**122** (entry 5)

**123** (entry 4)

**124** (80 °C, entry 3)

**125** (entry 5)

X = S **126** (entry 4)
X = N **127** (entry 4)

**128** (60 °C, entry 5)

**129** (entry 4)

**130** (Ar = 3,5-($t$-Bu)$_2$C$_6$H$_3$)
(entry 4)

**131** (80 °C, entry 4)

**132** (100 °C, entry 4)

**133** (Ar = 3,5-($t$-Bu)$_2$C$_6$H$_3$)
(entry 4)

**Figure 2.8** Selected heteroarylboronates synthesized using conditions of Table 2.8.

Functionalization of isoindoles (**135**) has been limited due to instability of the o-quinoid structure. 1-Boryl and 1,3-diborylisoindoles (**136**) are obtained in high yields via a dehydrogenation/C–H borylation sequence when isoindolines (**134**) are borylated with HBpin and a palladium catalyst (Equation 2.13) [129].

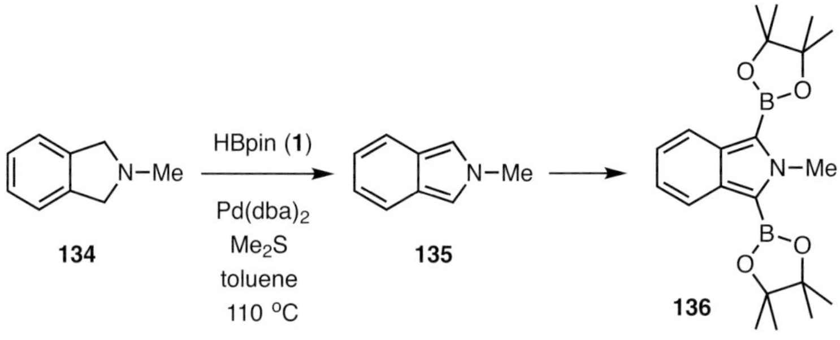

Trialkylsilanes, $R_3SiH$, undergoing oxidative addition to iridium(I) complexes works as an efficient directing group for *ortho*-selective borylation of arenes (Figure 2.9) [130]. A proposed mechanism involves an $Ir[(Si(Me)_2(X-Ar))(Bpin)_2(dtbpy)]$ ($X = CH_2$, NRH, O) intermediate (**138**) generated by oxidative addition of Si−H and B−B bonds. By this method, arenes possessing an *ortho*-$CH_2SiMe_2H$, $-N(R)SiMe_2H$, or $-OSiMe_2H$ group (**137**) are borylated selectively at the *ortho*-carbon.

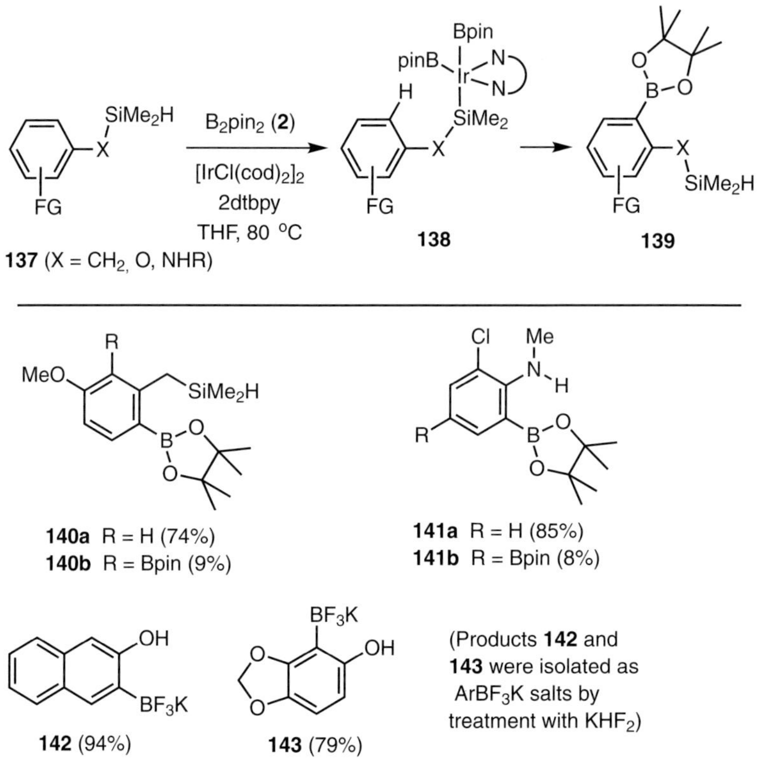

(Products **142** and **143** were isolated as $ArBF_3K$ salts by treatment with $KHF_2$)

**Figure 2.9** *ortho*-Selective borylation of arenes directed by $SiMe_2H$.

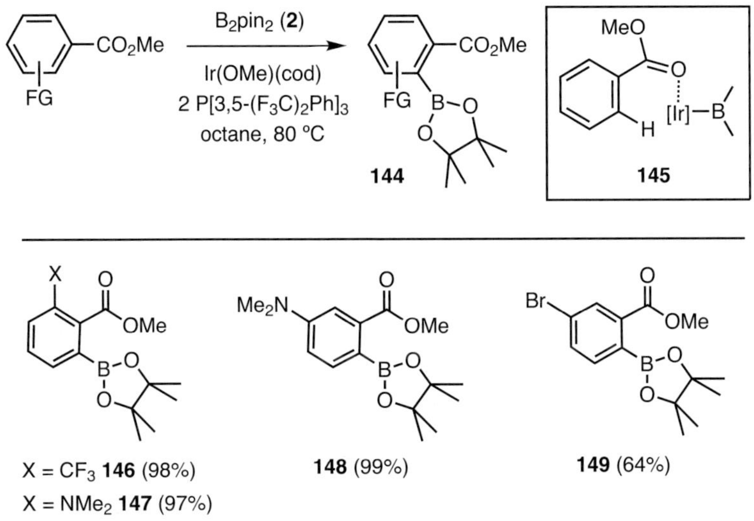

**Figure 2.10** ortho-Selective borylation of benzoates.

C—H activation controlled by a chelation to a heteroatom such as in **145** allows *ortho*-selective coupling among other possible C—H bonds. A π-acidic 3,5-bis(trifluoromethyl)phenylphosphine was recognized to be the most selective ligand for *ortho*-borylation of benzoates (Figure 2.10) [131]. Diboron selectively couples with benzoates substituted by a bromine, trifluoromethyl, or dimethyl amino group at *ortho*-, *meta*-, and *para*-carbons that have been synthesized by an *o*-lithiation–borylation protocol [132]. HBpin is inert; thus, one boron atom of $B_2pin_2$ participates in the coupling.

## 2.4
## Catalytic Cycle

Interaction between M-X (M = Rh(I), Ir(I); X = halogen, alkyl, OR); and HBpin or $B_2pin_2$ yields M-Bpin (**7**) as an intermediate leading to M(Bpin)$_3$ (**9**) (Scheme 2.1) [19, 22, 30]. The (methoxo)iridium(I) complex [Ir(OMe)(cod)]$_2$ is a better precursor than [Ir(Cl)(cod)]$_2$ for formation of **7** at room temperature due to the higher bond energy of the resulting B—O bond than that of the B—Cl bond. Thus, the catalyst activity of Ir(X)(cod) parallels the order of basic strength of the anionic ligand, for example, X = MeO > HO > PhO > AcO ≫ Cl [112a]. Although both Ir(I)- and Ir(III) boryl species (**7** and **9**) are viable for aromatic C—H borylation [110b], mechanistic studies by NMR have shown that Ir(boryl)$_3$ complex (**9**) is an active component involved in the catalytic cycle [22b]. Thus, the catalytic cycle proceeds through oxidative addition of an arene C—H bond to **151**, yielding an 18-electron Ir(V) intermediate (**152**) that reductively eliminates ArBpin (Figure 2.11). Oxidative addition of $B_2pin_2$ to **153**, giving tetraboryl **150**, can be followed by reductive elimination of HBpin to regenerate **151**. The resulting HBpin then participates in the catalytic cycle

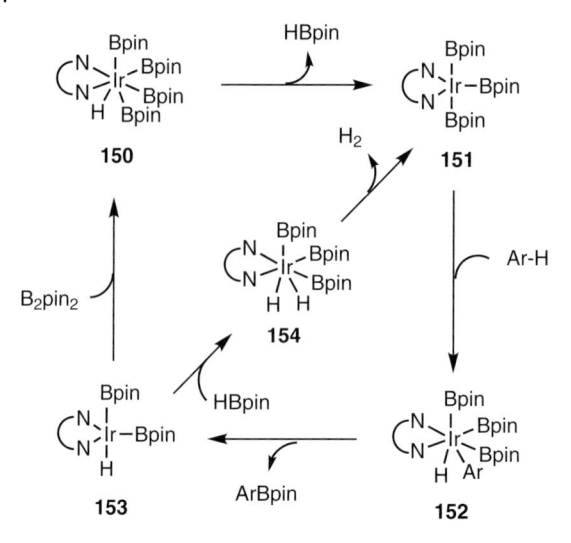

**Figure 2.11** Proposed catalytic cycle for aromatic C—H borylation catalyzed by Ir(OMe)(cod)/dtbpy.

via a sequence of oxidative addition and hydrogen reductive elimination from **154**. The reaction of B$_2$pin$_2$ is significantly faster than that of HBpin since the catalytic reaction shows a two-step process: fast borylation by B$_2$pin$_2$ followed by slow borylation by HBpin [22a]. A small steric hindrance from the planar bipyridine ligand as well as its property of electron donation to the metal center may allow smooth oxidative addition of an arene C—H bond at room temperature. The planar structures of dioxaboryl rings (Bpin) and an arene substrate (Ar) can also be critical for the formation of such sterically hindered hepta-coordinated Ir(V) intermediates (**152**). These processes have been supported by theoretical studies by Sakaki and coworkers [30].

The steric effect of substituents and the inductive effect of substituents or that of heteroatoms in aromatic rings play a role in the control of orientation, and the resonance effect of substituents is very small. A space-filling model depicted on the

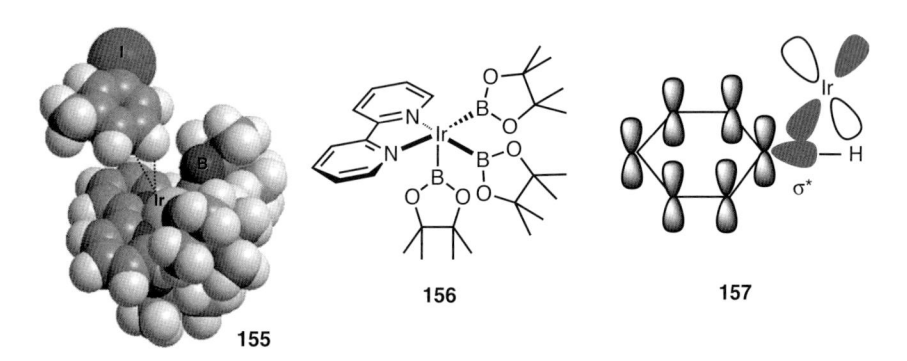

**Figure 2.12** Steric and electronic effects on orientation of aromatic C—H borylation.

basis of X-ray analysis of an active component (**156**) (cf. **151** in Figure 2.11) [22] and theoretical calculation [30] show a very small free space at the upper left area (**155**) (Figure 2.12). Thus, the coupling preferentially occurs at the *meta*- or *para*-carbon than at the *ortho*-carbon of substituted arenes due to this steric hindrance of the catalyst. Theoretical calculations suggested an interaction between a $\sigma^*$ antibonding of C−H bond and a filled d-orbital of iridium metal for the initial process of C−H activation (**157**) [30]. Thus, the reaction behaves as a nucleophilic substitution of aromatic C−H bonds. Trifluoromethylbenzene reacts six times faster than does anisole [22a] and a small and highly electronegative F and CN group increases the *ortho*-coupling products [121]. Although such electronic properties of the substituents are generally not very effective to control orientations of typical arenes, the inductive effect of heteroatoms in aromatic rings significantly accelerates the reaction at the α-carbon as is discussed in Figures 2.7 and 2.8. The regioselectivity in aliphatic borylation in Figure 2.5 also shows selectivities owing to the inductive effect of heteroatoms [100b].

## 2.5
## Summary

Recent studies on the developments of the transition metal-catalyzed new borylation reactions by B−B or B−H compounds are described. Cross-coupling reactions of B−B or B−H reagents with aryl, vinyl, allyl, and benzyl halides or triflates have provided a direct method for the borylation of organic electrophiles without using lithium or magnesium intermediates. Direct C−H borylation of hydrocarbons with B−B or B−H reagents catalyzed by a transition metal complex is an economical, efficient, elegant, and environmentally benign protocol for the synthesis of a variety of organoboron compounds.

## References

1 Nesmeyanov, A.N. and Sokolik, R.A. (1967) *Methods of Elemento-Organic Chemistry*, vol. 1, North-Holland, Amsterdam.
2 Matteson, D.S. (1995) *Stereodirected Synthesis with Organoboranes*, Springer, Berlin.
3 Miyaura, N. (2005) *Science of Synthesis*, vols 6.1.7 and 6.1.26 (eds D.E. Kaufmann and D.S. Matteson), Georg Thieme Verlag, Stuttgart.
4 Ishiyama, T. and Miyaura, N. (2005) Chapter 2, in *Boronic Acids* (ed. D.G. Hall), Wiley-VCH Verlag GmbH, Weinheim.
5 Miyaura, N. and Yamamoto, Y. (2007) *Comprehensive Organometallic Chemistry III*, vol. 9 (eds R.H. Crabtree, D.M. Mingos, and P. Knochel), Elsevier, Amsterdam, pp. 145–244.
6 (a) Burgess, K. and Ohlmeyer, M.J. (1991) *Chem. Rev.*, **91**, 1179–1191; (b) Beletskaya, I. and Pelter, A. (1997) *Tetrahedron*, **53**, 4957–5026.
7 Miyaura, N. (2001) Chapter 1, in *Catalytic Heterofunctionalization* (eds A. Togni and H. Grützmacher), Wiley-VCH Verlag GmbH, Weinheim.
8 (a) Reviews for catalyzed reactions of B−B compounds: Marder, T.B. and

Norman, N.C. (1998) *Top. Catal.*, **5**, 63–73; (b) Irvine, G.J., Lesley, M.J.G., Marder, T.B., Norman, N.C., Rice, C.R., Robins, E.G., Roper, W.R., Whittell, G.R., and Wright, L.J. (1998) *Chem. Rev.*, **98**, 2685–2722; (c) Beletskaya, I. and Moberg, C. (1999) *Chem. Rev.*, **99**, 3435–3462; (d) Ishiyama, T. and Miyaura, N. (2000) *J. Organomet. Chem.*, **611**, 392–402; (e) Ishiyama, T. and Miyaura, N. (2004) *Chem. Rec.*, **3**, 271–280; (f) Miyaura, N. (2008) *Bull. Chem. Soc. Jpn.*, **81**, 1535–1553.

9  Ishiyama, T., Nishijima, K.-i., Miyaura, N., and Suzuki, A. (1993) *J. Am. Chem. Soc.*, **115**, 7219–7225.

10  Review for addition reactions of B–Si compounds: Suginome, M. and Ito, Y. (2000) *Chem. Rev.*, **100**, 3221–3256.

11  Suginome, M., Yamamoto, A., and Murakami, M. (2003) *J. Am. Chem. Soc.*, **125**, 6358–6359.

12  Review for addition reactions of B–Sn compounds: Han, L.-B. and Tanaka, M. (1999) *Chem. Commun.*, 395–402.

13  (a) Reviews for borylation of RX with B–B or H–B compounds: Miyaura, N. (2002) *Top. Curr. Chem.*, **219**, 11–59; (b) Miyaura, N. (2004) *Metal-Catalyzed Cross-Coupling Reactions*, 2nd edn, vol. 1 (eds A. Meijere and F. Diederich), Wiley-VCH, pp. 41–124 and Refs [3, 5, 7, 8f].

14  (a) Reviews for borylation of RH with B–B or H–B compounds: Ishiyama, T. and Miyaura, N. (2003) *J. Organomet. Chem.*, **680**, 3–11; (b) Ishiyama, T. and Miyaura, N. (2006) *Pure and Appl. Chem.*, **78**, 1369–1375; (c) Mkhalid, I.A.I., Barnard, J.H., Marder, T.B., Murphy, J.M., and Hartwig, J.F. (2010) *Chem. Rev.*, **110**, 890–931.

15  (a) Ishiyama, T., Matsuda, N., Miyaura, N., and Suzuki, A. (1993) *J. Am. Chem. Soc.*, **115**, 11018–11019; (b)Ishiyama, T., Matsuda, N., Murata, M., Ozawa, F., Suzuki, A., and Miyaura, N. (1996) *Organometallics*, **15**, 713–720.

16  (a) Iverson, C.N. and Smith, M.R., III (1995) *J. Am. Chem. Soc.*, **117**, 4403–4404; (b) Iverson, C.N. and Smith, M.R., III (1996) *Organometallics*, **15**, 5155–5165.

17  (a) Lesley, G., Nguyen, P., Taylor, N.J., Marder, T.B., Scott, A.J., Clegg, W., and Norman, N.C. (1996) *Organometallics*, **15**, 5137–5154; (b) Kerr, A., Norman, N.C., Orpen, A.G., Quayle, M.J., Rice, C.R., Timms, P.L., Whittll, G.R., and Marder, T.B. (1998) *Chem. Commun.*, 319–320.

18  (a) Knorr, J.R. and Merola, J.S. (1990) *Organometallics*, **9**, 3008–3010; (b) The reported regioselectivities are opposite due to misprint: A private communication. Westcott, S.A., Taylor, N.J., Marder, T.B., Baker, R.T., Jones, N.J., and Calabrese, J.C. (1991) *J. Chem. Soc., Chem. Commun.*, 304–305; (c) Baker, R.T., Calabrese, J.C., Westcott, S.A., Nguyen, P., and Marder, T.B. (1993) *J. Am. Chem. Soc.*, **115**, 4367–4368.

19  Dai, C., Stringer, G., Marder, T.B., Scott, A.J., Clegg, W., and Norman, N.C. (1997) *Inorg. Chem.*, **36**, 272–273.

20  (a) Baker, R.T., Calabrese, J.C., Westcott, S.A., Nguyen, P., and Marder, T.B. (1993) *J. Am. Chem. Soc.*, **115**, 4367–4368; (b) Dai, C., Stringer, G., Marder, T.B., Baker, R.T., Scott, A.J., Clegg, W., and Norman, N.C. (1996) *Can. J. Chem.*, **74**, 2026–2031.

21  Nguyen, P., Blom, H.P., Westcott, S.A., Taylor, N.J., and Marder, T.B. (1993) *J. Am. Chem. Soc.*, **115**, 9329–9330.

22  (a) Ishiyama, T., Takagi, J., Ishida, K., Miyaura, N., Anastasi, N.R., and Hartwig, J.F. (2002) *J. Am. Chem. Soc.*, **124**, 390–391; (b) Boller, T.M., Murphy, J.M., Hapke, M., Ishiyama, T., Miyaura, N., and Hartwig, J.F. (2005) *J. Am. Chem. Soc.*, **127**, 14263–14278.

23  (a) Murata, M., Watanabe, S., and Masuda, Y. (1999) *Tetrahedron Lett.*, **40**, 2585–2588; (b) Murata, M., Kawakita, K., Asana, T., Watanabe, S., and Masuda, Y. (2002) *Bull. Chem. Soc. Jpn.*, **75**, 825–829.

24  (a) Westcott, S.A., Marder, T.B., and Baker, R.T. (1993) *Organometallics*, **12**, 975–979; (b) Vogels, C.M., Hayes, P.G., Shaver, M.P., and Westcott, S.A. (2000) *Chem. Commun.*, 51–52; (c) Mkhalid, I.A.I., Coapes, R.B., Edes, S.N., Coventry, D.N., Souza, F.E., Thomas, R.L., Hall, J.J., Bi, S.W., Lin, Z., and Marder, T.B. (2008) *Dalton Trans.*, **22**, 1055–1064.

25 Brown, J.M. and Lloyd-Jones, G.C. (1994) *J. Am. Chem. Soc.*, **116**, 866–878.
26 Ishiyama, T., Murata, M., and Miyaura, N. (1995) *J. Org. Chem.*, **60**, 7508–7510.
27 Ishiyama, T., Itoh, Y., Kitano, T., and Miyaura, N. (1997) *Tetrahedron Lett.*, **38**, 3447–3450.
28 (a) Takahashi, K., Ishiyama, T., and Miyaura, N. (2000) *Chem. Lett.*, 982–983; (b) Takahashi, K., Ishiyama, T., and Miyaura, N. (2001) *J. Organomet. Chem.*, **625**, 47–53.
29 (a) Cui, Q., Musaev, D.G., and Morokuma, K. (1997) *Organometallics*, **16**, 1355–1364; (b) Sakaki, S. and Kikuno, T. (1997) *Inorg. Chem.*, **36**, 226–229; (c) Cui, Q., Musaev, D.G., and Morokuma, K. (1998) *Organometallics*, **17**, 742–751; (d) Lam, W.H. and Lin, Z.Y. (2003) *Organometallics*, **22**, 473–480.
30 Tamura, H., Yamazaki, H., Sato, H., and Sakaki, S. (2003) *J. Am. Chem. Soc.*, **125**, 16114–16126.
31 (a) Murata, M., Watanabe, S., and Masuda, Y. (1997) *J. Org. Chem.*, **62**, 6458–6459; (b) Murata, M., Oyama, T., Watanabe, S., and Masuda, Y. (2000) *J. Org. Chem.*, **65**, 164–168.
32 Olsson, V.J., Sebelius, S., Selander, N., and Szabó, K.J. (2006) *J. Am. Chem. Soc.*, **128**, 4588–4589.
33 Sebelius, S., Olsson, V.J., and Szabó, K.J. (2005) *J. Am. Chem. Soc.*, **127**, 10478–10479.
34 Sebelius, S., Olsson, V.J., Wallner, O.A., and Szabó, K.J. (2006) *J. Am. Chem. Soc.*, **128**, 8150–8151.
35 (a) Aspley, C.J. and Williams, J.A.G. (2001) *New J. Chem.*, **25**, 1136–1147; (b) Thompson, A.L.S., Kabalka, G.W., Akula, M.R., and Huffman, J.W. (2005) *Synthesis*, 547–550.
36 Nakamura, H., Fujiwara, M., and Yamamoto, Y. (1998) *J. Org. Chem.*, **63**, 7529–7530.
37 Huang, Y., Chen, D., and Qing, F.-L. (2005) *Synlett*, 1740–1742.
38 Rosen, B.M., Huang, C., and Percec V., (2008) *Org. Lett.*, **10**, 2597–2600.
39 Euzenat, L., Horhant, D., Ribourdouille, Y., Duriez, C., Alcaraz, G., and Vaultier, M. (2003) *Chem. Commun.*, 2280–2281.
40 Iwadate, N. and Suginome, M. (2009) *J. Organomet. Chem.*, **694**, 1713–1717.
41 (a) Wang, Y.-C. and Georghiou, P.E. (2002) *Org. Lett.*, **4**, 2675–2678; (b) Yu, S., Saenz, J., and Srirangam, J.K. (2002) *J. Org. Chem.*, **67**, 1699–1702; (c) Zaidlewicz, M. and Wolan, A. (2002) *J. Organomet. Chem.*, **657**, 129–135; (d) Becht, J.M., Meyer, O., and Helmchen, G. (2003) *Synthesis*, 2805–2810; (e) Perttu, E.K., Arnold, M., and Iovine, P.M. (2005) *Tetrahedron Lett.*, **46**, 8753–8756.
42 (a) Malan, C. and Morin, C. (1998) *J. Org. Chem.*, **63**, 8019–8020; (b) Yuen, A.K.L. and Hutton, C.A. (2005) *Tetrahedron Lett.*, **46**, 7899–7903.
43 Zhu, L., Duquette, J., and Zhang, M. (2003) *J. Org. Chem.*, **68**, 3729–3732.
44 Ishiyama, T., Ishida, K., and Miyaura, N. (2001) *Tetrahedron*, **57**, 9813–9816.
45 Billingsley, L.L., Barder, T.E., and Buchwald, S.L. (2007) *Angew. Chem., Int. Ed.*, **46**, 5359–5363.
46 Fürstner, A. and Seidel, G. (2002) *Org. Lett.*, **4**, 541–543.
47 (a) Appukkuttan, P., Eycken, E.V., and Dehaen, W. (2003) *Synlett*, 1204–1206; (b) Metten, B., Smet, M., Boens, N., and Dehaen, W. (2005) *Synthesis*, 1838–1844.
48 (a) Brimble, M.A. and Lai, M.Y.H. (2003) *Org. Biomol. Chem.*, **1**, 2084–2095; (b) Skaff, O., Jolliffe, K.A., and Hutton, C.A. (2005) *J. Org. Chem.*, **70**, 7353–7363; (c) Han, F.S., Higuchi, M., and Kurth, D.G. (2007) *Org. Lett.*, **9**, 559–562.
49 Willis, D.M. and Strongin, R.M. (2000) *Tetrahedron Lett.*, **41**, 8683–8686.
50 Ma, Y., Song, C., Jiang, W., Xue, G., Cannon, J.F., Wang, X., and Andrus, M.B. (2003) *Org. Lett.*, **5**, 4635–4638.
51 (a) Nicolaou, K.C., Rao, P.B., Hao, J., Reddy, M.V., Rassias, G., Huang, X., Chen, D.Y.-K., and Snyder, S.A. (2003) *Angew. Chem., Int. Ed.*, **42**, 1753–1758; (b) Nicolaou, K.C., Hao, J., Reddy, M.V., Rao, P.B., Rassias, G., Snyder, S.A., Huang, X., Chen, D.Y.-K., Brenzovich, W.E., Giuseppone, N., Giannakakou, P., and O'Brate, A. (2004) *J. Am. Chem. Soc.*, **126**, 12897–12906.

52 (a) Bringmann, G., Rüdenauer, S., Götz, D.C.G., Gulder, T.M.A., and Reichert, M. (2006) *Org. Lett.*, **8**, 4743–4746; (b) Aratani, N. and Osuka, A. (2008) *Chem. Commun.*, 4067–4069.

53 Dietrich-Buchecker, C., Colasson, B., Jouvenot, D., and Sauvage, J.-P. (2005) *Chem. Eur. J.*, **11**, 4374–4386.

54 Abbott, B. and Thompson, P. (2003) *Aust. J. Chem.*, **56**, 1099–1106.

55 Araki, H., Katoh, T., and Inoue, M. (2006) *Synlett*, 555–558.

56 Read, M.W., Escobedo, J.O., Willis, D.M., Beck, P.A., and Strongin, R.M. (2000) *Org. Lett.*, **2**, 3201–3204.

57 (a) Elder, A.M. and Rich, D.H. (1999) *Org. Lett.*, **1**, 1443–1446; (b) Decicco, C.P., Song, Y., and Evans, D.A., (2001) *Org. Lett.*, **3**, 1029–1032; (c) Albrecht, B.K. and Williams, R.M. (2003) *Org. Lett.*, **5**, 197–200; (d) Bois-Choussy, M., Cristau, P., and Zhu, J. (2003) *Angew. Chem., Int. Ed.*, **42**, 4238–4241; (e) Inoue, M., Sakazaki, H., Furuyama, H., and Hirama, M. (2003) *Angew. Chem., Int. Ed.*, **42**, 2654–2657; (f) Lin, S., Yang, Z.-Q., Kwok, B.H.B., Koldobskiy, M., Crews, C.M., and Danishefsky, S.J. (2004) *J. Am. Chem. Soc.*, **126**, 6347–6355; (g) Gerstenberger, B.S. and Konopelski, J.P., (2005) *J. Org. Chem.*, **70**, 1467–1470; (h) Roberts, T.C., Smith, P.A., Cirz, R.T., and Romesberg, F.E. (2007) *J. Am. Chem. Soc.*, **129**, 15830–15838.

58 Obst, U., Betschman, P., Lerner, C., Seiler, P., Diederich, F., Gramlich, V., Weber, L., Banner, D.W., and Schönholzer, P. (2000) *Helv. Chim. Acta*, **83**, 855–1048.

59 Yasuda, S., Yorimitsu, H., and Oshima, K. (2008) *Organometallics*, **27**, 4025–4027.

60 Piettre, S.R. and Baltzer, S. (1997) *Tetrahedron Lett.*, **38**, 1197–1200.

61 (a) Koizumi, Y., Seki, S., Acharya, A., Saeki, A., and Tagawa, S., (2004) *Chem. Lett.*, **33**, 1290–1291; (b) Dudek, S.P., Pouderoijen, M., Abbel, R., Schenning, A.P.H.J., and Meijer, E.W. (2005) *J. Am. Chem. Soc.*, **127**, 11763–11768; (c) Hiroki, K. and Kijima, M. (2005) *Chem. Lett.*, **34**, 942–943; (d) Wong, W.W.H., Jones, D.J., Yan, C., Watkins, S.E., King, S., Haque, S.A., Wen, X., Ghiggino, K.P., and Holmes, A.B. (2009) *Org. Lett.*, **11**, 975–978.

62 (a) Zembower, D.E. and Zhang, H. (1998) *J. Org. Chem.*, **63**, 9300–9305; (b) Nicolaou, K.C., Huang, X., Giuseppone, N., Rao, P.B., Bella, M., Reddy, M.V., and Snyder, S.A. (2001) *Angew. Chem., Int. Ed.*, **40**, 4705–4709; (c) Kaiser, M., Groll, M., Renner, C., Huber, R., and Moroder, L. (2002) *Angew. Chem., Int. Ed.*, **41**, 780–783; (d) Kaiser, M., Siciliano, C., Assfalg-Machleidt, I., Groll, M., Milbradt, A.G., and Moroder, L. (2003) *Org Lett.*, **5**, 3435–3437; (e) Yoburn, J.C. and Vranken, D.L.V. (2003) *Org. Lett.*, **5**, 2817–2820; (f) Zheng, X., Meng, W.-D., and Qing, F.-L. (2004) *Tetrahedron Lett.*, **45**, 8083–8085; (g) Nicolaou, K.C., Snyder, S.A., Giuseppone, N., Huang, X., Bella, M., Reddy, M.V., Rao, P.B., Koumbis, A.E., Giannakakou, P., and O'Brate, A. (2004) *J. Am. Chem. Soc.*, **126**, 10174–10182; (h) Nicolaou, K.C., Snyder, S.A., Huang, X., Simonsen, K.B., Koumbis, A.E., and Bigot, A. (2004) *J. Am. Chem. Soc.*, **126**, 10162–10173; (i) Davies, J.R., Kane, P.D., and Moody, C.J. (2005) *J. Org. Chem.*, **70**, 7305–7316; (j) Lépine, R. and Zhu, J. (2005) *Org. Lett.*, **7**, 2981–2984; (k) Shinohara, T., Deng, H., Snapper, M.L., and Hoveyda, A.H. (2005) *J. Am. Chem. Soc.*, **127**, 7334–7336; (l) Jones, S.B., He, L., and Castle, S.L. (2006) *Org. Lett.*, **8**, 3757–3760; (m) DiMauro, E.F. and Vitullo, J.R. (2006) *J. Org. Chem.*, **71**, 3959–3962; (n) Jia, Y., Bois-Choussy, M., and Zhu, J. (2007) *Org. Lett.*, **9**, 2401–2404; (o) Young, R.J., Borthwick, A.D., Brown, D., Burns-Kurtis, C.L., Campbell, M., Chan, C., Charbaut, M., Chung, C.W., Convery, M.A., Kelly, H.A., King, N.P., Kleanthous, S., Mason, A.M., Pateman, A.J., Patikis, A.N., Pinto, I.L., Pollard, D.R., Senger, S., Shah, G.P., Toomey, J.R., Watson, N.S., and Weston, H.E. (2008) *Bioorg. Med. Chem. Lett.*, **18**, 23–27; (p) Lee, S.J., Gray, K.C., Paek, J.S., and Burke, M.D.

(2008) *J. Am. Chem. Soc.*, **130**, 466–468; (q) Bourdreux, Y., Nowaczyk, S., Billaud, C., Mallinger, A., Willis, C., Murr, M.D.-E., Toupet, L., Lion, C., Gall, T.L., and Mioskowski, C. (2008) *J. Org. Chem.*, **73**, 22–26; (r) Bringmann, G., Rüdenauer, S., Bruhn, T., Benson, L., and Brun, R. (2008) *Tetrahedron*, **64**, 5563–5568; (s) Lin, J., Gerstenberger, B.S., Stessman, N.Y.T., and Konopelski, J.P. (2008) *Org. Lett.*, **10**, 3969–3972; (t) Michaux, J., Retailleau, P., and Campagne, J.-M. (2008) *Synlett*, 1532–1536; (u) Nielsen, D.K., Nielsen, L.L., Jones, S.B., Toll, L., Asplund, M.C., and Castle, S.L. (2009) *J. Org. Chem.*, **74**, 1187–1199.

63 (a) Iovine, P.M., Kellett, M.A., Redmore, N.P., and Therien, M.J. (2000) *J. Am. Chem. Soc.*, **122**, 8717–8727; (b) Sakai, N., Gerard, D., and Matile, S. (2001) *J. Am. Chem. Soc.*, **123**, 2517–2524; (c) Kurotobi, K., Tabata, H., Miyauchi, M., Murafuji, T., and Sugihara, Y. (2002) *Synthesis*, 1013–1016; (d) Han, J.W., Castro, J.C., and Burgess, K. (2003) *Tetrahedron Lett.*, **44**, 9359–9362; (e) Ni, W., Fang, H., Springsteen, G., and Wang, B. (2004) *J. Org. Chem.*, **69**, 1999–2007; (f) Zheng, X., Mulcahy, M.E., Horinek, D., Galeotti, F., Magnera, T.F., and Michl, J. (2004) *J. Am. Chem. Soc.*, **126**, 4540–4542; (g) Chang, M.C.Y., Pralle, A., Isacoff, E.Y., and Chang, C.J. (2004) *J. Am. Chem. Soc.*, **126**, 15392–15393; (h) Bell, T.D.M., Jacob, J., Angeles-Izquierdo, M., Fron, E., Nolde, F., Hofkens, J., Müllen, K., and Schryver, F.C.D. (2005) *Chem. Commun.*, 4973–4975; (i) Miller, E.W., Albers, A.E., Pralle, A., Isacoff, E.Y., and Chang, C.J. (2005) *J. Am. Chem. Soc.*, **127**, 16652–16659; (j) Wang, S., Hong, J.W., and Bazan, G.C., (2005) *Org. Lett.*, **7**, 1907–1910; (k) Welter, S., Salluce, N., Benetti, A., Rot, N., Belser, P., Sonar, P., Grimsdale, A.C., Müllen, K., Lutz, M., Spek, A.L., and Cola, L.D. (2005) *Inorg. Chem.*, **44**, 4706–4718; (l) Cheung, K.-M., Zhang, Q.-F., Chan, K.-W., Lam, M.H.W., Williams, I.D., and Leung, W.-H. (2005) *J. Organomet. Chem.*, **690**, 2913–2921; (m) Thiemann, F., Piehler, T., Haase, D., Saak, W., and Lützen, A. (2005) *Eur. J. Org. Chem.*, 1991–2001; (n) Yuan, S.-C., Chen, H.-B., Zhang, Y., and Pei, J. (2006) *Org. Lett.*, **8**, 5701–5704; (o) Albers, A.E., Okreglak, V.S., and Chang, C.J. (2006) *J. Am. Chem. Soc.*, **128**, 9640–9641; (p) Kiehne, U., Bunzen, J., Staats, H., and Lützen, A. (2007) *Synthesis*, 1061–1069; (q) Yang, Y.J. and Nocera, D.G. (2007) *J. Am. Chem. Soc.*, **129**, 8192–8198; (r) Jacobsen, M.F., Andersen, C.S., Knudsen, M.M., and Gothelf, K.V. (2007) *Org. Lett.*, **9**, 2851–2854; (s) Bhayana, B. and Wilcox, C.S. (2007) *Angew. Chem., Int. Ed.*, **46**, 6833–6836; (t) Fendt, L.-A., Fang, H., Plonska-Brzezinska, M.E., Zhang, S., Cheng, F., Braun, C., Echegoyen, L., and Diederich, F. (2007) *Eur. J. Org. Chem.*, 4659–4673; (u) Trokowski, R., Akine, S., and Nabeshima, T. (2008) *Chem. Commun.*, 889–890; (v) Noguchi, H., Shioda, T., Chou, C.-M., and Suginome, M. (2008) *Org. Lett.*, **10**, 377–380; (w) Bolink, H.J., Santamaria, S.G., Sudhakar, S., Zhen, C., and Sellinger, A. (2008) *Chem. Commun.*, 618–620; (x) Su, S.-J., Tanaka, D., Li, Y.-J., Sasabe, H., Takeda, T., and Kido, J. (2008) *Org. Lett.*, **10**, 941–944.

64 (a) Deng, X. and Cai, C. (2003) *Tetrahedron Lett.*, **44**, 815–817; (b) Chi, C., Im, C., Enkelmann, V., Ziegler, A., Lieser, G., and Wegner, G. (2005) *Chem. Eur. J.*, **11**, 6833–6845; (c) Iyer, P.K. and Wang, S. (2006) *Tetrahedron Lett.*, **47**, 437–439; (d) Jiang, Y., Lu, Y.-X., Cui, Y.-X., Zhou, Q.-F., Ma, Y., and Pei, J. (2007) *Org. Lett.*, **9**, 4539–4542; (e) Chi, C., Mikhailovsky, A., and Bazan, G.C. (2007) *J. Am. Chem. Soc.*, **129**, 11134–11145; (f) Pol, C.V.D., Bryce, M.R., Wielopolski, M., Atienza-Castellanos, C., Guldi, D.M., Filippone, S., and Martin, N. (2007) *J. Org. Chem.*, **72**, 6662–6671; (g) Vives, G., Gonzalez, A., Jaud, J., Launay, J.-P., and Rapenne, G. (2007) *Chem. Eur. J.*, **13**, 5622–5631; (h) Araki, H., Katoh, T., and Inoue, M. (2007) *Tetrahedron Lett.*, **48**, 3713–3717; (i) Kulasi, A., Yi, H., and Iraqi, A. (2007) *J. Polymer. Sci. A*, **23**, 5957–5967; (j) Zhang, M., Yang, C., Mishra, A.K., Pisula, W., Zhou, G., Schmaltz, B., Baumgarten, M., and Müllen, K. (2007) *Chem. Commun.*,

1704–1706; (k) Liu, B. and Dishari, S.K. (2008) *Chem. Eur. J.*, **14**, 7366–7375; (l) Figueira-Duarte, T.M., Simon, S.C., Wagner, M., Druzhinin, S.I., Zachariasse, K.A., and Müllen, K. (2008) *Angew. Chem., Int. Ed.*, **47**, 10175–10178; (m) Fang, F.-C., Chu, C.-C., Huang, C.-H., Raffy, G., Guerzo, A.D., Wong, K.-T., and Bassani, D.M. (2008) *Chem. Commun.*, 6369–6371.

65. (a) Broutin, P.-E., Čerňa, I., Campaniello, M., Leroux, F., and Colobert, F. (2004) *Org. Lett.*, **6**, 4419–4422; (b) Murata, M., Sambommatsu, T., Watanabe, S., and Masuda, Y. (2006) *Synlett*, 1867–1870.

66. Stadlwieser, J.F. and Dambaur, M.E. (2006) *Helv. Chem. Acta*, **89**, 936–946.

67. (a) Baudoin, O., Guénard, D., and Guéritte, F. (2000) *J. Org. Chem.*, **65**, 9268–9271; (b) Billingsley, K.L., Anderson, K.W., and Buchwald, S.L. (2006) *Angew. Chem., Int. Ed.*, **45**, 3484–3488; (c) Billingsley, K.L. and Buchwald, S.L. (2007) *J. Am. Chem. Soc.*, **129**, 3358–3366; (d) Billingsley, K.L. and Buchwald, S.L. (2008) *J. Org. Chem.*, **73**, 5589–5591.

68. Christophersen, C., Begtrup, M., Ebdrup, S., Petersen, H., and Vedsø, P. (2003) *J. Org. Chem.*, **68**, 9513–9516.

69. (a) Melaimi, M., Mathey, F., and Floch, P.L. (2001) *J. Organomet. Chem.*, **640**, 197–199; (b) Melaimi, M., Thoumazet, C., Ricard, L., and Floch, P.L. (2004) *J. Organomet. Chem.*, **689**, 2988–2994.

70. (a) Zhu, W. and Ma, D. (2006) *Org. Lett.*, **8**, 261–263; (b) Kleeberg, C., Dang, L., Lin, Z., and Marder, T.B. (2009) *Angew. Chem., Int. Ed.*, **48**, 5350–5354.

71. Mentzel, U.V., Tanner, D., and Tønder, J.E. (2006) *J. Org. Chem.*, **71**, 5807–5810.

72. Gravett, E.C., Hilton, P.J., Jones, K., and Péron, J.M. (2003) *Synlett*, 253–255.

73. (a) Nakamura, H., Fujiwara, M., and Yamamoto, Y. (2000) *Bull. Chem. Soc. Jpn.*, **73**, 231–235; (b) Combs, A.P., Zhu, W., Crawley, M.L., Glass, B., Polam, P., Sparks, R.B., Modi, D., Takvorian, A., McLaughlin, E., Yue, E.W., Wasserman, Z., Bower, M., Wei, M., Rupar, M., Ala, P.J., Reid, B.M., Ellis, D., Gonneville, L., Emm, T., Taylor, N., Yeleswaram, S., Li, Y., Wynn, R., Burn, T.C., Hollis, G., Liu, P.C.C., and Metcalf, B. (2006) *J. Med. Chem.*, **49**, 3774–3789.

74. (a) Wolan, A. and Zaidlwicz, M. (2003) *Org. Biomol. Chem.*, **1**, 3274–3276; (b) Horn, J., Marsden, S.P., Nelson, A., House, D., and Weingarten, G.G. (2008) *Org. Lett.*, **10**, 4117–4120.

75. (a) Setsune, J., Toda, M., Watanabe, K., Panda, P.K., and Yoshida, T. (2006) *Tetrahedron Lett.*, **47**, 7541–7544; (b) Pinkerton, D.M., Banwell, M.G., and Willis, A.C. (2007) *Org. Lett.*, **9**, 5127–5130.

76. Heinrich, M.R., Steglich, W., Banwell, M.G., and Kashman, Y. (2003) *Tetrahedron*, **59**, 9239–9247.

77. (a) Chng, L.L., Chang, C.J., and Nocera, D.G. (2003) *J. Org. Chem.*, **68**, 4075–4078; (b) Felber, B. and Diederich, F. (2005) *Helv. Chim. Acta*, **88**, 120–153; (c) Zhang, T.-G., Zhao, Y., Asselberghs, I., Persoons, A., Clays, K., and Therien, M.J. (2005) *J. Am. Chem. Soc.*, **127**, 9710–9720; (d) Cheng, F., Zhang, S., Adronov, A., Echegoyen, L., and Diederich, F. (2006) *Chem. Eur. J.*, **12**, 6062–6070; (e) Dahms, K. and Senge, M.O. (2008) *Tetrahedron Lett.*, **49**, 5397–5399.

78. (a) Baudoin, O., Décor, A., Cesario, M., and Guéritte, F. (2003) *Synlett*, 2009–2012; (b) Joncour, A., Décor, A., Liu, J.-M., Dau, M.-E.T.H., and Baudoin, O. (2007) *Chem. Eur. J.*, **13**, 5450–5465.

79. Chan, J.M.W. and Swager, T.M. (2008) *Tetrahedron Lett.*, **49**, 4912–4914.

80. (a) Selected references: Murr, M.D.-E., Nowaczyk, S., Gall, T.L., Mioskowski, C., Amekraz, B., and Moulin, C. (2003) *Angew. Chem., Int. Ed.*, **42**, 1289–1293; (b) Penhoat, M., Levacher, V., and Dupas, G. (2003) *J. Org. Chem.*, **68**, 9517–9520; (c) Pla, D., Marchal, A., Olsen, C.A., Albericio, F., and Álvarez, M. (2005) *J. Org. Chem.*, **70**, 8231–8234; (d) Casimiro-Garcia, A. and Schultz, A.G. (2006) *Tetrahedron Lett.*, **47**, 2739–2742; (e) Babudri, F., Cardone, A., Gioffi, C.T., Farinola, G.M., Naso, F., and Ragni, R. (2006) *Synthesis*, 1325–1332;

(f) Altemöller, M., Podlech, J., and Fenske, D. (2006) *Eur. J. Org. Chem.*, 1678–1684; (g) Wipf, P. and Furegati, M. (2006) *Org. Lett.*, **8**, 1901–1904; (h) Kasahara, T. and Kondo, Y. (2006) *Chem. Commun.*, 891–893; (i) Miller, W.D., Fray, A.H., Quatroche, J.T., and Sturgill, C.D. (2007) *Org. Process Res. Dev.*, **11**, 359–364; (j) Liu, W., Buck, M., Chen, N., Shang, M., Taylor, N.J., Asoud, J., Wu, X., Hasinoff, B.B., and Dmitrienko, G.I. (2007) *Org. Lett.*, **9**, 2915–2918; (k) Roach, S.L., Higuchi, R.I., Adams, M.E., Liu, Y., Karanewsky, D.S., Marschke, K.B., Mais, D.E., Miner, J.N., and Zhi, L. (2008) *Bioorg. Med. Chem. Lett.*, **18**, 3504–3508; (l) Duong, H.A., Chua, S., Huleatt, P.B., and Chai, C.L.L. (2008) *J. Org. Chem.*, **73**, 9177–9180; (m) Morrison, M.D., Hanthorn, J.J., and Pratt, D.A. (2009) *Org. Lett.*, **11**, 1051–1054.

81 (a) Selected references: Krämer, C.S., Zeitler, K., and Müller, T.J.J. (2001) *Tetrahedron Lett.*, **42**, 8619–8624; (b) Krämer, C.S., Zimmermann, T.J., Sailer M., and Müller, T.J.J. (2002) *Synthesis*, 1163–1170; (c) Wegner, H.A., Scott, L.T., and Meijere, A.D. (2003) *J. Org. Chem.*, **68**, 883–887; (d) Yu, L., Muthukumaran, K., Sazanovich, I.V., Kirmaier, C., Hindin, E., Diers, J.R., Boyle, P.D., Bocian, D.F., Holten, D., and Lindsey, J.S. (2003) *Inorg. Chem.*, **42**, 6629–6647; (e) Schmittel, M. and Kishore, R.S.K. (2004) *Org. Lett.*, **6**, 1923–1926; (f) Tam, V.K., Liu, Q., and Tor, Y. (2006) *Chem. Commun.*, 2684–2686; (g) Montes, V.A., Pohl, R., Shinar, J., and Anzenbacher, P., Jr. (2006) *Chem. Eur. J.*, **12**, 4523–4535; (h) Ronan, D., Jeannerat, D., Pinto, A., Sakai, N., and Matile, S. (2006) *New J. Chem.*, **30**, 168–176; (i) Saito, S., Yamaguchi, H., Muto, H., and Makino, T. (2007) *Tetrahedron Lett.*, **48**, 7498–7501; (j) Takeuchi, T., Akeda, K., Murakami, S., Shinmori, H., Inoue, S., Lee, W.S., and Hishiya, T. (2007) *Org. Biomol. Chem.*, **5**, 2368–2374; (k) Schiek, M., Al-Shamery, K., and Lützen, A. (2007) *Synthesis*, 613–621; (l) Sakai, N., Sisson, A.L., Bhosale, S., Fürstenberg, A., Banerji, N., Vauthey, E., and Matile, S. (2007) *Org. Biomol. Chem.*, **5**, 2560–2563; (m) Mora, F., Tran, D.-H., Oudry, N., Hopfgartner, G., Jeannerat, D., Sakai, N., and Matile, S. (2008) *Chem. Eur. J.*, **14**, 1947–1953.

82 (a) Takahashi, K., Takagi, J. Ishiyama, T., and Miyaura, N. (2000) *Chem. Lett.*, **2**, 126–127; (b) Takagi, J., Takahashi, K., Ishiyama, T., and Miyaura, N. (2002) *J. Am. Chem. Soc.*, **124**, 8001–8006.

83 (a) Shimizu, T., Satoh, T., Murakoshi, K., and Sodeoka, M. (2005) *Org. Lett.*, **7**, 5573–5576; (b) Gopalarathnam, A. and Nelson, S.G. (2006) *Org. Lett.*, **8**, 7–10.

84 Takagi, J., Kamon, A., Ishiyama, T., and Miyaura, N. (2002) *Synlett*, 1880–1882.

85 (a) Eastwood, P.R. (2000) *Tetrahedron Lett.*, **41**, 3705–3708; (b) Jin, B., Liu, Q., and Sulikowski, G.A. (2005) *Tetrahedron*, **61**, 401–408; (c) Ghosh, S., Kinney, W.A., Gauthier, D.A., Lawson, E.C., Hudlicky, T., and Maryanoff, B.E. (2006) *Can. J. Chem.*, **84**, 555–560; (d) López, S., Montenegro, J., and Saá, C. (2007) *J. Org. Chem.*, **72**, 9572–9581; (e) Kohno, K., Azuma, S., Choshi, T., Nobuhiro, J., and Hibino, S. (2009) *Tetrahedron Lett.*, **50**, 590–592.

86 (a) Murata, M., Oyama, T., Watanabe, S., and Masuda, Y. (2000) *Synthesis*, 778–780; (b) Lessard, S., Peng, F., and Hall, D.G. (2009) *J. Am. Chem. Soc.*, **131**, 9612–9613.

87 Winkler, J.D., Londregan, A.T., Ragains, J.R., and Hamann, M.T. (2006) *Org. Lett.*, **8**, 3407–3409.

88 Högermeier, J. and Reissig, H.U. (2006) *Synlett*, 2759–2762.

89 (a) Ferrali, A., Guarna, A. Galbo, F.L., and Occhiato, E.G. (2004) *Tetrahedron Lett.*, **45**, 5271–5274; (b) Occhiato, E.G., Galbo, F.L., and Guarna, A. (2005) *J. Org. Chem.*, **70**, 7324–7330.

90 Ishiyama, T., Ahiko, T.-a., and Miyaura, N. (1996) *Tetrahedron Lett.*, **37**, 6889–6892.

91 Dutheuil, G., Selander, N., Szabó, K.J., and Aggarwal, V.K. (2008) *Synthesis*, 2293–2297.

92 Ahiko, T.-a., Ishiyama, T., and Miyaura, N. (1997) *Chem. Lett.*, 811–812.

93 (a) Ramachandran, P.V., Pratihar, D., Biswas, D., Srivastava, A., and Reddy, M.V.R. (2004) *Org. Lett.*, **6**, 481–484; (b) Kabalka, G.W., Venkataiah, B., and Dong, G. (2004) *J. Org. Chem.*, **69**, 5807–5809.

94 Mitra, S., Gurrala, S.R., and Coleman, R.S. (2007) *J. Org. Chem.*, **72**, 8724–8736.

95 (a) Ito, H., Kawakami, C., and Sawamura, M. (2005) *J. Am. Chem. Soc.*, **127**, 16034–16035; (b) Ito, H., Ito, S., Sasaki, Y., Matsuura, K., and Sawamura, M. (2007) *J. Am. Chem. Soc.*, **129**, 14856–14857; (c) Ito, H., Sasaki, Y., and Sawamura, M. (2008) *J. Am. Chem. Soc.*, **130**, 15774–15775.

96 Ishiyama, T., Oohashi, Z., Ahiko, T.-a., and Miyaura, N. (2002) *Chem. Lett.*, 780–781.

97 Giroux, A. (2003) *Tetrahedron Lett.*, **44**, 233–235.

98 (a) Waltz, K.M. and Hartwig, J.F. (1997) *Science*, **277**, 211–213; (b) Waltz, K.M., Muhoro, C.N., and Hartwig, J.F. (1999) *Organometallics*, **18**, 3383–3393.

99 Chen, H.Y. and Hartwig, J.F. (1999) *Angew Chem., Int. Ed.*, **38**, 3391–3393.

100 (a) Kawamura, K. and Hartwig, J.F. (2001) *J. Am. Chem. Soc.*, **123**, 8422–8423; (b) Lawrence, J.D., Takahashi, M., Bae, C., and Hartwig, J.F. (2004) *J. Am. Chem. Soc.*, **126**, 15334–15335; (c) Hartwig, J.F., Cook, K.S., Hapke, M.D., Incarvito, C., Fan, Y., Webster, C.E., and Hall, M.B. (2005) *J. Am. Chem. Soc.*, **127**, 2538–2552; (d) Boebel, T.A. and Hartwig, J.F. (2008) *Organometallics*, **27**, 6013–6019.

101 Murphy, J.M., Lawrence, J.D., Kawamura, K., Incarvito, C., and Hartwig, J.F. (2006) *J. Am. Chem. Soc.*, **128**, 13684–13685.

102 Kondo, Y., Garcia-Guadrado, D., Hartwig, J.F., Boaen, N.K., Wagner, N.L., and Hillmyer, M.A. (2002) *J. Am. Chem. Soc.*, **124**, 1164–1165.

103 Olsson, V.J. and Szabó, K.J. (2007) *Angew. Chem., Int. Ed.*, **46**, 6891–6893.

104 Shimada, S., Batsanov, A.S., Howard, J.A.K., and Marder, T.B. (2001) *Angew Chem., Int. Ed.*, **40**, 2168–2171.

105 Ishiyama, T., Ishida, K., Takagi, J., and Miyaura, N. (2001) *Chem. Lett.*, 1082–1083.

106 (a) Kikuchi, T., Takagi, J., Ishiyama, T., and Miyaura, N. (2008) *Chem. Lett.*, **37**, 664–665; (b) Kikuchi, T., Takagi, J., Isou, H., Ishiyama, T., and Miyaura, N. (2008) *Chem. Asian. J.*, **3**, 2082–2090.

107 Chen, H., Schlecht, S., Semple, T.C., and Hartwig, J.F. (2000) *Science*, **287**, 1995–1997.

108 Cho, J.-Y., Iverson, C.N., and Smith, M.R., III (2000) *J. Am. Chem. Soc.*, **122**, 12868–12869.

109 Tse, M.K., Cho, J.-Y., and Smith, M.R., III (2001) *Org. Lett.*, **3**, 2831–2833.

110 (a) Iverson, C.N. and Smith, M.R., III (1999) *J. Am. Chem. Soc.*, **121**, 7696–7697; (b) Cho, J.-Y., Tse, M.K., Holmes, D., Maleczka, R.E., Jr., and Smith, M.R., III (2002) *Science*, **295**, 305–307; (c) Holmes, D., Chotana, G.A., Maleczka, R.E., Jr., and Smith, M.R., III (2006) *Org. Lett.*, **8**, 1407–1410; (d) Lokare, K.S., Staples, R.J., and Odom, A.L. (2008) *Organometallics*, **27**, 5130–5138.

111 (a) Takagi, J., Sato, K., Hartwig, J.F., Ishiyama, T., and Miyaura, N. (2002) *Tetrahedron Lett.*, **43**, 5649–5651; (b) Liversedge, I.A., Higgins, S.J., Giles, M., Heeney, M., and McCulloch, I. (2006) *Tetrahedron Lett.*, **47**, 5143–5146; (c) Stockmann, V., Eriksen, K., and Fiksdahl, L.A. (2008) *Tetrahedron*, **64**, 11180–11184.

112 (a) Ishiyama, T., Takagi, J., Hartwig, J.F., and Miyaura, N. (2002) *Angew Chem., Int. Ed.*, **41**, 3056–3058; (b) Ishiyama, T., Takagi, J., Yonekawa, Y., Hartwig, J.F., and Miyaura, N. (2003) *Adv. Synth. Catal.*, **345**, 1103–1106; (c) Shi, F., Smith, M.R., III, and Maleczka, R.E., Jr. (2006) *Org. Lett.*, **8**, 1411–1414; (d) Tzschucke, C.C., Murphy, J.M., and Hartwig, J.F. (2007) *Org. Lett.*, **9**, 761–767; (e) Harrison, P., Morris, L., Marder, T.B., and Steel, P.G. (2009) *Org. Lett.*, **11**, 3586–3589.

113 Ishiyama, T., Nobuta, Y., Hartwig, J.F., and Miyaura, N. (2003) *Chem. Commun.*, 2924–2925.

114 Murata, M., Odajima, H., Watanabe, S., and Masuda, Y. (2006) *Bull. Chem. Soc. Jpn.*, **79**, 1980–1982.
115 Tagata, T. and Nishida, M. (2004) *Adv. Synth. Catal.*, **346**, 1655–1660.
116 Frey, G.D., Rentzsch, C.F., Preysing, D.V., Scherg, T., Muhlhofer, M., Herdtweck, E., and Herrmann, W.A. (2006) *J. Organomet. Chem.*, **691**, 5725–5738.
117 Tagata, T., Nishida, M., and Nishida, A. (2009) *Tetrahedron Lett.*, **50**, 6176–6179.
118 Finke, A.D. and Moore, J.S. (2008) *Org. Lett.*, **10**, 4851–4854.
119 Cordes, J., Wessel, C., Harms, K., and Koert, U. (2008) *Synthesis*, 2217–2220.
120 (a) Coventry, D.N., Batsanov, A.S., Goeta, A.E., Howard, J.A.K., Marder, T.B., and Perutz, R.N. (2005) *Chem. Commun.*, **22**, 2172–2174; (b) Wanninger-Weiß, B.C. and Wagenknecht, H.-A. (2008) *Eur. J. Org. Chem.*, 64–71.
121 Chotana, G.A., Rak, M.A., and Smith, M.R., III (2005) *J. Am. Chem. Soc.*, **127**, 10539–10544.
122 Datta, A., Kollhofer, A., and Plenio, H. (2004) *Chem. Commun.*, **21**, 1508–1509.
123 Kurotobi, K., Miyauchi, M., Takakura, K., Murafuji, T., and Sugihara, Y. (2003) *Eur. J. Org. Chem.*, 3663–3665.
124 Chotana, G.A., Kallepalli, V.A. Maleczka, R.E., Jr., and Smith, M.R., III (2008) *Tetrahedron*, **64**, 6103–6114.
125 Ishida, N., Narumi, M., and Murakami, M. (2008) *Org. Lett.*, **10**, 1279–1281.
126 Mkhalid, I.A.I., Coventry, D.N., Albesa-Jove, D., Batsanov, A.S., Howard, J.A.K., Perutz, R.N., and Marder, T.B. (2006) *Angew. Chem., Int. Ed.*, **45**, 489–491.
127 (a) Paul, S., Chotana, G.A., Holmes, D., Reichle, R.C., Maleczka, R.E., Jr., and Smith, M.R., III (2006) *J. Am. Chem. Soc.*, **128**, 15552–15553; (b) Lo, W.F., Kaiser, H.M., Spannenberg, A., Beller, M., and Tse, M.K. (2007) *Tetrahedron Lett.*, **48**, 371–375.
128 (a) Hata, H., Shinokubo, H., and Osuka, A. (2005) *J. Am. Chem. Soc.*, **127**, 8264–8265; (b) Hata, H., Yamaguchi, S., Mori, G., Nakazono, S., Katoh, T., Takatsu, K., Hiroto, S., Shinokubo, H., and Osuka, A. (2007) *Chem. Asian J.*, **2**, 849–859; (c) Mori, G., Shinokubo, H., and Osuka, A. (2008) *Tetrahedron Lett.*, **49**, 2170–2172.
129 Ohmura, T., Kijima, A., and Suginome, M. (2009) *J. Am. Chem. Soc.*, **131**, 6070–6071.
130 Boebel, T.A. and Hartwig, J.F. (2008) *J. Am. Chem. Soc.*, **130**, 7534–7535.
131 Ishiyama, T., Isou, H., Kikuchi, T., and Miyaura, N. (2010) *Chem. Commun.*, **46**, 159–161.
132 (a) Caron, S. and Hawkins, J.M. (1998) *J. Org. Chem.*, **63**, 2054–2055; (b) Kristensen, J., Lysén, M., Vedsø, P., and Begtrup, M. (2001) *Org. Lett.*, **3**, 1435–1437.

# 3
# Transition Metal-Catalyzed Element-Boryl Additions to Unsaturated Organic Compounds

*Michinori Suginome and Toshimichi Ohmura*

## 3.1
### Introduction

Increasing demands for highly functionalized, stereochemically defined organoboronic acid derivatives have prompted the development of new B−C bond forming reactions. In addition to their conventional synthesis on the basis of uncatalyzed hydroboration and transmetalation, that is, nucleophilic substitution on the boron atom, more attention is currently focused on the transition metal-catalyzed borylation reactions. In particular, catalytic additions of boron-element σ-bonds to unsaturated organic molecules have been studied intensively in recent years because such addition reactions provide new efficient synthetic routes to highly functionalized organoboronic acid derivatives. In addition, such catalytic borylations have advantages of high functional group compatibility and possibilities of efficient chemo-, regio-, and stereoselectivities, which are often switchable simply by the choice of catalysts.

We can trace back two important origins of the transition metal-catalyzed element-boryl addition reactions, which are covered in this chapter. One is transition metal-catalyzed bis-silylation chemistry, where palladium, nickel, and platinum catalysts activate silicon–silicon bonds and enable a variety of catalytic additions to unsaturated organic molecules [1, 2]. Although hydrosilylation had been known at the time, the possibility and synthetic applicability of the addition of σ-bonds between two nonhydrogen elements are clearly demonstrated by the chemistry. The other important origin is the finding on the catalytic hydroboration by Männig and Nöth [3]. The study demonstrated the possibility of using transition metal catalysts in hydroboration, which at the time was recognized as a "catalyst-free" reaction. Nöth demonstrated in his first paper on the catalytic hydroboration that use of a rhodium catalyst can change the chemoselectivity dramatically [3]. These research activities eventually led to the key finding on the Pt-catalyzed diboration reaction of alkynes by Miyaura and coworkers in 1993 [4], which is recognized as the first successful catalytic addition of σ-bonds between boron and a nonhydrogen element, bringing about rapid development of the element-boryl addition reactions.

This chapter focuses on the transition metal-catalyzed reactions in which boryl and nonhydrogen groups are introduced in a single catalytic cycle. Additions of B−B

*Boronic Acids: Preparation and Applications in Organic Synthesis, Medicine and Materials*, Second Edition.
Edited by Dennis G. Hall.
© 2011 Wiley-VCH Verlag GmbH & Co. KGaA. Published 2011 by Wiley-VCH Verlag GmbH & Co. KGaA.

(diboration), B—Sn (stannaboration), B—Si (silaboration), B—C (carboboration), and B—S (thioboration) to unsaturated organic molecules are included in this section [5]. Catalytic hydroboration is not treated in this chapter. In addition to the "direct additions," in which the two added groups come from a single reactant, the chapter includes some related element-boryl addition reactions in which the two incorporated groups are originated from two separate reactants (Scheme 3.1). The latter processes usually involve transmetalation in their catalytic cycle and therefore are classified as "transmetalative element-boryl addition" in this chapter.

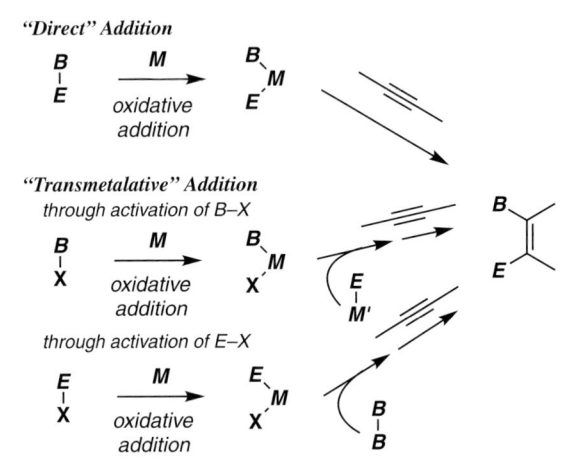

"Direct" Addition

"Transmetalative" Addition
through activation of B–X

through activation of E–X

**Scheme 3.1** Direct and transmetalative element-boryl additions (B, boryl groups; E, nonhydrogen elements; X, halogen; M′, main group metal; M, transition metal complexes).

## 3.2
## Diboration

### 3.2.1
### Diboron Reagents for Diboration

Diboration, that is, addition of a boron–boron bond across a carbon–carbon unsaturated bond, was originally recognized as a noncatalytic reaction of diboron tetrahalides with unsaturated hydrocarbons [6]. Although this reaction is potentially important as a method for simultaneous introduction of two boryl groups to organic frameworks, less attention has been paid to application in organic synthesis because of the highly Lewis acidic, unstable nature of such diborons. Development of transition metal-catalyzed methods utilizing much less reactive diboron tetraalkoxides made diboration chemistry a valuable tool for organic synthesis [7]. Air- and moisture-stable, commercially available bis(pinacolato)diboron ($B_2pin_2$) [8] is most frequently used for diborations (Figure 3.1). The diboration products are generally purified by distillation or silica gel column chromatography or used

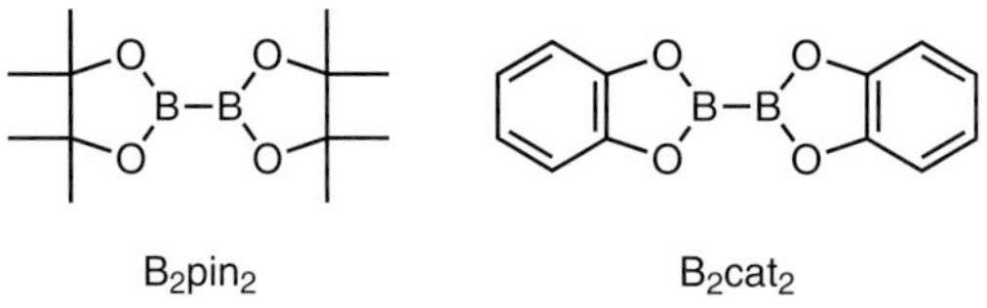

## 3.2 Diboration

**B₂pin₂**

**B₂cat₂**

**Figure 3.1** Representative tetraalkoxydiborons for catalytic diboration.

directly for further conversion of the carbon–boron bond to other functional groups. Bis(catecholato)diboron ($B_2cat_2$) [9] is also used for catalytic diboration. The Bcat groups in the diboration product are often converted to both Bpin by transesterification with pinacol and free boronic acids $B(OH)_2$ by hydrolysis.

### 3.2.2
### Diboration of Alkynes

Although diboron tetraalkoxides such as $B_2pin_2$ and $B_2cat_2$ are stable and easy to handle, the B–B bonds are totally inert for addition to C–C multiple bonds in the absence of transition metal catalysts. In 1993, the first catalytic diboration utilizing diboron tetraalkoxides was achieved in the reaction with alkynes using a platinum (0) catalyst (Scheme 3.2) [4]. In the presence of a catalytic amount of $Pt(PPh_3)_4$, $B_2pin_2$ adds to alkynes in DMF at 80 °C to give cis-1,2-diborylalkenes **1** through selective syn-1,2-addition. Other platinum complexes such as $Pt(CO)_2(PPh_3)_2$ and $Pt(CH_2=CH_2)(PPh_3)_2$ are also effective for catalytic alkyne diboration with $B_2pin_2$ under similar reaction conditions [10, 11]. Room-temperature diboration of alkynes is achieved by combined use of $B_2cat_2$ with a phosphine-free platinum catalyst such as $PtCl_2(cod)_2$ or $Pt(NHC)[O(SiMe_2CH=CH_2)_2]$ (**2**) [12, 13] and $B_2pin_2$ with a $Pt(norbornene)_3/PPh_3$ (Pt/L = 1/1) catalyst [14]. A wide array of alkynes including both terminal and internal alkynes bearing aromatic and aliphatic substituents are applicable to the diboration (entries 1 and 2, Table 3.1). Many functional groups are tolerable under the platinum-catalyzed diboration conditions: alkynes having alkenyl, chloro, cyano, epoxy, ester (entry 1, Table 3.1), and ketone groups can be

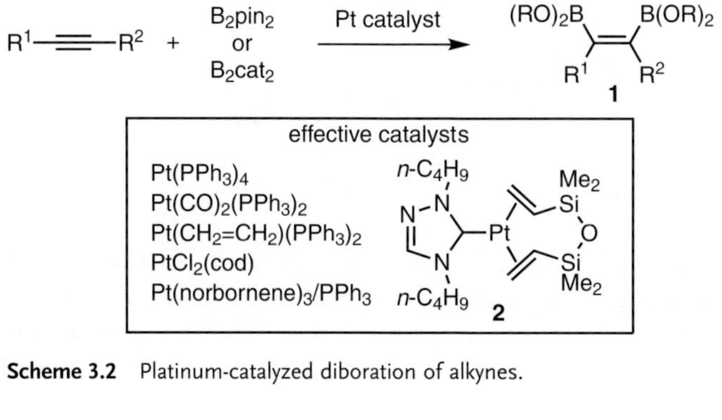

**Scheme 3.2** Platinum-catalyzed diboration of alkynes.

**Table 3.1** Selected examples of diboration reactions.

| Entry | Substrate | Conditions | Product | Reference |
|---|---|---|---|---|
| 1 | 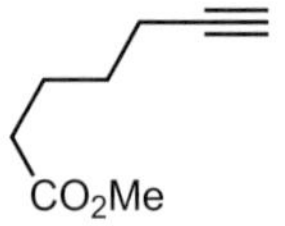 CH₂=CH–CH₂CH₂–CO₂Me (alkene with CO₂Me tether) | $B_2pin_2$, Pt(PPh$_3$)$_4$ (3 mol%), DMF, 80 °C, 24 h | 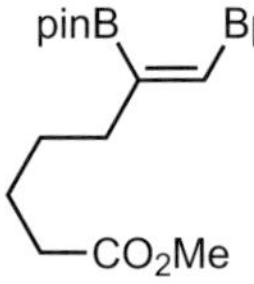 pinB, Bpin across double bond, CO₂Me tether (89%) | [10] |
| 2 |  Ph—≡—Ph | $B_2pin_2$, Pt(PPh$_3$)$_4$ (3 mol%), DMF, 80 °C, 24 h | 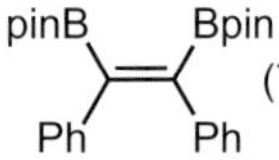 pinB, Bpin on Ph,Ph alkene (79%) | [4] |
| 3 |  catB—≡—Bcat | $B_2cat_2$, Pt(cod)$_2$ (4 mol%), toluene, 40 °C, 48 h | 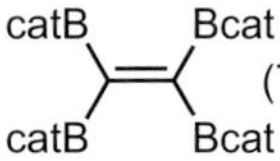 catB, Bcat / catB, Bcat alkene (70%) | [16] |
| 4 | 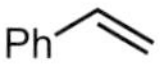 Ph–CH=CH$_2$ | $B_2pin_2$, Pt(dba)$_2$ (3 mol%), toluene, 50 °C, 1 h | 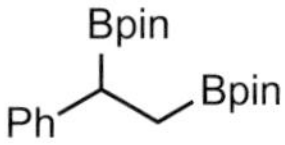 Ph–CH(Bpin)–CH$_2$Bpin (86%) | [19a] |
| 5 |  cyclopentene | $B_2pin_2$, Pt(dba)$_2$ (3 mol%), toluene, 50 °C, 1 h | 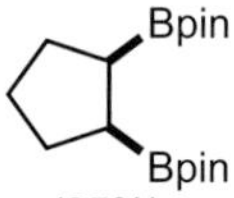 cis-1,2-bis(Bpin)cyclopentane (85%) | [19a] |
| 6 | 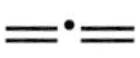 H$_2$C=C=CH$_2$ (allene) | $B_2pin_2$, Pt(PPh$_3$)$_4$ (3 mol%), toluene, 80 °C, 16 h | 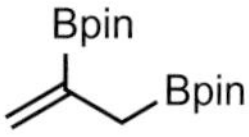 CH$_2$=C(Bpin)–CH$_2$Bpin (99%) | [29] |
| 7 | 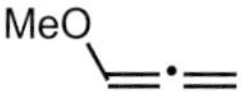 MeO–CH=C=CH$_2$ | $B_2pin_2$, Pt(dba)$_2$/PCy$_3$ (3 mol%), toluene, rt | 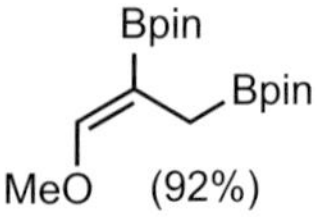 MeO–CH=C(Bpin)–CH$_2$Bpin (92%) | [29] |
| 8 | 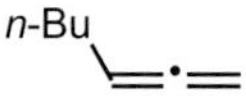 n-Bu–CH=C=CH$_2$ | $B_2pin_2$, Pt(dba)$_2$/PCy$_3$ (3 mol%), toluene, rt | 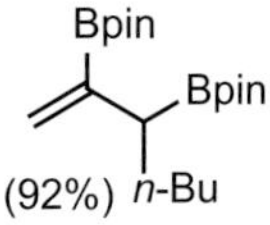 CH$_2$=C(Bpin)–CH(Bpin)(n-Bu) (92%) | [29] |
| 9 |  1,3-butadiene | $B_2pin_2$, Pt(PPh$_3$)$_4$ (3 mol%), toluene, 80 °C, 16 h |  pinB–CH$_2$–CH=CH–CH$_2$–Bpin (95%, Z > 99%) | [33] |

## 3.2 Diboration

**Table 3.1** (Continued)

| Entry | Substrate | Conditions | Product | Reference |
|---|---|---|---|---|
| 10 |  Me-CH=CH-C≡CH (but-1-en-3-yne with Me) | B₂pin₂<br>Pt(PPh₃)₄ (3 mol%)<br>toluene, 80 °C, 16 h | 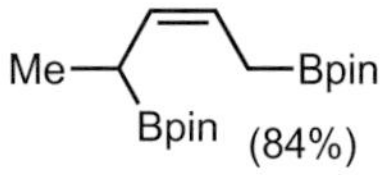 Me–C(=CH₂)–C(Bpin)=CH(Bpin) (84%) | [19a] |
| 11 | 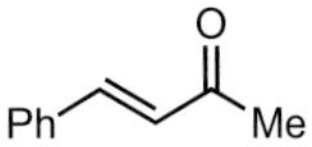 Ph–CH=CH–C(O)Me | B₂pin₂<br>Pt(CH₂=CH₂)(PPh₃)₂<br>(5 mol%)<br>toluene, 80 °C | 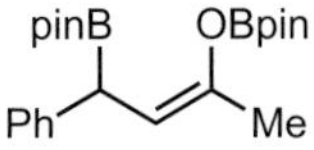 Ph–CH(pinB)–CH=C(OBpin)Me<br>**23** (90%) | [37a] |
| 12 | 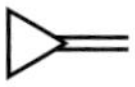 cyclopropylidene-methylene (allene with cyclopropyl) | B₂pin₂<br>Pt(PPh₃)₄ (3 mol%)<br>toluene, 80 °C, 5 h | 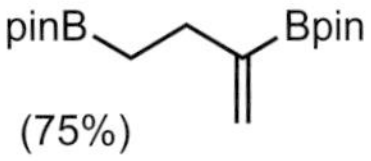 pinB–CH₂–C(=CH–Bpin) (75%) | [44] |
| 13 | 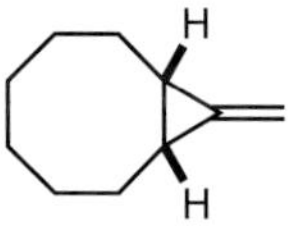 bicyclic methylenecyclopropane | B₂pin₂<br>Pt(PPh₃)₄ (3 mol%)<br>toluene, 80 °C, 2 h | 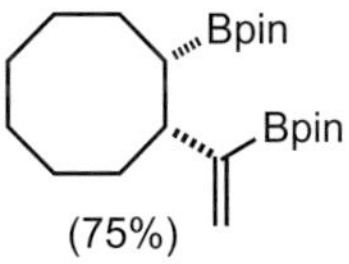 cyclooctane-fused bis(Bpin)methylene (75%) | [44] |
| 14 | 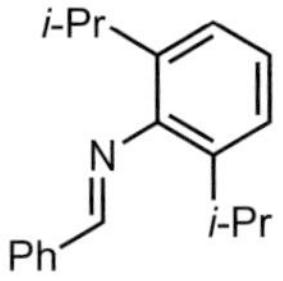 2,6-(i-Pr)₂C₆H₃–N=CH–Ph | B₂cat₂<br>Pt(cod)Cl₂ (3 mol%)<br>benzene, 25 °C | 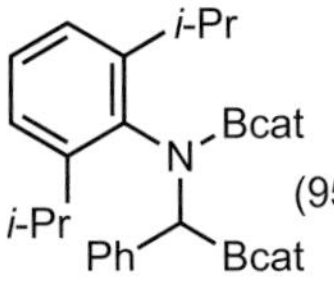 2,6-(i-Pr)₂C₆H₃–N(Bcat)–CH(Bcat)–Ph (95%) | [12] |
| 15 | H₂C=N₂ | B₂pin₂<br>Pt(PPh₃)₄ (3 mol%)<br>Et₂O, 0 °C | 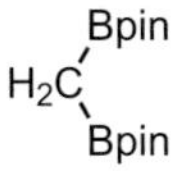 H₂C(Bpin)₂ (82%) | [45] |

used for reaction with B₂pin₂. Diborations of 1,3-diynes [11], 1-borylalkynes [15], 1,2-diborylalkynes (entry 3, Table 3.1) [16, 17], and alkynyl posphonates [15] are also reported.

A reaction mechanism starting from oxidative addition of the B−B bond to Pt(0) to form (boryl)₂Pt(II) complex **3** is proposed (Scheme 3.3). The complex **3** then undergoes insertion of the C−C triple bond to the Pt−B bond to afford (alkenyl)(boryl)Pt(II) **4**, which is followed by reductive elimination to give diborylated alkene with regeneration of Pt(0). The elementary steps were confirmed in stoichiometric reactions of the corresponding platinum complexes **5** and **6** (Schemes 3.4 and 3.5) [10, 18].

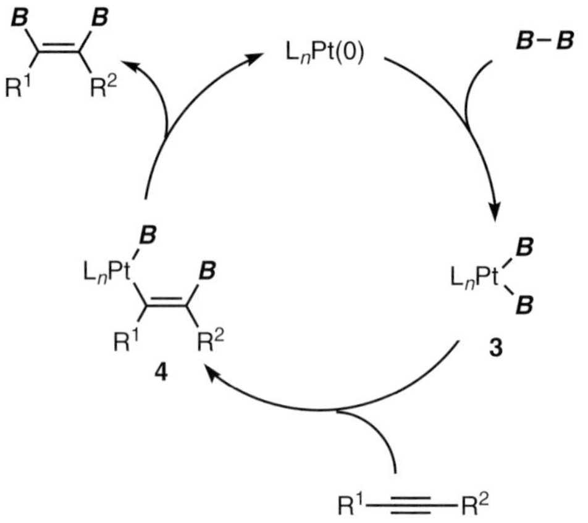

**Scheme 3.3** Reaction mechanism for platinum-catalyzed alkyne diboration.

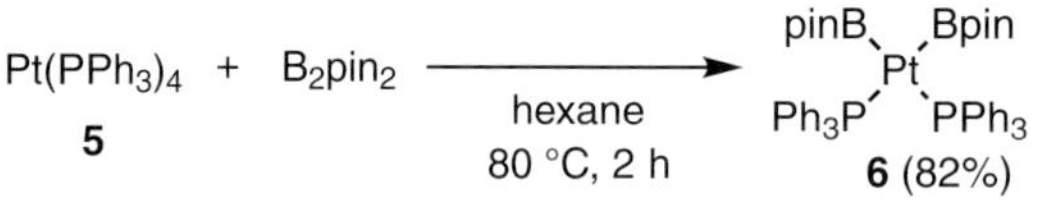

**Scheme 3.4** Stoichiometric reaction of diboron with a Pt(0) complex.

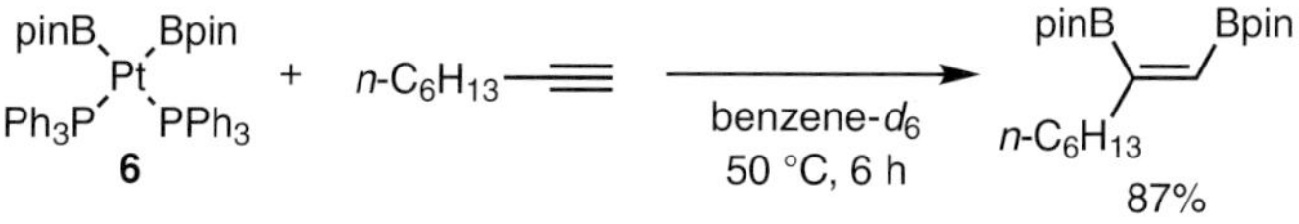

**Scheme 3.5** Stoichiometric reaction of Pt(Bpin)$_2$ with 1-octyne.

Representative catalytic borylation reactions are summarized in Table 3.1 with typical reaction conditions and products.

### 3.2.3
### Diboration of Alkenes, Allenes, 1,3-Dienes, and Methylenecyclopropanes

In sharp contrast to alkyne diboration, in which only platinum complexes are effective, alkene diboration is achieved with various transition metal complexes such as platinum [12, 19], rhodium [20], palladium [21], copper [22], silver [23], and gold [20a, 24] complexes (Scheme 3.6 and entries 4 and 5 in Table 3.1). The alkene diboration catalyzed by such transition metal catalysts proceeds with *cis*-1,2-addition to give 1,2-diborylalkanes. The order of reactivity of alkenes in catalytic diboration is as follows: terminal alkenes > strained cyclic alkenes > internal alkenes. The platinum-catalyzed diboration is typically carried out at elevated

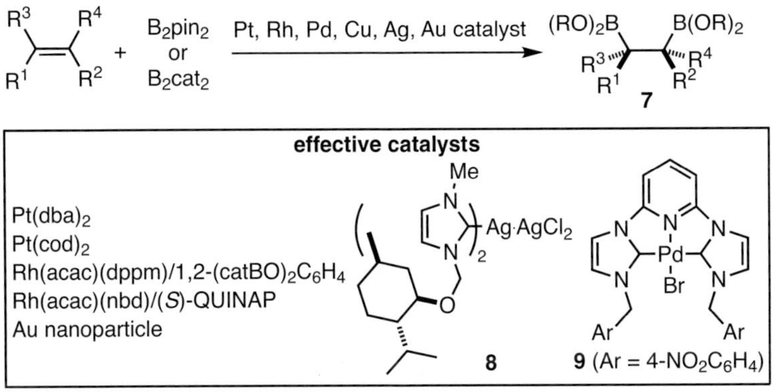

## 3.2 Diboration

**Scheme 3.6** Catalytic diboration of alkenes.

temperatures with $B_2pin_2$ using $Pt(dba)_2$ catalyst [19a] or at room temperature with $B_2cat_2$ in the presence of a $Pt(cod)_2$ catalyst [19b]. The platinum-catalyzed diboration is limited to terminal alkenes and strained cyclic alkenes. Use of a Ag (NHC)$_2$ **8** [23] or CuCl(SIPr) [22] catalysts with $B_2cat_2$ is also effective for certain alkenes. On the contrary, rhodium catalysts such as Rh(acac)(dppm)/1,2-(catBO)$_2C_6H_4$ [20b] or Rh(acac)(nbd)/(S)-QUINAP [20d] show higher reactivity in diboration of alkenes with $B_2cat_2$, resulting in successful expansion of the substrate scope to both aromatic and aliphatic internal alkenes. The combination of rhodium catalysts and $B_2cat_2$ is crucial for selective diboration; using $B_2pin_2$, dehydrogenative borylation becomes the major reaction pathway [25]. The C−N−C pincer-type palladium **9** [21] and gold nanoparticle catalysts [24] efficiently catalyze diboration of 1-phenyl-1-propane with $B_2cat_2$.

Asymmetric diboration of alkenes provides an attractive route to enantioenriched organoboron compounds. Attempts at asymmetric induction were initially made in diboration of styrenes with optically active chiral diboron reagents in the presence of an achiral platinum complex, affording moderate diastereoselection [26]. Two efficient systems for enantioselective diboration with achiral diborons have been established to date: $B_2cat_2$/Rh(acac)(nbd)/(S)-QUINAP system [20d, 27] and $B_2pin_2$/Pt$_2$(dba)$_3$/chiral phosphoramidite **11** system (Scheme 3.7) [28]. The former rhodium-catalyzed system is especially effective for 1,2-disubstituted (E)-alkenes, affording syn-1,2-diborylated alkanes in good yields with up to 98% enantiomeric excesses [20d, 27]. As for terminal monosubstituted alkenes, only those having a tertiary alkyl group give high enantioselectivity, whereas sterically less demanding alkenes, such as 1-octene and styrene, give moderate to poor enantioselectivities. On the other hand, the $B_2pin_2$/Pt$_2$(dba)$_3$/chiral phosphoramidite system is effective for a broad range of terminal monosubstituted alkenes regardless of substituents, although internal alkenes are not reactive [28].

Diboration of allenes is highly attractive because the reaction provides β-borylallylboranes that can be useful synthetic intermediates bearing both allylborane and alkenylborane moieties. Platinum complexes such as Pt(PPh$_3$)$_4$ and Pt(dba)$_2$/PCy$_3$

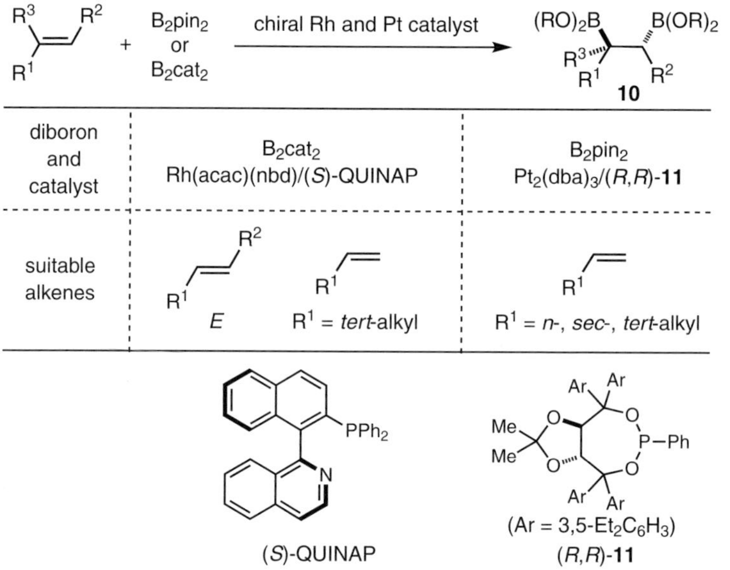

**Scheme 3.7** Enantioselective diboration of alkenes.

are effective catalysts for addition of $B_2pin_2$ to allene ($C_3H_4$) and various substituted allenes (Scheme 3.8 and entries 6–8 in Table 3.1) [29]. In the platinum-catalyzed diboration of substituted allenes, the addition takes place at the internal or terminal C−C double bond, depending on the substituents on the allene and the ligands on platinum. In contrast, regioselectivity of diboration of terminal allenes is effectively controlled by the choice of palladium catalysts. Selective diboration of the terminal C−C double bond takes place with $Pd(dba)_2$/3-iodo-2-methyl-2-cyclohexene as a catalyst [30], whereas a $Pd_2(dba)_3$/$PCy_3$ (Pd/P = 1/1) catalyst promotes selective diboration at the internal C−C double bond [31]. The internal addition under the latter conditions gives chiral β-borylallylboranes **12**. A palladium catalyst bearing a chiral phosphoramidite **14** efficiently catalyzes enantioselective diboration to give

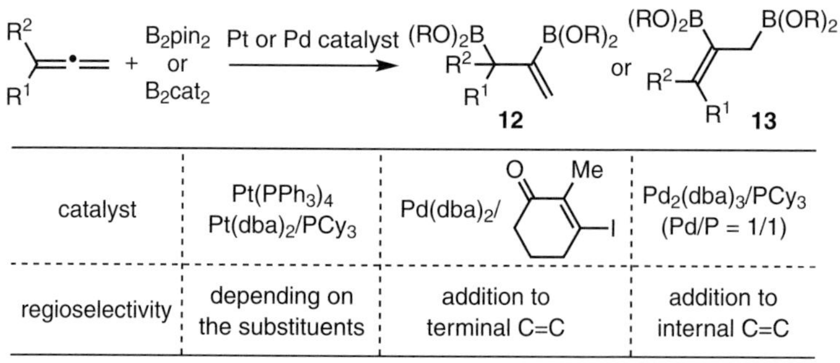

**Scheme 3.8** Catalytic diboration of allenes.

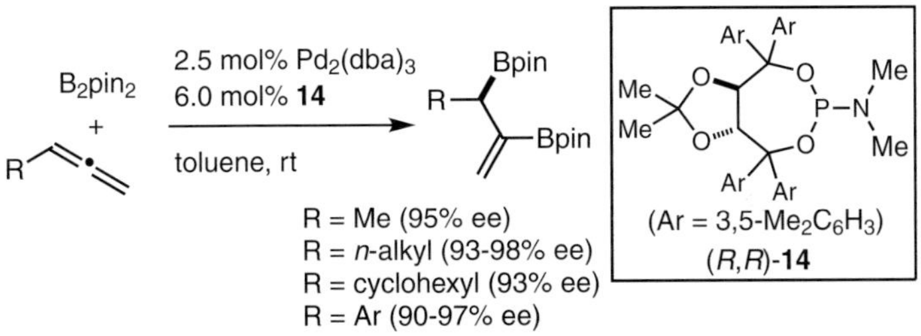

**Scheme 3.9** Enantioselective diboration of terminal allenes.

enantioenriched β-borylallylboranes with up to 98% ee (Scheme 3.9) [31]. The reaction mechanism of the platinum- and palladium-catalyzed diboration of allenes was investigated by deuterium labeling experiment and DFT calculations [31b, 32].

Diboration of 1,3-dienes provides allylboron compounds difficult to synthesize by other methods. Platinum catalysts such as Pt(PPh$_3$)$_4$ and Pt(dba)$_2$/PR$_3$ are effective for the addition of B$_2$pin$_2$ to 1,3-butadiene, isoprene, 2,3-dimethyl-1,3-butadiene, and other 1-substituted 1,3-butadienes, giving (Z)-1,4-diboryl-2-butenes **15** through selective 1,4-addition (Scheme 3.10, entries 9 and 10 in Table 3.1) [19a, 33, 34]. Regioselective 1,2-diboration takes place to give **16** in the reaction of B$_2$pin$_2$ with 1,3-pentadiene in the presence of a phosphine-free Pt(dba)$_2$ catalyst (Scheme 3.11) [19a]. Under the phosphine-free conditions, reaction of isoprene with B$_2$pin$_2$ affords a diborative dimerization product **17** in a regio- and stereoselective manner (Scheme 3.12) [33]. Diborative C—C bond formation has also been achieved in the reaction of 1,3-diene, aldehyde, and B$_2$pin$_2$ in the presence of a Ni(cod)$_2$/PCy$_3$ catalyst (Scheme 3.13) [35]. The three-component coupling may proceed through formation of an oxanickelacyclopentane intermediate **18**. Other diborons such as B$_2$cat$_2$ and a chiral diboron prepared from diethyl L-tartrate are used in the platinum-catalyzed diboration [36]. 1,4-Diboration of 1-substituted 1,3-butadienes affords allylboranes having a boron-substituted stereogenic carbon center. Enantioselective diboration has been achieved with up to 96% ee using a platinum catalyst bearing a chiral TADDOL-derived phosphonite **20** (Scheme 3.14) [34].

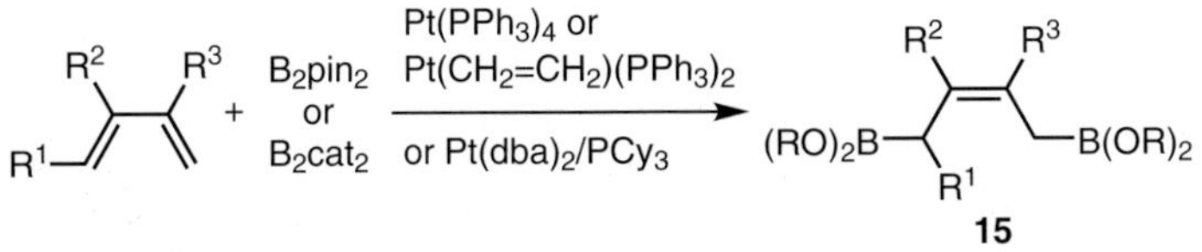

**Scheme 3.10** Catalytic diboration of 1,3-dienes.

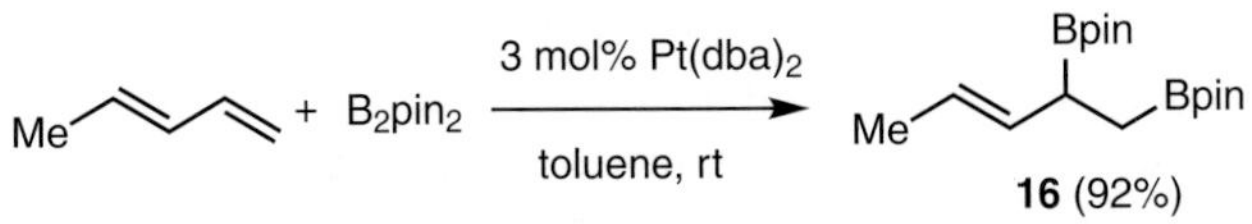

**Scheme 3.11** 1,2-Diboration of 1,3-pentadiene.

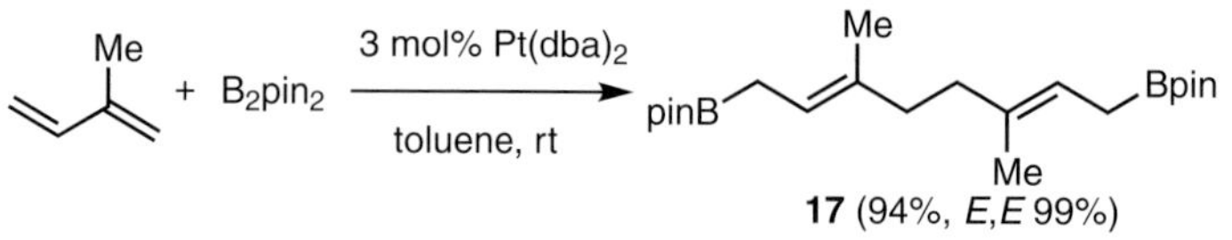

**Scheme 3.12** Diborative dimerization of isoprene.

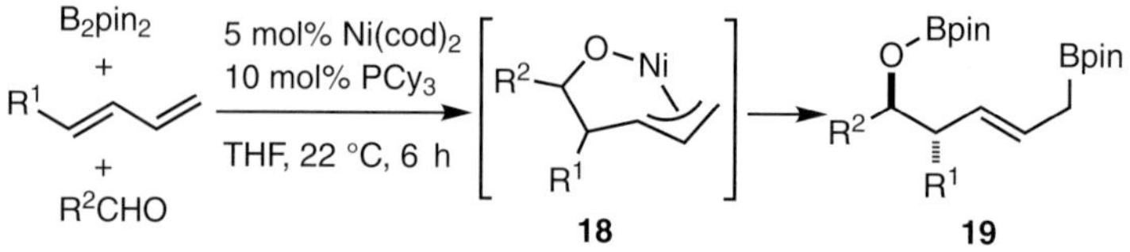

**Scheme 3.13** Diborative coupling of 1,3-dienes with aldehydes.

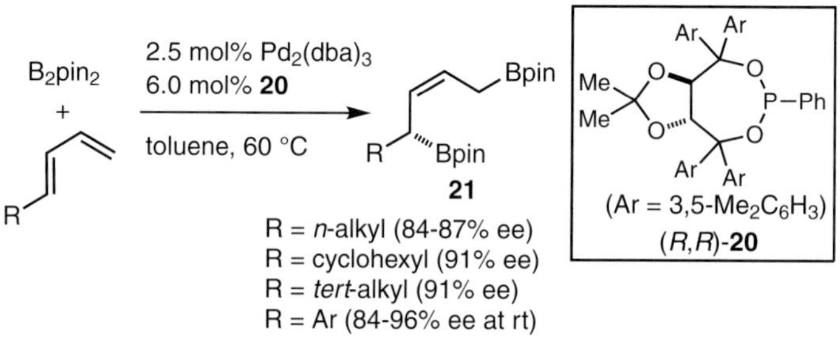

**Scheme 3.14** Enantioselective diboration of 1,3-dienes.

Addition of diboron reagents to α,β-unsaturated carbonyl compounds has received increasing attention because it provides an efficient synthetic route to β-boryl carbonyl compounds 22 (Scheme 3.15). Platinum [37], copper [38], rhodium [39], and nickel [40] complexes promote the catalytic reaction that proceeds through activation of the B—B bond through oxidative addition or transmetalation. Formation of diborylated compounds 23 is observed in a platinum-catalyzed reaction of α,β-unsaturated ketones and esters (entry 11 in Table 3.1) [37a, 37c]. Instead of isolating the unstable diborylated products, β-monoborylated carbonyl compounds 22 are isolated. Platinum complexes bearing PPh$_3$ or diimine ligand, bis(phenylimino)acenaphthene, show good catalyst activity for the reactions of cyclic and acyclic α,β-unsaturated ketones, acrylates, and cinnamaldehyde derivatives [37].

Conjugate borylation of α,β-unsaturated carbonyl compounds is also catalyzed by copper salts, such as CuCl/AcOK [38c] and CuOTf/P(n-Bu)$_3$ [38b] in DMF. An alternative catalyst system, CuCl/DPEphos/t-BuONa/MeOH in THF, is reported to be more effective for a wider range of substrates, including α,β-unsaturated nitriles, amides, and phosphonates [38d, 38e]. In these reactions, a large acceleration of reaction rate is observed by addition of MeOH. A similar enhancement of reaction

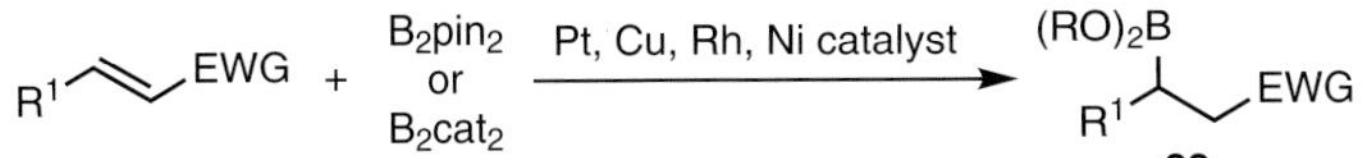

## 3.2 Diboration

$$R^1\diagup\!\!\!\diagdown\text{EWG} + \begin{array}{c}B_2pin_2\\ \text{or}\\ B_2cat_2\end{array} \xrightarrow{\text{Pt, Cu, Rh, Ni catalyst}} \underset{\mathbf{22}}{\underset{R^1}{(RO)_2B}\diagup\!\!\!\diagdown\text{EWG}}$$

E = CHO, COR$^2$, CO$_2$R$^2$, CONR$^2$R$^3$, CN, P(O)R$^2{}_2$

| | | | |
|---|---|---|---|
| catalyst and conditions | Pt(CH$_2$=CH$_2$)(PPh$_3$)$_2$ Pt(PPh$_3$)$_4$ 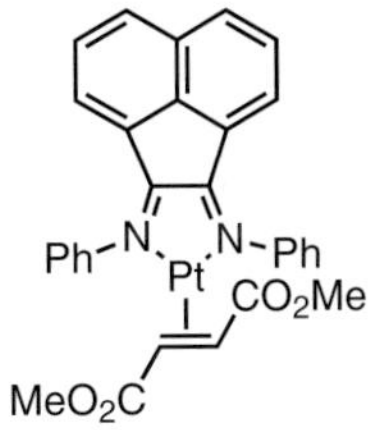 [Pt complex with Ph-N, N-Ph, CO$_2$Me, MeO$_2$C] | CuCl/AcOK or CuOTf/P(n-Bu)$_3$ in DMF | CuCl/DPEphos/ t-BuONa/MeOH in THF |
| suitable substrates (EWG) | CHO, COR$^2$, CO$_2$R$^2$ | COR$^2$ | COR$^2$, CO$_2$R$^2$, CONR$^2$R$^3$, CN, P(O)(OEt)$_2$ |
| catalyst and conditions | Cu(IPr)(OMe) in MeOH/THF | Ni(cod)$_2$/PCy$_3$ with Cs$_2$CO$_3$, H$_2$O in toluene/MeOH | 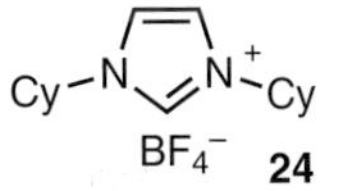 Cy-N⌒N-Cy BF$_4^-$ **24** t-BuONa in THF |
| suitable substrates (EWG) | CHO | CO$_2$R$^2$, CONR$^2$R$^3$ | COR$^2$, CO$_2$R$^2$ |

**Scheme 3.15** Catalytic conjugate borylation of α,β-unsaturated carbonyl compounds.

efficiency by protic additives is also observed in Ni(cod)$_2$/PCy$_3$-catalyzed conditions, in which smooth borylation takes place even to less reactive α,β-unsaturated amides having multiple substituents on the C–C double bond [40]. Selective conjugate addition of B$_2$pin$_2$ to crotonaldehydes proceeds by a copper catalyst bearing a N-heterocyclic carbene ligand, Cu(IPr)(OMe), in the presence of MeOH to give β-boryl aldehydes [38f]. The 1,4-addition of diboron to α,β-unsaturated ketones is also catalyzed by organocatalysts in the absence of transition metals. The B–B bond of B$_2$pin$_2$ is activated by nucleophilic attack of N-heterocyclic carbene generated from **24**, undergoing insertion of electron-deficient C–C double bond to give the corresponding β-borylketones and esters [41].

Enantioselective addition of diborons to α,β-unsaturated carbonyl compounds has been achieved with chiral copper [38e, 42] and rhodium [43] catalysts (Scheme 3.16). Optically active chiral bidentate ferrocenylphosphines, for example, (R)-(S)-Josiphos and (R,S)-Taniaphos, are effective in the CuCl/t-BuONa/MeOH catalyst system, leading to highly enantioselective addition of B$_2$pin$_2$ to α,β-unsaturated nitriles [42a],

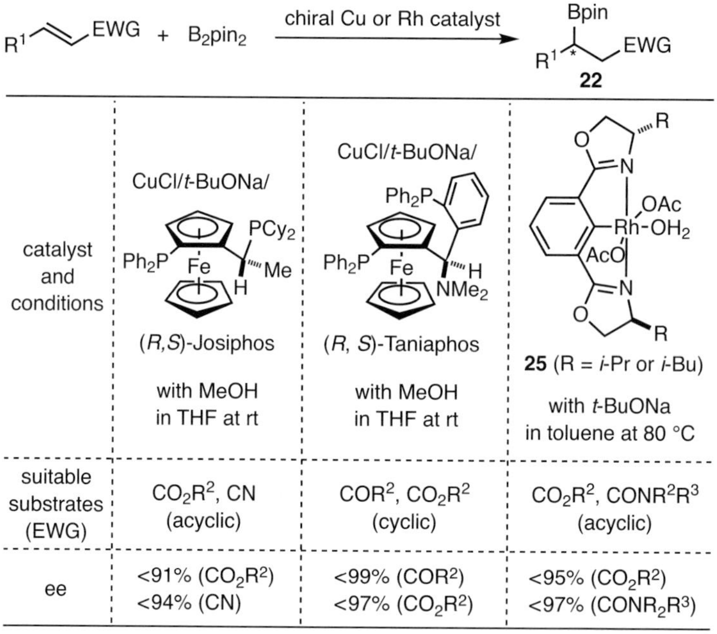

**Scheme 3.16** Enantioselective conjugate borylation of α,β-unsaturated carbonyl compounds.

| catalyst and conditions | CuCl/t-BuONa/ (R,S)-Josiphos (Ph$_2$P-Fe-PCy$_2$, Me) with MeOH in THF at rt | CuCl/t-BuONa/ (R,S)-Taniaphos (Ph$_2$P-Fe-Ph$_2$P, NMe$_2$) with MeOH in THF at rt | 25 (R = i-Pr or i-Bu) Rh-bisoxazoline/OAc with t-BuONa in toluene at 80 °C |
|---|---|---|---|
| suitable substrates (EWG) | CO$_2$R$^2$, CN (acyclic) | COR$^2$, CO$_2$R$^2$ (cyclic) | CO$_2$R$^2$, CONR$^2$R$^3$ (acyclic) |
| ee | <91% (CO$_2$R$^2$) <br> <94% (CN) | <99% (COR$^2$) <br> <97% (CO$_2$R$^2$) | <95% (CO$_2$R$^2$) <br> <97% (CONR$_2$R$^3$) |

Reaction: R$^1$CH=CH-EWG + B$_2$pin$_2$ → (chiral Cu or Rh catalyst) → R$^1$CH(Bpin)-CH$_2$-EWG (**22**)

esters [42a], ketones [42b], and amides [38e]. Chiral rhodium–bisoxazolinylphenyl acetate complexes **25** also catalyze the reaction with high enantioselectivities [43].

Platinum-catalyzed reaction of methylenecyclopropanes **26** with B$_2$pin$_2$ proceeds through cleavage of the proximal C—C bond of the cyclopropane ring, giving 2,4-diboryl-1-alkenes **27** (Scheme 3.17 and entries 12 and 13 in Table 3.1) [44]. B$_2$cat$_2$ adds to aldimines **28** in the presence of PtCl$_2$(cod) catalyst to afford α-aminoalkylboronic esters **29** that serve as a precursor of the boron analogues of α-amino acids (Scheme 3.18 and entry 14 in Table 3.1) [12]. Diazoalkanes **30** also undergo catalytic

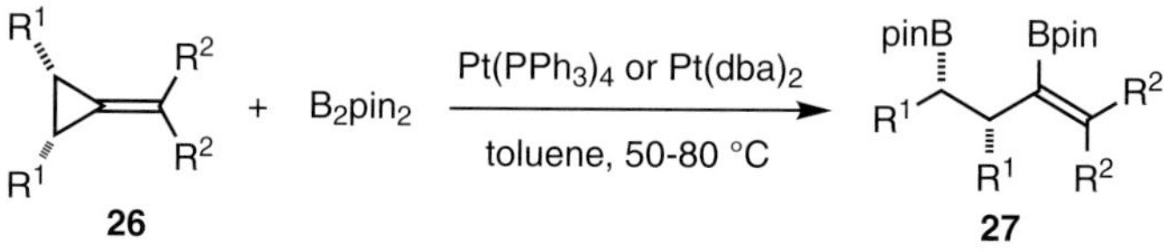

**Scheme 3.17** Catalytic diboration of methylenecyclopropanes.

**26** + B$_2$pin$_2$ → (Pt(PPh$_3$)$_4$ or Pt(dba)$_2$, toluene, 50–80 °C) → **27**

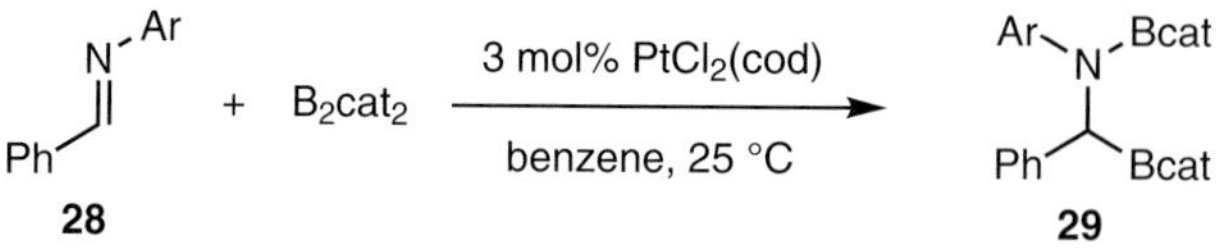

**Scheme 3.18** Catalytic diboration of imines.

**28** (PhCH=N-Ar) + B$_2$cat$_2$ → (3 mol% PtCl$_2$(cod), benzene, 25 °C) → **29** (PhCH(Bcat)-N(Ar)(Bcat))

diboration: $B_2pin_2$ reacts with diazomethane ($CH_2N_2$) and aryl-substituted diazomethanes in the presence of $Pt(PPh_3)_4$ to give 1,1-diborylmethanes **31** in high yields (Scheme 3.19 and entry 15 in Table 3.1) [37b, 45]. Uncatalyzed 1,1-diboration with alkylidene carbenoids **32** is also reported (Scheme 3.20) [46].

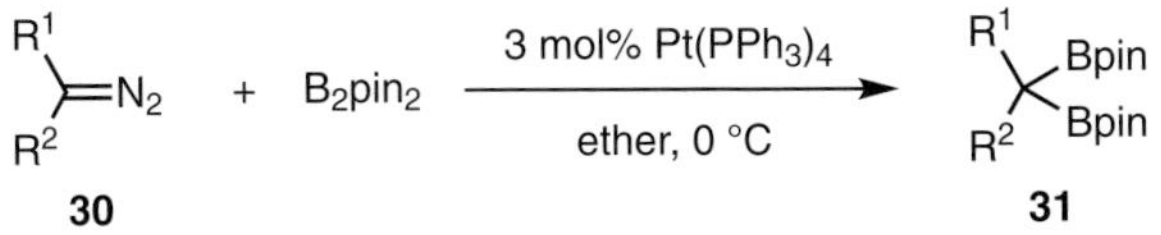

**Scheme 3.19** Catalytic diboration of diazomethanes.

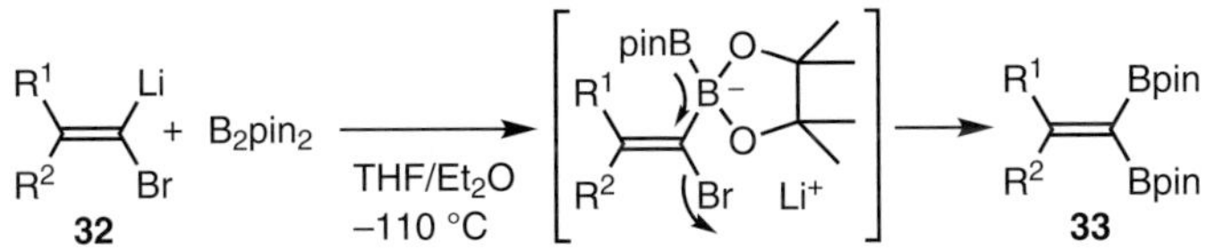

**Scheme 3.20** Diboration of alkylidene carbenoids.

### 3.2.4
### Synthetic Applications of Diboration Products

Catalytic diboration has found some synthetic applications, of which representative examples are shown next.

Products **34** of platinum-catalyzed alkyne diboration were used in stepwise cross-coupling at the two nonequivalent Bpin groups (Scheme 3.21) [47]. The first coupling takes place selectively at the less sterically demanding terminal Bpin group, followed by the second coupling at the internal Bpin group. This selective sequential coupling was utilized for the synthesis of conjugated enyne derivatives **36**.

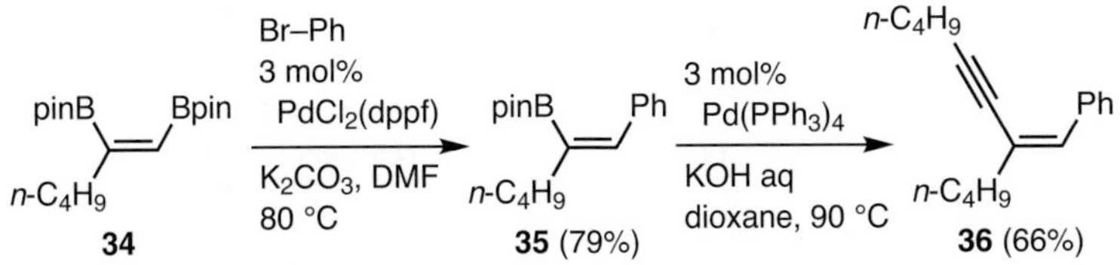

**Scheme 3.21** Conversion of 1,2-diboryl-1-hexene **34** via sequential cross-coupling reactions.

Enantioenriched 2-substituted 1,4-butanediol **39** was prepared via asymmetric alkene diboration, followed by a double one-carbon homologation in combination with $H_2O_2$ oxidation (Scheme 3.22) [28].

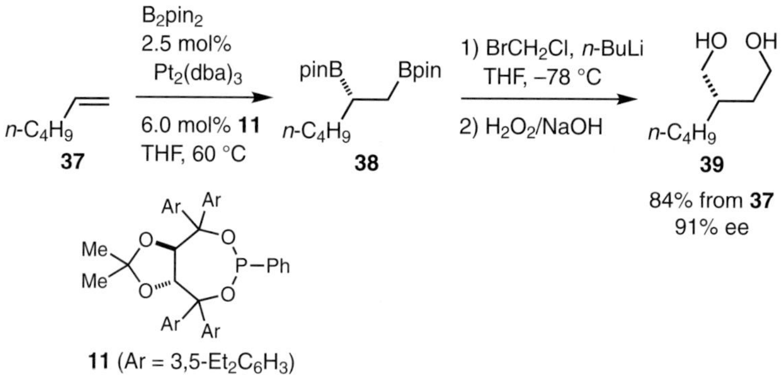

**Scheme 3.22** Conversion of enantioenriched chiral 1,2-diborylhexane **38** via a double one-carbon homologation.

β-Borylallylborane **40**, that is, a diboration product of allene, reacts with aldehydes via allylboration, which is followed by Suzuki–Miyaura coupling, giving 3-aryl homoallyllic alcohols **42** (Scheme 3.23) [29].

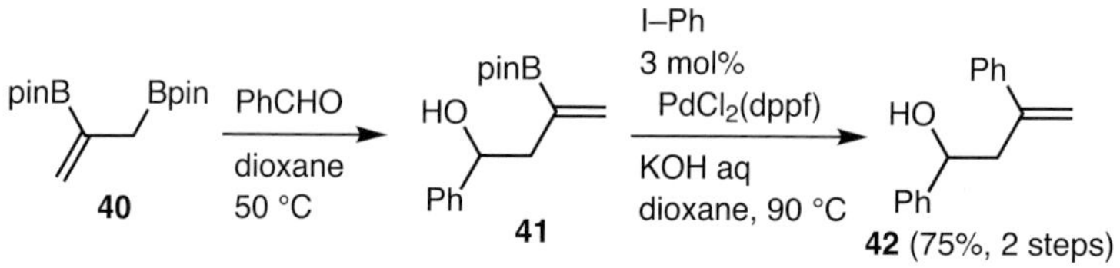

**Scheme 3.23** Conversion of 2,3-diboryl-1-propene **40** via allylboration followed by cross-coupling.

Platinum-catalyzed 1,4-diboration of dienal affords an aldehyde **44** bearing an allylborane moiety in the molecule that undergoes regio- and stereoselective six-membered ring formation through intramolecular allylation (Scheme 3.24) [48].

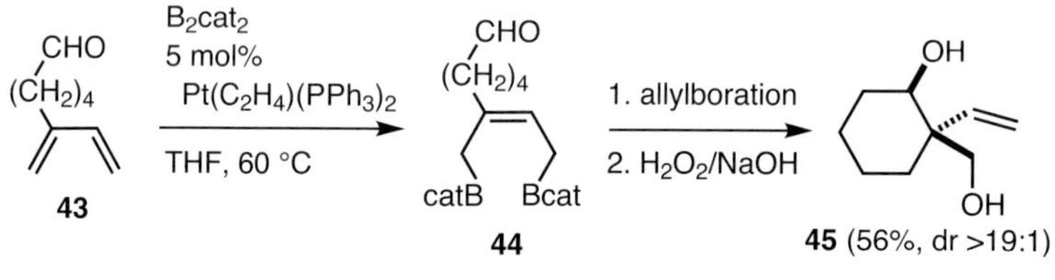

**Scheme 3.24** Conversion of 1,4-diboryl-2-butene **44** via intramolecular allylation followed by oxidation.

Using a diboration product **46** of methylenecyclopropane, a competitive Suzuki–Miyaura coupling between $sp^2$ and $sp^3$ organoboron compound was examined (Scheme 3.25). The boryl group connected to an $sp^2$ carbon center is more prone to undergoing cross-coupling. Oxidation of the remaining $sp^3$ C–B bond afforded a *cis*-2-alkenylcyclohexane derivative **48** selectively [44].

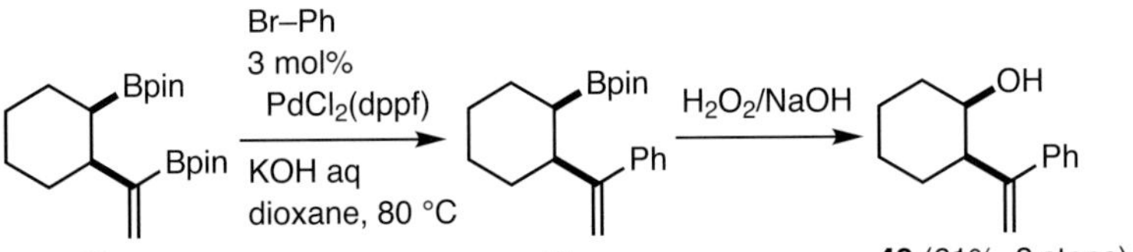

Scheme 3.25 Conversion of 2,4-diboryl-1-butene **46** via cross-coupling followed by oxidation.

## 3.3
## Silaboration

Silylboronic acid derivatives have been known since 1960, but silaboration affording organoboronic acid derivatives as isolable products had never been reported until 1996. Although addition of a silylborate (Me$_2$PhSiBEt$_3$Li) to alkynes proceeds in MeOH in the presence of a copper catalyst, the reaction gives formal hydrosilylation products via *in situ* protodeborylation of the initially formed 2-silylalkenylborates, which shifts the equilibrium to the product side [49]. A reaction of phenyldimethylsilyl(catechol)borane (**49**) with ethyl diazoacetate was reported in 1995 to give ethyl 1-boryl-1-silylacetate [50]. This 1,1-silaboration product, however, was not isolated, but was converted into 1-silylacetate via protodeborylation during workup. An efficient catalyst system for the activation of the boron–silicon bond in a neutral silylborane was then reported in 1996, allowing high-yield formation of 1,2-*cis*-addition products [51, 52]. It should be noted that 1,1-additions of silylboranes to carbenoids [46, 53] and isocyanides [54], having a carbene-like isocyanide carbon, have been developed later on. In this chapter, only 1,1-silaborations catalyzed by transition metal complexes are covered.

### 3.3.1
### Silylborane Reagents for Silaboration

Compounds containing a silicon–boron bond are readily accessible via reaction of (triorganosilyl)lithiums with boron electrophiles. For example, bis(diethylamino)(dimethylphenylsilyl)borane (**52**) is prepared via reaction of (dimethylphenylsilyl)lithum (**50**) with chlorobis(diethylamino)borane (**51**) (Scheme 3.26) [50].

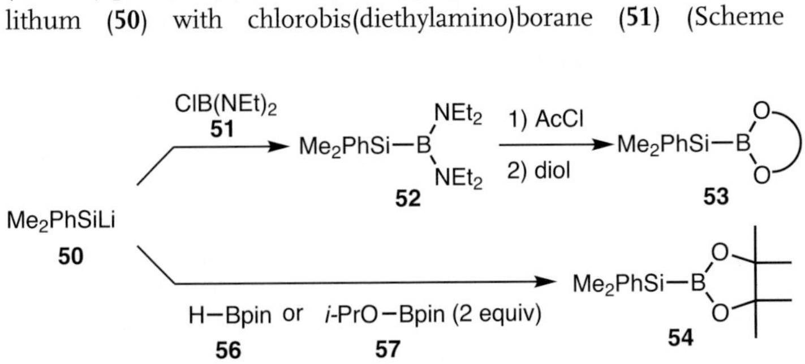

Scheme 3.26 Synthetic routes to (triorganosilyl)boronic esters.

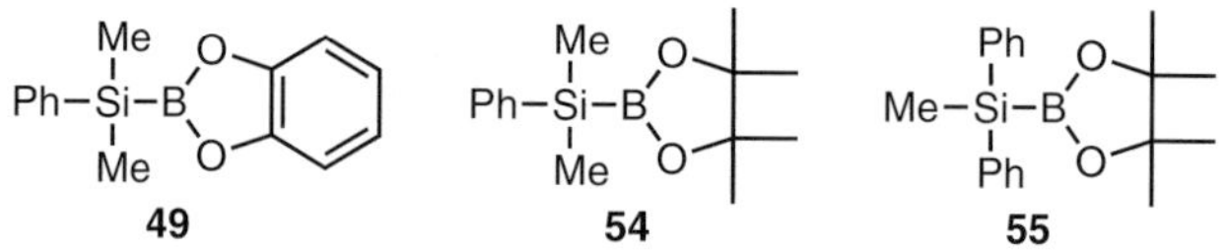

**Figure 3.2** Examples of (triorganosilyl)boronic ester for catalytic silaboration.

This silylborane (**52**) is a useful precursor of (dimethylphenylsilyl)boronic esters **53**: the diethylamino groups are substituted by various alkoxy groups through treatment with acetyl chloride followed by diols such as pinacol [49a], catecohol [48], and pinanediol [55]. A pinacol derivative (Me$_2$PhSi–Bpin) **54** is air and moisture stable, and thus is a standard silylboronic ester for catalytic silaborations (Figure 3.2). The pinacol derivatives of silylboronic esters are synthesized more practically by the reaction of (triorganosilyl)lithium with pinacolborane (HBpin, **56**) or isopropoxypinacolborane (*i*-PrOBpin, **57**) (Scheme 3.26) [56].

Silylboronic esters bearing heteroatoms on the silicon atom are attractive in that the functional groups can change the reactivity, applicability of the reaction products, and even reaction course. Amino-substituted silylborane (Et$_2$N)Ph$_2$Si–Bpin **59**, prepared by reaction of (diethylamino)diphenylsilyllithium **58** with *i*-PrOBpin **57** [57], serves as a useful precursor for the synthesis of silylboranes functionalized on silicon (Scheme 3.27). The diethylamino group is converted to a chlorine group by reaction with hydrogen chloride. Thus, synthesized chlorine-substituted silylborane ClPh$_2$Si–Bpin (**60**) reacts with various alcohols in the presence of pyridine to afford silylboronic esters **61** having alkoxy groups on the silicon atoms. The corresponding dimethylsilyl derivative ClMe$_2$Si–Bpin **62** is prepared by chlorodephenylation of Me$_2$PhSi–Bpin **54** by reaction with hydrogen chloride in the presence of AlCl$_3$ (Scheme 3.28) [57]. Subsequent amination with dialkylamine in the presence of triethylamine gives (R$_2$N)Me$_2$Si–Bpin **64**.

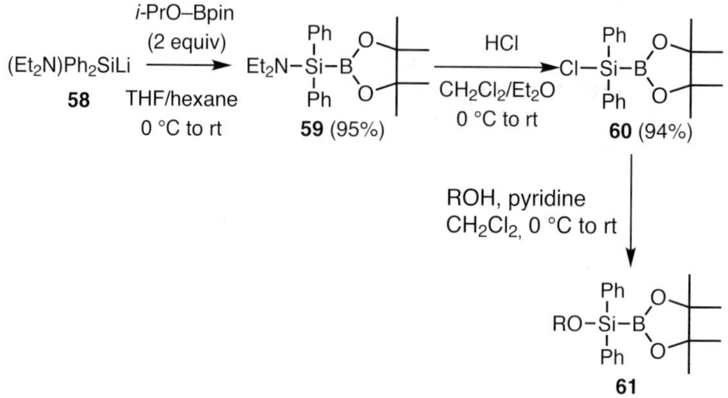

**Scheme 3.27** Preparation of XPh$_2$Si–Bpin (X = Cl, NR$_2$, and OR).

Scheme 3.28 Preparation of XMe$_2$Si–Bpin (X = Cl, NR$_2$, and OR).

### 3.3.2
### Silaboration of Alkynes

The Si–B bonds of silylboranes are inert toward addition to C–C multiple bonds in the absence of transition metal catalysts. In the initial report on the silaboration, a palladium catalyst generated *in situ* from Pd(OAc)$_2$ with *t*-BuCH$_2$CMe$_2$NC catalyzes addition of Me$_2$PhSi–Bpin (**54**) to alkynes in toluene at 50–110 °C (Scheme 3.29) [51]. Palladium complexes bearing phosphorus ligands, such as Pd$_2$(dba)$_3$ with 4-ethyl-2,6,7-trioxa-1-phosphabicyclo[2.2.2]octane (etpo) [58], Pd$_2$(dba)$_3$ with P(OEt)$_3$ [51b], and PdCl$_2$(PPh$_3$)$_2$ [51b], are also effective for alkyne silaboration. It is noteworthy that the palladium-catalyzed silaboration of terminal alkynes proceeds with high regio- and stereoselectivities. The boryl and silyl groups were introduced to the terminal and

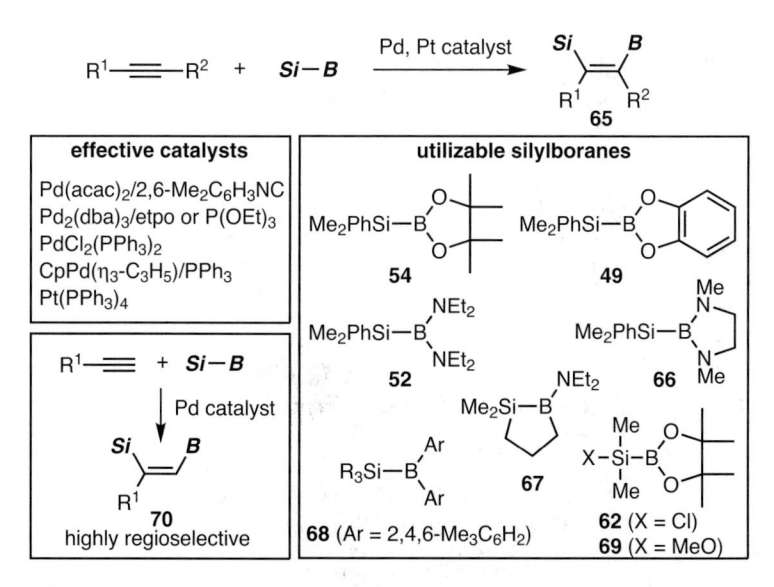

Scheme 3.29 Catalytic silaboration of alkynes.

internal carbon atoms, respectively, in a *cis*-fashion to give (Z)-1-boryl-2-silyl-1-alkenes **70** (Scheme 3.29). A platinum complex, Pt(PPh$_3$)$_4$, also catalyzes the silaboration of alkyne, although the regioselectivity in the reaction of 1-octyne is slightly lower than that with palladium catalysts. The palladium-catalyzed silaboration is applicable to a broad range of alkynes, including terminal and internal alkynes bearing aromatic, aliphatic, silyl, and ester substituents. Gaseous acetylene (C$_2$H$_2$, 1 atm) also undergoes silaboration. Functional groups such as chloro, cyano, silyloxy, acetal, and hydroxy groups are tolerable in the silaboration reactions.

In addition to Me$_2$PhSi–Bpin (**54**), the palladium catalysts promote addition of some other silylboranes, including Me$_2$PhSi–Bcat (**49**) [51b], (dimethylphenylsilyl)bis(dialkylamino)boranes **52** and **66** [51b, 58], Me$_2$PhSi–B(mesityl)$_2$ (**68**) [59], and five-membered cyclic silylborane **67** [60] (Scheme 3.29). A large rate acceleration in comparison to Me$_2$PhSi–Bpin (**54**) was observed in the silaboration of 1-octyne with ClMe$_2$Si–Bpin (**62**) in the presence of a catalyst generated *in situ* from CpPd($\eta^3$-C$_3$H$_5$) with PPh$_3$ (Pd/P = 1/1.2) (Scheme 3.30) [61]. The reaction is completed within 15 min at room temperature using 1.0 mol% of the catalyst, whereas the silaboration with Me$_2$PhSi–Bpin takes 70 h under identical reaction conditions. Methoxy-substituted MeOMe$_2$Si–Bpin (**69**) shows a moderate rate acceleration. The reactivity in alkyne silaboration decreases in the following order: ClMe$_2$Si–Bpin (**62**) > MeOMe$_2$Si–Bpin (**69**) > Me$_2$PhSi–Bpin (**54**). A remarkable change in the reaction course was observed in the palladium-catalyzed reaction of silylboronic esters **59** and **71** bearing dialkylamino groups on the silicon atoms (Scheme 3.31). In the presence of palladium–triarylphosphine catalyst, (Et$_2$N)Me$_2$Si–Bpin **71** reacts with 2 equiv of terminal alkynes to give 2,4-disubstituted siloles **72** and Et$_2$N–Bpin through silylene-alkyne [2 + 2 + 1] cycloaddition [61]. Generation of silylene-palladium species is presumed in the formation of silole derivatives.

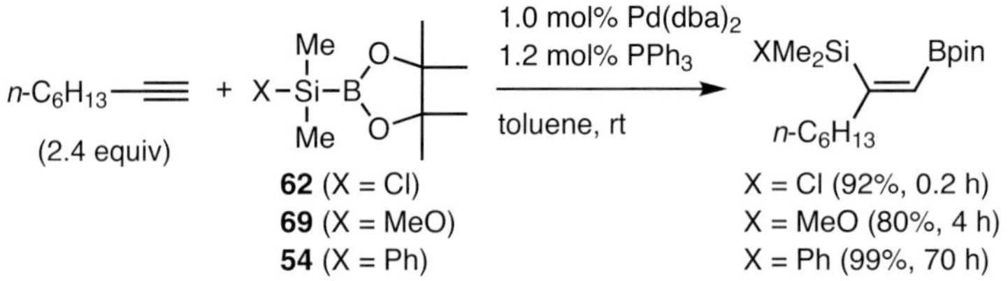

**Scheme 3.30** Comparison of reactivity.

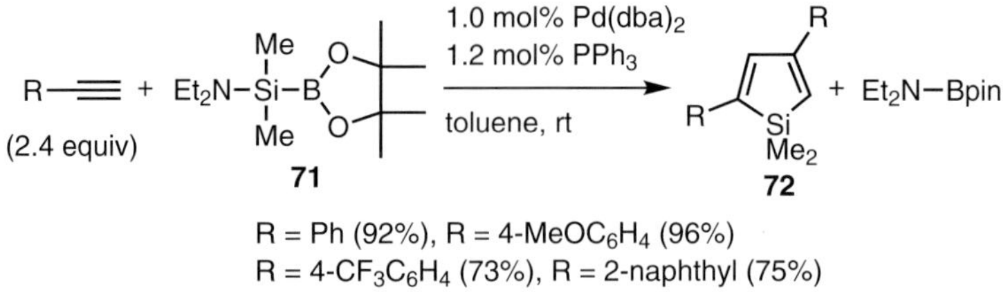

**Scheme 3.31** Silole formation in the reaction of (Et$_2$N)Me$_2$Si–Bpin with terminal alkynes.

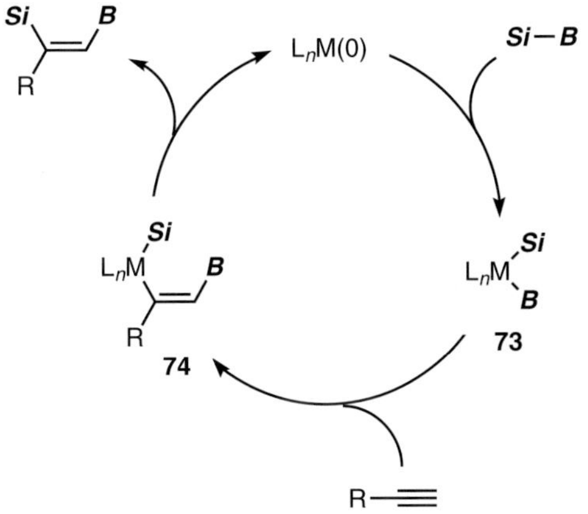

**Scheme 3.32** Catalytic cycle of alkyne silaboration.

A reaction mechanism for the palladium- and platinum-catalyzed silaboration of alkynes was proposed (Scheme 3.32). The catalytic cycle may consist of (a) oxidative addition of the Si–B bond to M(0) (M = Pd and Pt) giving (boryl)(silyl)M(II) **73**, (b) insertion of alkyne into M–B bond of **73** affording (β-borylalkenyl)(silyl)M(II) **74**, and (c) product formation via reductive elimination from **74**. All the elementary processes were observed in stoichiometric reactions of the silylborane with a platinum complex [62].

As mentioned previously, silaboration with ClMe$_2$Si–Bpin (**62**) proceeds rapidly to give the *cis*-addition product selectively. In the subsequent reaction in the same reaction vessel, the Cl group on the silicon atom of the silaboration products is converted into *i*-PrO group for isolation by silica gel chromatography [63]. When the conversion was applied to the silaboration that was carried out with a slight excess of silylborane (silylborane:alkyne = 1.2 : 1.0), stereoselective formation of *trans*-silaboration products (*E*)-**75** took place (*Z*:*E* = 7 : 93–11 : 89) (Scheme 3.33). This stereochemical outcome is in sharp contrast to the formation of *Z* isomer in the corresponding silaboration using excess alkyne (silylborane:alkyne = 1.0 : 1.2) under the otherwise identical conditions. This stereoselective formation of the *E* isomer is

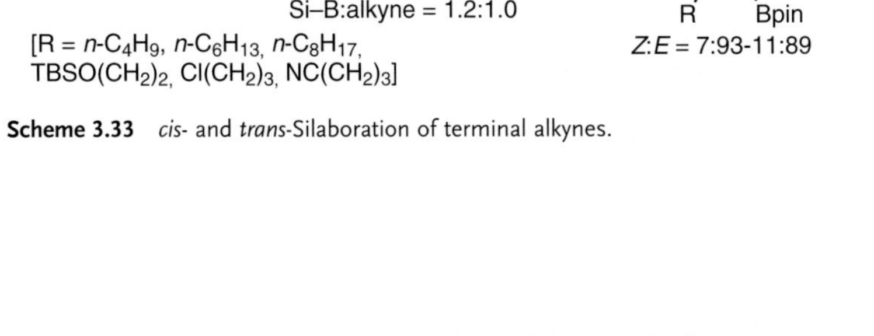

[R = n-C$_4$H$_9$, n-C$_6$H$_{13}$, n-C$_8$H$_{17}$, TBSO(CH$_2$)$_2$, Cl(CH$_2$)$_3$, NC(CH$_2$)$_3$]

**Scheme 3.33** *cis*- and *trans*-Silaboration of terminal alkynes.

ascribed to Z-to-E isomerization in the second step of Cl–i-PrO exchange, in which some highly active species for alkene isomerization is generated from i-PrOH, silylborane, and palladium complex.

A C—C bond formation accompanied by alkyne silaboration is attractive for the synthesis of boron- and silicon-containing 1,3-dienes **77**. Silaborative carbocyclization takes place in the reaction of 1,6-heptadiyne (**76**) in the presence of Pd$_2$(dba)$_3$/ etpo catalyst (Scheme 3.34) [58]. A phosphine-free nickel catalyst generated *in situ* by the reaction of Ni(acac)$_2$ with diisobutylalminium hydride (DIBAH) promotes silaborative dimerization of internal alkynes, giving 1-boryl-4-silyl-1,3-butadiene derivatives **79** (Scheme 3.34) [64].

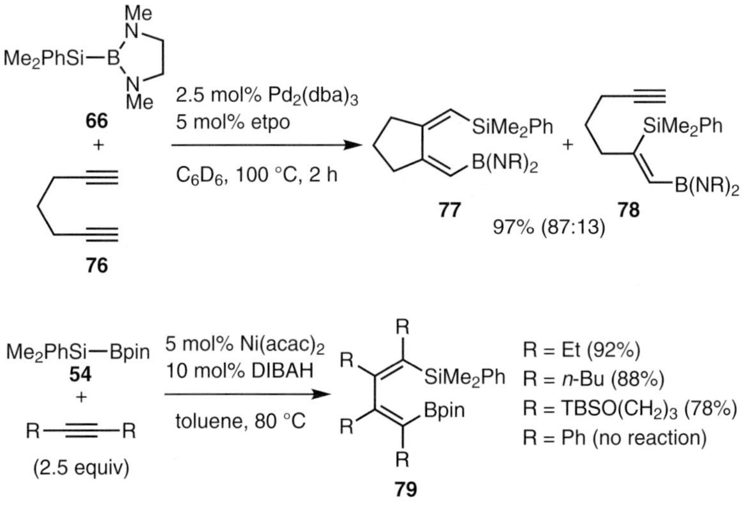

**Scheme 3.34** Silaborative coupling of alkynes.

Silaboration of 4-substituted but-1-en-3-ynes **80** is carried out in the presence of nickel, palladium, and platinum catalysts (Scheme 3.35) [65]. When the conjugated 1,3-enynes have sterically less demanding *n*-alkyl groups at 4-position, 1,2-silaboration takes place to give conjugated enynes **81** and **81′**. On the other hand, for enynes

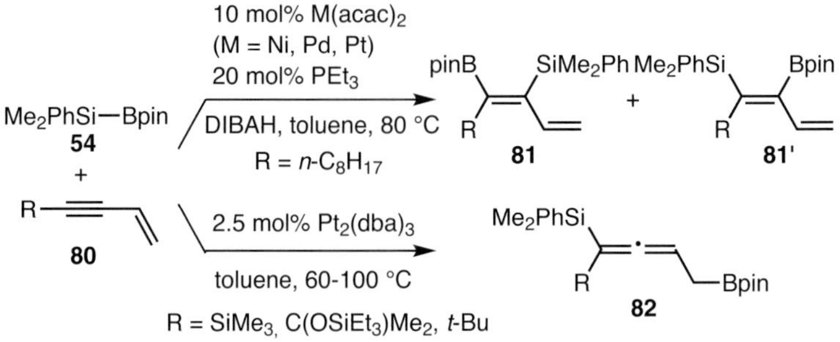

**Scheme 3.35** Catalytic silaboration of conjugated enynes with terminal C=C bonds.

with bulky 4-substituents such as $Me_3Si$ or $Me_2C(OSiEt_3)$, silaboration takes place in a 1,4-fashion, leading to the formation of allene products **82**.

### 3.3.3 Silaboration of Alkenes, Allenes, 1,3-Dienes, and Methylenecyclopropanes

Intermolecular silaboration of alkenes takes place under reflux in dioxane in the presence of platinum catalysts such as $Pt(PPh_3)_4$ and $Pt(CH_2{=}CH_2)(PPh_3)_2$ (Scheme 3.36) [66]. Aliphatic alkenes give 2-boryl-1-silylalkanes **83** in moderate yields along with the formation of a 1,1-addition product **84**, 1-boryl-1-silylalkanes. Platinum-catalyzed intramolecular silaboration of homoallylic alcohol derivatives **85** gives five-membered cyclic silyl ethers **86** in high yields through 5-exo cyclization (Scheme 3.37) [67]. The stereoselectivity for secondary homoallylic substrates highly depends on the phosphorus ligands on the platinum catalyst. In the presence of Pt $(dba)_2/PCyPh_2$ ($Pt/P = 1/2$) catalyst, *trans*-cyclic products are selectively formed, whereas $Pt(dba)_2/P[O(2,4\text{-}t\text{-}Bu_2C_6H_3)]_3$ ($Pt/P = 1/2$) catalyst gives *cis*-products. Using similar silicon-tethered substrate **87** having a 2-substituent on the internal alkenyl carbon atoms, stereoselective dehydroborylation has been established (Scheme 3.38) [68]. The product **88** still possesses a reactive tethered Si−H bond in the molecule and can undergo stereoselective intramolecular hydrosilylation when a dienenol derivative **87** is used as the starting material.

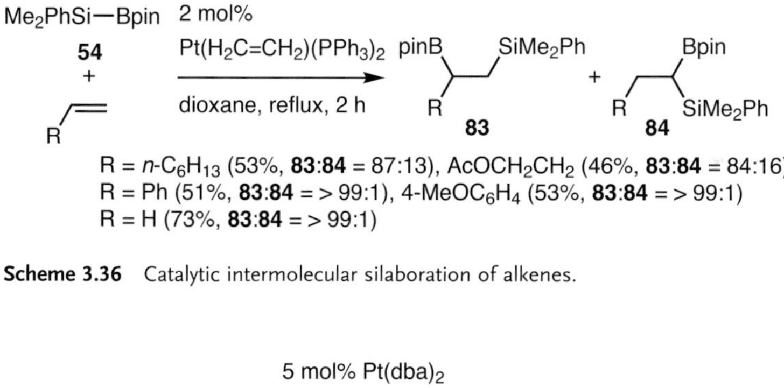


R = $n$-$C_6H_{13}$ (53%, **83**:**84** = 87:13), $AcOCH_2CH_2$ (46%, **83**:**84** = 84:16)
R = Ph (51%, **83**:**84** = > 99:1), 4-$MeOC_6H_4$ (53%, **83**:**84** = > 99:1)
R = H (73%, **83**:**84** = > 99:1)

**Scheme 3.36** Catalytic intermolecular silaboration of alkenes.

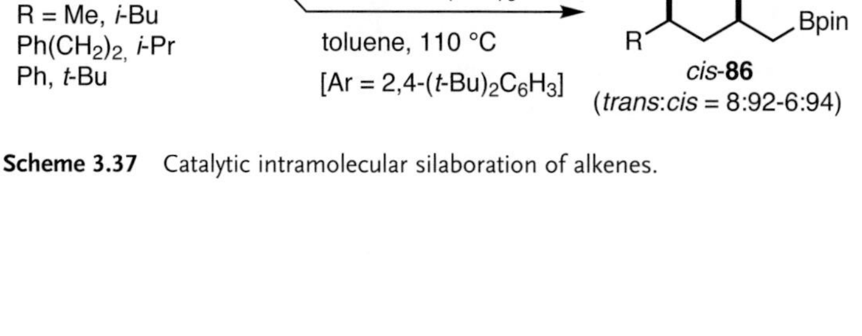


**Scheme 3.37** Catalytic intramolecular silaboration of alkenes.

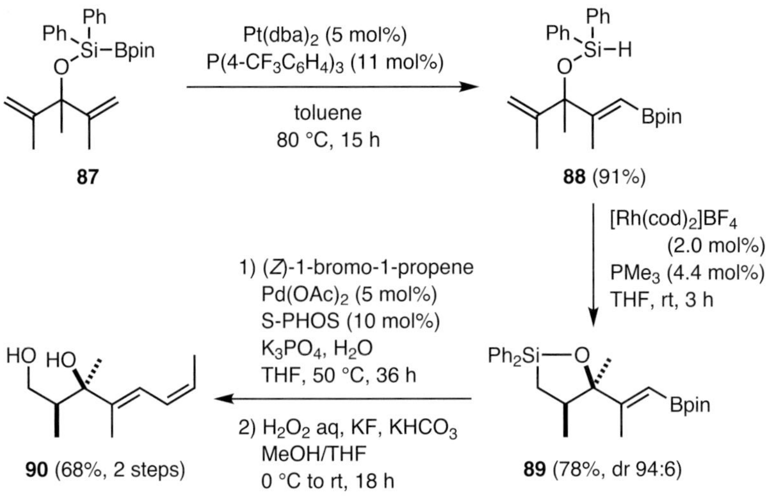

**Scheme 3.38** Dehydrogenative borylation of silicon-tethered diene **87**.

Silaboration of conjugated 1,3-dienes is catalyzed by nickel and platinum catalysts (Schemes 3.39 and 3.40) [69, 70]. Palladium catalysts are totally inactive in this reaction. Derivatives of 4-silylallylborane are obtained via 1,4-silaboration of acyclic dienes with high stereoselectivity for *cis*-products in the presence of a nickel

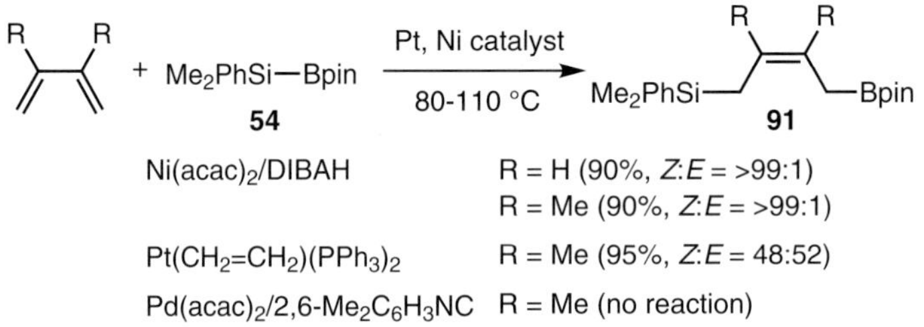

**Scheme 3.39** Catalytic silaboration of acyclic 1,3-dienes.

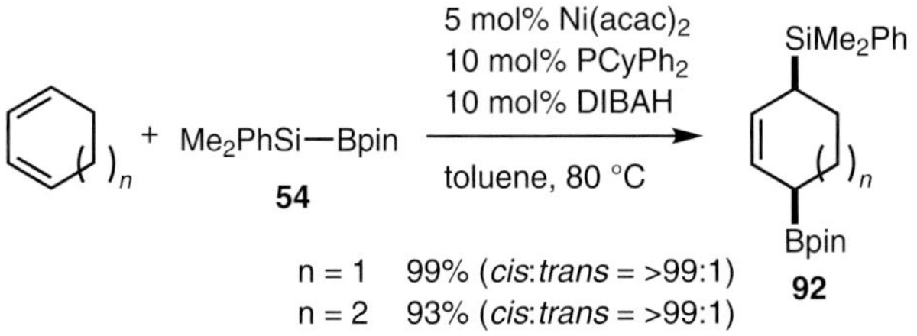

**Scheme 3.40** Catalytic silaboration of cyclic 1,3-dienes.

catalyst [69]. Use of a platinum catalyst resulted in the formation of a 1 : 1 mixture of cis- and trans-products [70]. Nickel-catalyzed silaboration of 1,3-cyclohexadiene affords cis-addition product selectively in the presence of PCyPh$_2$ as a ligand. Use of other phosphines results in the formation of a significant amount of trans-product.

Reaction of 1,3-dienes, aldehydes, and silylborane in the presence of a platinum catalyst affords a three-component coupling reaction that may involve nucleophilic attack of an intermediary formed allylplatinum species **95** to aldehydes (Scheme 3.41) [70]. The product **93** cannot be obtained by the corresponding sequential reaction, in which the allylborane is formed before the allylboration of aldehyde takes place. Hence, it is elucidated that a (silyl)(4-borylallyl)platinum intermediate **95** is formed and reacts with aldehydes. Enantioselective silaboration of 1,3-cyclohexadienes has been reported by using a platinum catalyst having a chiral phosphoramidite ligand **98** (Scheme 3.42) [71].

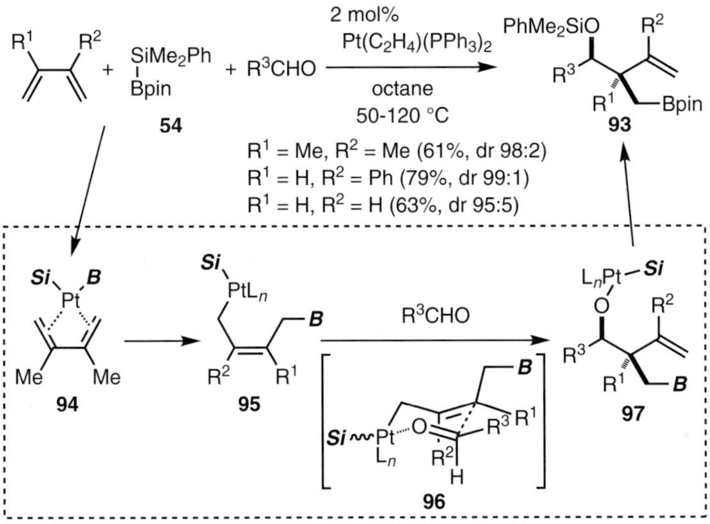

**Scheme 3.41** Silaborative coupling of 1,3-dienes with aldehydes.

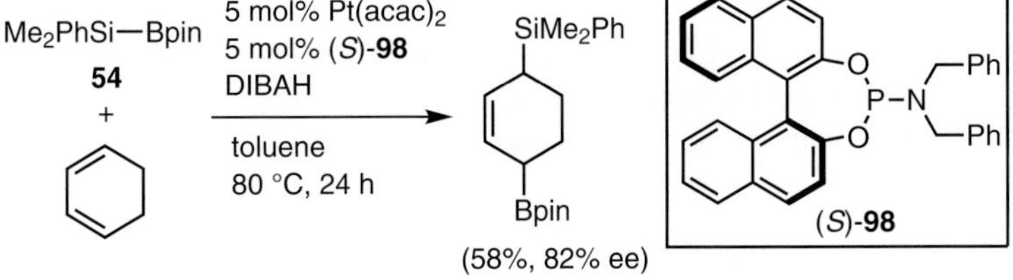

**Scheme 3.42** Enantioselective silaboration of 1,3-cyclohexadiene.

Silaboration of allenes have received much attention because the reaction gives β-borylallylsilanes **99** and **100** that would be useful synthetic intermediates through sequential conversion of both alkenylboron and allylsilane moieties (Scheme 3.43).

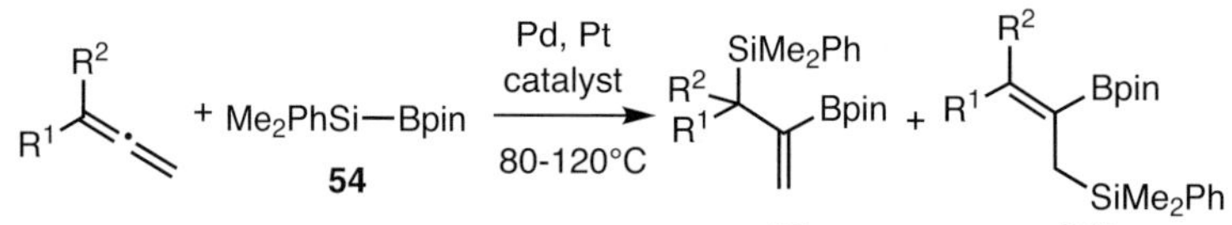

**Scheme 3.43** Catalytic silaboration of allenes.

| effective catalysts | selected examples |
|---|---|
| Pd(acac)$_2$/2,6-Me$_2$C$_6$H$_3$NC<br>Pd$_2$(dba)$_3$/etpo<br>PdCl$_2$(PPh$_3$)$_2$<br>Pd$_2$(dba)$_3$/PPh$_3$ or PMe$_3$<br>Pt(CH$_2$=CH$_2$)(PPh)$_2$ | catalyst: Pd(acac)$_2$/2,6-Me$_2$C$_6$H$_3$NC<br>R$^1$ = PhCH$_2$CH$_2$, R$^2$ = H (99%, **99**:**100** = 100:0)<br>R$^1$ = MeO, R$^2$ = H (92%, **99**:**100** = 100:0)<br>R$^1$ = 4-MeOC$_6$H$_4$, R$^2$ = H (76%, **99**:**100** = 94:6)<br>R$^1$ = 4-F$_3$CC$_6$H$_4$, R$^2$ = H (81%, **99**:**100** = 36:64)<br>R$^1$ = n-C$_6$F$_{13}$, R$^2$ = H (94%, **99**:**100** = 0:100) |

Palladium complexes such as Pd(acac)$_2$/2,6-Me$_2$C$_6$H$_3$NC and Pd$_2$(dba)$_3$/etpo were initially found to be effective for addition of Me$_2$PhSi–Bpin (**54**) to terminal and internal allenes as well as 1,2-propadiene (C$_3$H$_4$) [72, 73]. The allene silaboration proceeds with introduction of the boryl group to the central carbon atom of the allene to give β-borylallylsilanes. In the palladium-isocyanide-catalyzed silaboration, regioselectivity of the reaction depends on the electronic nature of the substituents on the allenes: the addition prefers to proceed at the double bond having more electron-donating substituents. For example, silylborane adds to the internal C−C double bond to give 2-boryl-3-silyl-1-alkenes **99** in the reaction of terminal allenes bearing alkyl, 4-methoxyphenyl, and methoxy groups, whereas the silaboration of 4-trifluoromethylphenyl- and perfluoroalkyl-substituted 1,2-propadiene takes place at the terminal C−C double bond.

Catalysts generated from Pd(0) precursors, such as CpPd(η$^3$-C$_3$H$_5$) and Pd(dba)$_2$, and monodentate phosphorous ligands with a Pd/PR$_3$ ratio of 1:1 exhibit high catalyst activity toward silaboration of terminal allenes, allowing silaboration to proceed even below room temperature (Scheme 3.44). Highly enantioface-selective silaboration of terminal allenes has been achieved through double asymmetric induction using pinanediol-derived optically active silylborane (−)-**102** with a palladium catalyst bearing a chiral monodentate phosphine (Scheme 3.45) [55]. The double asymmetric induction system was applied to highly effective reagent-controlled asymmetric induction with terminal allenes **105** and **106** having stereogenic carbon centers α to the double bonds (Scheme 3.46) [74]. Enantioselective silaboration of achiral allenes with an achiral silylborane was achieved with up to 93% ee by a

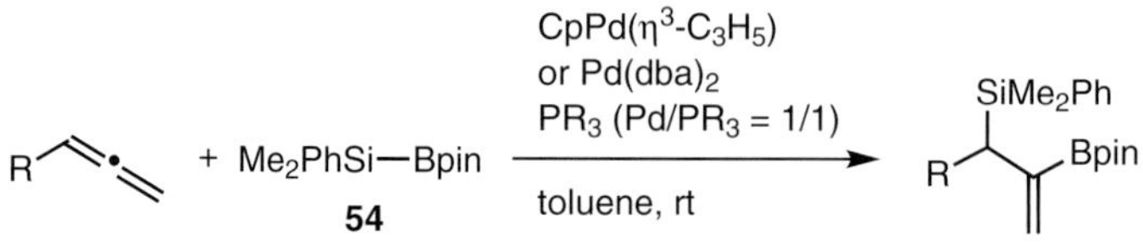

**Scheme 3.44** Room-temperature silaboration of allenes.

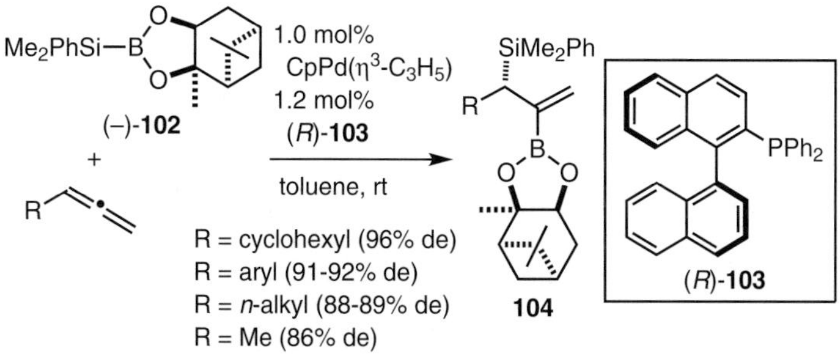

**Scheme 3.45** Enantioface-selective silaboration of terminal allenes with chiral silylborane.

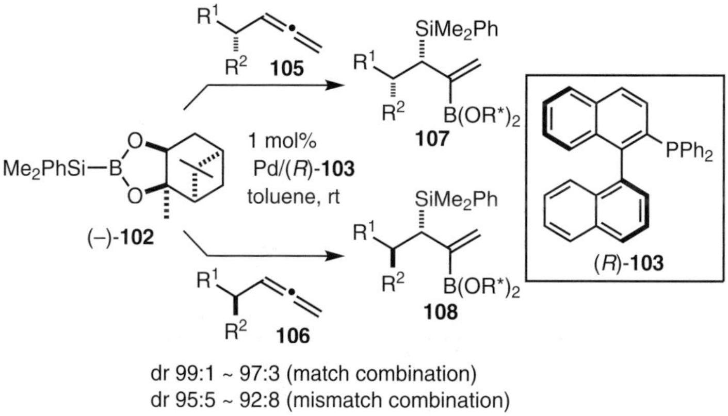

**Scheme 3.46** Asymmetric silaboration of terminal allenes bearing α-stereogenic carbon centers.

palladium catalyst bearing a chiral monodentate binaphthylphosphine (R)-**109** (Scheme 3.47) [75]. An interesting change in regioselectivity in palladium-catalyzed silaboration of terminal allenes was observed when the reaction was carried out with a catalytic amount of organic iodide, such as 3-iodo-2-methyl-2-cyclohexanone (Scheme 3.48) [76]. This reaction is classified as transmetalative silaboration because

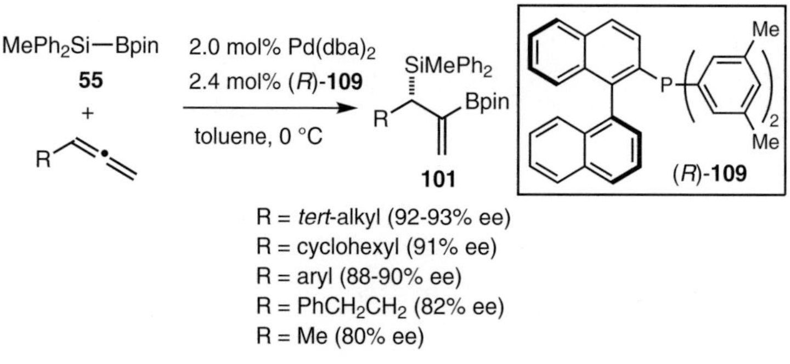

**Scheme 3.47** Enantioselective silaboration of terminal allenes.

**Scheme 3.48** Silaboration of terminal allenes in the presence of organic iodide.

it is supposed that the reaction proceed through formation of silyl iodide, which undergoes oxidative addition to the palladium(0) complex.

A mechanism that correctly rationalizes the observed regioselectivity has been proposed [72b] and studied theoretically (Scheme 3.49) [77]. The proposed mechanism involves addition of the B–Pd bond to the terminal C=C bond with Pd–C bond formation at the terminal allylic carbon atom of the allene (**111**). It was initially supposed and then supported by the theoretical study that thus formed σ-allylpalladium intermediate **111** does not undergo reductive elimination, because of the *trans*-orientation of the allyl and silyl groups, but forms π-allylpalladium intermediate **112**, in which the more substituted carbon atom is located *cis* to the silyl group. It is supposed that only the internal addition product is formed because the reductive elimination step is much more facile than the isomerization of the π-allylpalladium intermediate.

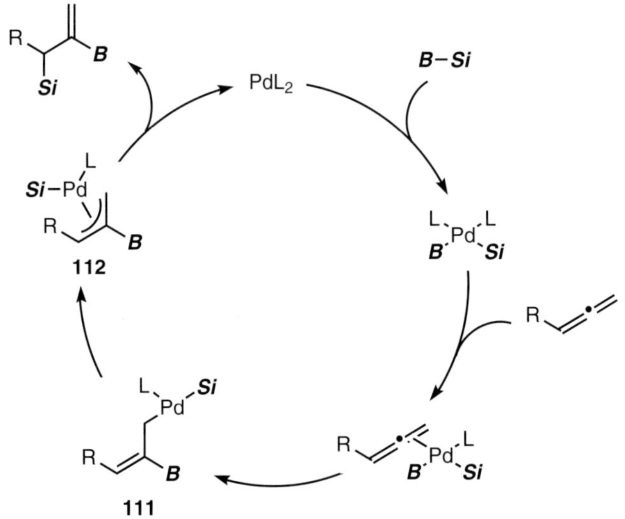

**Scheme 3.49** Catalytic cycle for allene silaboration.

Catalytic silaboration has also been applied to cyclopropane derivatives. Reaction of methylenecyclopropanes (MCPs) proceeds with cleavage of the ring C—C bonds in the presence of palladium or platinum catalysts (Scheme 3.50) [78]. The position of the C—C bond cleavage relies on the choice of the catalyst. For instance, selective cleavage of either of the two nonequivalent proximal C—C bonds of benzylidenecyclopropane **115** affords (*E*)- and (*Z*)-alkenylboranes **116** selectively with the use of palladium and platinum catalysts, respectively (Scheme 3.51) [78]. In the palladium-catalyzed reactions of cyclohexylidenecyclopropane **117**, the distal or the proximal C—C bond is cleaved selectively by use of different phosphorus ligands (Scheme 3.52) [78]. In addition, a palladium-catalyzed reaction of cyclohexane-fused MCP **120** resulted in the formation of simple silaborative C—C cleavage product **121**, whereas the corresponding platinum-catalyzed reaction afforded regioisomeric 1,3-disubstituted cyclohexane **122** in a stereoselective fashion (Scheme 3.53) [78]. Palladium-catalyzed silaboration of bicyclopropylidene **123** affords silaborative C—C cleavage product in which the silyl group adds to the alkenyl carbon atom in contrast to the examples shown above (Scheme 3.54) [79].

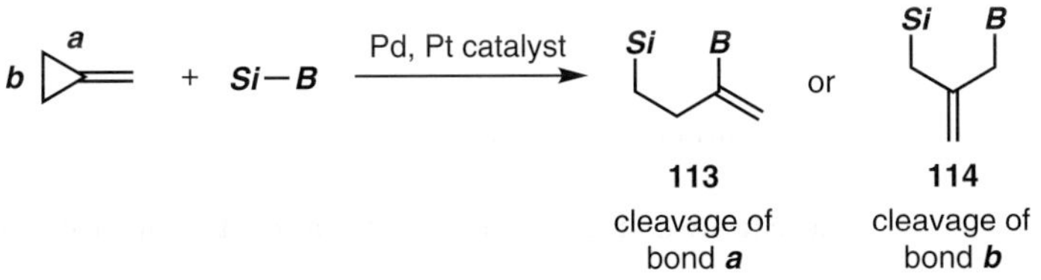

**Scheme 3.50** Catalytic silaborative C—C cleavage of methylenecyclopropanes.

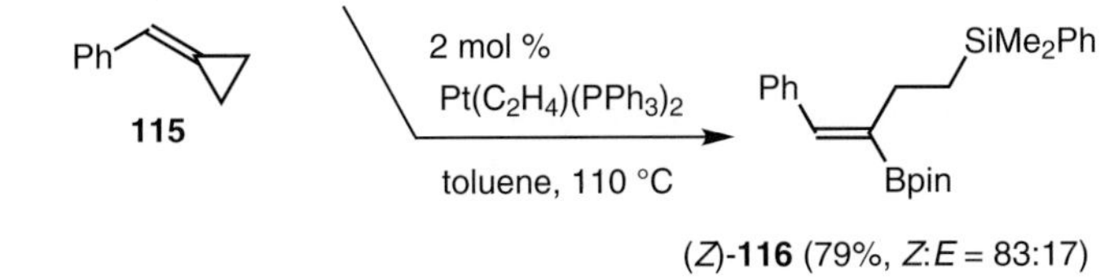

**Scheme 3.51** Silaboration of benzylidenecyclopropane **115**.

Asymmetric silaborative C—C bond cleavage of *meso*-MCP has been established by applying the same chiral catalyst system as that of terminal allenes (Scheme 3.55) [80]. Various *meso*-MCPs **125** undergo cleavage of one of the two enantiotopic proximal C—C bonds, leading to the formation of enantioenriched alkenylboronic acid

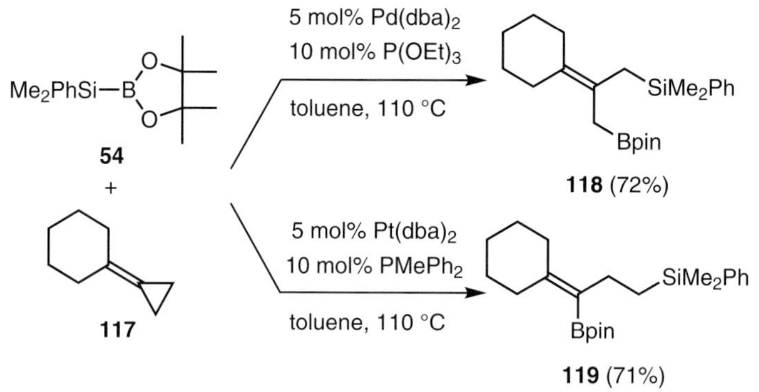

**Scheme 3.52** Silaboration of cyclohexylidenecyclopropane **117**.

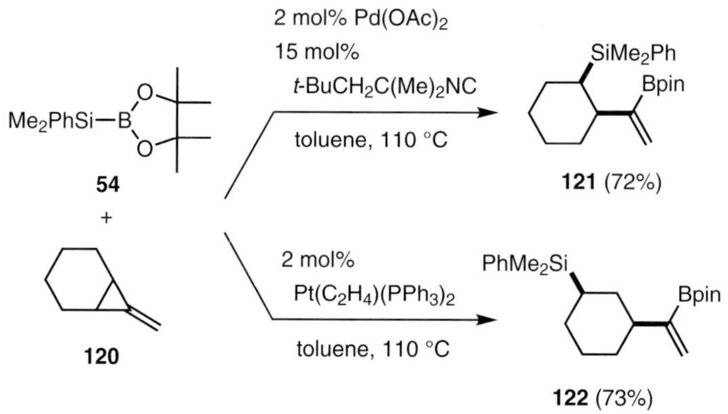

**Scheme 3.53** Silaboration of cyclohexane-fused methylenecyclopropane **120**.

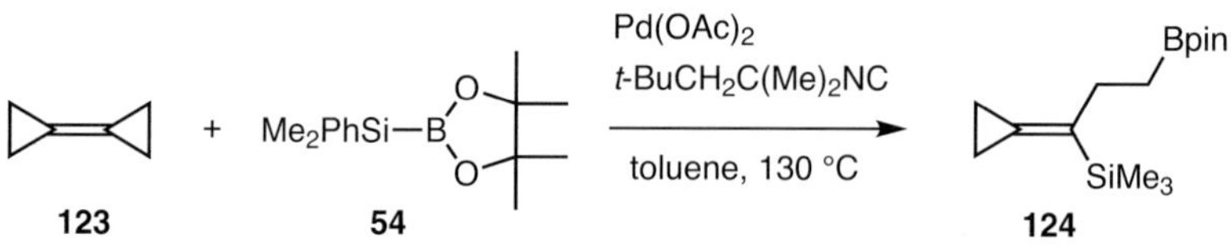

**Scheme 3.54** Silaboration of bicyclopropylidene **123**.

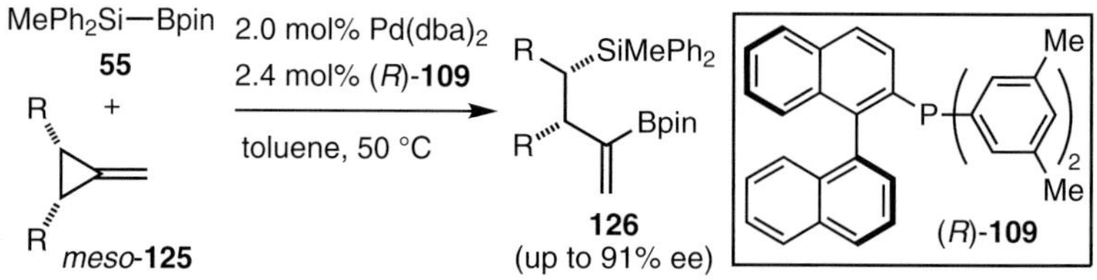

**Scheme 3.55** Enantioselective silaborative C−C bond cleavage of methylenecyclopropanes.

derivatives **126** with up to 91% ee. Asymmetric kinetic resolution of 1-alkyl-2-methylenecyclopropanes has also been achieved (Scheme 3.56) [81]. This kinetic resolution may be classified as "parallel kinetic resolution," in which two enantiomeric starting materials are selectively converted into constitutional isomers. For instance, racemic 2-substituted methylenecyclopropane **127** (R = CH$_2$OSiMe$_2$-t-Bu, 3 equiv) affords constitutional isomers **129** and **130** in a 86 : 14 ratio, of which the former exhibited 92% ee.

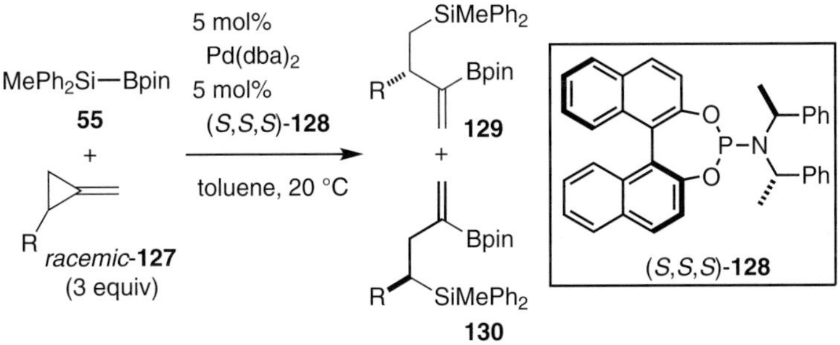

R = n-C$_6$H$_{13}$      90% (**129**:**130** = 78:22, 91% ee for **129**)
R = CH$_2$CH$_2$Ph      97% (**129**:**130** = 80:20, 90% ee for **129**)
R = CH$_2$OSiMe$_2$(t-Bu)  85% (**129**:**130** = 86:14, 92% ee for **129**)

**Scheme 3.56** Kinetic resolution of methylenecyclopropanes via silaborative C—C cleavage.

Nickel-catalyzed reaction of vinylcyclopropanes (VCPs) with silylboranes proceeds with cleavage of the proximal C—C bond in the three-membered ring (Scheme 3.57) [82]. The reaction provided borylated allylsilane **132** with high stereoselectivity for the (E)-products. It should be noted that vinylcyclobutanes **133** also undergo the silaborative C—C bond cleavage reaction in the presence of the nickel catalyst (Scheme 3.58) [82].

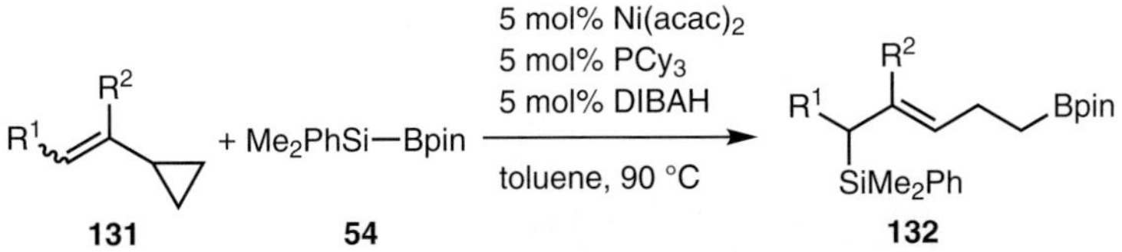

R$^1$ = H, R$^2$ = Ph (89%), R$^1$ = H, R$^2$ = Me (74%)
R$^1$ = Ph, R$^2$ = H (82%), R$^1$ = n-Bu, R$^2$ = H (87%)
R$^1$ = Ph, R$^2$ = Me (56%)

**Scheme 3.57** Silaborative C—C cleavage of vinylcyclopropanes.

Rhodium-catalyzed reactions of silylboranes silylboronates with α,β-unsaturated carbonyl compounds were reported. These reactions provide new methods for the synthesis of enantioenriched β-silyl carbonyl compounds (Scheme 3.59) [83].

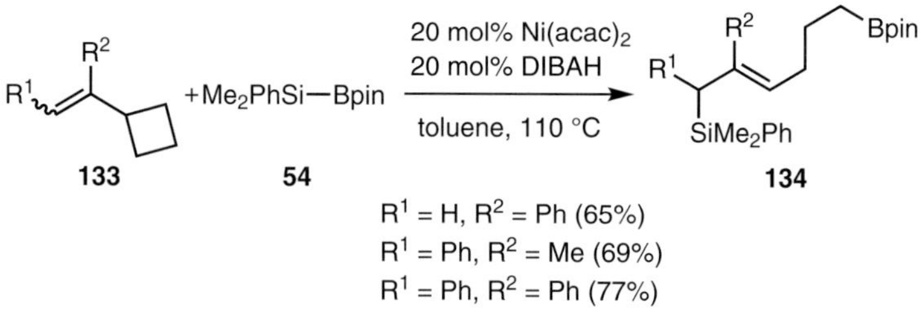

**Scheme 3.58** Silaborative C—C cleavage of vinylcyclobutanes.

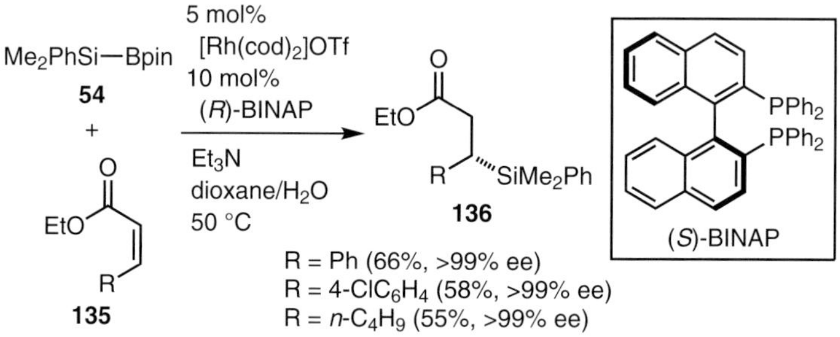

R¹ = H, R² = Ph (65%)
R¹ = Ph, R² = Me (69%)
R¹ = Ph, R² = Ph (77%)

**Scheme 3.59** Catalytic silylation of α,β-unsaturated carbonyl compounds.

R = Ph (66%, >99% ee)
R = 4-ClC$_6$H$_4$ (58%, >99% ee)
R = n-C$_4$H$_9$ (55%, >99% ee)

### 3.3.4
### Synthetic Application of Silaboration Products

The products of catalytic silaboration have found wide synthetic applications, in which the reactivity difference between silyl and boryl groups is utilized.

Stereoisomeric alkyne silaboration products (Z)- and (E)-**137** undergo sequential cross-coupling with two aryl iodides, leading to stereoselctive synthesis of unsymmetrical (Z)- and (E)-stilbene derivatives **139** (Scheme 3.60) [63]. Initial coupling with an aryl iodide proceeds under Suzuki–Miyaura coupling conditions (Pd(OAc)$_2$, S-Phos, K$_3$PO$_4$), leaving the isopropoxydimethylsilyl group untouched. Thus, obtained alkenylsilanes are subjected to Hiyama coupling conditions (Pd(dba)$_2$, TBAF) for the synthesis of (Z)- and (E)-stilbene derivatives **139**.

β-Borylallylsilanes react with various electrophiles with retention of the boryl group. Reaction of **140** with benzaldehyde dimethyl acetal (**141**) gives boryl-substituted homoallylic ether **142** in high yield in the presence of TiCl$_4$ (Scheme 3.61) [84]. Allylative cyclization takes place in the reaction of **143**, which have siloxy groups, with benzaldehyde to give boron-substituted cyclic unsaturated ether **144** in good yield with complete chirality transfer (Scheme 3.62) [75]. The corresponding cyclization of enantioenriched **145** with benzyl trimethylsilyl ether also proceeds with highly efficient chirality transfer, leading to the formation of enantioenriched cyclopentene product **146** (Scheme 3.63) [75]. β-Borylallylsilane **147** reacts sequentially with two

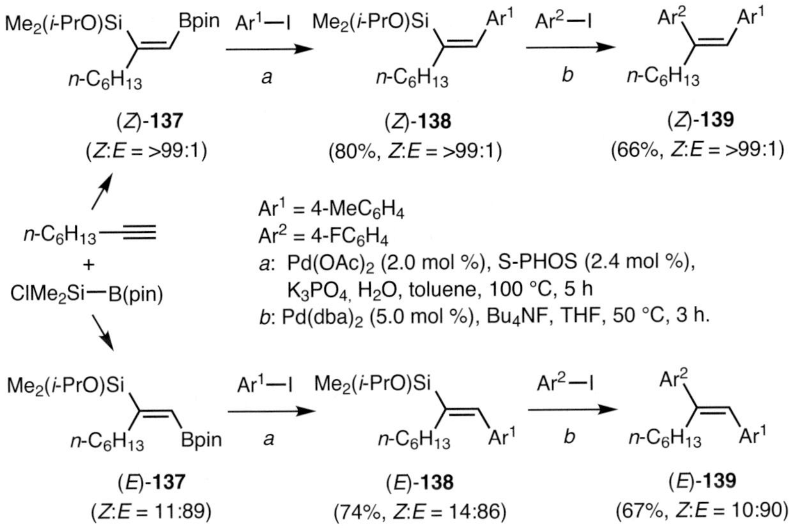

**Scheme 3.60** Stereoselective synthesis of *cis*- and *trans*-1,2-diaryloct-1-enes via *cis*- and *trans*-selective silaborations.

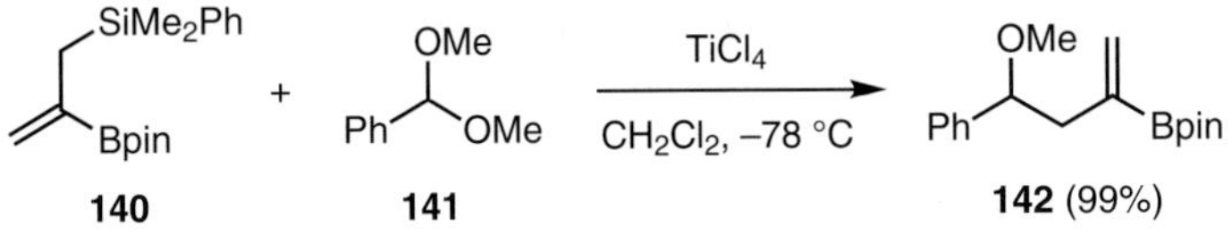

Ar$^1$ = 4-MeC$_6$H$_4$
Ar$^2$ = 4-FC$_6$H$_4$
a: Pd(OAc)$_2$ (2.0 mol %), S-PHOS (2.4 mol %), K$_3$PO$_4$, H$_2$O, toluene, 100 °C, 5 h
b: Pd(dba)$_2$ (5.0 mol %), Bu$_4$NF, THF, 50 °C, 3 h.

(Z)-137 (Z:E = >99:1)
(Z)-138 (80%, Z:E = >99:1)
(Z)-139 (66%, Z:E = >99:1)

(E)-137 (Z:E = 11:89)
(E)-138 (74%, Z:E = 14:86)
(E)-139 (67%, Z:E = 10:90)

**Scheme 3.61** TiCl$_4$-promoted allylation of acetals with β-borylallylsilane **140**.

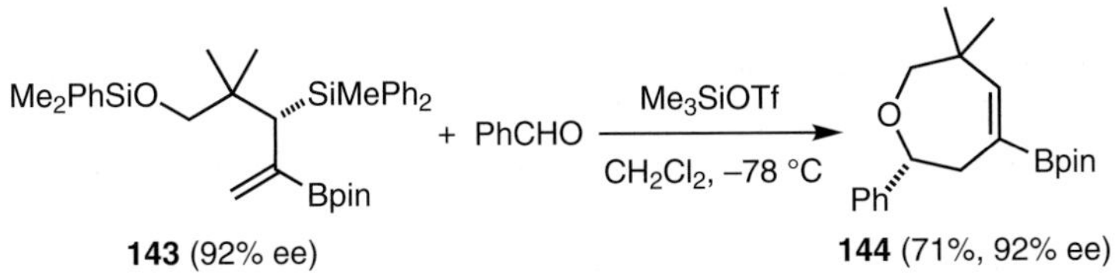

140, 141, 142 (99%)

**Scheme 3.62** Seven-membered cyclic ether formation with β-borylallylsilane **143** and aldehyde with efficient chirality transfer.

143 (92% ee), 144 (71%, 92% ee)

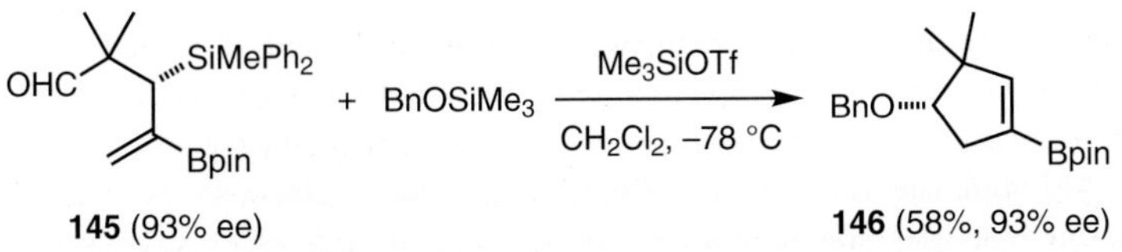

**Scheme 3.63** Intramolecular allylation of β-borylallylsilane **145** with efficient chirality transfer.

145 (93% ee), 146 (58%, 93% ee)

aldehydes in the presence of TiCl$_4$, leading to the formation of tricyclic organoboronate **148** in good yield as a single diastereomer (Scheme 3.64) [85].

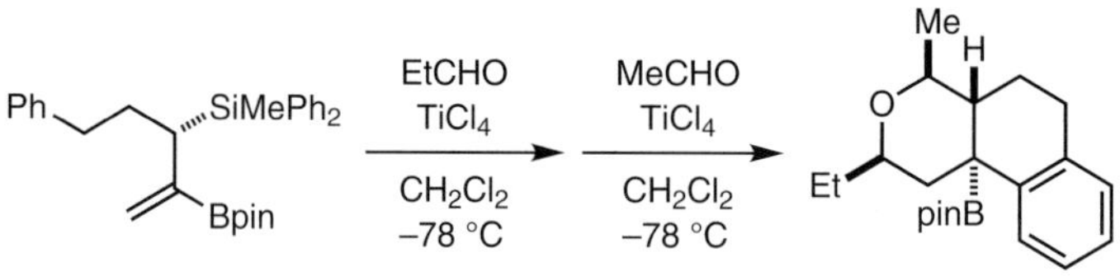

**Scheme 3.64** Sequential reactions of β-borylallylsilane **147** with two aldehydes.

Alkenylboronate **149**, which was obtained by silaborative C–C cleavage of cyclohexane-fused methylenecyclopropane, was converted into allylic organoboronate **150** by one-carbon homologation with lithium carbenoid (Scheme 3.65) [80]. The allylboron compound **150** reacts with benzaldehydes with high diastereoselectivity, giving homoallylic alcohol **151** in good yield. It should be noted that the stereochemistry of the newly formed stereogenic center is efficiently controlled by the chiral group at the β-position of the allylic boron compound **150**.

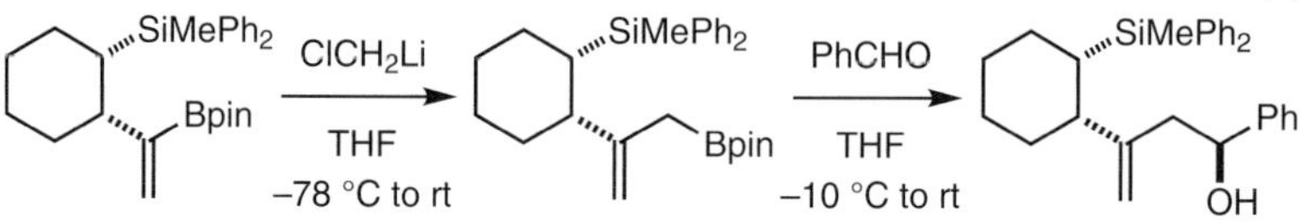

**Scheme 3.65** Diastereoselective conversion of alkenylboronic ester **149** through one-carbon homologation followed by allylboration.

Reaction of ClCH$_2$Li with *trans*-**152**, which was obtained by intramolecular silaboration of **85** (R = *i*-Pr), followed by Tamao oxidation gave 1,3,5-triol *anti*-**153** in good yield (Scheme 3.66) [67]. The corresponding diastereomer *syn*-**153** was also obtained stereoselectively via the identical conversion of *cis*-**152**.

## 3.4
## Carboboration

Carboboration can be the most straightforward and efficient method for the preparation of organoboronic acids. It provides new synthetic routes to rather complex and functionalized organoboronic acids through concomitant formation of C–C and C–B bonds. Although the uncatalyzed direct carboboration of alkynes was reported using the highly reactive triallylborane [86], no catalytic variants using more stable organoboron compounds appeared before 2000. As mentioned in the introduction, there are two distinctive classes of carboboration. One is direct carboboration, in

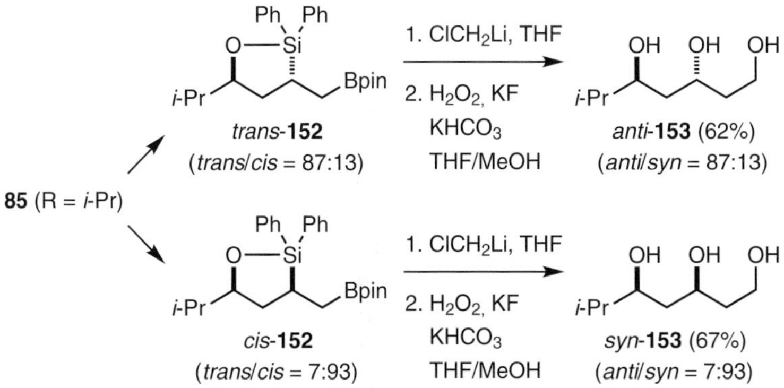

**Scheme 3.66** Stereoselective synthesis of diastereomeric 1,3,5-triols utilizing intramolecular silaboration.

which B−C bond in a reactant is activated by a catalyst and both organic and boryl groups are introduced into the product molecule. The other is transmetalative carboboration, in which the boryl and organic groups come from separate reactants. The former strategy is advantageous from the viewpoint of atom economy and the latter has an advantage of wider substrate scope because the organic parts can be varied easily by changing the organometallic reagents or organic halides used as the carbon source.

### 3.4.1
### Direct Addition: Cyanoboration and Alkynylboration

Although a number of cyanoboranes have been prepared since 1959 [87], there has been no application of these interesting compounds in organic synthesis. On the basis of the study on the use of these unique reagent in Strecker-type aminative cyanation of aldehydes [88], the reactivity of cyano bis(dialkylamino)boranes has been investigated in the presence of transition metal catalysts. An initial report on the catalytic addition of the B−CN bond across carbon–carbon triple bonds deals with an intramolecular cyclization of cyanoboranes **154** tethered onto carbon–carbon triple bonds by boryl homopropargyl ether linkages, which are easily cleavable after the reaction (Scheme 3.67) [89]. Palladium and nickel catalysts with or without phosphorus ligands showed high activity for the intramolecular cyanoboration. Intramolecular cyanoboration of allenes was also reported (Scheme 3.68) [90].

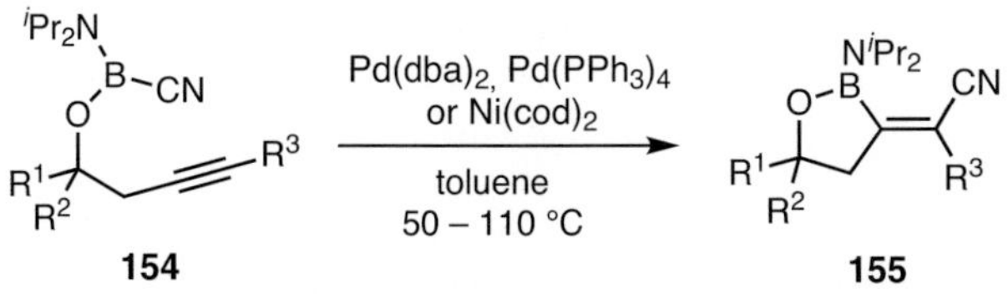

**Scheme 3.67** Catalytic intramolecular cyanoboration of alkynes.

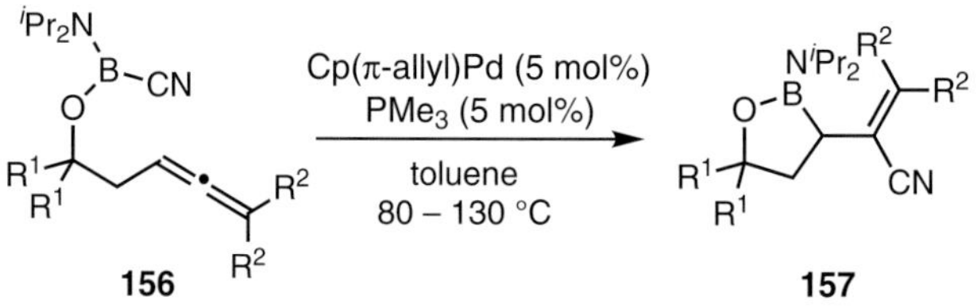

**Scheme 3.68** Catalytic intramolecular cyanoboration of allenes.

An intermolecular variant was reported using cyclic cyanoboranes **158** having 1,2-ethylenediamine-type ligands with a Pd/PMe$_3$ catalyst (Scheme 3.69) [91]. It is interesting to note that cyanoboration of 1-aryl-1-alkynes ($R^1$ = aryl) proceeds with high regioselectivities for the product in which the cyano group is attached α to the aryl group.

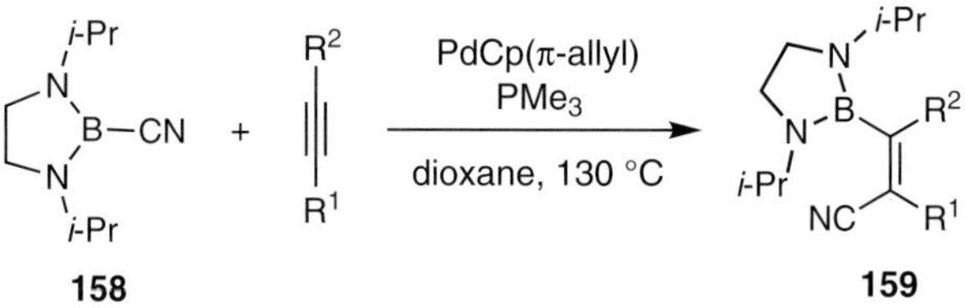

**Scheme 3.69** Catalytic intermolecular cyanoboration of alkynes.

A mechanistic study on the basis of stoichiometric reactions revealed that the oxidative addition of B—CN bond to palladium is reversible [92]. Insertion of alkynes takes place at the B—Pd bond rather than Pd—CN bond, followed by the rate-determining reductive elimination step.

Another example of the direct carboboration is the reaction of alkynylboronates **160** with alkynes (Scheme 3.70) [93]. Although alkynylboranes have been widely utilized in organic synthesis, no such addition reaction is reported. In the presence of a PCy$_3$-based nickel catalyst, *cis* addition of the boryl–alkynyl bond across carbon–carbon triple bonds takes place in good yields. So far only internal alkynes are suitable, and unsymmetrical 1-aryl-1-alkynes ($R^1$ = aryl) undergo the alkynylboration in a regioselective manner, in which the alkynyl group is introduced α to the aryl group.

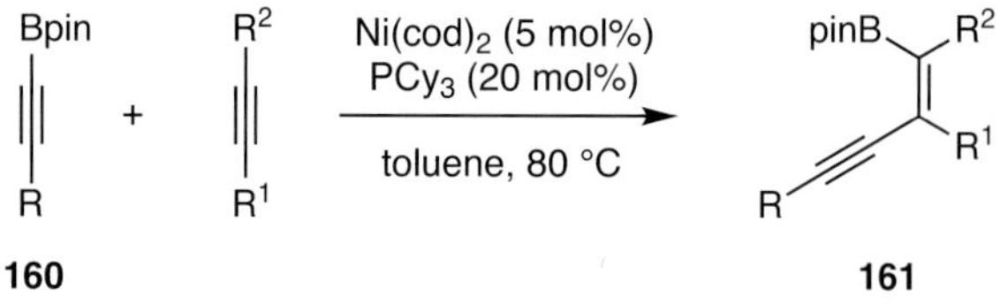

**Scheme 3.70** Catalytic alkynylboration of alkynes.

## 3.4.2
### Transmetalative Carboboration

Although the direct carboboration is highly attractive in terms of atom economy, it has been difficult to introduce organic groups other than cyano and alkynyl groups. This is mainly because other organoboron derivatives have more stable B—C bonds than B—CN and B—alkynyl bonds. Therefore, transmetalative carboboration seems to be a more feasible approach. A transmetalative carboboration of allenes utilizing acyl halides with $B_2pin_2$ was reported in 2000 (Scheme 3.71) [94]. The reaction provided a new synthetic route to β-borylmethyl-α,β-unsaturated ketones. This is the only report on the transmetalative carboboration that involves oxidative addition of carbon–halogen bond.

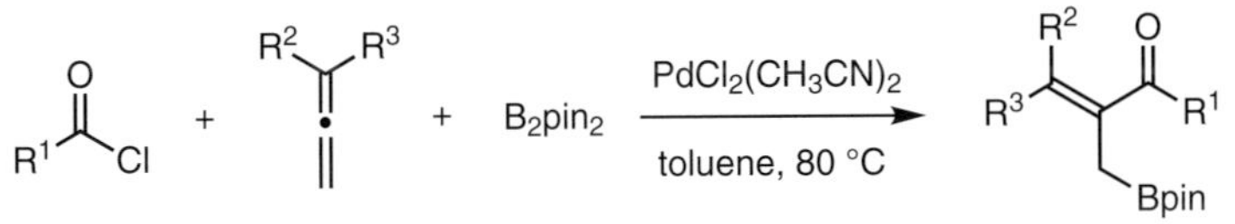

**Scheme 3.71** Transmetalative acylboration of allenes.

On the other hand, the other types of transmetalative carboboration involving oxidative addition of the boron–halogen bonds have shown rapid progress in these years. On the basis of mechanistic studies on the activation of B—CN bonds, in which the CN group can be regarded as a pseudohalide, activation of closely related B—Cl bonds has been pursued. A stoichiometric reaction of chloroboranes with a palladium(0) complex, leading to the formation of borylpalladium chloride species, was reported [95]. The boryl complex undergoes insertion with alkynes at the B—Pd bond, but no reductive elimination proceeds.

In an initial report on the transmetalative carboboration, tethered starting materials were used (Scheme 3.72) [96]. The chloroborane-alkynol conjugates **162** undergo intramolecular borylation followed by coupling with alkynylstannane **163**, leading to the formation of conjugated enynes **164** as the alkynylboration products in good yields. It should be noted that the carboboration takes place with a *trans*-addition mode with high selectivity. The primary products **164** were treated with pinacol and acetic anhydride with base, affording chromatographically stable pinacolborane derivatives **166**.

Following the transmetalative carboboration using alkynylstannane reagents, more versatile systems for the introduction of various organic groups have appeared. The use of organozirconium reagents **168** as transmetalation reagents is the key for a successful carboboration (Scheme 3.73) [97]. The system allows introduction of alkenyl, aryl, and even alkyl substituents into the products. It also allows the use of various organic groups on the *sp* carbon of the starting alkynes **167**, including H, alkyl, and aryl groups. More important, the stereochemical mode of addition is switchable: the use of $PMe_3$ as a ligand provides *cis*-addition products with high

**Scheme 3.72** Catalytic transmetalative alkynylboration.

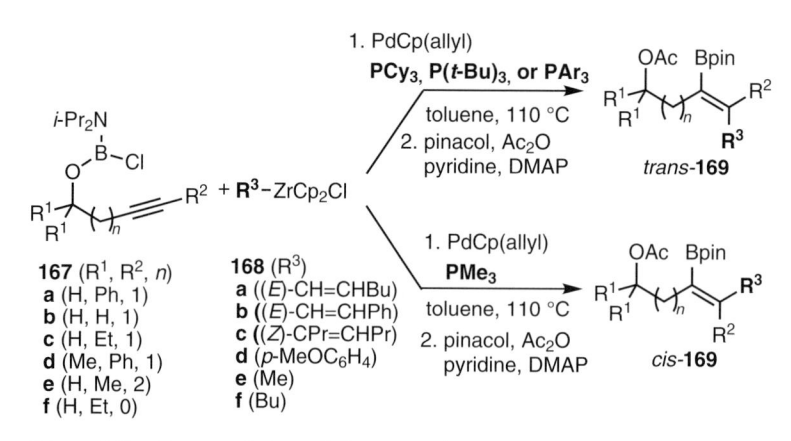

**Scheme 3.73** Transmetalative carboboration utilizing organozirconium reagents.

167 (R$^1$, R$^2$, n)
a (H, Ph, 1)
b (H, H, 1)
c (H, Et, 1)
d (Me, Ph, 1)
e (H, Me, 2)
f (H, Et, 0)

168 (R$^3$)
a ((E)-CH=CHBu)
b ((E)-CH=CHPh)
c ((Z)-CPr=CHPr)
d (p-MeOC$_6$H$_4$)
e (Me)
f (Bu)

selectivities, whereas use of more bulky ligands such as PPh$_3$, PCy$_3$, and tri-2-furylphosphine gives *trans*-addition products in high selectivities.

The transmetalative carboboration approach has been extended to an intermolecular carboboration system, in which chloroborane **170** bearing a diamine ligand and organozirconium reagents **168** are used [98] (Scheme 3.74). In the reactions of terminal alkynes (R$^1$ = H), the boryl group becomes attached to the terminal carbon atom in a highly regioselective manner. Note that only *cis*-addition products are obtained in the intermolecular reaction regardless of the phosphine ligand used.

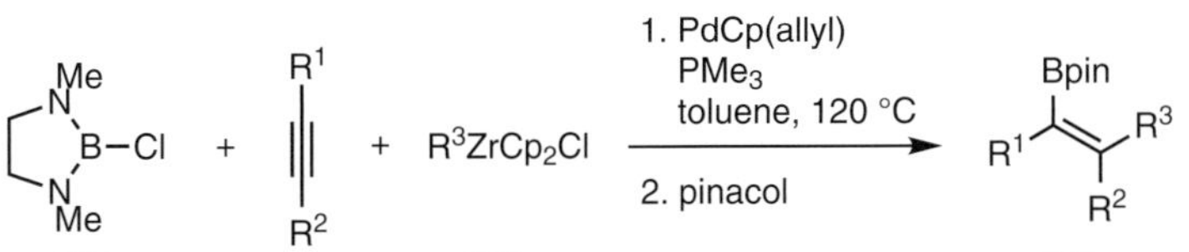

**Scheme 3.74** Intermolecular transmetalative carboboration.

Only alkenyl- and arylzirconium reagents rather than alkylzirconium reagents can be utilized in this reaction.

## 3.5
## Miscellaneous Element-Boryl Additions

In addition to the above-mentioned element-boryl addition reactions, some other direct additions including thioboration, stannaboration, and germaboration have been reported. The thioboration is not discussed here in detail because the particular reaction uses phenylthio-9-BBN, which is beyond the scope of this book on boronate derivatives (Scheme 3.75) [99].

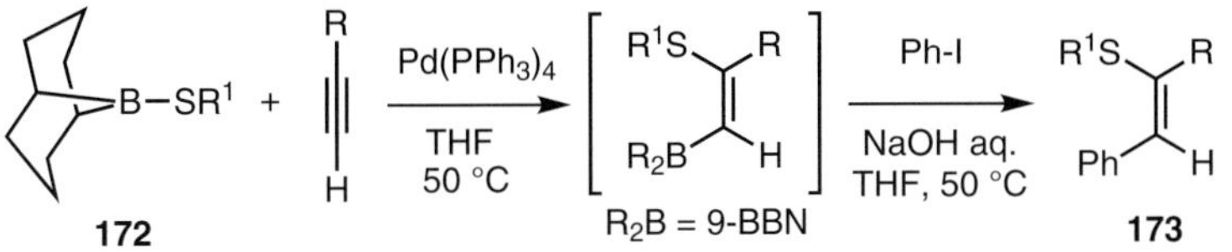

**Scheme 3.75** Catalytic thioboration of alkynes.

Trialkylstannyl(pinacol)borane **174** is used with palladium catalysts in the stannaboration of alkynes (Scheme 3.76) [100]. The reaction of terminal alkynes proceeds at room temperature with high regioselectivity for the attachment of the boryl group at the terminal position. A (stannyl)(boryl)palladium intermediate was isolated and characterized by a single-crystal X-ray analysis [100].

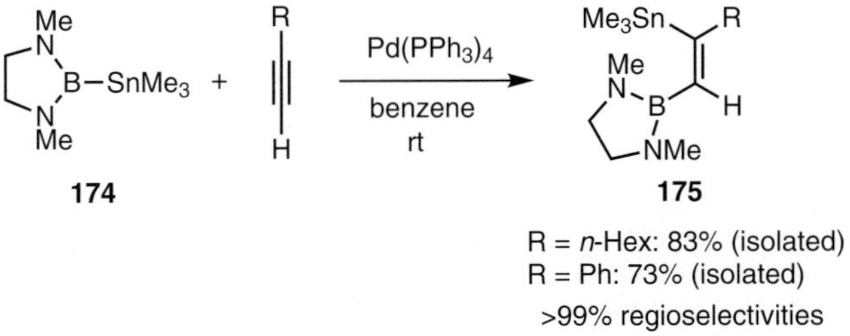

R = n-Hex: 83% (isolated)
R = Ph: 73% (isolated)
>99% regioselectivities

**Scheme 3.76** Catalytic stannaboration of alkynes.

Germaboration has been examined in the course of mechanistic investigation on the silaborative alkyne dimerization (Scheme 3.77) [64]. A platinum catalyst afforded simple germaboration product **177** selectively in high yield, whereas a nickel catalyst selectively afforded germaborative dimerization product **178**. Using a palladium catalyst, a mixture of these two products is obtained in good total yields.

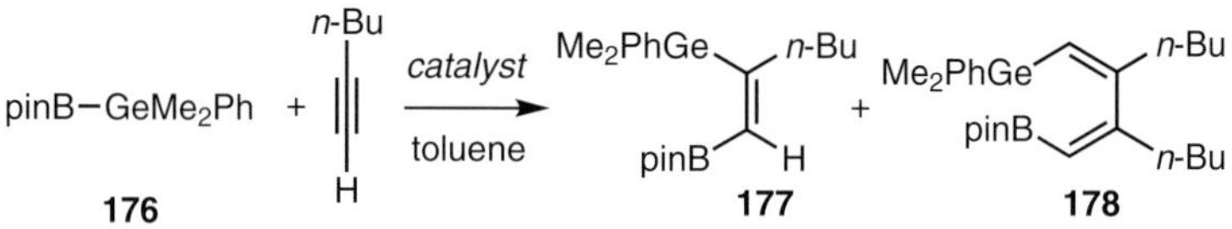

| catalyst | %yield of 177 (ratio) | %yield of 178 (ratio) | 177/178 |
|---|---|---|---|
| Pt(CH$_2$=CH$_2$)(PPh$_3$)$_2$ | 87 (91/9) | 0 | >99:1 |
| Ni(acac)$_2$/DIBAH | trace | 74 (74:26) | 4:96 |
| Pd(acac)$_2$/$t$-OcNC | 46 (>99:1) | 39 (96:4) | 53:47 |

**Scheme 3.77** Catalytic germaboration and germaborative dimerization of 1-hexyne.

## 3.6
## Conclusion

Catalytic element-boryl additions are powerful strategies for the synthesis of highly functionalized organoboronic acid derivatives. Boron-containing σ-bonds such as B−B, B−Si, B−Ge, B−Sn, B−S, B−CN, and boron–alkynyl bonds are activated by transition metal complexes in the direct element-boryl addition reactions. In addition to these direct additions, transmetalative addition reactions are rapidly explored to expand the reaction scope. The transmetalative element-boryl additions so far found involves activation of B−Cl bond. In the next decade, new, more efficient element-boryl additions will be exploited on the basis of the development of new catalysts and reagents.

## References

1 (a) Okinoshima, H., Yamamoto, K., and Kumada, M. (1972) *J. Am. Chem. Soc.*, **94**, 9263–9264; (b) Okinoshima, H., Yamamoto, K., and Kumada, M. (1975) *J. Organomet. Chem.*, **86**, C27–C30; (c) Sakurai, H., Kamiyama, Y., and Nakadaira, Y. (1975) *J. Am. Chem. Soc.*, **97**, 931–932.
2 Review on the element-silyl additions: Suginome, M. and Ito, Y. (2000) *Chem. Rev.*, **100**, 3221–3256.
3 Männig, D. and Nöth, H. (1985) *Angew. Chem., Int. Ed.*, **24**, 878–879.
4 Ishiyama, T., Matsuda, N., Miyaura, N., and Suzuki, A. (1993) *J. Am. Chem. Soc.*, **115**, 11018–11019.
5 Reviews on the element–element σ-bond additions: (a) Beletskaya, I. and Moberg, C. (1999) *Chem. Rev.*, **99**, 3435–3462; (b) Beletskaya, I. and Moberg, C. (2006) *Chem. Rev.*, **106**, 2320–2354.

6 Ceron, P., Frey, A.F.J., Kerrigan, J., Parsons, T., Urry, G., and Schlesinger, H.I. (1959) *J. Am. Chem. Soc.*, **81**, 6368–6371.

7 Reviews on the B–B additions: (a) Marder, T.B. and Norman, N.C. (1998) *Top. Catal.*, 63–73; (b) Ishiyama, T. and Miyaura, N. (2000) *J. Organomet. Chem.*, **611**, 392–402; (c) Ishiyama, T. and Miyaura, N. (2004) *Chem. Rec.*, **3**, 271–280; (d) Dembitsky, V.M., Abu Ali, H., and Srebnic, M. (2003) *Appl. Organomet. Chem.*, **17**, 327–345; (e) Dembitsky, V.M., Abu Ali, H., and Srebnic, M. (2004) *Adv. Organomet. Chem.*, **51**, 193–250.

8 (a) Nöth, H. (1984) *Z. Naturforsch.*, **39b**, 1463–1466; (b) Ishiyama, T., Murata, M., Ahiko, T., and Miyaura, N. (2000) *Org. Synth.*, **77**, 176–181.

9 Welch, C.N. and Shore, S.G. (1968) *Inorg. Chem.*, **7**, 225.

10 Ishiyama, T., Matsuda, N., Murata, M., Ozawa, F., Suzuki, A., and Miyaura, N. (1996) *Organometallics*, **15**, 713–720.

11 Lesley, G., Nguyen, P., Taylor, N.J., Marder, T.B., Scott, A.J., Clegg, W., and Norman, N.C. (1996) *Organometallics*, **15**, 5137–5154.

12 Mann, G., John, K.D., and Baker, R.T. (2000) *Org. Lett.*, **2**, 2105–2108.

13 Lillo, V., Mata, J., Ramírez, J., Peris, E., and Fernandez, E. (2006) *Organometallics*, **25**, 5829–5831.

14 Thomas, R.L., Souza, F.E.S., and Marder, T.B. (2001) *J. Chem. Soc., Dalton Trans.*, 1650–1656.

15 Ali, H.A., Quntar, A.E.A.A., Goldberg, I., and Srebnik, M. (2002) *Organometallics*, **21**, 4533–4539.

16 Maderna, A., Pritzkow, H., and Siebert, W. (1996) *Angew. Chem., Int. Ed. Engl.*, **35**, 1501–1503.

17 Bluhm, M., Maderna, A., Pritzkow, H., Bethke, S., Gleiter, R., and Siebert, W. (1999) *Eur. J. Inorg. Chem.*, 1693–1700.

18 Iverson, C.N. and Smith, M.R., III (1995) *J. Am. Chem. Soc.*, **117**, 4403–4404.

19 (a) Ishiyama, T., Yamamoto, M., and Miyaura, N. (1997) *Chem. Commun.*, 689–690; (b) Iverson, C.N. and Smith, M.R., III (1997) *Organometallics*, **16**, 2757–2759; (c) Carter, C.A.G., Vogels, C.M., Harrison, D.J., Gagnon, M.K.J., Norman, D.W., Langler, R.F., Baker, R.T., and Westcott, S.A. (2001) *Organometallics*, **20**, 2130–2132.

20 (a) Baker, R.T., Nguyen, P., Marder, T.B., and Westcott, S.A. (1995) *Angew. Chem., Int. Ed. Engl.*, **34**, 1336–1338; (b) Dai, C., Robins, E.G., Scott, A.J., Clegg, W., Yufit, D.S., Howard, J.A.K., and Marder, T.B. (1998) *Chem. Commun.*, 1983–1984; (c) Nguyen, P., Coapes, R.B., Woodward, A.D., Taylor, N.J., Burke, J.M., Howard, J.A.K., and Marder, T.B. (2002) *J. Organomet. Chem.*, **652**, 77–85; (d) Morgan, J.B., Miller, S.P., and Morken, J.P. (2003) *J. Am. Chem. Soc.*, **125**, 8702–8703.

21 Lillo, V., Mas-Marzá, E., Segarra, A.M., Carbó, J.J., Bo, C., Peris, E., and Fernández, E. (2007) *Chem. Commun.*, 3380–3382.

22 Lillo, V., Fructos, M.R., Ramírez, J., Braga, A.A.C., Maseras, F., Díaz-Requejo, M.M., Pérez, P.J., and Fernández, E. (2007) *Chem. Eur. J.*, **13**, 2614–2621.

23 Ramírez, J., Corberán, R., Sanaú, M., Peris, E., and Fernández, E. (2005) *Chem. Commun.*, 3056–3058.

24 Ramírez, J., Sanaú, M., and Fernández, E. (2008) *Angew. Chem., Int. Ed.*, **47**, 5194–5197.

25 Coapes, R.B., Souza, F.E.S., Thomas, R.L., Hall, J.J., and Marder, T.B. (2003) *Chem. Commun.*, 614–615.

26 (a) Marder, T.B., Norman, N.C., and Rice, C.R. (1998) *Tetrahedron Lett.*, **39**, 155–158; (b) Mkhalid, I.A.I., Coapes, R.B., Edes, S.N., Coventry, D.N., Souza, F.E.S., Thomas, R.L., Hall, J.J., Bi, S.-W., Lin, Z., and Marder, T.B. (2008) *Dalton Trans.*, 1055–1064.

27 (a) Miller, S.P., Morgan, J.B., Nepveux, V.F.J., and Morken, J.P. (2004) *Org. Lett.*, **6**, 131–133; (b) Trudeau, S., Morgan, J.B., Shrestha, M., and Morken, J.P. (2005) *J. Org. Chem.*, **70**, 9538–9544; (c) Ramírez, J., Segarra, A.M., and Fernández, E. (2005) *Tetrahedron Asymmetry*, **16**, 1289–1294.

28 Kliman, L.T., Mlynarski, S.N., and Morken, J.P. (2009) *J. Am. Chem. Soc.*, **131**, 13210–13211.

29 Ishiyama, T., Kitano, T., and Miyaura, N. (1998) *Tetrahedron Lett.*, **39**, 2357–2360.

30 Yang, F.-Y. and Cheng, C.-H. (2001) *J. Am. Chem. Soc.*, **123**, 761–762.

31 (a) Pelz, N.F., Woodward, A.R., Burks, H.E., Sieber, J.D., and Morken, J.P. (2004) *J. Am. Chem. Soc.*, **126**, 16328–16329; (b) Burks, H.E., Liu, S., and Morken, J.P. (2007) *J. Am. Chem. Soc.*, **129**, 8766–8773.

32 Wang, M., Cheng, L., and Wu, Z. (2008) *Organometallics*, **27**, 6464–6471.

33 Ishiyama, T., Yamamoto, M., and Miyaura, N. (1996) *Chem. Commun.*, 2073–2074.

34 Burks, H.E., Kliman, L.T., and Morken, J.P. (2009) *J. Am. Chem. Soc.*, **131**, 9134–9135.

35 Cho, H.Y. and Morken, J.P. (2008) *J. Am. Chem. Soc.*, **130**, 16140–16141.

36 (a) Morgan, J.B. and Morken, J.P. (2003) *Org. Lett.*, **5**, 2573–2575; (b) Ballard, C.E. and Morken, J.P. (2004) *Synthesis*, 1321–1324.

37 (a) Lawson, Y.G., Lesley, M.J.G., Marder, T.B., Norman, N.C., and Rice, C.R. (1997) *Chem. Commun.*, 2051–2052; (b) Ali, H.A., Goldberg, I., and Srebnik, M. (2001) *Organometallics*, **20**, 3962–3965; (c) Bell, N.J., Cox, A.J., Cameron, N.R., Evans, J.S.O., Marder, T.B., Duin, M.A., Elsevier, C.J., Baucherel, X., Tulloch, A.A.D., and Tooze, R.P. (2004) *Chem. Commun.*, 1854–1855.

38 (a) Takahashi, K., Ishiyama, T., and Miyaura, N. (2000) *Chem. Lett.*, 982–983; (b) Ito, H., Yamanaka, H., Tateiwa, J., and Hosomi, A. (2000) *Tetrahedron Lett.*, **41**, 6821–6825; (c) Takahashi, K., Ishiyama, T., and Miyaura, N. (2001) *J. Organomet. Chem.*, **625**, 47–53; (d) Mun, S., Lee, J.-E., and Yun, J. (2006) *Org. Lett.*, **8**, 4887–4889; (e) Chea, H., Sim, H.-S., and Yun, J. (2009) *Adv. Synth. Catal.*, **351**, 855–858; (f) Bonet, A., Lillo, V., Ramírez, J., Díaz-Requejo, M.M., and Fernández, E. (2009) *Org. Biomol. Chem.*, **7**, 1533–1535.

39 Kabalka, G.W., Das, B.C., and Das, S. (2002) *Tetrahedron Lett.*, **43**, 2323–2325.

40 Hirano, K., Yorimitsu, H., and Oshima, K. (2007) *Org. Lett.*, **9**, 5031–5033.

41 Lee, K., Zhugralin, A.R., and Hoveyda, A.H. (2009) *J. Am. Chem. Soc.*, **131**, 7253–7255.

42 (a) Lee, J.-E. and Yun, J. (2008) *Angew. Chem., Int. Ed.*, **47**, 145–147; (b) Feng, X. and Yun, J. (2009) *Chem. Commun.*, 6577–6579; (c) Fleming, W.J., Müller-Bunz, H., Lillo, V., Fernández, E., and Guiry, P.J. (2009) *Org. Biomol. Chem.*, **7**, 2520–2524.

43 Shiomi, T., Adachi, T., Toribatake, K., Zhou, L., and Nishiyama, H. (2009) *Chem. Commun.*, 5987–5989.

44 Ishiyama, T., Momota, S., and Miyaura, N. (1999) *Synlett*, 1790–1791.

45 Ali, H.A., Goldberg, I., Kaufmann, D., Burmeister, C., and Srebnik, M. (2002) *Organometallics*, **21**, 1870–1876.

46 (a) Hata, T., Kitagawa, H., Masai, H., Kurahashi, T., Shimizu, M., and Hiyama, T. (2001) *Angew. Chem., Int. Ed.*, **40**, 790–792; (b) Kurahashi, T., Hata, T., Masai, H., Kitagawa, H., Shimizu, M., and Hiyama, T. (2002) *Tetrahedron*, **58**, 6381–6395; (c) Shimizu, M., Schelper, M., Nagao, I., Shimono, K., Kurahashi, T., and Hiyama, T. (2006) *Chem. Lett.*, **35**, 1222–1223.

47 Ishiyama, T., Yamamoto, M., and Miyaura, N. (1996) *Chem. Lett.*, 1117–1118.

48 Ballard, C.E. and Morken, J.P. (2004) *Synthesis*, 1321–1324.

49 Nozaki, K., Wakamatsu, K., Nonaka, T., Tückmantel, W., Oshima, K., and Utimoto, K. (1986) *Tetrahedron Lett.*, **27**, 2007–2010.

50 Buynak, J.D. and Geng, B. (1995) *Organometallics*, **14**, 3112–3115.

51 (a) Suginome, M., Nakamura, H., and Ito, Y. (1996) *Chem. Commun.*, 2777–2778; (b) Suginome, M., Matsuda, T., Nakamura, H., and Ito, Y. (1999) *Tetrahedron*, **55**, 8787–8800.

52 Ohmura, T. and Suginome, M. (2009) *Bull. Chem. Soc. Jpn.*, **82**, 29–49.

53 (a) Shimizu, M., Kitagawa, H., Kurahashi, T., and Hiyama, T. (2001) *Angew. Chem., Int. Ed.*, **40**, 4283–4286; (b) Shimizu, M., Kurahashi, T., Kitagawa, H., and Hiyama, T. (2003) *Org. Lett.*, **5**, 225–227; (c) Shimizu, M., Kitagawa, H., Kurahashi, T., Shimono, K., and Hiyama, T. (2003) *J. Organomet. Chem.*, **686**, 286–293; (d) Murakami, M., Usui, I., Hasegawa, M., and Matsuda, T.

(2005) *J. Am. Chem. Soc.*, **127**, 1366–1367.

54 (a) Suginome, M., Fukuda, T., Nakamura, H., and Ito, Y. (2000) *Organometallics*, **19**, 719–721; (b) Suginome, M., Fukuda, T., and Ito, Y. (2002) *J. Organomet. Chem.*, **643-644**, 508–511.

55 Suginome, M., Ohmura, T., Miyake, Y., Mitani, S., Ito, Y., and Murakami, M. (2003) *J. Am. Chem. Soc.*, **125**, 11174–11175.

56 Suginome, M., Matsuda, T., and Ito, Y. (2000) *Organometallics*, **19**, 4647–4649.

57 Ohmura, T., Masuda, K., Furukawa, H., and Suginome, M. (2007) *Organometallics*, **26**, 1291–1294.

58 Onozawa, S., Hatanka, Y., and Tanaka, M. (1997) *Chem. Commun.*, 1229–1230.

59 Da Silva, J.C.A., Birot, M., Pillot, J.-P., and Pétraud, M. (2002) *J. Organomet. Chem.*, **646**, 179–190.

60 Suginome, M., Noguchi, H., Hasui, T., and Murakami, M. (2005) *Bull. Chem. Soc. Jpn.*, **78**, 323–326.

61 Ohmura, T., Masuda, K., and Suginome, M. (2008) *J. Am. Chem. Soc.*, **130**, 1526–1527.

62 Sagawa, T., Asano, Y., and Ozawa, F. (2002) *Organometallics*, **21**, 5879–5886.

63 Ohmura, T., Oshima, K., and Suginome, M. (2008) *Chem. Commun.*, 1416–1418.

64 Suginome, M., Matsuda, T., and Ito, Y. (1998) *Organometallics*, **17**, 5233–5235.

65 Lüken, C. and Moberg, C. (2008) *Org. Lett.*, **10**, 2505–2508.

66 Suginome, M., Nakamura, H., and Ito, Y. (1997) *Angew. Chem., Int. Ed.*, **36**, 2516–2518.

67 Ohmura, T., Furukawa, H., and Suginome, M. (2006) *J. Am. Chem. Soc.*, **128**, 13366–13367.

68 Ohmura, T., Takasaki, Y., Furukawa, H., and Suginome, M. (2009) *Angew. Chem., Int. Ed.*, **48**, 2372–2375.

69 Suginome, M., Matsuda, T., Yoshimoto, T., and Ito, Y. (1999) *Org. Lett.*, **1**, 1567–1569.

70 Suginome, M., Nakamura, H., Matsuda, T., and Ito, Y. (1998) *J. Am. Chem. Soc.*, **120**, 4248–4249.

71 (a) Gerdin, M. and Moberg, C. (2005) *Adv. Synth. Catal.*, **347**, 749–753; (b) Gerdin, M., Penhoat, M., Zalubovskis, R., Pétermann, C., and Moberg, C. (2008) *J. Organomet. Chem.*, **693**, 3519–3526; (c) Durieux, G., Gerdin, M., Moberg, C., and Jutand, A. (2008) *Eur. J. Inorg. Chem.*, 4236–4241.

72 (a) Suginome, M., Ohmori, Y., and Ito, Y. (1999) *Synlett*, 1567–1568; (b) Suginome, M., Ohmori, Y., and Ito, Y. (2000) *J. Organomet. Chem.*, **611**, 403–413.

73 Onozawa, S., Hatanaka, Y., and Tanaka, M. (1999) *Chem. Commun.*, 1863–1864.

74 Ohmura, T. and Suginome, M. (2006) *Org. Lett.*, **8**, 2503–2506.

75 Ohmura, T., Taniguchi, H., and Suginome, M. (2006) *J. Am. Chem. Soc.*, **128**, 13682–13683.

76 Chang, K.-J., Rayabarapu, D.-K., Yang, F.-Y., and Cheng, C.-H. (2005) *J. Am. Chem. Soc.*, **127**, 126–131.

77 Abe, Y., Kuramoto, Y., Ehara, M., Nakatsuji, H., Suginome, M., Murakami, M., and Ito, Y. (2008) *Organometallics*, **27**, 1736–1742.

78 Suginome, M., Matsuda, T., and Ito, Y. (2000) *J. Am. Chem. Soc.*, **122**, 11015–11016.

79 Pohlmann, T. and de Meijere, A. (2000) *Org. Lett.*, **2**, 3877–3879.

80 Ohmura, T., Taniguchi, H., Kondo, Y., and Suginome, M. (2007) *J. Am. Chem. Soc.*, **129**, 3518–3519.

81 Ohmura, T., Taniguchi, H., and Suginome, M. (2009) *Org. Lett.*, **11**, 2880–2883.

82 Suginome, M., Matsuda, T., Yoshimoto, T., and Ito, Y. (2002) *Organometallics*, **21**, 1537–1539.

83 (a) Walter, C., Auer, G., and Oestreich, M. (2006) *Angew. Chem., Int. Ed.*, **45**, 5675–5677; (b) Walter, C., and Oestreich, M. (2008) *Angew. Chem., Int. Ed.*, **47**, 3818–3820; (c) Walter, C., Fröhlich, R., and Oestreich, M. (2009) *Tetrahedron*, **65**, 5513–5520.

84 Suginome, M., Ohmori, Y., and Ito, Y. (2001) *J. Am. Chem. Soc.*, **123**, 4601–4602.

85 Suginome, M., Ohmori, Y., and Ito, Y. (2001) *Chem. Commun.*, 1090–1091.

86 Mikhailov, B.M. and Bubnov, Yu N. (1971) *Tetrahedron Lett.*, **12**, 2127–2130.

87 (a) Evers, E.C., Freitag, W.O., Kriner, W.A., and MacDiarmid, A.G. (1959) *J. Am. Chem. Soc.*, **81**, 5106–5108;

(b) Spielvogel, B.F., Bratton, R.F. and Moreland, C.G. (1972) *J. Am. Chem. Soc.*, **94**, 8597–8598; (c) Bessler, V.E. and Goubeau, J.Z. (1967) *Anorg. Allg. Chem.*, **352**, 67–76; (d) Meller, A., Maringgele, W., and Sicker, U. (1977) *J. Organomet. Chem.*, **141**, 249–255.

88 Suginome, M., Yamamoto, A., and Ito, Y. (2002) *Chem. Commun.*, 1392–1393.

89 (a) Suginome, M., Yamamoto, A., and Murakami, M. (2003) *J. Am. Chem. Soc.*, **125**, 6358–6359; (b) Ohmura, T., Awano, T., Suginome, M., Yorimitsu, H., and Oshima, K. (2008) *Synlett*, 423–427.

90 Yamamoto, A., Ikeda, Y., and Suginome, M. (2009) *Tetrahedron Lett.*, **50**, 3168–3170.

91 Suginome, M., Yamamoto, A., and Murakami, M. (2005) *Angew. Chem., Int. Ed.*, **44**, 2380–2382.

92 Suginome, M., Yamamoto, A., Sasaki, T., and Murakami, M. (2006) *Organometallics*, **25**, 2911–2913.

93 Suginome, M., Shirakura, M., and Yamamoto, A. (2006) *J. Am. Chem. Soc.*, **128**, 14438–14439.

94 Yang, F.-Y., Wu, M.-Y., and Cheng, C.-H. (2000) *J. Am. Chem. Soc.*, **122**, 7122–7123.

95 Onozawa, S.-Y. and Tanaka, M. (2001) *Organometallics*, **20**, 2956–2958.

96 Yamamoto, A. and Suginome, M. (2005) *J. Am. Chem. Soc.*, **127**, 15706–15707.

97 Daini, M., Yamamoto, A., and Suginome, M. (2008) *J. Am. Chem. Soc.*, **130**, 2918–2919.

98 Daini, M. and Suginome, M. (2008) *Chem. Commun.*, 5224–5226.

99 Ishiyama, T., Nishijima, K., Miyaura, N., and Suzuki, A. (1993) *J. Am. Chem. Soc.*, **115**, 7219–7225.

100 Onozawa, S.-y., Hatanaka, Y., Sakakura, T., Shimada, S., and Tanaka, M. (1996) *Organometallics*, **15**, 5450–5452.

# 4
# The Contemporary Suzuki–Miyaura Reaction
*Cory Valente and Michael G. Organ*

## 4.1
## Introduction

### 4.1.1
### Preamble and Outlook

The "tried, tested, and true" transformations that are relied upon most heavily by synthetic chemists all have one underlying commonality – they are general. It is no wonder, then, that much effort is continuously dedicated to realizing new thresholds for reactions that aim to further improve their general application. This quest is largely fueled by an ever-evolving understanding of reaction mechanisms in parallel with the discovery of novel chemicals and methodologies. Since its discovery, Pd-mediated (and to a lesser extent Ni-mediated) cross-coupling has evolved into one of the most reliable means to construct C–C and C–N bonds; indeed, many synthetic routes contain at least one step that relies on this methodology [1–3]. Sifting out all the nuances that distinguish one cross-coupling from another makes it clear that the reaction conditions that are the most general unsurprisingly constitute the most widely utilized cross-coupling, namely, the Suzuki–Miyaura reaction [2–10].

Over the past decade, the Suzuki–Miyaura reaction has been pushed to new limits by way of the rational design of highly active and efficient catalysts [11–13]. These endeavors have paid off handsomely and have garnered enhanced scope and generality for this reaction. This chapter will serve to ledger these recent achievements, placing focus on (i) new ligands that yield highly active catalysts capable of coupling (ii) aryl chlorides and (iii) sterically hindered substrates and (iv) unactivated alkyl electrophiles, with advances in the last two areas making possible (v) the asymmetric Suzuki–Miyaura reaction. The chapter will close with a synopsis of (vi) iterative and orthogonal Suzuki–Miyaura cross-couplings. Although the focus of this chapter is on boronic acids, a variety of organoboron derivatives will be discussed for the sake of completeness.

---

*Boronic Acids: Preparation and Applications in Organic Synthesis, Medicine and Materials*, Second Edition.
Edited by Dennis G. Hall.
© 2011 Wiley-VCH Verlag GmbH & Co. KGaA. Published 2011 by Wiley-VCH Verlag GmbH & Co. KGaA.

## 4.1.2
## A Brief History

In 1978, Negishi and coworkers discovered that 2-iodotoluene could be coupled with lithium 1-heptynyltributylborate, establishing for the first time that organoboranes were effective transmetalating agents in Pd-mediated cross-couplings [14, 15]. Neutral alkenylboranes were next applied by Suzuki and Miyaura, where they found that the addition of an exogenous base was necessary to effect the cross-coupling of these organometallic species [16, 17]. The Suzuki–Miyaura reaction has since evolved into the most commonly applied cross-coupling reaction in both academia and industry alike [5–7]. This stems primarily from the many attractive attributes of boronic acids and their derivatives, including trifluoroborates (discussed in detail in Chapter 14), such as their air and moisture insensitivity, thermal stability, functional group tolerance, and negligible toxicity, with myriad boronic acids being commercially available [18–20].

## 4.1.3
## Mechanistic Aspects

The general scheme for the Suzuki–Miyaura reaction is presented in Scheme 4.1. The catalytic cycle begins and ends as it does for most other cross-couplings, with oxidative addition and reductive elimination, respectively (Figure 4.1). The Suzuki–Miyaura reaction is unique in that it is the only cross-coupling that requires the addition of a stoichiometric excess of base. Base has been found to drastically accelerate the transmetalation step of the catalytic cycle [1]. Two proposed pathways for metal–metal exchange have been put forward to rationalize the observed enhancement in rate (Path A and B) [21]. In Path A, the boronic acid and base react to form a borate that is sufficiently electron-rich to transmetalate with the Pd(II) oxidative addition adduct. The rate of transmetalation also depends on the cation of the base in addition to the halide (X) of the electrophile, that is, the rate decreases in the order $X = Cl > Br > I$, suggesting that transmetalation is more facile with more electrophilic Pd species. In Path B, a transient (oxo)palladium(II) intermediate is formed prior to transmetalation with a neutral organoborane. Path B is consistent with experimental results showing that preformed (oxo)palladium(II) intermediates undergo transmetalation with neutral organoboranes in the absence of added base. In most

$$R^1-X + R^2-BY_n \xrightarrow[\text{Base}]{\text{Pd or Ni (cat.)}} R^1-R^2 + BXY_n$$

$R^1, R^2$ = Aryl, Alkenyl, Benzyl, Allyl, Alkyl
X = I, Br, Cl, OTf, OTs, OPiv
$BY_n$ = $B(OH)_2$, $B(OC(Me_2)C(Me_2)O)$, 9-BBN, $BF_3^-$
Base = $^-OH$, $CO_3^{2-}$, $PO_4^{3-}$, $F^-$

**Scheme 4.1** The general reaction scheme for the Suzuki–Miyayra reaction. BBN = borabicyclo[3.3.1]nonane; Piv = pivalate; Tf = trifluoromethanesulfonate; Ts = 4-toluenesulfonate.

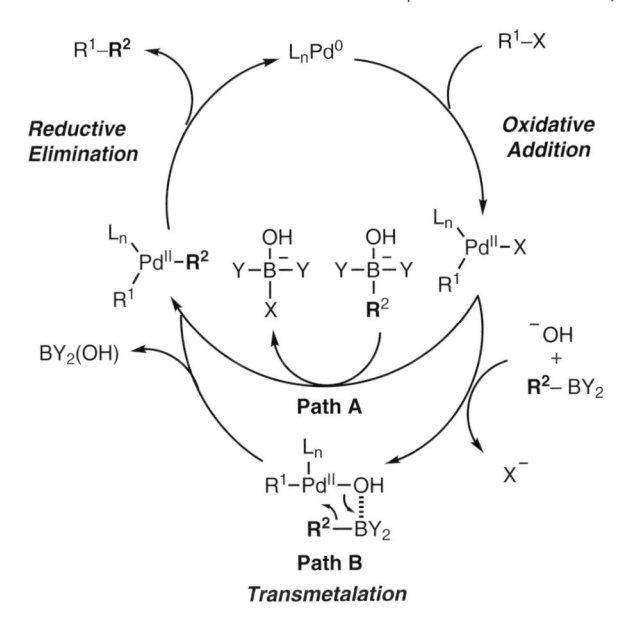

**Figure 4.1** The general catalytic cycle of the Suzuki–Miyaura reaction. Refer to Scheme 4.1 for the definition of $R^1$, $R^2$, Y, and X, and the text for elaboration.

Suzuki–Miyaura cross-couplings, both Path A and B are believed to be operational concurrently, and it is not immediately obvious as to why one pathway predominates under a particular set of reaction conditions.

The choice of base and solvent in the Suzuki–Miyaura reaction is still largely empirical; however, ethereal and aromatic hydrocarbon solvents tend to be optimal as are carbonate, phosphate, hydroxide, and fluoride bases. Water can have a beneficial effect on the reaction, and much work has gone into translating the Suzuki–Miyaura reaction to a green platform [22, 23]. The Suzuki–Miyaura reaction is most commonly associated with the formation of a $C_{sp2}$–$C_{sp2}$ bond between an aryl or alkenyl iodide or bromide and an aryl or alkenylboronic acid and many reviews dealing with this subject matter have been written, and the reader is directed to these materials for further information [2–10].

## 4.2
## Developments Made in the Coupling of Nontrivial Substrates

### 4.2.1
### Rational Design of Ligands for Use in the Suzuki–Miyaura Reaction

In being able to fully understand what has led to effectively a "second wind" for not only the Suzuki–Miyaura reaction but also cross-coupling reactions in general, it is important to first describe the basis upon which such advancements have been made. The barriers that once excluded sterically and electronically demanding substrates [24–30] from undergoing cross-coupling have been resolved primarily through advanced ligand design; this has been aided by greater mechanistic insight into the catalytic cycle, leading to more finely tuned catalysts [11–13, 31, 32]. The most effective ligands utilized in cross-couplings can be grouped into two main categories, namely, organophosphines possessing at least two alkyl fragments (Figure 4.2) [32–34] and N-heterocyclic carbenes (NHCs) [35–41], specifically

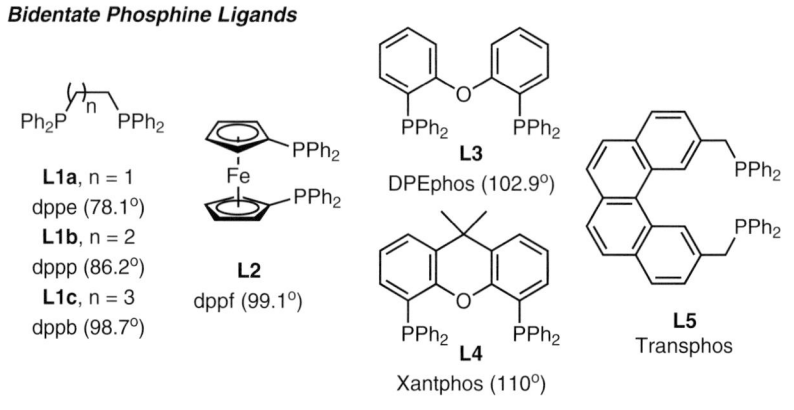

**Bidentate Phosphine Ligands**

**L1a**, n = 1
dppe (78.1°)
**L1b**, n = 2
dppp (86.2°)
**L1c**, n = 3
dppb (98.7°)

**L2**
dppf (99.1°)

**L3**
DPEphos (102.9°)

**L4**
Xantphos (110°)

**L5**
Transphos

**Monodentate Phosphine Ligands**

PPh$_3$      P(o-tolyl)$_3$   P(n-Bu$_3$)   PCy$_3$      P(t-Bu)$_3$    P(t-Bu)$_2$Me
(145°) {2.7}  (194°) {3.1}   (132°) {8.4}  (170°) {9.7}  (182°) {11.4}  (161°)

PAd$_2$(n-Bu)

**L6**

**L7**

**L8**

**L9a**, R = Cy
**L9b**, R = t-Bu
JohnPhos

**L9c**
DavePhos

**L9d**, R = H
SPhos
**L9f**, R = SO$_3$Na
SPhos(SO$_3$Na)

**L9e**, R = i-Pr
XPhos
**L9g**, R = SO$_3$Na
XPhos(SO$_3$Na)

**Figure 4.2** A selection of organophosphines used as ligands in the Suzuki–Miyaura reaction. Selected cone and bite angles are given in parentheses for bi- and monodentate ligands, respectively. Selected $pK_a$'s of the conjugate acids are given in braces [44, 45]. Ad = adamantyl; Cy = cyclohexyl.

N, N'-disubstituted imidazolylidines (Figure 4.3). A select number of amine-based ligands have also proven very useful, especially in Ni-catalyzed couplings of secondary alkyl halides (see below) [26, 27]. Both main classes of ligands are excellent σ-donors that increase the electron density on Pd, which benefits the oxidative addition step, while possessing suitable steric bulk that enhances reductive elimination. There is a fine balance between ligand structure and the optimal steric topography around the metal center. For example, the use of two bulky monodentate ligands is only capable of complexing Pd *trans* so as to minimize ligand–ligand steric repulsion, the consequence being that reductive elimination is arrested. By optimizing the ligand sterics, monoligated Pd-adducts become favorable and reductive elimination can proceed [31, 42, 43]. Second, the increased steric topography around the monoligated metal center (i.e., [(L)$R^1PdR^2$]) forces the coupling fragments $R^1$ and $R^2$ *cis*, which is a prerequisite for reductive elimination, and provides an added "push" to expel the cross-coupled product and relieve the steric congestion.

### 4.2.1.1 Organophosphine Ligands and Properties

Organophosphines are the most routinely employed ligands in cross-coupling reactions. Diphosphines of the type **L1–L4** (Figure 4.2) form a variety of bisligated Pd and Ni species, coordinating *cis* positions on the transition metal. This ensures that $R^1$ and $R^2$ in the $R^1R^2Pd{\zeta}L$ intermediate are also oriented *cis*. Varying the bite angle (see Figure 4.2 for selected bite angles) of these bis(diphenylphosphino) ligands has a substantial effect on the rate of reductive elimination [46–48]; dppf (**L2**), dppp (**L1b**), dppb (**L1c**), and DPEphos (**L3**) have been shown to be superior to dppe (**L1a**) and Xantphos (**L4**) in a variety of examples [1, 46, 47, 49, 50]. For example, the rate of reductive elimination from $Me_2Ni{\zeta}L$ complexes has been shown to be 46 times faster when L = dppp (**L1b**, bite angle 86.2°) compared to when L = dppe (**L1a**, bite angle 78.1°) [51]. This trend occurs until a maximum bite angle is reached, above which the geometric constraints become counterproductive for the catalytic cycle [1, 46–54]. For example, Transphos (**L5**) prevents reductive elimination from occurring, as the two coupling fragments cannot adopt a *cis* orientation on Pd [53]. Ligand electronic properties also have been found to significantly influence the rate of reductive elimination [55, 56]. More recently, organophosphines harnessing sterically laden, electron-donating secondary and tertiary *alkyl* groups as "activating" ligands in Pd-catalyzed cross-couplings have been demonstrated to be greatly superior to triarylphosphines and less bulky trialkylphosphines [32–34, 57]. These benefits source from optimal cone angles and better overall σ-donation to Pd via inductive effects from the secondary and tertiary alkyl groups of these ligands [57, 58]. In particular, dialkylbiaryl phosphines **L7–L9** [32] and trialkylphosphines $PCy_3$ and $P(t\text{-}Bu)_3$ [33] have emerged as optimal ligands, providing highly active catalysts that have expanded considerably the scope of the Suzuki–Miyaura reaction. Buchwald and Fu have largely led this charge, and each have recently published accounts of their endeavors from the past decade [32, 33]. In terms of ease of use, trialkylphosphines tend to be pyrophoric, and some organophoshines can be oxidized to their corresponding phosphine oxide upon prolonged exposure to air. To circumvent this impediment,

## Imidazolium Salts

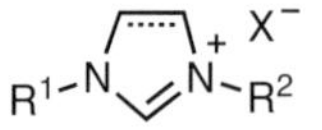

| | | IAd<br>L10a | IPr<br>L10b | IEt<br>L10c | IMes<br>L10d | SIPr<br>L10e | SIPrEt<br>L10f | SIPrMes<br>L10g | SIEt<br>L10h | SIMes<br>L10i |
|---|---|---|---|---|---|---|---|---|---|---|
| | SI/I | I | I | I | I | SI | SI | SI | SI | SI |
| | R$^1$ | Ad | Pr | Et | Mes | Pr | Pr | Pr | Et | Mes |
| | R$^2$ | Ad | Pr | Et | Mes | Pr | Et | Mes | Et | Mes |
| | % BV | 37 | 29 | – | 26 | 30 | – | – | – | 27 |

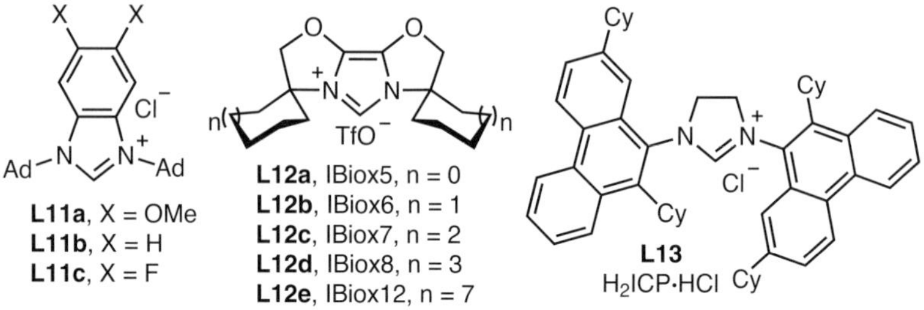

L10a–i
Imidazoliums
Unsaturated (I)
Saturated (SI)

L11a, X = OMe
L11b, X = H
L11c, X = F

L12a, IBiox5, n = 0
L12b, IBiox6, n = 1
L12c, IBiox7, n = 2
L12d, IBiox8, n = 3
L12e, IBiox12, n = 7

L13
H$_2$ICP·HCl

## NHC–Pd Complexes

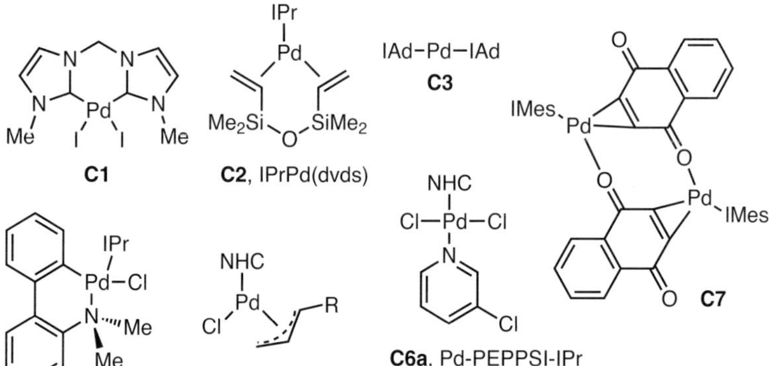

C1

C2, IPrPd(dvds)

C3, IAd–Pd–IAd

C4

C5a, R = H, NHC = IMes
C5b, R = H, NHC = IPr
C5c, R = Ph, NHC = IPr

C6a, Pd-PEPPSI-IPr
C6b, Pd-PEPPSI-IBu
C6c, Pd-PEPPSI-Ic-Pent
C6d, Pd-PEPPSI-IPent

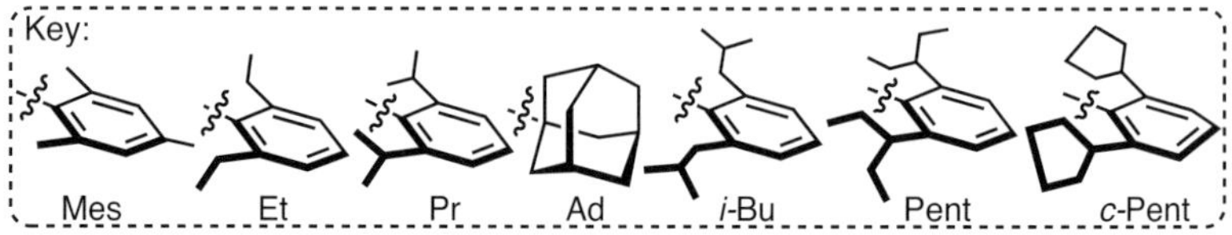

C7

Key: Mes, Et, Pr, Ad, i-Bu, Pent, c-Pent

**Figure 4.3** A selection of imidazolium salts and NHC–Pd complexes used in the Suzuki–Miyaura reaction. Selected % buried volumes (%BV) are given for imidazolium salts L10a–i. Biox = bioxazoline; Cy = cyclohexyl; dvds = 1,1,3,3-tetramethyl-1,3-divinyldisiloxane; H$_2$ICP = N,N-bis-(2,9-dicyclohexyl-10-phenanthryl)-4,5-dihydroimidazolium; PEPPSI = pyridine enhanced precatalyst preparation, stabilization, and initiation.

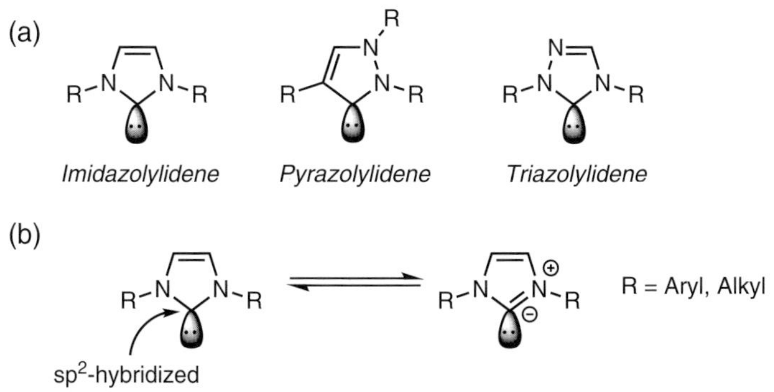

**Figure 4.4** (a) Selected structures of N-heterocyclic carbenes and (b) the stabilization of the carbene carbon by neighboring nitrogen atoms.

air-stable trialkyl phosphonium salts have been employed that are converted into their neutral organophosphine counterpart *in situ* under the basic reaction conditions [59].

#### 4.2.1.2 N-Heterocyclic Carbene Ligands and their Properties

Nitrogen-stabilized singlet carbenes (Figure 4.4a) were first described in the early 1960s [60] and by the end of that decade Wanzlick [61] and Öfele [62] had independently prepared transition metal complexes of these carbenes. The $sp^2$-hybridized singlet carbene is stabilized by the σ-electron-withdrawing neighboring nitrogen atom(s) whose electron lone pairs interact with the vacant p-orbital of the carbene. As a result, these carbenes are electron-rich, nucleophilic entities (Figure 4.4b) [63, 64]. Not until 1991 when Arduengo isolated and characterized for the first time the free and crystalline IAd carbene [65] were NHCs more seriously considered for transition metal-catalyzed reactions [35, 41, 66–70], among other applications [71, 72]. Overall, NHCs are better σ-donors than the most basic trialkylphosphines and relief of electron density on Pd via backbonding is not as prominent as it is on Pd-organophosphine adducts [41, 73–79]. Moreover, NHCs do not dissociate readily from their ligated transition metal. These attributes combined, the metal center is more electron-rich in NHC complexes than it is in organophosphine complexes. NHCs are also unique in that alterations made to substituents on the imidazolium ring do not appreciably alter the level of σ-donation of the carbene [13, 80, 81]. This facet renders the electronic component independent of the steric component of NHCs, thereby permitting the fine-tuning of the steric topography around its ligated metal center without deleterious effects to the electronic properties of the metal. Experimentally, it has been shown that sterically similar/electronically dissimilar benzimidazolium ligands **L11a–c** (Figure 4.3) are comparable in their ability to "activate" Pd toward the Suzuki–Miyaura reaction involving aryl chlorides [13, 82]. As such, the carbene carbon is sufficiently electron-rich to overshadow substituent alterations made to the backbone of the NHC.

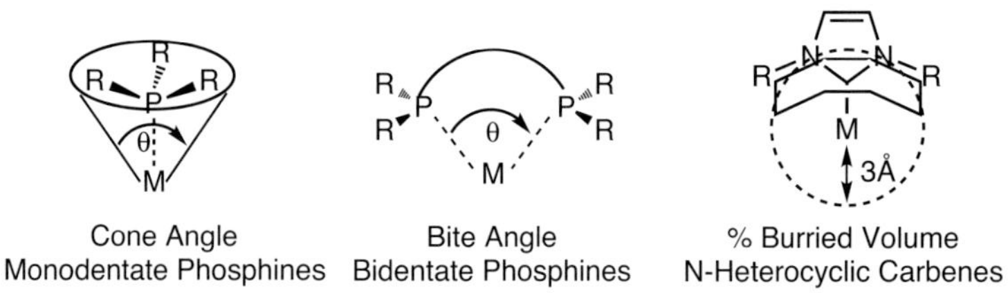

| Cone Angle | Bite Angle | % Burried Volume |
|---|---|---|
| Monodentate Phosphines | Bidentate Phosphines | N-Heterocyclic Carbenes |

**Figure 4.5** Steric descriptors for organophosphine and NHC ligands.

The steric topography around the metal center imposed by NHCs is very different from that of organophosphines (Figure 4.5). In the case of NHCs, the substituents on the imidazolium ring are directed toward the metal center; for organophosphines, the substituents point away. As such, the steric contribution of an NHC to a metal center cannot be measured by Tolman's cone angle descriptor [83–85]. Instead, the buried volume of the ligand is calculated as the percentage of occupied space by the NHC that lies within a sphere of a 3 Å radius centered at the metal [75, 79, 81, 86]. A selection of these values is included in Figure 4.3. As is in the case for organophosphines, the rate of reductive elimination is affected by alterations made to the structure of the NHC [36, 82, 86]. For example, the effect of variably substituted NHC ligands on the room temperature alkyl–alkyl Negishi reaction has been related indirectly to the reductive elimination step, where azolium salt/$Pd_2(dba)_3$ catalyst systems perform in the order of decreasing sterics, where IPr (**L10b**) ≈ SIPr (**L10e**) > SIPr-Et (**L10f**) > SIPr-Mes (**L10g**) > IEt (**L10c**) ≈ SIEt (**L10h**) > IMes (**L10d**) ≈ SIMes (**L10i**) [13].

The use of bulky, electron-rich organophosphines and NHCs as ligands in the Pd-catalyzed Suzuki–Miyaura reaction is reviewed in the following sections. Given intense interest in this field, only pertinent examples that have most significantly contributed to the advancement of the field will be reviewed, and these examples best portray the present state of the art.

## 4.2.2
### The Suzuki–Miyaura Cross-Coupling of Challenging Aryl Halides

#### 4.2.2.1 Overview of Challenges
Aryl bromides and iodides are the electrophiles used most routinely in the Suzuki–Miyaura reaction, despite the fact that aryl chlorides are considerably more commercially abundant and economical. In part, this constraint is the direct consequence of bond dissociation energies, such that C—Br and C—I bonds are reduced more readily relative to that of a C—Cl bond during oxidative addition to Pd(0) [24, 25, 30]. Di-, tri-, and tetra-*ortho*-substituted biaryls have also been elusive products as sterically hindered aryl halide and arylboronic acid precursors are prone to protodehalogenation and protodeboration, respectively. Functionalized heterocycles can be challenging substrates in cross-couplings given they are prone to protodeboration and can serve as catalyst poisons [87, 88]. With the above challenges in mind, most new catalyst systems are designed for and evaluated in the cross-coupling of one or more of these difficult substrate classes.

### 4.2.2.2 Organophosphine-Derived Catalysts

**4.2.2.2.1 Coupling of Carbocyclic Substrates** Seminal work by the groups of Fu [89] and Buchwald [90] revealed that P(t-Bu)₃ and DavePhos (**L9c**, Figure 4.2) were effective ligands for coupling aryl chlorides with arylboronic acids. With preliminary results from these experiments in hand, both groups refined the reaction conditions to be milder, hence more general to permit a broader range of functionalized substrates (Scheme 4.2). For example, a variety of "activated" (electron-deficient) aryl chlorides were coupled with arylboronic acids to provide functionalized products

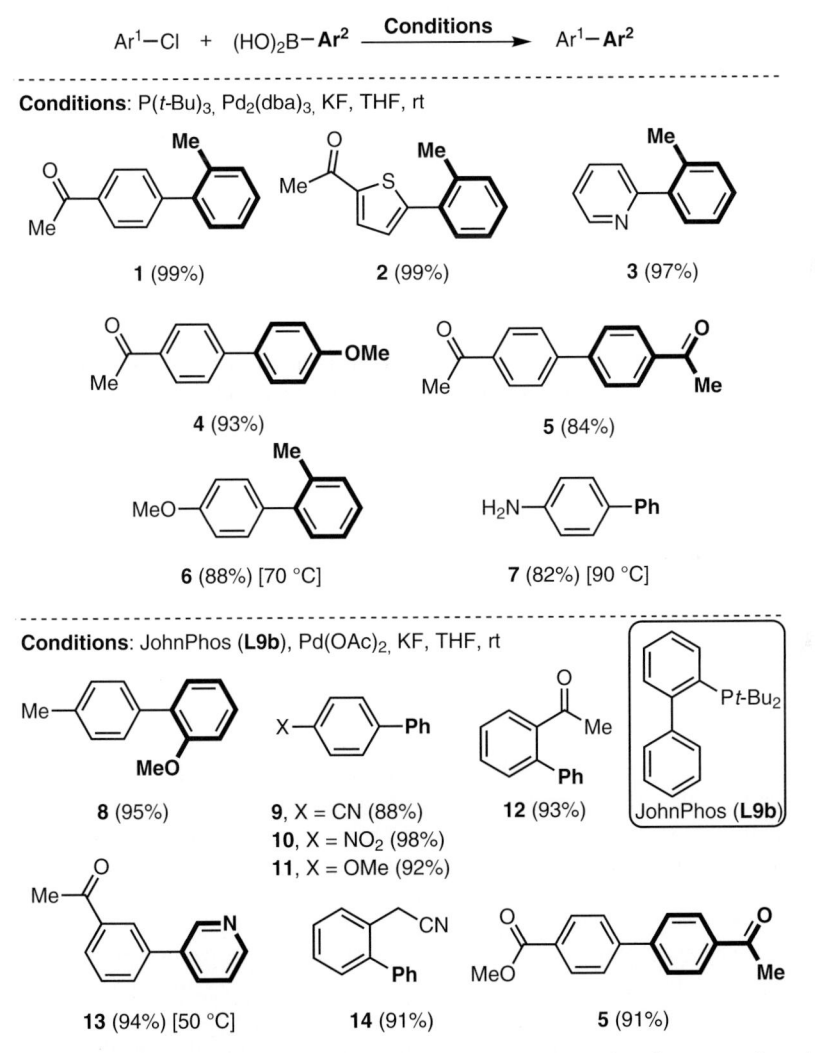

$$Ar^1-Cl + (HO)_2B-Ar^2 \xrightarrow{\text{Conditions}} Ar^1-Ar^2$$

**Conditions**: P(t-Bu)₃, Pd₂(dba)₃, KF, THF, rt

1 (99%)  2 (99%)  3 (97%)
4 (93%)  5 (84%)
6 (88%) [70 °C]  7 (82%) [90 °C]

**Conditions**: JohnPhos (**L9b**), Pd(OAc)₂, KF, THF, rt

8 (95%)
9, X = CN (88%)
10, X = NO₂ (98%)
11, X = OMe (92%)
12 (93%)
JohnPhos (**L9b**)
13 (94%) [50 °C]  14 (91%)  5 (91%)

**Scheme 4.2** The Suzuki–Miyaura cross-coupling of aryl chlorides with arylboronic acids in the presence of bulky, electron-rich organophosphines.

(1–5) using the P(t-Bu)$_3$/Pd$_2$(dba)$_3$ catalyst system at room temperature [91]. Electron-rich aryl chlorides were also coupled effectively (leading to **6** and **7**) by heating the reaction at or above 70 °C. Similarly, the JohnPhos (**L9b**)/Pd(OAc)$_2$ catalyst system was demonstrated to be general in substrate scope [42, 92], and a range of functionalized biaryls (**5** and **8–14**) were prepared in excellent yield from both electron-deficient and electron-rich aryl chlorides alike. Aryl bromides and iodides were also found to be suitable electrophiles under these very mild reaction conditions.

Subsequently, bulky trialkylphosphine and dialkylbiaryl phosphine ligands were evaluated in the Suzuki–Miyaura coupling of sterically hindered aryl substrates (Scheme 4.3). Using either P(t-Bu)$_3$ or PCy$_3$ in the presence of Pd$_2$(dba)$_3$, di- and tri-*ortho*-methylbiphenyls (**15–17**) were provided at *room temperature* from their corresponding *ortho*-substituted aryl bromides and arylboronic acids [91]; substituting aryl bromides with aryl chlorides required elevated temperatures to achieve comparable product yields. Tetra-*ortho*-substituted biaryls were not obtained under the specified reaction conditions. However, in the presence of **L8**/Pd$_2$(dba)$_3$, di-*ortho*-substituted aryl bromides and chlorides were effectively coupled with di-*ortho*-substituted arylboronic acids to give the corresponding tetra-*ortho*-substituted biaryl products (**18–21**) in impressive yields at 110 °C [93]. Stepwise optimization of the pendant groups on the dialkylbiphenyl phosphine ligand backbone eventually resulted in SPhos (**L9d**, Figure 4.2), which has since emerged as a fairly general ligand both for Pd-catalyzed Suzuki–Miyaura reactions [12, 32, 94] and for aryl aminations [34]. The application of SPhos in the Pd-catalyzed coupling of di-*ortho*-substituted aryl bromides, where the *ortho* substituents are methyl or *i*-propyl or a *t*-butyl/methyl combination, provided excellent yields of the corresponding tri- and tetra-*ortho*-substituted biaryls **22–25**. One *ortho*-*t*-butyl group is tolerated, while the presence of two *ortho*-*t*-butyl groups is not. Specifically, oxidative addition of 2,4,6-tri-*t*-butylbromobenzene to Pd(0) was found to proceed smoothly; however, the substantial sterics imparted by the two *ortho*-*t*-butyl groups discourages subsequent transmetalation with phenylboronic acid (Scheme 4.4) [94]. Instead, an alternative pathway takes over wherein the oxidative addition adduct undergoes cyclometalation to provide the corresponding palladacycle. Subsequent protonation of the palladacycle provides an alkyl palladium halide that is incapable of undergoing β-hydride elimination, and is sufficiently stable to transmetalate with arylboronic acids. This alternate catalytic cycle directs the formation of a new $C_{alkyl}$–$C_{aryl}$ bond upon reductive elimination to give cross-coupled products **26–28**. These examples are among the reported few alkyl–aryl Suzuki–Miyaura cross-couplings utilizing dialkylbiaryl phosphine ligands – a handful of examples of aryl halides being coupled with alkyl-9-BBN or methylboronic acid reagents have been reported [12].

#### 4.2.2.2.2 Coupling of Heterocyclic Substrates
The cross-couplings discussed in the previous section primarily involved carbocycles. However, cross-couplings involving functionalized heterocycles are paramount given their ubiquitous presence in natural products, pharmaceuticals, and agrochemicals [95]. Overcoming the aforementioned difficulties in coupling such substrates, the application of the PCy$_3$/Pd$_2$(dba)$_3$ catalyst system to the coupling of a variety of nitrogen-containing hetero-

Ar¹–X + (HO)₂B–Ar² →[Conditions] Ar¹–Ar²

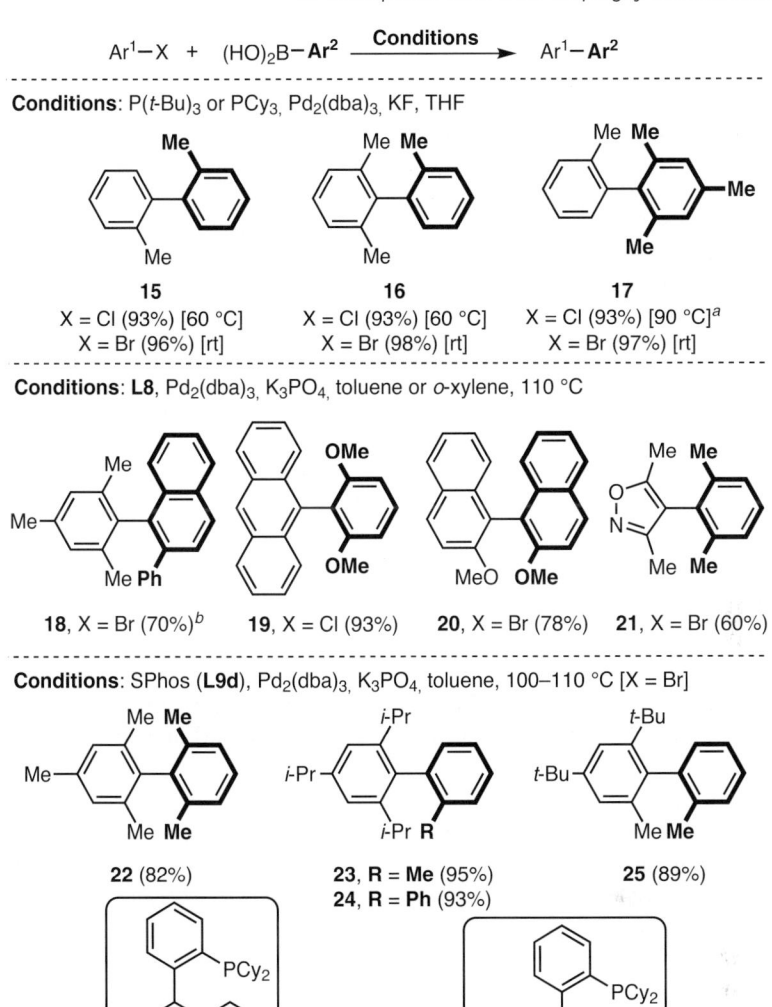

**Conditions**: P(t-Bu)₃ or PCy₃, Pd₂(dba)₃, KF, THF

**15**
X = Cl (93%) [60 °C]
X = Br (96%) [rt]

**16**
X = Cl (93%) [60 °C]
X = Br (98%) [rt]

**17**
X = Cl (93%) [90 °C][a]
X = Br (97%) [rt]

**Conditions**: L8, Pd₂(dba)₃, K₃PO₄, toluene or o-xylene, 110 °C

**18**, X = Br (70%)[b]
**19**, X = Cl (93%)
**20**, X = Br (78%)
**21**, X = Br (60%)

**Conditions**: SPhos (**L9d**), Pd₂(dba)₃, K₃PO₄, toluene, 100–110 °C [X = Br]

**22** (82%)
**23**, R = Me (95%)
**24**, R = Ph (93%)
**25** (89%)

L8

SPhos (**L9d**)

**Scheme 4.3** The Suzuki–Miyaura cross-coupling of sterically hindered aryl chlorides and bromides with arylboronic acids in the presence of bulky, electron-rich organophosphines.

[a]Reaction was completed using K₃PO₄ and toluene in place of KF and THF, respectively.
[b]Reaction was conducted in o-xylene at 120 °C in place of toluene at 110 °C.

cyclic chlorides and bromides with pyridine-based boronic acids, which tend to have a slow rate of transmetalation [96], provided excellent yields of the corresponding products (Scheme 4.5) [97]. Notably, unprotected aryl alcohols and amines (leading to **29** and **31**) were compatible substrates under these reaction conditions. The application of dialkylbiaryl phosphine ligands has provided a more general route

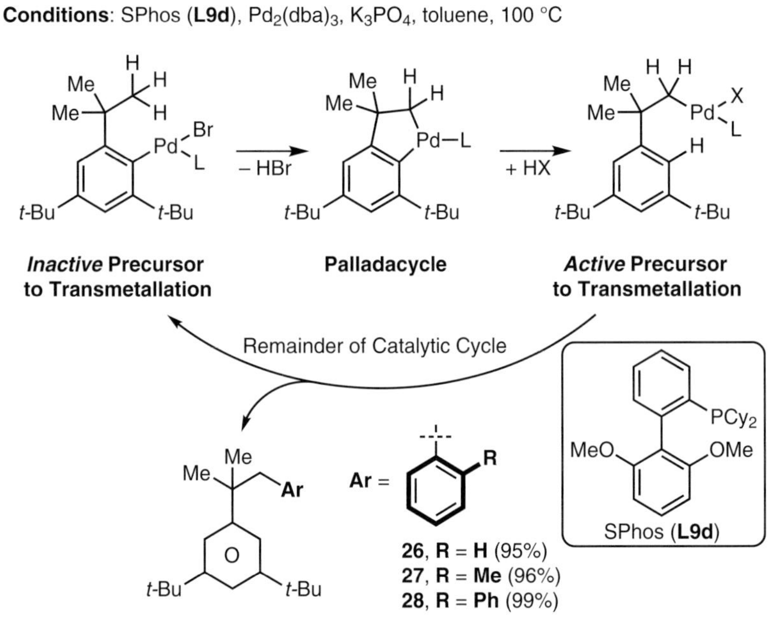

**Conditions**: SPhos (**L9d**), Pd$_2$(dba)$_3$, K$_3$PO$_4$, toluene, 100 °C

**Inactive Precursor to Transmetallation** → (– HBr) **Palladacycle** → (+ HX) **Active Precursor to Transmetallation**

Remainder of Catalytic Cycle

SPhos (**L9d**)

**26**, R = H (95%)
**27**, R = Me (96%)
**28**, R = Ph (99%)

**Scheme 4.4** This alternate catalytic cycle is proposed to account for the formation of **26–28** when coupling aryl halides that possess *t*-butyl substituents at both *ortho* positions.

to the preparation of heterobiaryls, as illustrated by the variety of functionalized products in Scheme 4.5 [96]. SPhos (**L9d**, Figure 4.2) was found to be an effective ligand for coupling unprotected chloroaminopyridines and -pyrimidines with arylboronic acids (leading to **35–37**) [98]. Pyrrole and indole boronic acids are scarcely reported in the Suzuki–Miyaura reaction for a variety of reasons [96]; however, SPhos (**L9d**)/Pd(OAc)$_2$ was found to be a fairly general catalyst system for the coupling of these transmetalating species with heterocyclic chlorides providing access to functionalized heterocycles **38–40**. The coupling of deactivated aryl and heteroaryl chlorides were sluggish when using SPhos as the ancillary ligand, whereas the sterically more hindered XPhos (**L9e**, Figure 4.2) ligand enabled the coupling of such electrophiles (leading to **41–43**) [96]. The authors attribute the pronounced catalyst activity to the increased sterics of XPhos (**L9e**), which raises the relative concentration of the monoligated oxidative addition intermediate (e.g., [(XPhos)Pd(Ar)Cl]).

**4.2.2.2.3 Suzuki–Miyaura Reactions in Water** Replacing organic solvent for water has clear implications in the presence of mounting environmental concerns. Treatment of SPhos (**L9d**) or XPhos (**L9e**) with H$_2$SO$_4$ in CH$_2$Cl$_2$ provided their monosulfonated sodium salt derivatives (**L9f** and **L9g**, respectively, Figure 4.2) [99] that are as active as the parent ligands. A collection of hydrophilic functionalized heterocycles (**44–52**) were prepared from their heteroaryl chloride and boronic acid precursors using SPhos(SO$_3$Na) (**L9f**) as the spectator ligand in water (Scheme 4.6).

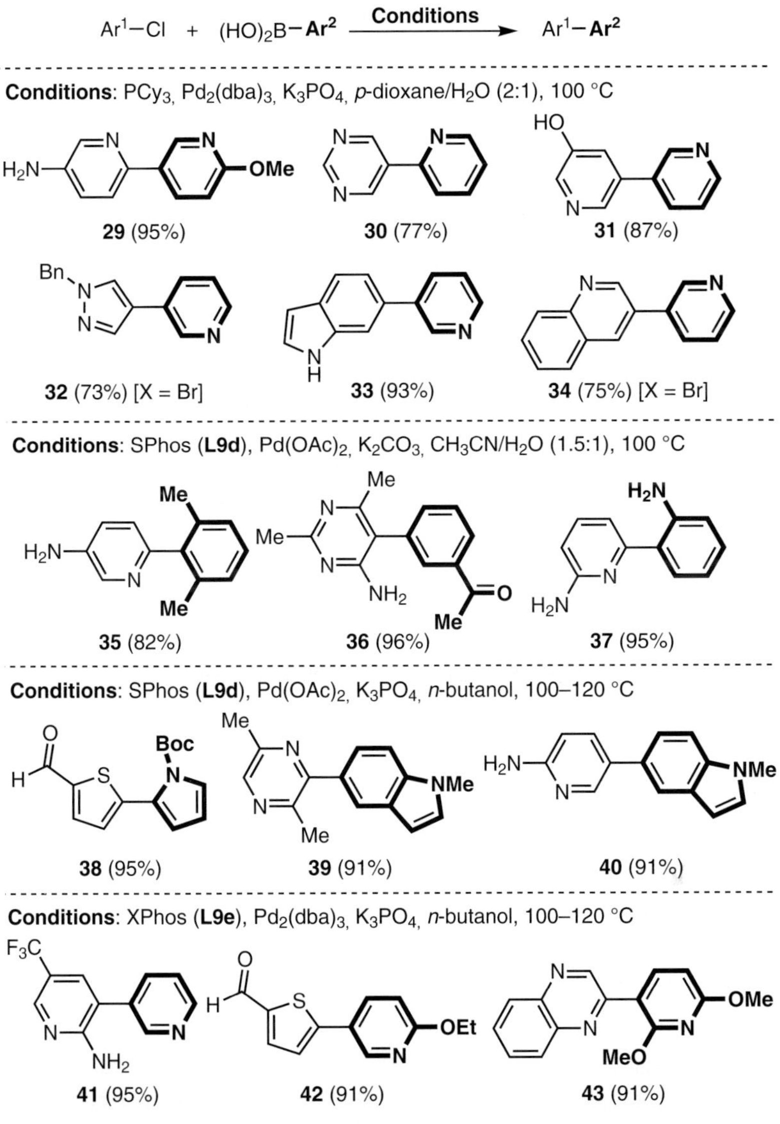

## 4.2 Developments Made in the Coupling of Nontrivial Substrates | 225

Ar¹—Cl + (HO)₂B—Ar² —Conditions→ Ar¹—Ar²

---

**Conditions**: PCy₃, Pd₂(dba)₃, K₃PO₄, p-dioxane/H₂O (2:1), 100 °C

**29** (95%)   **30** (77%)   **31** (87%)

**32** (73%) [X = Br]   **33** (93%)   **34** (75%) [X = Br]

---

**Conditions**: SPhos (**L9d**), Pd(OAc)₂, K₂CO₃, CH₃CN/H₂O (1.5:1), 100 °C

**35** (82%)   **36** (96%)   **37** (95%)

---

**Conditions**: SPhos (**L9d**), Pd(OAc)₂, K₃PO₄, n-butanol, 100–120 °C

**38** (95%)   **39** (91%)   **40** (91%)

---

**Conditions**: XPhos (**L9e**), Pd₂(dba)₃, K₃PO₄, n-butanol, 100–120 °C

**41** (95%)   **42** (91%)   **43** (91%)

**Scheme 4.5** The Suzuki–Miyaura cross-coupling of functionalized heterocyclic aryl chlorides and bromides and aryl- and heteroarylboronic acids in the presence of bulky, electron-rich organophosphines.

There is a clear advantage over similar couplings carried out in anhydrous organic solvents where solubility is a concern. Selected couplings were achievable at room temperature, but for the most part these cross-couplings were carried out at 100 °C. Heating to 150 °C in a microwave reactor accelerated the reaction to provide the cross-coupled products in 10 min.

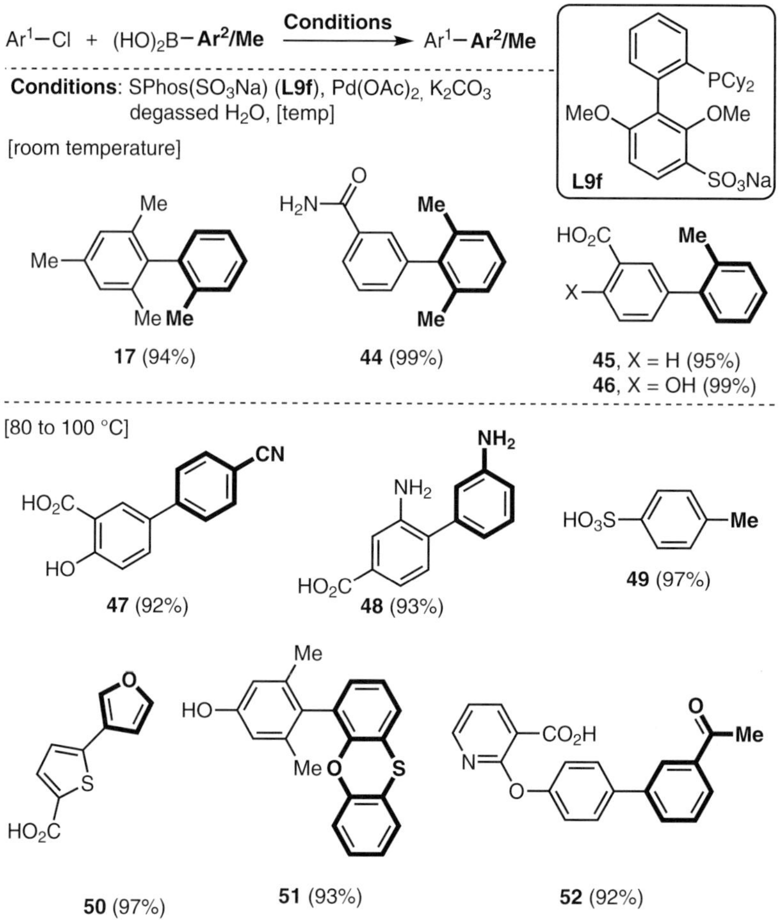

Ar¹—Cl + (HO)₂B—**Ar²/Me** $\xrightarrow{\text{Conditions}}$ Ar¹—**Ar²/Me**

**Conditions**: SPhos(SO₃Na) (**L9f**), Pd(OAc)₂, K₂CO₃
degassed H₂O, [temp]

[room temperature]

**17** (94%)   **44** (99%)   **45**, X = H (95%)
                              **46**, X = OH (99%)

[80 to 100 °C]

**47** (92%)   **48** (93%)   **49** (97%)

**50** (97%)   **51** (93%)   **52** (92%)

**Scheme 4.6** The Suzuki–Miyaura cross-coupling in water of functionalized heterocyclic aryl chlorides and arylboronic in the presence of **L9f** (Figure 4.2), a water-soluble monosulfonated derivative of SPhos (**L9d**).

**4.2.2.2.4 Low Catalyst Loadings** The majority of Suzuki–Miyaura cross-couplings require catalyst loadings in the range of 0.5–5 mol% Pd. However, there are examples with catalyst loadings at or below 0.05 mol% (Table 4.1). Low catalyst loadings have obvious implications when considering large-scale transformations. The catalyst turnover number (TON) is a quantitative measure for evaluating a catalyst's activity. Using bulky trialkylphosphines, in particular P($t$-Bu)₃, as the supporting ligand, catalyst TONs on the order of approximately 10 000 have been achieved for electron-deficient aryl chlorides (entry 1); catalyst TONs for electron-neutral aryl chlorides fall on average an order of magnitude below that (entry 2) [91]. Relatively low catalyst loadings were also possible when using the $N$-aryl-2-(dialkylphosphino)pyrrole **L7** (Figure 4.2) [100] in the coupling of a variety of aryl chlorides with phenylboronic acid (entries 3–5 and 7). In particular, catalyst TONs approached 10 000 with **L7** for the

**Table 4.1** The Suzuki–Miyaura cross-coupling at low loadings of Pd in the presence of bulky, electron-rich organophosphines.

$$Ar^1-X \ + \ (HO)_2B-Ar^2 \ \xrightarrow[\text{Base}]{\text{Conditions}} \ Ar^1-Ar^2$$

| Entry | Product | X | Pd (mol %) | Ligand | Temp. (°C) (Time, h) | Yield (%) | TON |
|---|---|---|---|---|---|---|---|
| 1 | 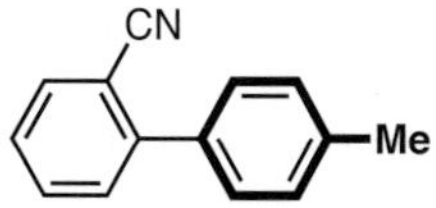 2-CN-C₆H₄–C₆H₄–4-Me | Cl | 0.01 | P(t-Bu)₃ | 90 (25) | 97 | 9700 |
| 2 |  Ph–C₆H₄–4-Me | Cl | 0.1 | P(t-Bu)₃ | 100 (43) | 92 | 920 |
| 3 | R–C₆H₄–Ph | R = Me, Cl | 0.01 | L7 | 60 (24) | >99 | 9900 |
|   |   | Me Cl | 0.005 | L7 | 100 (24) | 98 | 19 600 |
|   |   | Me Cl | 0.05 | JohnPhos | 100 (25) | 93 | 1860 |
|   |   | n-Bu Cl | 0.003 | SPhos | 100 (24) | 93 | 31 000 |
| 4 | 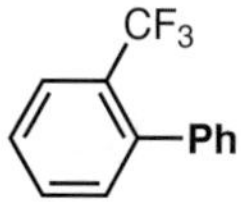 2-CF₃-C₆H₄–Ph | Cl | 0.05 | L7 | 60 (24) | 90 | 1800 |
| 5 | 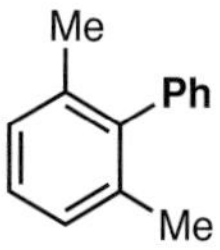 2,6-Me₂-C₆H₃–Ph | Cl | 0.01 | L7 | 60 (24) | 91 | 9100 |
|   |   | Cl | 0.005 | L7 | 60 (24) | 16 | 320 |
| 6 | 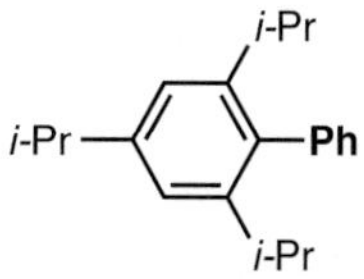 2,4,6-(i-Pr)₃-C₆H₂–Ph | Br | 0.01 | SPhos | 100 (16) | 97 | 9700 |
| 7 | 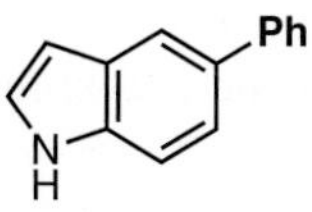 5-Ph-indole | Cl | 0.05 | L7 | 100 (24) | 97 | 1940 |
| 8 | 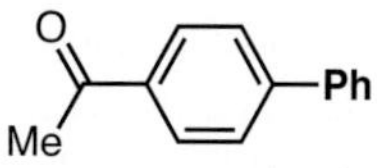 4-MeC(O)-C₆H₄–Ph | Br | 0.000001 | DavePhos | 100 (24) | 91 | 91 × 10⁶ |
|   |   | Br | 0.001 | — | 100 (19) | >99 | 99 000 |
|   |   | Cl | 0.02 | DavePhos | 100 (23) | 92 | 4600 |

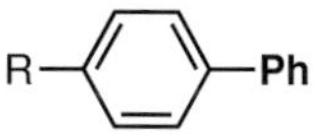

coupling of 4-chlorotoluene at 60 °C; increasing the temperature to 100 °C allowed for lower catalyst loadings (TON ~20 000, entry 3). While JohnPhos (**L9b**) was less effective for the same coupling [42, 92], SPhos (**L9d**) yielded catalyst TONs above 30 000 (entry 3) [12, 94]. In the case of **L7**, the presence of *ortho* substituents on the electrophile lowered catalyst activity (entry 3 versus 5). Although the same is true for SPhos (**L9d**), relatively high TONs are still achieved for the very sterically hindered 2,4,6-tri-*i*-propylbromobenzene (entry 6). Catalyst TONs are comparatively low in the coupling of heterocyclic aryl chlorides (entry 7). The use of DavePhos (**L9c**) as the supporting ligand for the coupling of 4-acetylbromobenzene improved the catalyst TON by three orders of magnitude compared to that where no ligand is present (entry 8) – as this was an "activated" aryl bromide, both catalysts were able to achieve quantitative conversions. As expected, the derivative 4-acetylchlorobenzene was more sluggish, leading to catalyst TONs approximately four orders of magnitude lower than the coupling of the corresponding aryl bromide in the presence of DavePhos (**L9c**). Aside from the ligand/Pd combinations in Table 4.1, the use of $PAd_2(n\text{-}Bu)/Pd(OAc)_2$ at 100 °C has also proven effective for the coupling of a variety of aryl chlorides with arylboronic acids at 0.001–0.005 mol% Pd loading, achieving catalyst TONs ranging from 11 600 to 69 000 [101].

#### 4.2.2.3 NHC-Derived Catalysts

##### 4.2.2.3.1 *In-Situ*-Generated Catalysts from Imidazolium Salt Precursors
Imidazolium salts are shelf-stable, free-flowing crystalline materials that can be prepared on a large-scale, stored for long periods of time, and conveniently weighed out in air as needed. Their conversion to the free carbene is mediated by deprotonation of the imidazolium salt. This process typically occurs under the basic conditions of the Suzuki–Miyaura reaction, where capture by Pd generates the active catalyst *in situ*. Alternatively, the free carbene may be generated as a solution that is then added to a Pd source to generate the catalytically active NHC(Pd) species. An early report by Nolan and coworkers demonstrated the use of IMes•HCl (**L10d**, Figure 4.3) in the presence of $Pd_2(dba)_3$ and $Cs_2CO_3$ to generate *in situ* the (IMes)Pd catalyst that was able to couple relatively simple electron-rich aryl chlorides with arylboronic acids in *p*-dioxane at 80 °C in near quantitative yields [102]. A follow-up study showed that the success of the cross-coupling highly depended on the sterics of the NHC used; among a variety of *unsaturated* NHC ligands evaluated, IMes (**L10d**) and IPr (**L10b**) were found to be optimal and were equally effective [103].

Various *saturated* imidazolium salts have also been evaluated as ligands in the coupling of both electron-rich and electron-deficient aryl chlorides and arylboronic acids at room temperature [104]. Under the conditions specified in Scheme 4.7, $H_2$ICP•HCl (**L13**, Figure 4.3) was superior to both SIMes•HCl (**L10i**) and SIPr•HCl (**L10e**) in the preparation of biaryls **15** and **53–66**. Removal of two or all of the pendant cyclohexyl groups from $H_2$ICP•HCl (**L13**) significantly attenuated the activity of the *in situ* generated (NHC)Pd complex, again demonstrating the strict dependence of cross-couplings on the precise sterics of the ancillary ligand. Di-*ortho*-substituted

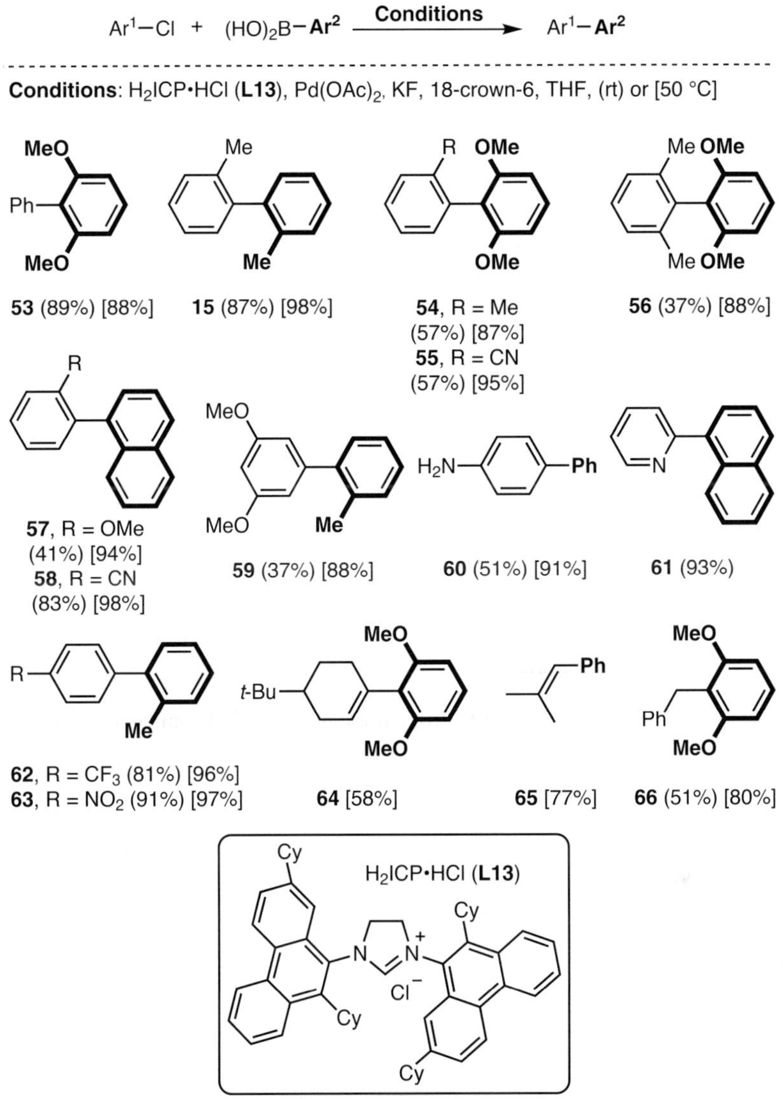

## 4.2 Developments Made in the Coupling of Nontrivial Substrates

$Ar^1-Cl + (HO)_2B-Ar^2 \xrightarrow{\text{Conditions}} Ar^1-Ar^2$

**Conditions**: H₂ICP·HCl (**L13**), Pd(OAc)₂, KF, 18-crown-6, THF, (rt) or [50 °C]

**53** (89%) [88%]

**15** (87%) [98%]

**54**, R = Me (57%) [87%]
**55**, R = CN (57%) [95%]

**56** (37%) [88%]

**57**, R = OMe (41%) [94%]
**58**, R = CN (83%) [98%]

**59** (37%) [88%]

**60** (51%) [91%]

**61** (93%)

**62**, R = CF₃ (81%) [96%]
**63**, R = NO₂ (91%) [97%]

**64** [58%]

**65** [77%]

**66** (51%) [80%]

H₂ICP·HCl (**L13**)

**Scheme 4.7** The Suzuki–Miyaura cross-coupling of aryl and alkenyl chlorides with arylboronic acids using the NHC precursor H₂ICP•HCl (**L13**, Figure 4.3).

products (**15**, **53**, **57**, and **58**) were obtained in mostly excellent yields at room temperature. Tri- and tetra-*ortho*-substituted products (**54–56**) were also coupled in moderate yields; however, warming the reaction to 50 °C improved the yields in shorter reaction times. Aryl chlorides functionalized with trifluoromethyl, nitro, and nitrile groups as well as primary amines were well tolerated, as was 2-chloropyridine (leading to **60–63**). Alkenyl and benzylic chlorides (leading to **64–66**) were also suitable substrates. In addition, boronic acid derivatives phenyl pinacolatoborane and trimethylboroxine were evaluated and found to be suitable transmetalating agents [104].

NHCs possessing "flexible" steric bulk give rise to highly active catalysts that are able to adapt their conformation to best suit various stages of the catalytic cycle [86]. Ir-complexes of bioxazoline (IBiox) (**L12**, Figure 4.3) and cyclic (alkyl)(amino) carbenes (CAAC) [105] are more electron-rich than their IPr and IMes derivatives making these NHCs among the best-known σ-donor ligands for transition metals. This enhanced σ-donor ability of these ligands is an artifact of the alkylated quaternary carbons neighboring the carbene carbon.

The IBiox6 (**L12b**, Figure 4.3)/Pd(OAc)$_2$ catalyst system is effective for coupling aryl chlorides at room temperature (Scheme 4.8) [31]. Sterically hindered arylboronic acids leading to di- and tri-*ortho*-substituted biphenyls (**16, 67,** and **68**) were coupled with aryl chlorides in good yield. Following these initial results, a series of IBiox•HOTf imidazolium salts were prepared with five (**L12a**), seven (**L12c**), eight (**L12d**), and twelve-membered (**L12e**) aliphatic rings branching off from the quaternary carbon. They were each evaluated along with IBiox6 (**L12b**) in the cross-coupling of 1-chloro-2,6-dimethylbenzene with mesitylboronic acid to give tetra-*ortho*-substituted biphenyl **22** [106]. IBiox7 (**L12c**), IBiox8 (**L12d**), and IBiox12 (**L12e**) ligands outperformed IBiox5 (**L12a**) and IBiox6 (**L12b**), with IBiox12 (**L12e**) being optimal. For comparison, the use of either IMes (**L10d**) or IAd (**L10a**) as spectator ligands

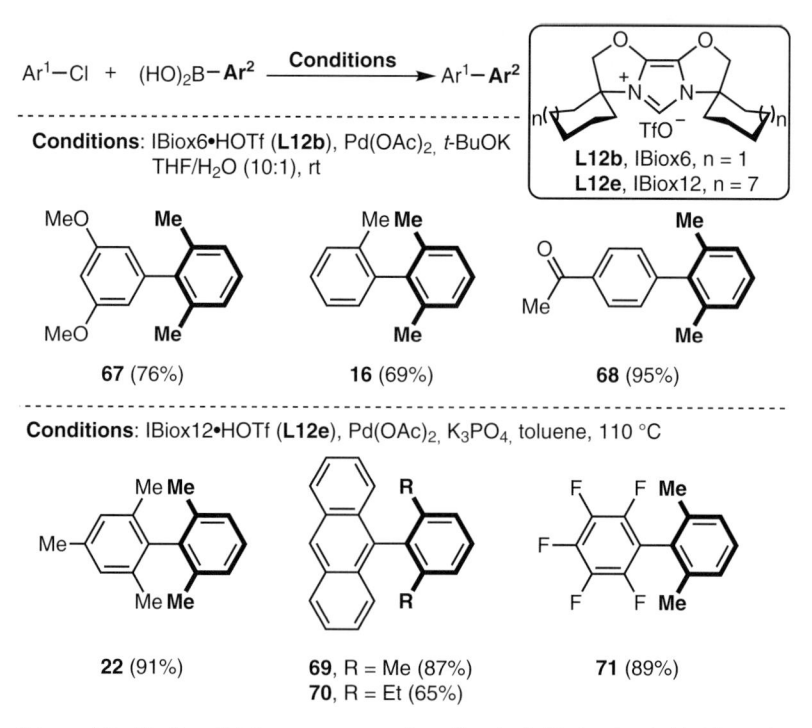

**Scheme 4.8** The Suzuki–Miyaura cross-coupling of sterically hindered aryl chlorides with arylboronic acids using NHC ligands possessing "flexible" steric bulk.

provided none of the desired products. The IBiox12 (**L12e**) ligand was found to be effective for a variety of challenging cross-couplings of sterically hindered substrates leading to tetra-*ortho*-substituted biaryls **22** and **69–71** (Scheme 4.8).

**4.2.2.3.2 Preformed NHC–Pd Complexes** The *in situ* generation of free carbenes from their imidazolium salt precursors is highly sensitive to moisture, reaction conditions, and the technical skill of the practitioner. These inherent factors lead to poor reproducibility from one experiment to the next when using *in-situ*-derived carbene-based catalysts. Although the active catalytic species is believed to be monoligated (NHC)Pd, twofold excess to Pd of the NHC or imidazolium salt is often employed in order to attain optimal results. It has been demonstrated that only a fraction of the active (NHC)Pd(0) catalyst is actually formed *in situ* from a 2:1 mixture of IPr•HCl (**L10b**) and a Pd(0) source [107]. Thus, there exists an uncertainty surrounding the stoichiometry and composition of the active catalytic species for *in situ* prepared NHC catalysts [108]. To bypass these drawbacks and concerns, discrete (NHC)Pd(0) and (NHC)Pd(II) complexes (Figure 4.3, **C1–C7**) have been prepared [109]. (NHC)Pd complexes are typically stabilized with noncarbene-based coligands; however, examples of stable (NHC)$_2$Pd complexes are known [108, 110].

Bischelated (NHC)Pd(II) complex **C1** was first reported for use in the Heck and Suzuki–Miyaura reactions [43]. For the latter, aryl bromides and a single aryl chloride were coupled with phenylboronic acid at 120 °C. (IAd)$_2$Pd(0) (**C3**) has also been prepared and was found to be effective for the coupling of aryl chlorides with phenyl and *p*-anisylboronic acids at room temperature [110]. In this system, one IAd ligand is shed *in situ* to generate the catalytically active monoligated species.

A variety of (NHC)Pd complexes that utilize noncarbene-based coligands have been prepared and evaluated. The (IMes)Pd(dvds) complex (**C2**, Figure 4.3) was found to be capable in the coupling of simple aryl chlorides with phenylboronic acid in moderate-to-good yields; however, heating to 100 °C was required [111]. The diminished catalytic activity stems from the strong binding of the olefins in dvds to Pd, effectively poisoning the catalyst [112]. Naphthoquinone (NQ) coordinates Pd via both the carbonyl oxygens and the α,β-unsaturated ketone olefins to give the bridged [(IMes)Pd(NQ)]$_2$ dimer (**C7**). This catalyst was evaluated in the coupling of aryldiazonium salts with arylboronic acids in MeOH at 50 °C with good results [113].

The most successful and generally applicable NHC-based Pd catalysts that have been developed are IPr-palladacycle **C4** [114], (IPr)Pd(allyl)Cl (**C5b**) [115], and Pd-PEPPSI-IPr (**C6a**, structures in Figure 4.3) [38]. All these (NHC)Pd(II) precatalysts are reduced *in situ* to presumably generate the same catalytically active species, namely, (IPr)Pd(0). However, IPr–palladacycle **C4** was found to be superior to (IPr)Pd(allyl)Cl (**C5b**) in room-temperature Suzuki–Miyaura cross-couplings [116], indicating that either (1) there are substantial differences in the rate of *in situ* reduction of these two precatalysts or (2) there is a coligand "memory effect," which includes the possibility of catalyst poisoning by the coligand under the reaction conditions. Both **C4** and Pd-PEPPSI-IPr (**C6a**) were comparably effective precatalysts for the room temperature Suzuki–Miyaura reaction in technical grade *i*-propanol (Scheme 4.9), providing access to biaryls **3**, **11**, **16**, and **72–78** [107, 114]. Sterically hindered aryl

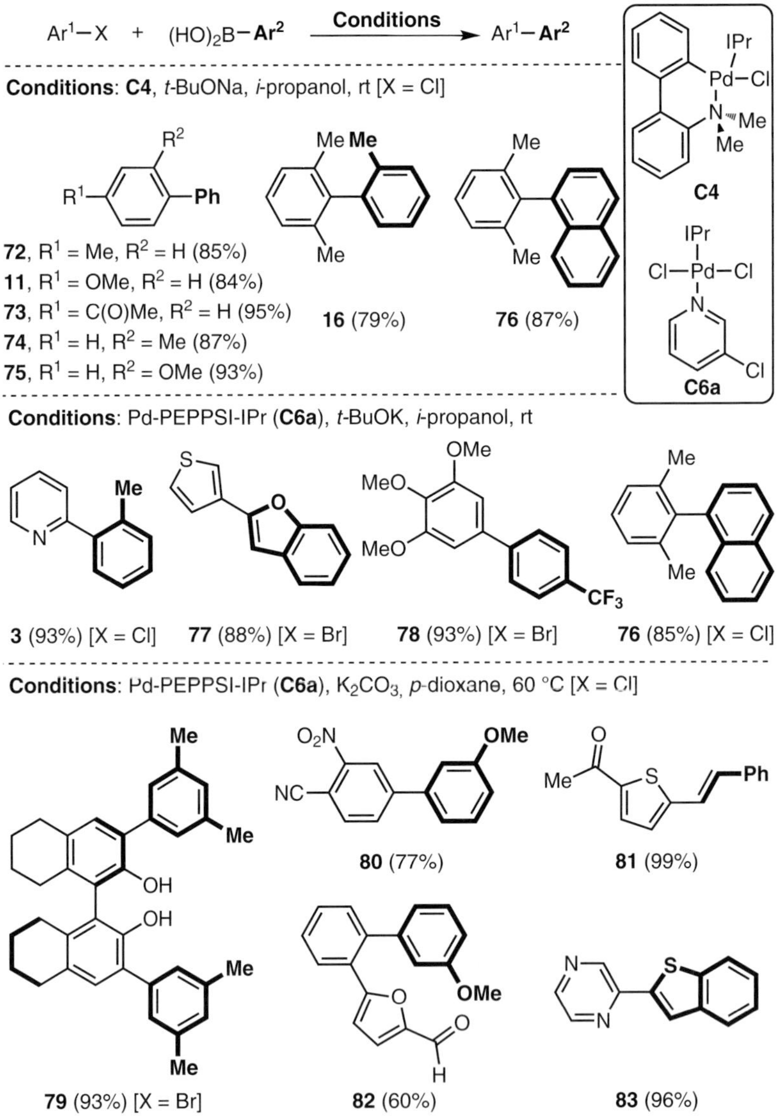

Ar¹–X + (HO)₂B–Ar² $\xrightarrow{\text{Conditions}}$ Ar¹–Ar²

**Conditions: C4**, *t*-BuONa, *i*-propanol, rt [X = Cl]

72, $R^1$ = Me, $R^2$ = H (85%)
11, $R^1$ = OMe, $R^2$ = H (84%)
73, $R^1$ = C(O)Me, $R^2$ = H (95%)
74, $R^1$ = H, $R^2$ = Me (87%)
75, $R^1$ = H, $R^2$ = OMe (93%)

16 (79%)   76 (87%)

C4

C6a

**Conditions:** Pd-PEPPSI-IPr (**C6a**), *t*-BuOK, *i*-propanol, rt

3 (93%) [X = Cl]   77 (88%) [X = Br]   78 (93%) [X = Br]   76 (85%) [X = Cl]

**Conditions:** Pd-PEPPSI-IPr (**C6a**), $K_2CO_3$, *p*-dioxane, 60 °C [X = Cl]

79 (93%) [X = Br]   80 (77%)   81 (99%)   82 (60%)   83 (96%)

**Scheme 4.9** The Suzuki–Miyaura cross-coupling of aryl halides and arylboronic acids using NHC–Pd precatalysts at room temperature. Functionalized and heteroaryl substrates were coupled in the presence of a less nucleophilic base at 60 °C.

chlorides leading to **16** and **76** were well tolerated. Select-few heteroaryl substrates (leading to **3** and **77**) were shown to be compatible using the precatalyst Pd-PEPPSI-IPr (**C6a**) under these reaction conditions. For the most part, *t*-BuO⁻ is too harsh of a base rendering it only moderately functional group tolerant. As such, milder reaction conditions consisting of $K_2CO_3$ base in *p*-dioxane were developed that permitted the cross-coupling of a variety of functionalized and heterocyclic substrates (leading to

## 4.2 Developments Made in the Coupling of Nontrivial Substrates

79–83) with Pd-PEPPSI-IPr (**C6a**), albeit at a slightly elevated temperature. Nitrile, nitro, and unprotected hydroxyl, ketone, and aldehyde functionality on heteroaryl frameworks was well tolerated. Notably, each of the functionalized and heteroaryl precursors leading to these products were not tolerated under the $i$-propanol/$t$-BuO$^-$ conditions and led to decomposition products.

Generally speaking, IPr-based catalysts are considerably more active than their IMes counterparts. This "trend" prompted the question as to whether further increasing the sterics of the substituents on the $N$-aryl moiety of the NHC would prove beneficial. As such, IBu, I$c$-Pent, and IPent-based Pd-PEPPSI complexes (**C6b, 6c,** and **6d**, respectively, Figure 4.3) were prepared and evaluated by Organ and coworkers [117]. While IBu and I$c$-Pent were ineffective ancillary ligands for cross-couplings leading to tetra-*ortho*-substituted biaryls, IPent proved to be very effective (Scheme 4.10). A variety of tetra-*ortho*-substituted biaryls (**22, 84,** and **86–88**) were produced under one of the lowest reaction temperatures reported to date for this transformation. In addition to aryl chlorides, aryl bromides were equally effective electrophiles. Functional group tolerance of this catalyst is demonstrated through cross-coupling leading to **85**. In all cases, Pd-PEPPSI-IPent (**C6d**) greatly outperformed Pd-PEPPSI-IPr (**C6a**). The major side reaction when using Pd-PEPPSI-IPr (**C6a**) is hydroxydeborylation, which was found to consume the balance of the boronic acid and thus stall the cross-coupling reaction. The increased sterics around the metal

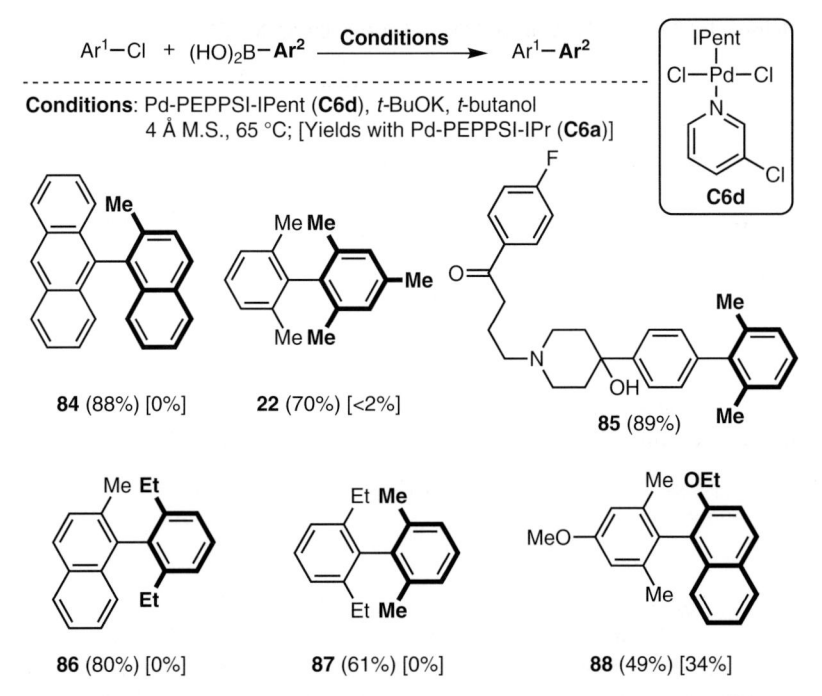

**Scheme 4.10** The Suzuki–Miyaura cross-coupling of sterically hindered aryl chlorides and arylboronic acids using Pd-PEPPSI-IPent (**C6d**, Figure 4.3). Yields in square brackets are for cross-couplings carried out using Pd-PEPPSI-IPr (**C6a**) in place of Pd-PEPPSI-IPent (**C6d**).

center is thought to have little effect on the rate of oxidative addition but instead significantly influences transmetalation and reductive elimination.

A variety of (NHC)Pd catalysts have been developed that are effective at low-catalyst loadings. In general, (NHC)Pd complexes are not effective on their own, and an auxiliary ligand is usually required, presumably to prolong active catalyst lifetime. Most commonly, organophosphines are added for this purpose [118–120]. Among the most active and general catalysts is (IPr)Pd(cinnamyl)Cl (**C5c**), which is effective at Pd loadings in the range of 50 ppm [121, 122].

All told, NHC-based catalysts provide conversions on par with electron-rich organophosphine ligated species. NHCs have an advantage over organophosphines in that they form air- and moisture-stable Pd adducts, and so are more attractive from a user standpoint in terms of ease of use, safety, and practicality. Furthermore, the large number of (NHC)Pd precatylst complexes that are commercially available make for their convenient use.

### 4.2.3
### The Suzuki–Miyaura Reaction Involving Unactivated Alkyl Halides

#### 4.2.3.1 Associated Difficulties

The use of unactivated alkyl halides (i.e., nonbenzylic, allylic, or α-carbonyl halides) as the electrophilic cross-coupling partner has only recently been realized [26, 28, 123]. For the most part, these substrates are poor candidates for Pd and Ni-catalyzed cross-couplings due to the comparatively low propensity of the unactivated $C_{sp3}$–X bond to undergo oxidative addition and the tendency for the resulting adduct to encounter *intra*molecular β-hydride elimination or hydrodehalogenation side reactions in preference to *inter*molecular transmetalation (Figure 4.6) [26, 28]. These obstacles have been significantly marginalized through the use of bulky, electron-rich ligands

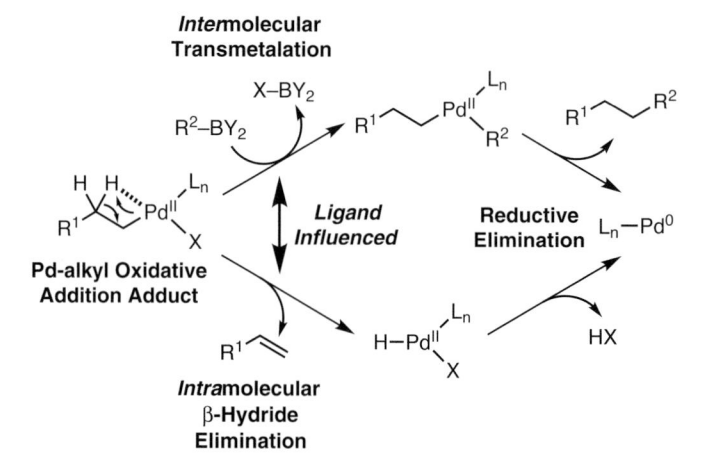

**Figure 4.6** β-Hydride elimination is a facile intramolecular process in cross-couplings. Rational ligand design has provided ligands that disfavor β-hydride elimination so that intermolecular transmetalation can proceed.

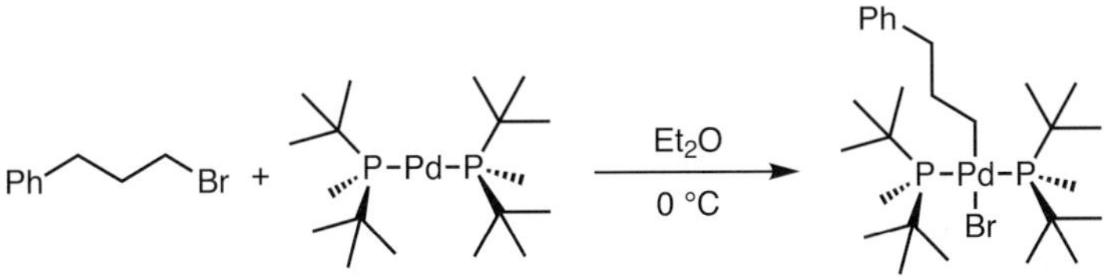

**Scheme 4.11** Fu and coworkers characterized using X-ray crystallography the alkyl–Pd complex [89•C8] that possesses β-hydrogens and that is stabilized by bulky, electron-rich phosphine ligands.

that prevent Pd-alkyl species from reaching geometries required for β-hydride elimination. In addition, the imparted electron density on the metal center renders it more nucleophilic and thus the energetic barrier to oxidative addition is attenuated. This increased electron density on the metal, in conjunction with the imposed steric topography from the ligand, lessens the agostic and anagostic [124] interactions of the metals d-orbitals with β-hydrogen(s) in the alkyl moiety. The ligand's sterics also favor reductive elimination of these less bulky alkyl fragments (relative to aryl groups), thereby shortening the lifetime of the transient $R^1R^2PdL_n$ species (refer to Figure 4.1) that provides less opportunity for deleterious side reactions to occur. The enhanced reactivity of Pd(0) using bulky, electron-rich ligands and the stability of the alkyl-Pd intermediates is evident in the elegant near-quantitative isolation of the oxidative addition adduct [89•C8] that was sufficiently stable to characterize by X-ray crystallography (Scheme 4.11) [125].

### 4.2.3.2 Cross-Couplings Promoted by Phosphines and Amine-Based Ligands

#### 4.2.3.2.1 Cross-Couplings of Primary Alkyl Halides

The early part of 2000 witnessed the first reports of the $C_{sp3}$–$C_{sp3}$ Suzuki–Miyaura cross-coupling of unactivated primary alkyl bromides [126], chlorides [127], and tosylates [128] with alkyl-9-BBN reagents (Scheme 4.12). The sterically demanding organophosphine PCy$_3$ was optimal for the coupling of primary alkyl bromides (leading to **90–94** at room temperature) and chlorides (leading to **95–98** at 90 °C) in the presence of Pd(OAc)$_2$ or Pd$_2$(dba)$_3$, respectively. While organophosphines P(t-Bu)$_2$Et and P(t-Bu)$_2$i-Pr provided only a trace amount of product, P(t-Bu)$_2$Me was optimal for coupling primary alkyl tosylates (leading to **99–102**). This result suggests that there exists a strict dependence on the precise sterics of the ligand for this cross-coupling. It is not immediately obvious what stage(s) of the catalytic cycle is(are) so highly dependent on such subtle changes to ligand structure or more specifically how the halide/pseudohalide influences this fine balance. Computational analysis has indicated that the transmetalation step is greatly affected by ligand sterics in the alkyl–alkyl Negishi cross-coupling [38]. Of course, the electronic contribution from the organophosphine cannot be ruled out as minor structural changes do attenuate their basicity. Regardless of the mechanistic underpinnings, a variety of functionalized products were prepared, with tertiary amines (**90** and **97**), esters (**92**, **94**, and **98**), nitriles (**94**), silyl

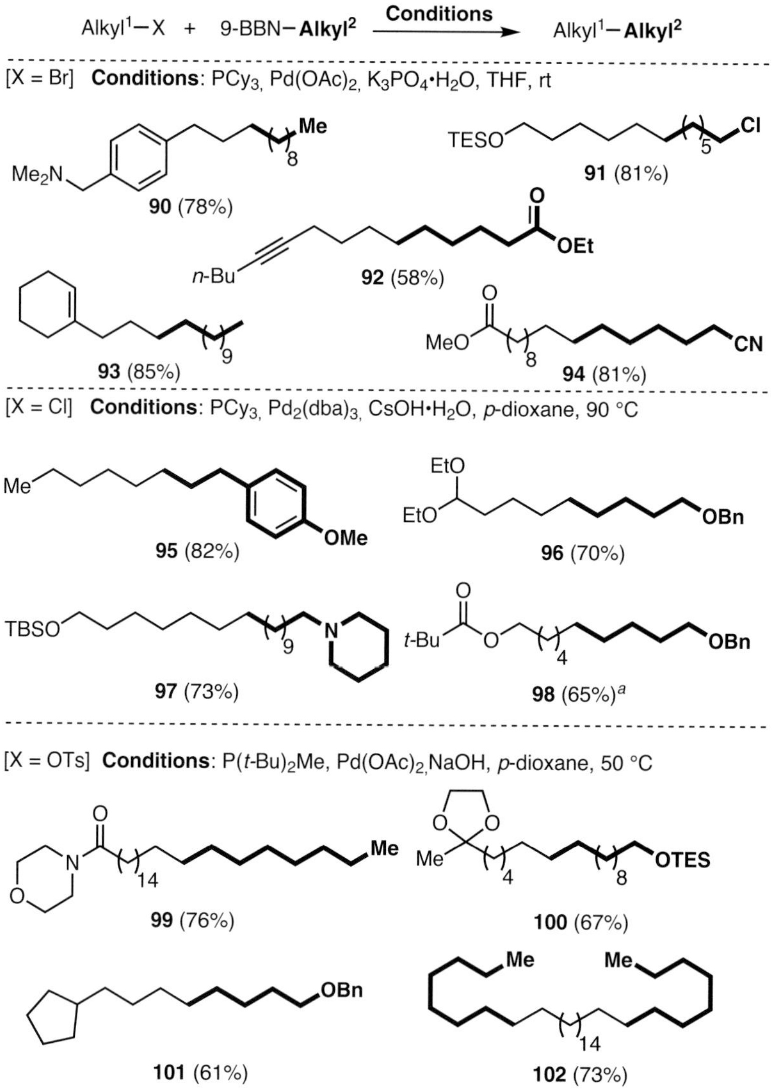

# 4 The Contemporary Suzuki–Miyaura Reaction

Alkyl$^1$—X + 9-BBN—Alkyl$^2$ →(Conditions) Alkyl$^1$—Alkyl$^2$

---

**[X = Br] Conditions**: PCy$_3$, Pd(OAc)$_2$, K$_3$PO$_4$·H$_2$O, THF, rt

- **90** (78%)
- **91** (81%)
- **92** (58%)
- **93** (85%)
- **94** (81%)

---

**[X = Cl] Conditions**: PCy$_3$, Pd$_2$(dba)$_3$, CsOH·H$_2$O, p-dioxane, 90 °C

- **95** (82%)
- **96** (70%)
- **97** (73%)
- **98** (65%)$^a$

---

**[X = OTs] Conditions**: P(t-Bu)$_2$Me, Pd(OAc)$_2$, NaOH, p-dioxane, 50 °C

- **99** (76%)
- **100** (67%)
- **101** (61%)
- **102** (73%)

**Scheme 4.12** The Suzuki–Miyaura cross-coupling of unactivated primary alkyl chlorides, bromides, and tosylates with alkyl-9-BBN reagents carried out in the presence of PCy$_3$ and P(t-Bu)$_2$Me. $^a$KOH was used in place of CsOH·H$_2$O. Bn = benzyl; TBS = t-butyldimethylsilyl; TES = triethylsilyl.

ethers (**91**, **97**, and **100**), and amides (**99**) being compatible with the reaction conditions. Ketones and aldehydes may not be tolerated, as inferred from the use of acetal (**96**) and ketal (**100**) masking groups. In addition, alkyl bromides are selectively coupled in the presence of alkyl chlorides, as seen in the coupling leading to **91**, and monosubstituted alkenes can be selectively hydroborated in the presence of disubstituted alkenes, providing access to **93**.

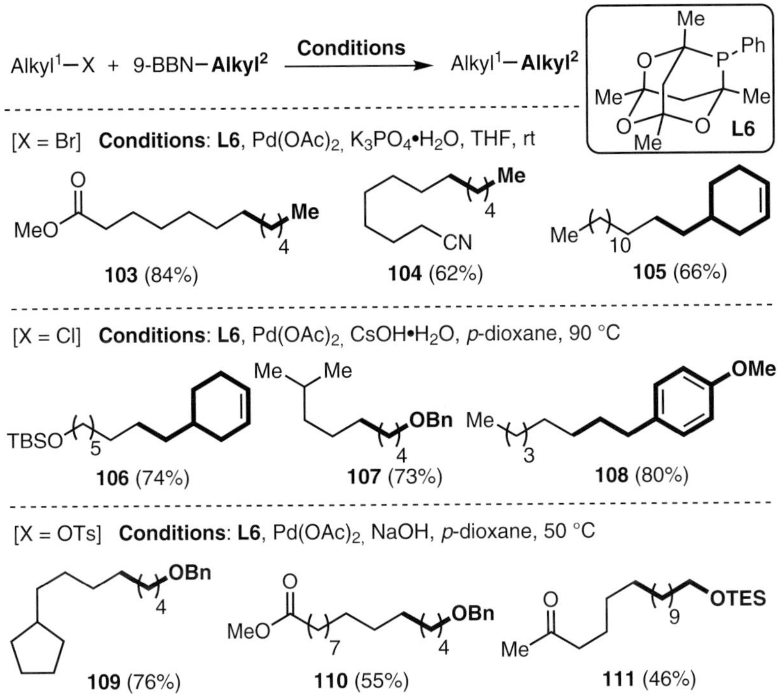

## 4.2 Developments Made in the Coupling of Nontrivial Substrates

Alkyl¹—X + 9-BBN—Alkyl² $\xrightarrow{\text{Conditions}}$ Alkyl¹—Alkyl²

[X = Br] **Conditions**: L6, Pd(OAc)$_2$, K$_3$PO$_4$•H$_2$O, THF, rt

**103** (84%)  **104** (62%)  **105** (66%)

[X = Cl] **Conditions**: L6, Pd(OAc)$_2$, CsOH•H$_2$O, p-dioxane, 90 °C

**106** (74%)  **107** (73%)  **108** (80%)

[X = OTs] **Conditions**: L6, Pd(OAc)$_2$, NaOH, p-dioxane, 50 °C

**109** (76%)  **110** (55%)  **111** (46%)

**Scheme 4.13** The Suzuki–Miyaura cross-coupling of unactivated primary alkyl chlorides, bromides, and tosylates with alkyl-9-BBN reagents carried out in the presence of phosphaadamantane-derived ligand **L6** (Figure 4.2).

Under the reaction conditions that were optimal for coupling using PCy$_3$ and P(t-Bu)$_2$Me (Scheme 4.12), phosphaadamantane **L6** (Figure 4.2) has been shown to be a general ligand for the coupling of unactivated primary alkyl bromides, chlorides, and tosylates providing a means to access aliphatic products **103–111** (Scheme 4.13) [129]. Functional group compatibility is on par with couplings carried out in the presence of PCy$_3$ and P(t-Bu)$_2$Me. That a single organophosphine ligand can couple each of these three electrophiles is an advantage of this system. Phosphaadamantane **L6** (Figure 4.2) was also effective at coupling unactivated primary alkyl bromides and chlorides with arylboronic acids [130]. As such, this ligand is as effective as trialkylphosphines for a variety of cross-couplings; however, it has received less attention presumably as a consequence of it being synthetically more laborious to prepare, limited studies on its use in C$_{sp2}$–C$_{sp2}$ cross-couplings, and being commercially unavailable at the time this chapter was composed.

**4.2.3.2.2 Coupling of Secondary Alkyl Halides** Unactivated secondary alkyl halides are more challenging electrophiles than primary ones due to the more electron-rich C—X bond and increased sterics surrounding the reactive site, as well as the presence of additional β-hydrogens that can interact with the metals d-orbitals. The employment of more reactive Ni-based catalysts has bypassed some of these barriers, as Ni(0)

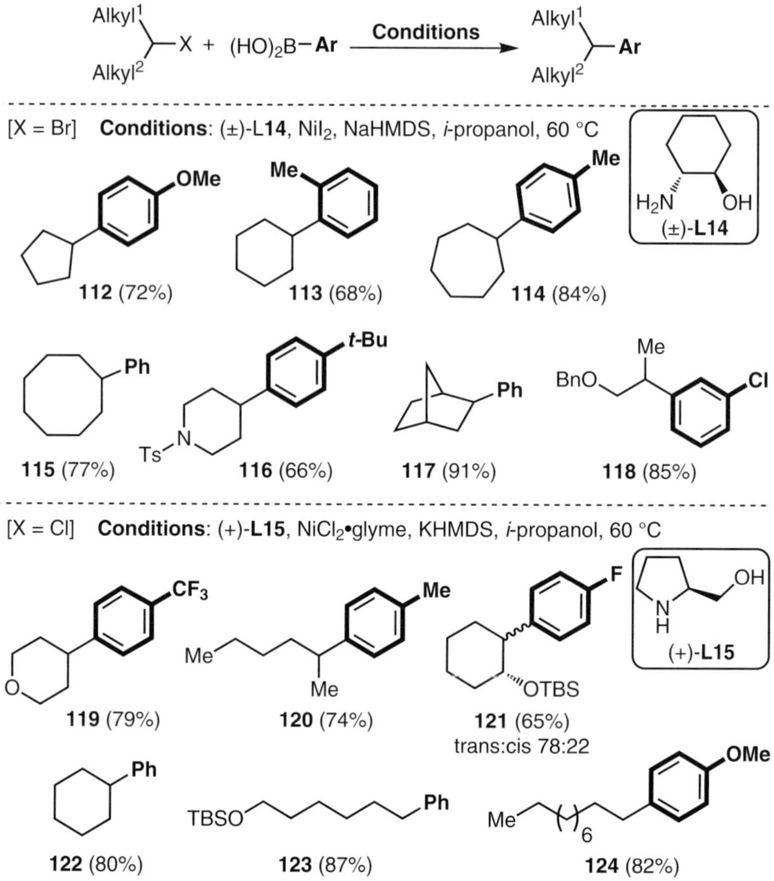

**Scheme 4.14** The Suzuki–Miyaura cross-coupling of unactivated secondary alkyl bromides and chlorides with arylboronic acids carried out in the presence of amino alcohol-based ligands **L14** and **L15**.

has a lower oxidation potential than does Pd(0) and proceeds via a radical mechanism (see below). Suzuki–Miyaura cross-couplings of secondary alkyl bromides and iodides with arylboronic acids were first demonstrated using Ni(cod)$_2$ (cod = cyclooctadiene) and bathophenanthroline as the spectator ligand in the presence of *t*-BuOK and *s*-butanol at 60 °C [131]. Yields of cross-coupled products ranged from 44 to 90%; however, as acknowledged by the authors the substrate scope is limited. This prompted a follow-up study that elucidated highly active Ni/amino alcohol-based catalyst systems (ligands **L14** and **L15**, Scheme 4.14) [132]. A variety of unactivated secondary bromides (leading to **112–118**) and chlorides (leading to **119–124**) were coupled with arylboronic acids in high yields. The use of heteroaryl-, alkenyl-, and alkylboronic acids were less effective substrates than arylboronic acids, often resulting in <30% yields. Electron-rich arylboronic acids were required in excess due to protodeborylation, which may stem from the use of *i*-propanol as solvent that

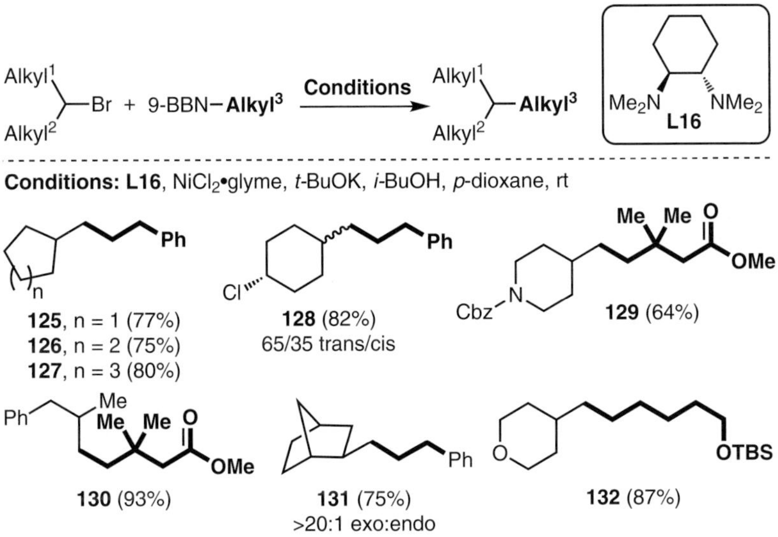

## 4.2 Developments Made in the Coupling of Nontrivial Substrates | 239

Alkyl¹\\Alkyl²C−Br + 9-BBN−Alkyl³ →[Conditions] Alkyl¹\\Alkyl²C−Alkyl³

L16: Me₂N–(cyclohexane-1,2-diyl)–NMe₂ (trans)

Conditions: L16, NiCl₂·glyme, *t*-BuOK, *i*-BuOH, *p*-dioxane, rt

**125**, n = 1 (77%)
**126**, n = 2 (75%)
**127**, n = 3 (80%)
(cyclopentyl/cyclohexyl/cycloheptyl-CH₂CH₂CH₂Ph)

**128** (82%), 65/35 trans/cis (4-chlorocyclohexyl-CH₂CH₂CH₂Ph)

**129** (64%) (Cbz-piperidinyl with –CH₂CH₂CH(Me)CH(Me)C(O)OMe side chain)

**130** (93%) (Ph–CH(CH₃)–CH₂CH(Me)CH(Me)C(O)OMe)

**131** (75%) >20:1 exo:endo (norbornyl-CH₂CH₂CH₂Ph)

**132** (87%) (tetrahydropyranyl-(CH₂)₅-OTBS)

**Scheme 4.15** The Suzuki–Miyaura cross-coupling of unactivated secondary alkyl bromides with primary alkyl-9-BBN reagents in the presence of the diamino ligand **L16**.

undergoes β-hydride elimination to give acetone and the corresponding Pd(II) hydride species [114, 133]. As well, *ortho*-substituted arylboronic acids were poor substrates in the coupling of secondary alkyl chlorides. Notwithstanding these shortcomings, in terms of cross-coupling protocols this method is relatively general, and a variety of secondary alkyl benzene derivatives were prepared that otherwise would be difficult to prepare in a single synthetic transformation. The authors did not comment on whether the use of (S)-(+)-proline (**L15**) provides asymmetric induction in the coupling of racemic secondary alkyl chlorides.

The cross-coupling of secondary alkyl bromides with *alkyl*-9-BBN reagents has also been achieved (Scheme 4.15) [134]. The optimal Ni-based catalyst systems utilizing amino alcohol-based ligands **L14** or **L15** that were viable for coupling secondary alkyl halides with arylboronic acids provided only trace (<5%) amounts of product in the coupling of bromocyclohexane with 9-(3-phenylpropyl)-9-BBN. Instead *trans-N,N'*-dimethyl-1,2-cyclohexanediamine (**L16**) was found to be an effective ligand for this Ni-catalyzed transformation. A range of unactivated secondary alkyl bromides were coupled with various primary alkyl-9-BBN reagents (leading to **125–132**) in good-to-excellent yields at *room temperature*! Protic solvent was necessary for effective coupling of these secondary electrophiles. Unactivated primary and secondary alkyl iodides and primary alkyl bromides were also suitable substrates under these reaction conditions. Interestingly, secondary alkyl bromides were coupled selectively in the presence of primary alkyl bromides in a competition experiment, and both greatly outperformed tertiary alkyl bromides. The two-electron oxidative addition commonly associated with aryl halides and Pd-mediated cross-couplings gives way to a radical mechanism for the coupling of secondary alkyl halides with low-valent Ni catalysts [132, 135, 136]. Ligand-dependant one-electron oxidation of Ni by the secondary alkyl bromide provides an sp$^2$-hybridized alkyl radical alongside the formation of X$^-$ [137]. The alkyl radical is in the vicinity of the metal and undergoes oxidative radical addition to Ni so long as radical dimerization and other side processes are slower kinetically. This perhaps accounts for the observed preference of secondary alkyl bromides over primary alkyl bromides as their radicals are more stable. It is likely that the sterics of tertiary alkyl electrophiles impede one or more steps of this revised catalytic cycle.

### 4.2.3.3 Cross-Couplings Promoted by NHC Ligands

*N*-Heterocyclic carbenes have also proven effective as ligands in the Suzuki–Miyaura $C_{sp3}$–$C_{sp3}$ reactions; however, their range of use is less advanced relative to organophosphine and amine-based ligands. An *in-situ*-prepared catalyst from the imidazolium salt IPr•HCl and Pd$_2$(dba)$_3$ generated yields of cross-coupled aliphatic products ranging in 28–56% yields from primary alkyl bromides and alkyl-9-BBN reagents at 40 °C [138]. The preformed complex Pd-PEPPSI-IPr (**C6a**, Figure 4.3) was used in similar couplings and was found to be superior to the *in-situ*-generated (NHC) Pd catalyst [139]. At room temperature, a range of primary $C_{sp3}$–$C_{sp3}$ cross-couplings were carried out to provide the corresponding products **(133–138)** in good-to-excellent yields (Scheme 4.16). Functional group compatibility was quite good under the mild reaction conditions. The alkylation of aryl bromides and chlorides was

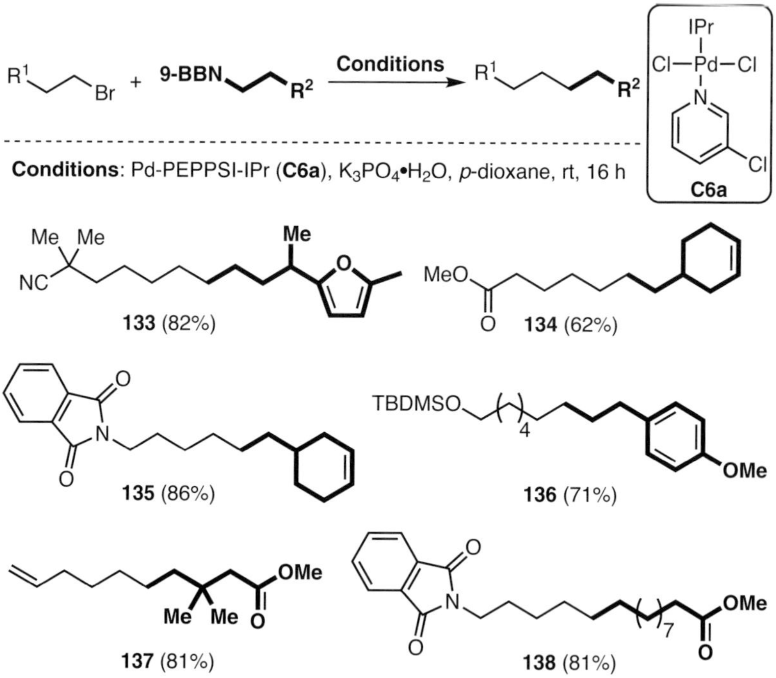

**Scheme 4.16** The Suzuki–Miyaura cross-coupling of unactivated primary alkyl bromides with primary alkyl-9-BBN reagents in the presence of Pd-PEPPSI-IPr (**C6a**, Figure 4.3).

also reported using the same reaction conditions, with free aniline and phenol derivatives being compatible substrates, thus removing the need to implement protecting group chemistry [139].

## 4.3
## Asymmetric Suzuki–Miyaura Cross-Couplings

Recently developing apace is the asymmetric Suzuki–Miyaura reaction wherein asymmetric induction by way of chrial ligands and/or substrates results in the stereoselective formation of a new C—C bond [140, 141]. Advanced ligand design to facilitate more effective cross-coupling has aided in overcoming the steric and electronic barriers that once marginalized the use of challenging *ortho*-substituted haloarenes or secondary aliphatic halides, species capable of achieving *axial* and *point* chirality in the coupling step, respectively. The vast majority of research in this realm has been focused on preparing biaryl compounds that possess hindered rotation about the newly formed aryl–aryl (axial) bond such that if the magnitude of the barrier to rotation is large enough [141], two isomers, more specifically atropisomers, are possible. Approaches to achieve atroposelective Suzuki–Miyaura cross-couplings rely on asymmetric induction via (1) chiral phosphine and nitrogen-based ligands for the metal catalyst (Section 4.3.1.1), (2) coordination of the metal catalyst by a stereogenic *ortho*-substituent on the electrophilic coupling partner (Section 4.3.1.2), and (3) (arene)chromium complexes that possess an axis of planar chirality (Section 4.3.1.3) [140, 141]. All methods provide a chiral environment for the metal center, with the first of these techniques being the most general and thus most exploited approach. Cross-couplings where point chirality is introduced into the product are much less developed due to the inherent difficulty in effectively coupling secondary alkyl halides via transition metal-mediated catalysis. However, some success has been made and approaches include the use of chiral catalyst systems or "advanced" substrates in which the point chirality is preestablished in the reactive C—X or C—B bonds that is transferred to the forming C—C bond (i.e., stereoretention). Each of these approaches will be discussed in the following sections.

### 4.3.1
### Achieving Axial Chirality in the Suzuki–Miyaura Reaction

#### 4.3.1.1 Axial Chirality Induced by Chiral Ligands/Catalysts
The use of chiral ligands to prepare atropisomers is desirable as it is potentially quite general and avoids the need to employ advanced enantiopure precursors with functionality that must later be removed. Progress has been slow and largely unfruitful in this area; however, a handful of recent investigations have achieved excellent atroposelectivities for this transformation (i.e., $>85\%$ *ee*). The atroposelective formation of the $C_{aryl}-C_{aryl}$ bond in 1-phenylnaphthalene and 1,1′-binaphthalene derivatives has become the benchmark for the evaluation of chiral ligands (i.e.,

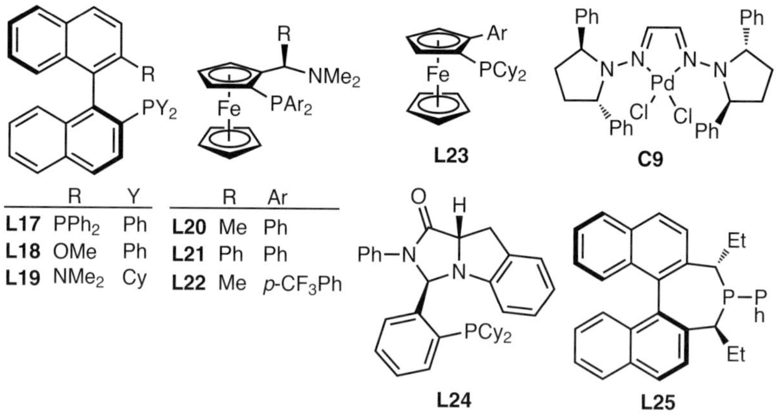

R
Y
L17 PPh2 Ph
L18 OMe Ph
L19 NMe2 Cy
R
Ar
L20 Me Ph
L21 Ph Ph
L22 Me p-CF3Ph
L23
C9
L24
L25


**Figure 4.7** Chiral ligands used in the asymmetric Suzuki–Miyaura reaction.

**L17**–**L25** and **C9**, Figure 4.7) developed for use in the asymmetric Suzuki–Miyaura reaction [140, 141]. Most couplings leading to these products are plagued by low *ee*'s and/or low yields that stems in part from protodeborylation due to increased sterics in the boronic acid substrate [142–150].

The use of chiral binaphthyl ligand **L19** in combination with $Pd_2(dba)_3$ under the conditions outlined in Scheme 4.17 was one of the earliest developed conditions to achieve consistently high atroposelectivities [151]. Long reaction times were typically required; however, excellent ee's and yields were obtained for a variety of 1-aryl-2-naphthylphosphonates (**139**) produced from their precursor 1-bromo-2-naphthylphosphonates (**140**) and *ortho*-substituted phenylboronic acids (**141**).

More recently, some of the highest ee's achieved to date were reported using the $C_2$-symmetric bis-hydrazone–$PdCl_2$ complex **C9** (Table 4.2) [152]. Coupling of substituted 1-bromobenzene or 1-bromonaphthalenes (**142**) with 1-naphthyl- or 1-dihydroacenaphthylboronic acid (**143**) derivatives possessing *ortho* substituents provided their corresponding chiral biaryl products (**144**) in excellent ee's and yields (entries

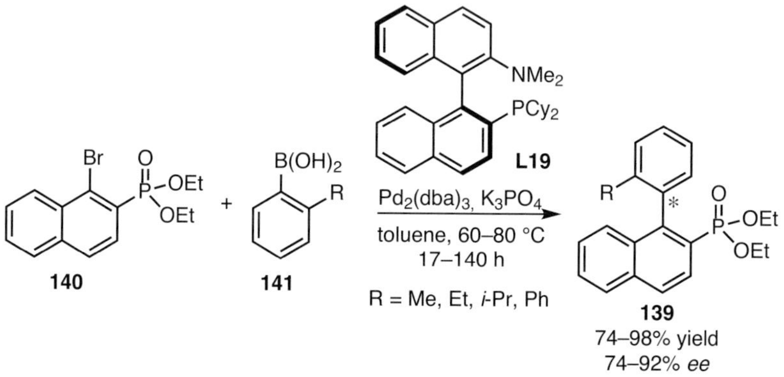

L19
Pd2(dba)3, K3PO4
toluene, 60–80 °C
17–140 h
R = Me, Et, i-Pr, Ph
140
141
139
74–98% yield
74–92% ee


**Scheme 4.17** Atroposelective Suzuki–Miyaura cross-couplings using chiral ligand **L19**.

Table 4.2 The Suzuki–Miyaura cross-coupling employing chiral catalyst **C9** for the preparation of biaryl atropisomers.

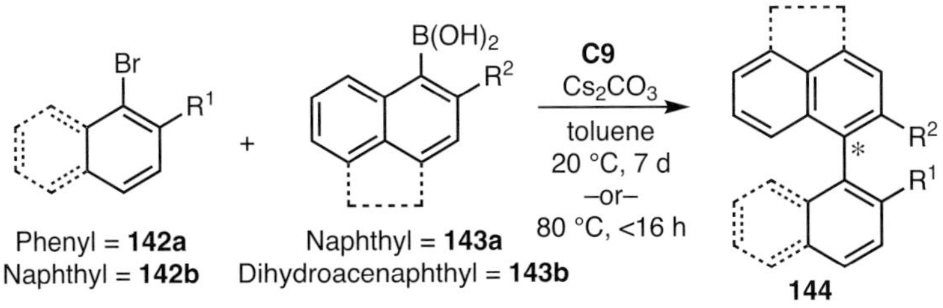

Phenyl = **142a**
Naphthyl = **142b**

Naphthyl = **143a**
Dihydroacenaphthyl = **143b**

C9, $Cs_2CO_3$, toluene, 20 °C, 7 d –or– 80 °C, <16 h → **144**

| Entry | Ar–R$^1$ | Ar–R$^2$ | 20 °C yield% (ee) | 80 °C yield% (ee) | Config. |
|---|---|---|---|---|---|
| 1 | 142b–OMe | 143a–H | 97 (86) | 99 (75) | S |
| 2 | 142b–Me | 143a–H | 80 (95) | 98 (90) | R |
| 3 | 142b–H | 143a–Me | 71 (98) | 99 (86) | R |
| 4 | 142b–Me | 143b–H | 40 (>98) | 98 (70) | R |
| 5 | 142a–Ph | 143a–Me | 97 (84) | — | S |

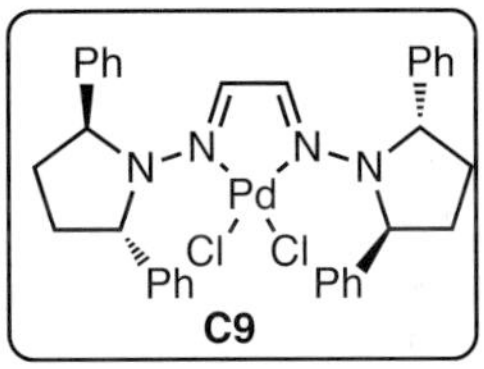

1–5). Although no tetra-*ortho*-substituted biaryls could be formed, tri-*ortho*-substituted biaryls were easily accessible. The reaction can be accelerated from 7 days to less than 16 h by heating to 80 °C. Yields are generally improved; however, there is a concomitant erosion in *ee*.

Remarkable reactivity and atroposelectivity come by way of ligand **L24** and Pd(OAc)$_2$ in toluene at 100 °C using K$_3$PO$_4$ base [153]. 1-Iodo-2-methylnaphthalene and 2-methyl-1-naphthylboronic acid were coupled to give the corresponding tetra-*ortho*-substituted binaphthyl derivative in 98% yield and 92% *ee* in just 5 h! The halide derivative 1-chloro-2-methylnaphthalene was also coupled in excellent *ee* (88%) and moderate yield (57%), making this protocol unprecedented in terms of its reactivity and atroposelectivity. Chiral ligand **L24** was subsequently attached to a polystyrene-poly(ethyleneglycol) copolymer (PS-PEG) resin (Scheme 4.18). In the presence of Pd(OAc)$_2$ and TBAF in H$_2$O at 80 °C, a wide variety of chiral biaryls (**145**, from the cross-coupling of **146** and **147**) were prepared in excellent *ee* and yield. The obvious environmental benefits from using solid-supported reagents and H$_2$O as solvent greatly add to the attractiveness of this approach. The PS-PEG-**L24** resin was reused up to four times without loss in *ee* and only marginal erosion in yield. At the time of writing, this study stands as the "state of the art" in the asymmetric Suzuki–Miyaura reaction for the atroposelective preparation of biaryls.

Palladium nanoparticles (1.2–1.7 nm) stabilized by (*S*)-BINAP (**L17**) have also been utilized in the asymmetric Suzuki–Miyaura reaction (Table 4.3) [154], establishing a new approach for the design and development of catalyst systems for use in this reaction. Cross-couplings of substituted 1-bromonaphthylenes (**148**) with naphthyl-

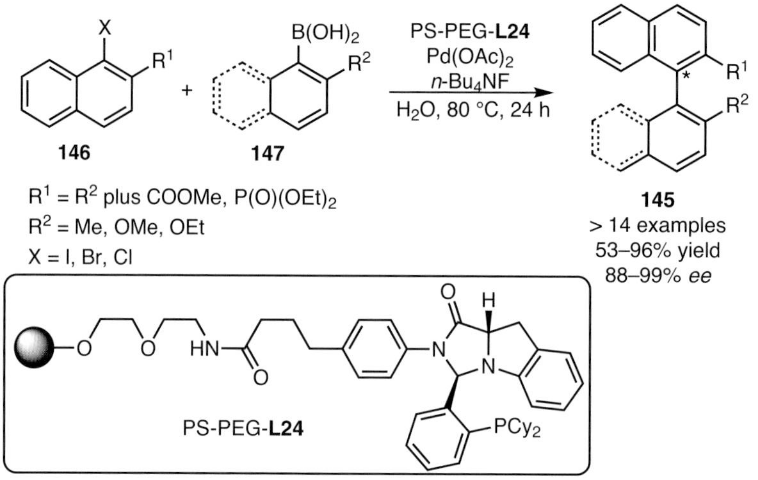

**Scheme 4.18** Atroposelective Suzuki–Miyaura cross-couplings using a solid-supported chiral ligand **PS-PEG-L24**.

**Table 4.3** The Suzuki–Miyaura cross-coupling employing chiral Pd nanoparticles for the preparation of biaryl atropisomers.

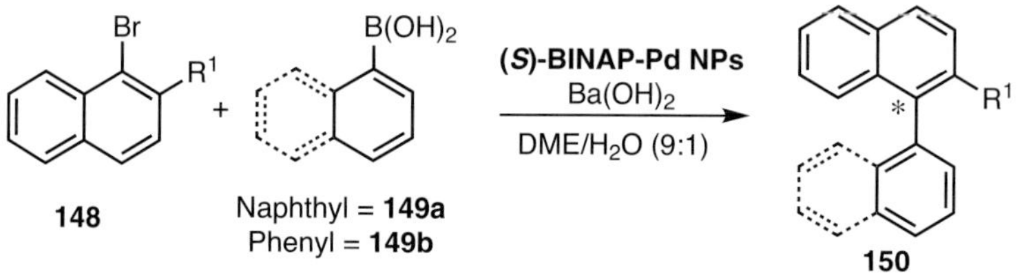

| Entry | Ar–R$^1$ | Ar–B(OH)$_2$ | Temp. (°C) | Time (h) | Yield, % (ee, %) |
|---|---|---|---|---|---|
| 1 | 148–OMe | 149a | 25 | 3 | 96 (69) |
| 2 | 148–OMe | 149a | −7 | 72 | 42 (74) |
| 3 | 148–OEt | 149a | 25 | 3 | 90 (70) |
| 4 | 148–OMe | 149b | 25 | 24 | 89 (55) |

(**149a**) and phenylboronic acids (**149b**) leading to substituted 1-arylnaphthalenes (**150**) were achieved in excellent yields and moderate atroposelectivities (entries 1, 3, and 4). Impressively, all couplings were carried out at room temperature; cooling to −7 °C cut the yield roughly by half while improving the *ee* only slightly (entry 2).

#### 4.3.1.2 Axial Chirality Induced by Point Chirality

Installing metal chelating functionality that is stereogenic *ortho* to the oxidative addition site on the electrophile induces a transient asymmetric topography around the metal center that can bias the atropstereoselectivity of the forming C$_{aryl}$–C$_{aryl}$

**Scheme 4.19** Atroposelective Suzuki–Miyaura cross-couplings using Pd-chelating substituents that possess point chirality on one of the coupling partners. TIPS = tri-*i*-propylsilyl; BHT = 3,5-di-*t*-butyl-4-methylphenol.

bond. In essence, the electrophile becomes an asymmetric ligand for the metal. This approach was utilized in the coupling of the boronic ester **151** with derivatives of **152** (Scheme 4.19) [155]. The silyl derivative (**152a**, weak Pd ligation) provides poor atropdiastereoselectivity, whereas the phosphine derivative (**152b**, good Pd ligation) provides a single atropdiastereomer of **153**, a precursor to *korupensamine A*, which is a component of the naturally occurring Michellamine alkaloids.

Another example involves the coupling of a single diastereomer of the sulfinyl derivative **154** with naphthylboronic acids and esters of the type **155** (Scheme 4.20) [156, 157]. Again, the induced chiral environment around Pd provides sulfinyl-containing biaryls **156** in good-to-excellent atropdiastereoselectivities.

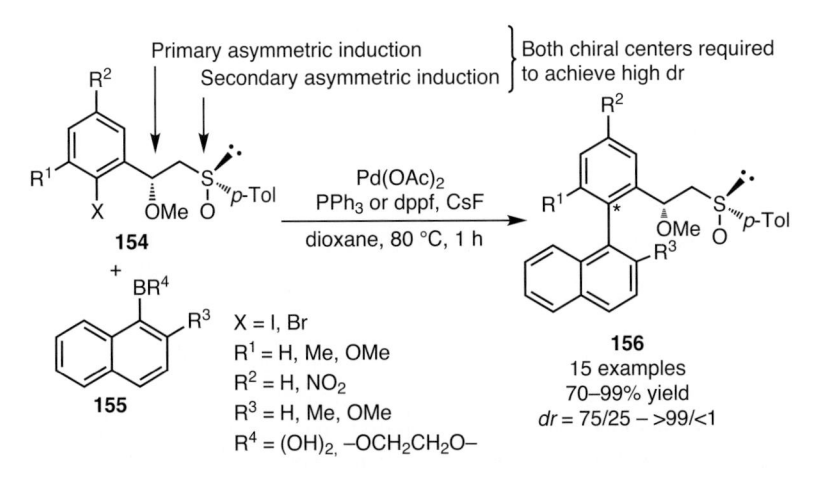

**Scheme 4.20** Suzuki–Miyaura cross-couplings using point chirality in *ortho* substituents of aryl bromides and iodides to achieve atroposelectivity.

### 4.3.1.3 Axial Chirality Induced by Planar Chirality

Axially chiral *ortho*-substituted biphenyls and phenylnaphthalenes have been prepared from arylboronic acids and (haloarene)chromium complexes (**157**) that possess a plane of chirality (Scheme 4.21) [158–163]. The atroposelectivity is highly dependent on the *ortho* substituent on the boronic acid (**158**). *Syn*-**159** atropisomers are formed exclusively in these couplings; however, the product can isomerize to the thermodynamically more stable isomer *anti*-**159** if the energetic barrier to axial rotation is sufficiently low. For the reverse reaction, (arylboronic acid)chromium complexes were found to be ineffective coupling partners with aryl halides forming only trace quantities of cross-coupled product. This approach has limited applications due to the use of a stoichiometric amount of toxic chromium and the difficulty associated with isolating single enantiomers of the chiral(arene)tricarbonylchromium complexes, which is typically accomplished by chromatography or fractional crystallization of their diastereoisomeric salts [164–167].

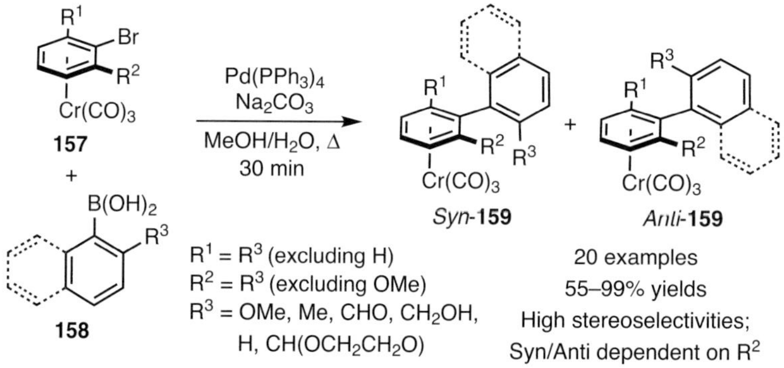

**Scheme 4.21** Suzuki–Miyaura cross-couplings using planar chirality in the substrate to achieve atroposelectivity.

## 4.3.2
### Achieving Point Chirality in the Suzuki–Miyaura Reaction

Unactivated secondary alkyl halides have been elusive substrates in transition metal-mediated couplings as a result of both the increased sterics and unfavorable electronics of these electrophiles. Only recently have examples emerged in which Ni-based catalysts, among others, have shown appreciable reactivity (Section 4.2.3.2.2) [136, 168]. Impressively, an asymmetric variant of this challenging reaction using the chiral diamine (*R,R*)-**L26** in the presence of Ni(cod)$_2$ provided access to products possessing a new stereogenic tertiary carbon (**160–165**) in good yields and enantioselectivities at or below room temperature (Scheme 4.22) [169]. There is a strict homobenzylic structural requirement for the alkyl halides as extending the chain length between the "bulky" aryl moiety and the $C_{sp3}$–Br bond leads to drastically reduced enantioselectivities. The observed stereoconvergence can be accounted for by the radical mechanism for oxidative addition (described in

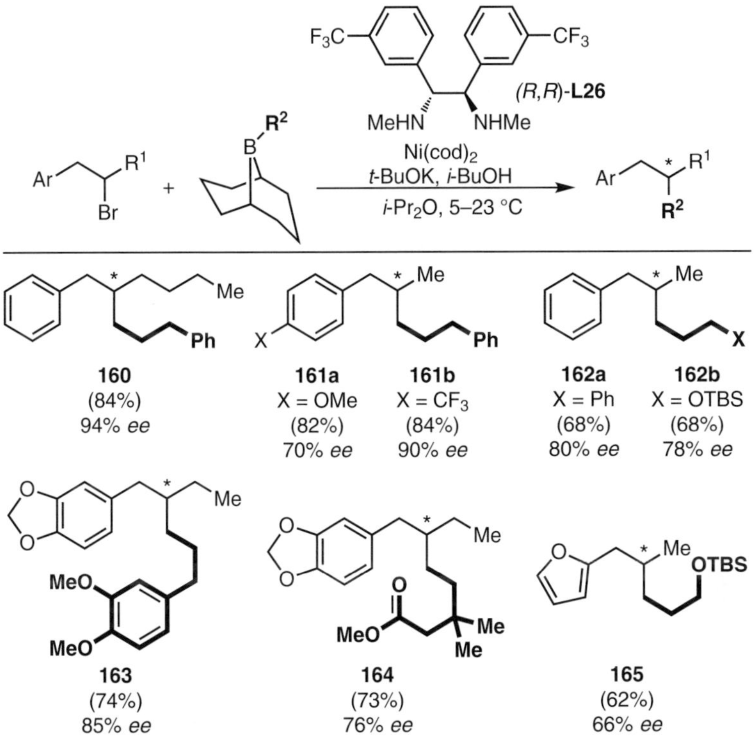

**Scheme 4.22** The asymmetric Suzuki–Miyaura cross-coupling of unactivated racemic secondary alkyl bromides with primary alkyl-9-BBN reagents in the presence of the diamino ligand (R,R)-**L26**.

Section 4.2.3.2.2) for Ni-catalyzed couplings of secondary alkyl bromides; the planar sp$^2$-hybridized alkyl radical is under the influence of the asymmetric environment imposed by the chiral ligand and induces enantioselective oxidative radical addition to Ni [137].

A different approach to the asymmetric Suzuki–Miyaura coupling has been reported that uses chiral organoboranes (**166**, Scheme 4.23). These are prepared in high levels of regio- and enantioselectivties by the Rh-catalyzed asymmetric hydroboration of styrene (**167**) and its analogues. These chiral secondary benzylic pinacol boranes are then coupled with aryl iodides in the presence of Pd$_2$(dba)$_3$, PPh$_3$, and AgO to yield the corresponding α-substituted phenylethanes (**168**) in good yields with excellent stereoretention of the stereogenic center [170]. Interestingly, achiral primary alkylboranes are not reactive under these reaction conditions.

To date, the above examples are the only known asymmetric Suzuki–Miyaura reactions of alkyl halides and alkylboranes. Although both couplings require stringent placement of aromatic motifs in one of the reacting substrates to achieve high levels of enantioselectivity, they nonetheless succeed in the long-sought goal of bringing the most widely used cross-coupling reaction into the realm of asymmetric catalysis.

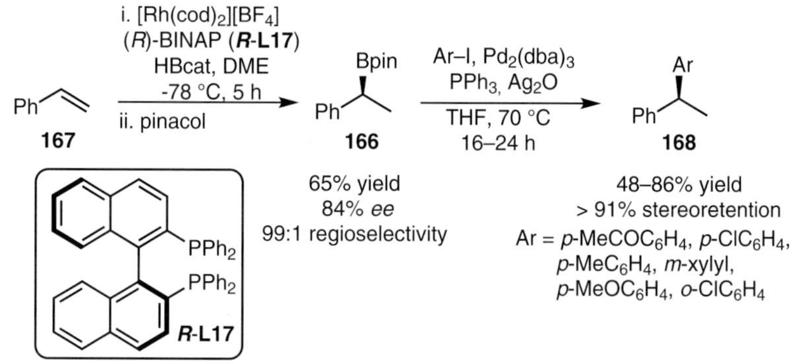

Scheme 4.23 The Suzuki–Miyaura cross-coupling of chiral organoboranes (**166**) to provide the stereoconserved chiral products (**168**). Yields of cross-coupled product based on $^1$H NMR spectroscopy versus an internal standard. HBcat = catecholborane.

## 4.4
## Iterative Suzuki–Miyaura Cross-Couplings

Robust methods for the facile preparation of oligoarenes are of interest as these are key structural components both for molecules of biological relevance, including enzyme mimics, and for use in molecular electronics and self-assembly [171, 172]. The Suzuki–Miyaura reaction has been applied to the preparation of such materials [173].

### 4.4.1
### *ortho* Metalation–Cross-coupling Iterations

*ortho* Metalation is a powerful method for selectively functionalizing arenes that possess a directed metalation group. Such methodology has been applied to the preparation of tetraphenylenes as a means of realizing the iterative Suzuki–Miyaura reaction (Scheme 4.24) [174]. Biphenylboronic acid derivative **169** was cross-coupled with *o*-bromophenyl diethylcarbamate **170** to provide **171a** along with its unmasked derivative **171b**. Subsequent *ortho* metalation of **171a** and quenching the intermediate phenylide with trimethylborate provided boronic acid **172** after acid hydrolysis. Iteration of the cross-coupling step provided tetraphenyl **174** in good overall yield. Successive iterations of (i) *ortho* metalation and (ii) cross-coupling should be possible given the presence of a directed metalation group in **174**.

### 4.4.2
### Triflating–Cross-Coupling Iterations

Another strategy involves the iteration of (i) the cross-coupling of aryl triflates functionalized with either a free or a methyl-protected hydroxyl group and (ii) triflation of the unmasked hydroxyl group in the product [171, 172, 175]. This

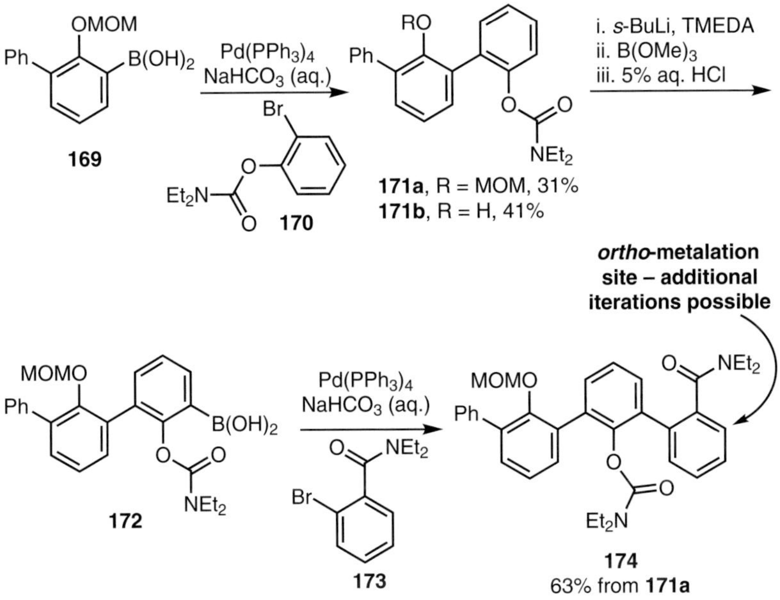

**Scheme 4.24** Iterative functionalization by *ortho*-lithiation and Suzuki–Miyaura cross-couplings. MOM = methoxymethyl; TMEDA = N,N,N′,N′-tetraethylenediamine.

approach (Figure 4.8a) was used to prepare the functionalized oligoarene **175** (Figure 4.8b) in excellent yields for each iteration [171, 172].

### 4.4.3
### Iterative Cross-Couplings via Orthogonal Reactivity

#### 4.4.3.1 Bifunctional Electrophiles

**4.4.3.1.1 Organohalides** Other iteration strategies exploit chemo- and regioselectivity in coupling bromoiodoarenes or dibromoarenes followed by conversion of the remaining bromide functionality to a boronic acid [176–180]. An example is presented in Scheme 4.25. The boronic acid **176** was coupled to bromoiodobiphenyl **177** chemoselectively to give the corresponding bromotetraarene **178a**. Conversion of the TMS group to iodide provided **178b** that was then selectively coupled to a second equivalent of **176** to give **179a**. Lithium–bromide exchange followed by quenching with a source of iodine provided **179b** that was then converted to pinacol boronate **180**. Dimerization of **180** provided macrocyclic oligophenylene **181** [177]. A similar strategy was applied for the preparation of a cyclotetraicosaphenylene, a macrocycle possessing 24 phenylene units [176].

**4.4.3.1.2 Alternate Electrophiles** The scope of tolerated electrophiles in the Suzuki–Miyaura reaction has recently been expanded to include ArOR derivatives, including aryl methyl ethers [181], allylic ethers [182], carboxylates [183, 184],

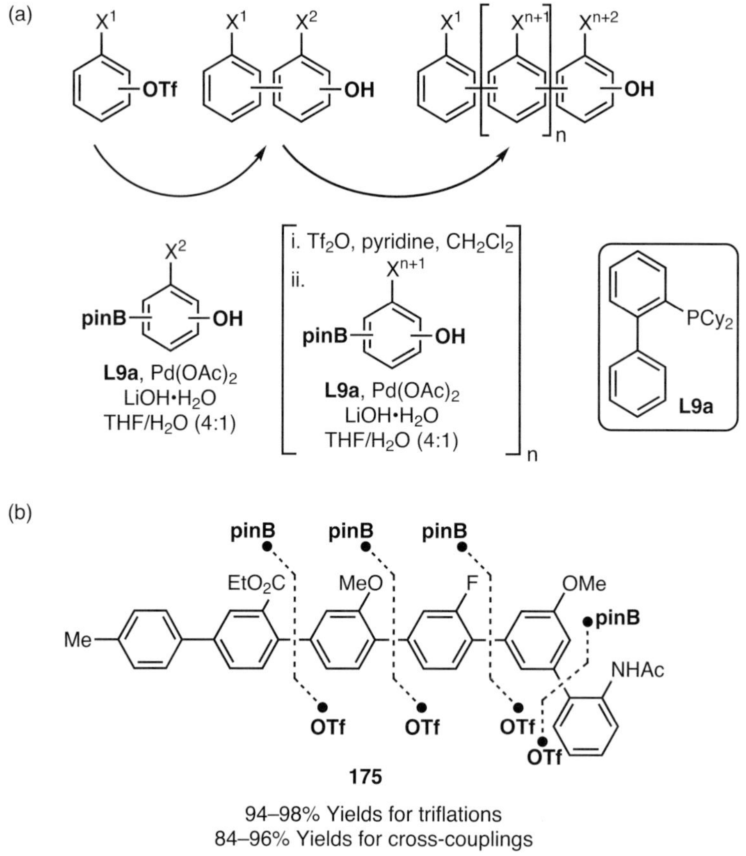

**Figure 4.8** (a) An iterative Suzuki–Miyaura cross-coupling and triflation strategy that (b) has been applied to the synthesis of polyarene **175**.

carbamates, carbonates, and sulfamates [185]. This allows one to use the relatively dormant $C_{aryl}-O$ bond, and in doing so allows for the development of new sequential strategies that play off of the cross-coupling step. While the $C_{aryl}-$halide bond is susceptible to oxidative addition by both Pd and Ni catalysts, the $C_{aryl}-O$ bond is, for the main part, inert toward Pd. This has also opened the door for iterative processes. For example, 2-bromo-6-methoxynaphthalene (**182**) has been shown to undergo a double cross-coupling sequence that begins with a Pd-catalyzed cross-coupling of the $C_{aryl}-Br$ (providing **183**) that is followed up with a Ni-catalyzed cross-coupling of the $C_{aryl}-OMe$ bond to provide the polyaryl scaffold present in **184** (Scheme 4.26) [181].

In addition to aryl methyl ethers, the more reactive aryl carbonates [186] have also been used as electrophiles successfully in the Suzuki–Miyaura reaction. The inert behavior of boronic acids toward carbonyl groups renders these functionalized electrophiles well matched to the Suzuki–Miyaura reaction, with the $C_{aryl}-O$ bond undergoing chemoselective oxidative addition in the presence of the relatively weaker

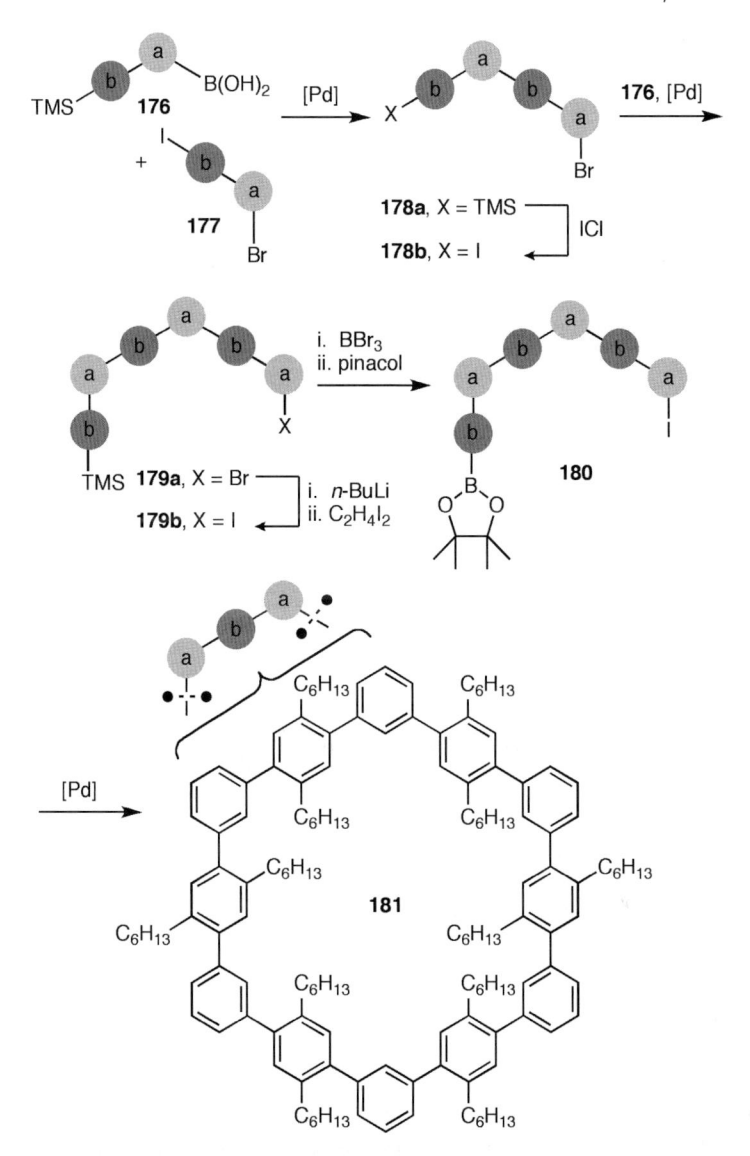

**Scheme 4.25** Iteration of Suzuki–Miyaura cross-couplings to provide macrocyclic oligophenylenes by means of orthogonal reactivity of aryl–I and aryl–Br bonds.

$C_{acyl}$–O bond (~106 and 80 kcal/mol, respectively). Aryl pivalates are the carbonate of choice, as they undergo relatively slow hydrolytic cleavage, and their application to iterative cross-coupling strategies was demonstrated in two seminal reports [183, 184]. One of these approaches (Scheme 4.27a) makes use of 4-acetylphenyl pivalate (**185**), wherein the $C_{aryl}$–OPiv bond is cross-coupled in the presence of commercially available $NiCl_2(PCy_3)_2$, followed by transformation of the remaining acetyl group in

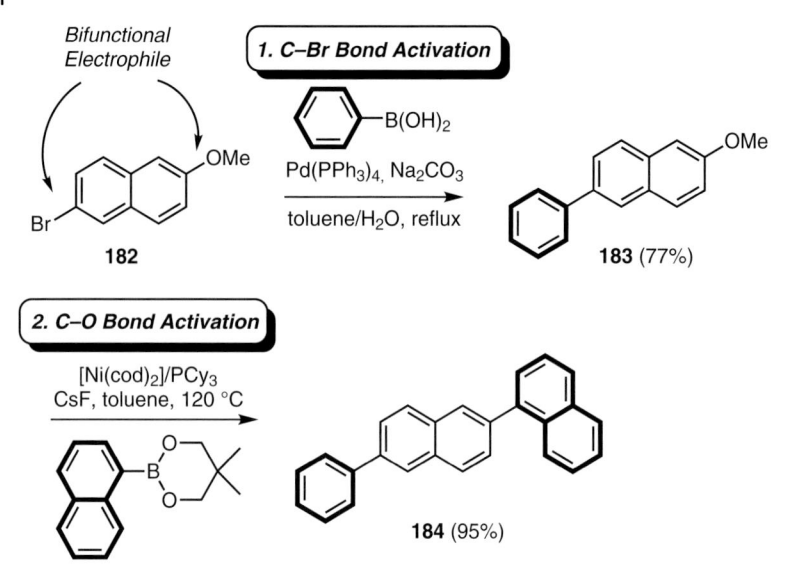

**Scheme 4.26** Iterative cross-couplings/orthogonal reactivity of aryl methyl ethers, and aryl bromides.

**186** into a second pivalate group via a three-step process (to provide **187**) that is rounded off with a second cross-coupling to provide the triphenylene **188**. A second approach (Scheme 4.27b) begins with the bifunctional bromonaphthyl pivalate **189** that undergoes successive cross-couplings of the $C_{aryl}$–Br with **190** followed by $C_{aryl}$–OPiv with phenylboronic acid using Pd and Ni catalysts, respectively, to provide **191** via intermediate **192**.

Perhaps the most useful strategy lies in the directing functionalization potential of aryl carbamates and sulfamates [185]. These readily available substrates can be *ortho*- and *para*-functionalized by *ortho* lithiation/quenching strategies [187, 188] and/or electrophilic aromatic substitution chemistry (Scheme 4.28). Hence, densely functionalized aromatic lynchpins with unique substitution patterns are readily available for subsequent derivatization via, among other processes, cross-couplings.

#### 4.4.3.2 Bifunctional Organoboranes

Molander and Sandrock have demonstrated the orthogonal cross-coupling reactivity in 9-BBN/BF$_3$K substrates (Scheme 4.29) [189]. These diboryl substrates are prepared by the hydroboration of alkenyl-containing BF$_3$K salts with 9-BBN and are used directly in a one-pot sequential cross-coupling strategy with high efficiencies.

Boron-masking groups such as 1,8-diaminonaphthalene (dan) [190–193] and *N*-methyliminodiacetic acid (MIDA) [194–198] have been exploited as a means to "deactivate" the C–B bond. This has made possible the sequential coupling of *lynchpins* via orthogonal reactivity that contain both an "active" electrophilic site and an "inactive" nucleophilic site that can subsequently be "activated" through acid- or base-catalyzed hydrolysis [199]. Both dan and MIDA function to deactivate the

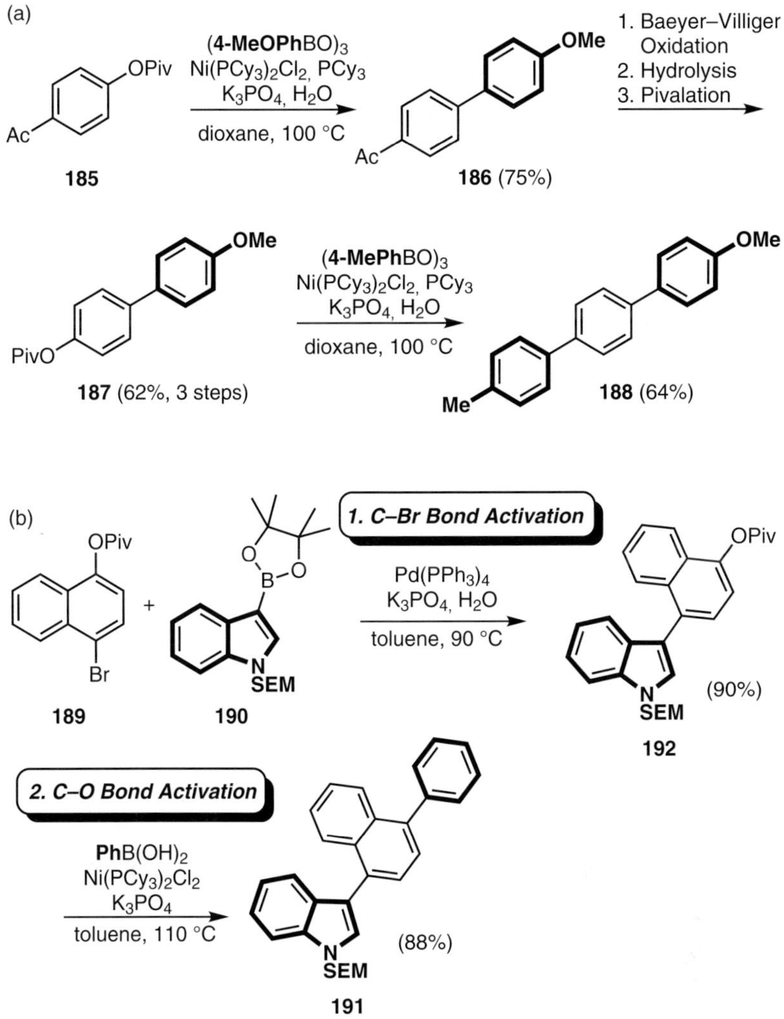

**Scheme 4.27** Iterative cross-coupling/orthogonal reactivity strategies involving aryl pivalates as an alternative electrophile.

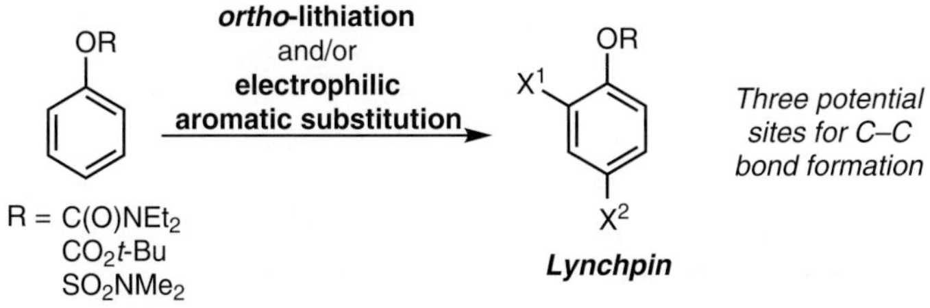

**Scheme 4.28** Aryl sulfamates and carbamates are valuable synthons in cross-couplings as they can be *ortho-* and/or *para-*functionalized, lending these species to synthetic strategies unavailable to aryl halides.

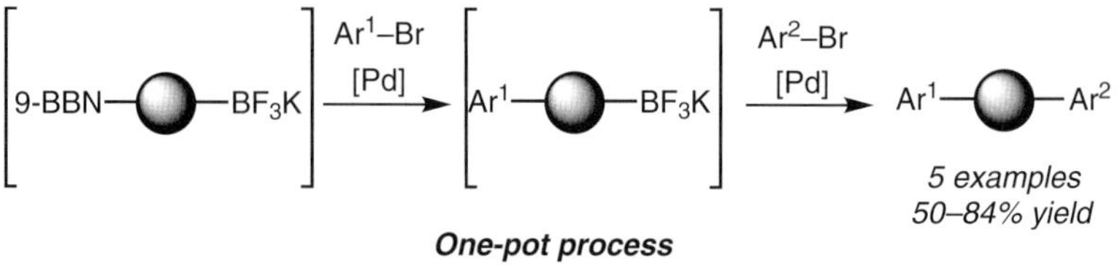

**Scheme 4.29** Differentially activated diboron reagents as substrates capable of orthogonal reactivity and iterative cross-couplings.

organoboronic ester by reducing the Lewis acidity of boron via stabilization of the vacant p-orbital on boron with that of the neighboring nitrogen lone pairs. This renders the organoborane inert toward transmetalation by effectively masking the C−B bond. The synthesis of oligophenylene **193** (Scheme 4.30) by Suginome and coworkers demonstrates how this principle can be applied to the iterative Suzuki–Miyaura reaction [192]. *p*-Tolylboronic acid (**194**) was coupled with haloboronamide **195** to provide the substituted biphenyl intermediate **196**. Acid-catalyzed hydrolysis of the boronamide provides the boronic acid derivative that can then participate in the

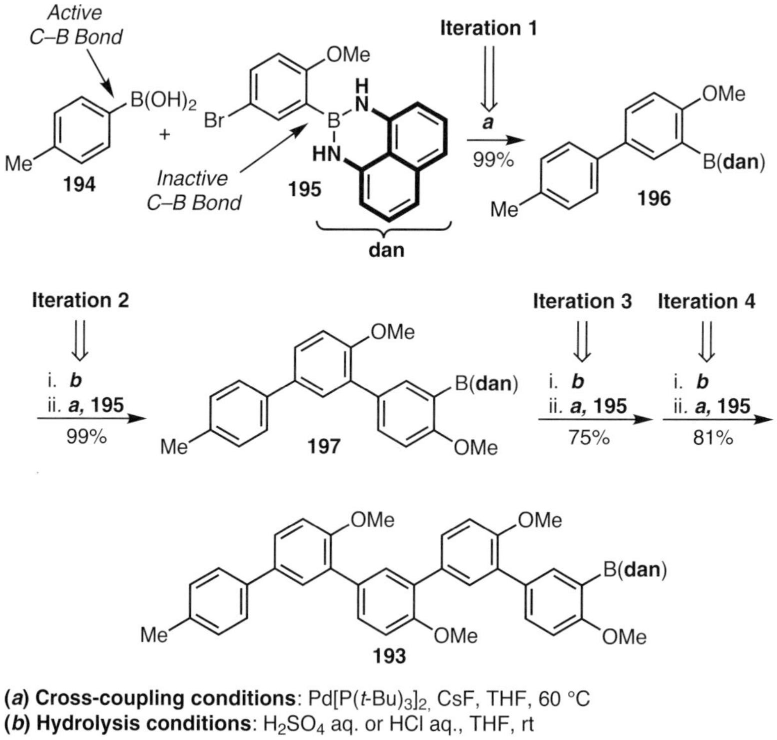

(*a*) **Cross-coupling conditions**: Pd[P(*t*-Bu)$_3$]$_2$, CsF, THF, 60 °C
(*b*) **Hydrolysis conditions**: H$_2$SO$_4$ aq. or HCl aq., THF, rt

**Scheme 4.30** Iterative Suzuki–Miyaura reactions (step *a*) that utilize masked bifunctional substrates (**195**) containing a "deactivated" boronamide that can be "activated" by acid-catalyzed hydrolysis (step *b*).

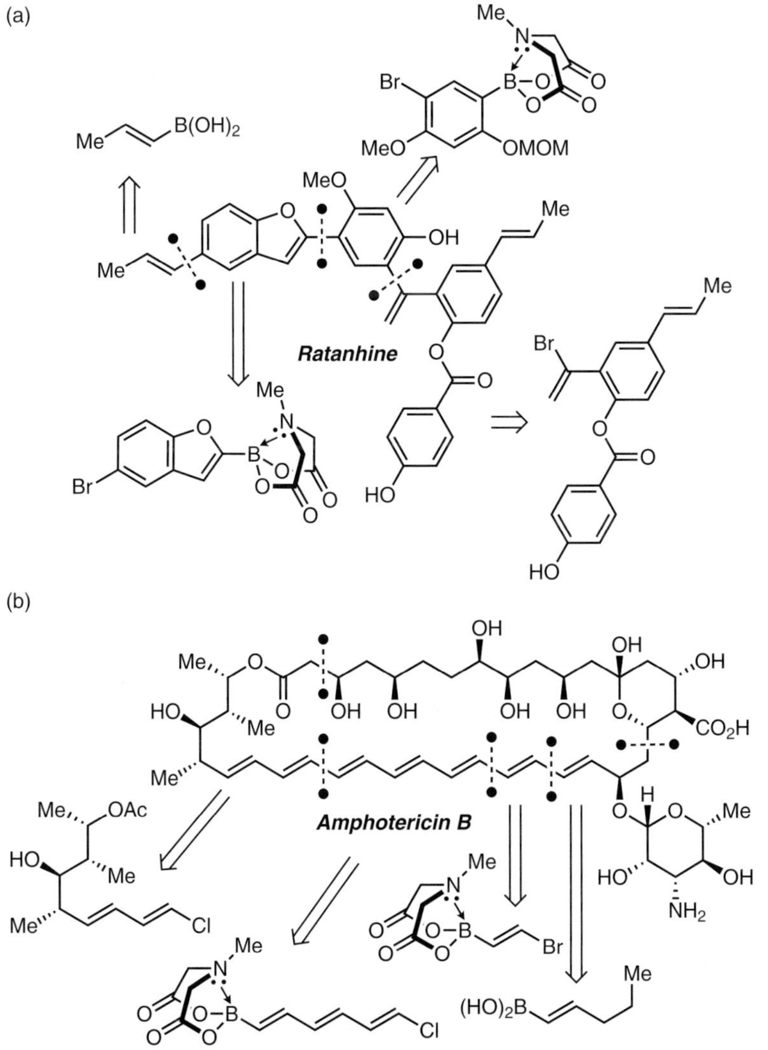

**Figure 4.9** Retrosynthetic breakdown to functionalized precursors for the total and partial synthesis of (a) *ratanhine* and (b) *amphotericin B*, respectively.

subsequent cross-coupling reaction with **195** to provide triarylboronamide **197**. Iteration of this sequence extends the phenylene chain to provide **193**.

Iterative Suzuki–Miyaura reactions have also been applied for the total synthesis of *ratanhine* [194] and a partial synthesis of *amphotericin B* (Figure 4.9) [196]. This work by Burke and coworkers was accomplished through the use of MIDA as the boronic acid-masking group. Notably, while polyenylboronic acids are unstable entities, haloalkenyl and polyenyl MIDA boronate esters are shelf stable. Conversion of the MIDA group back to the boronic acid was achieved under mild basic conditions (1M aq. NaOH/THF, 10 min; NaHCO$_3$/MeOH, 6 h) at room temperature.

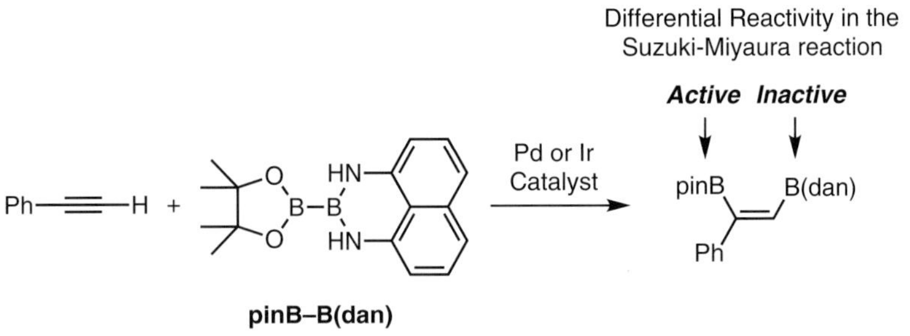

Ph≡H + pinB–B(dan) [structure with naphthalene-diamine] → (Pd or Ir Catalyst) → pinB / B(dan) alkene with Ph

**Differential Reactivity in the Suzuki-Miyaura reaction**
Active ↓    Inactive ↓

**pinB–B(dan)**

**Scheme 4.31** The reagent (pin)B–B(dan) can regioselectively diborylate teminal alkynes in the presence of a Pd or Ir catalyst. The 1-alkene-1,2-diboronic acid derivative contains both an active and an inactive C–B bond that can be selectively manipulated.

A variety of MIDA boronates are now commercially available. They are both air and chromatographically stable and are easily handled as free-flowing powders, as are the dan-protected derivatives. Although the latter arylboronamides can be prepared readily via condensation of 1,8-diaminonaphthalene with the desired boronic acid, a few recent studies illustrate their preparation via the stereoselective iridium-catalyzed (i) hydroboration of terminal alkynes [191], (ii) C–H borylation of functionalized arenes with (dan)BH [190], and (iii) diboration of alkynes with (pin)B–B(dan) to provide differentially protected 1,2-diboron alkenes (Scheme 4.31) [200]. Recent reports demonstrate that MIDA organoboronate derivatives are stable to a variety of chemical reactions that allows for their elaboration and functionalization. MIDA-protected haloalkenylboronic acids are easily derivatized via the Suzuki–Miyaura, Negishi, Heck–Mizoroki, and Sonogashira cross-couplings. In addition, MIDA boronates tolerate a wide variety of reactions including, for example, Swern–Moffat and Jones oxidations, Horner–Wadsworth–Emmons, Takai, Evans aldol, cyclopropanation, epoxidation, olefin metathesis reactions, silylation, and HF•Py-mediated desilylations, acid-catalyzed *p*-methoxybenzylation and subsequent unmasking with DDQ, and reductive amination [195]. As such, the preparation of advanced, functionalized MIDA boronate synthons is possible, making these masked boronic acid derivatives well suited for application in complex synthesis.

## 4.5
### Conclusions and Future Outlook

The design, synthesis, and application of bulky, electron-rich ligands has made possible the facile coupling of nontrivial substrates, including a myriad of aryl chlorides, unactivated alkyl halides, and both sterically hindered aryl halides and arylboronic acids. It now remains to the rest of the synthetic community to leave behind "traditional" cross-coupling protocols and more widely embrace these "contemporary" methods by applying them to challenging cross-couplings in synthesis.

The community has cracked the surface in terms of asymmetric Suzuki–Miyaura couplings, iterative processes, and alternate electrophiles, and further development of these areas will occupy the next decade of research in this field. The results from this future research will be of paramount importance as the application of these transformations in industry and academia seems limitless.

Still, despite the great advances that have been made over the past decade, the predictability of success in a given cross-coupling is not yet a certainty. With a virtually unlimited set of cross-coupling conditions now available to the synthetic chemist, it has become rather difficult to assign a particular set of reaction conditions *a priori*. Ensuing optimization of the Pd-catalyzed cross-coupling for a particular substrate pairing is to a large extent a long and tedious process as variables include the Pd source, ligand, organometallic reagent, solvent, additive(s), and temperature. The goal of developing a truly universal catalyst and global set of reaction conditions for this coupling may likely never be reached. Time and again, it has been observed that a strict balance between ligand and substrate structure is at play, notwithstanding the high dependence of the reaction on solvent polarity and additives. Thus, further elucidation of the intricacies of the cross-coupling mechanism and its reliance on substrate electronics and structure is requisite for the further rational development of this field. Once better understood, it will become more likely to be able to rationally choose a catalyst and conditions for any given Suzuki–Miyaura cross-coupling.

## References

1  Negishi, E.-I. (ed.) (2002) *Handbook of Organopalladium Chemistry for Organic Synthesis*, John Wiley & Sons, Inc., New York.
2  Chemler, S.R., Trauner, D., and Danishefsky, S.J. (2001) *Angew. Chem. Int. Ed.*, **40**, 4544–4568.
3  Nicolaou, K.C., Bulger, P.G., and Sarlah, D. (2005) *Angew. Chem. Int. Ed.*, **44**, 4442–4489.
4  Bellina, F., Carpita, A., and Rossi, R. (2004) *Synthesis*, **15**, 2419–2440.
5  Miyaura, N. and Suzuki, A. (1995) *Chem. Rev.*, **95**, 2457–2483.
6  Miyaura, N. (2002) *Top. Curr. Chem.*, **219**, 11–59.
7  Suzuki, A. (1999) *J. Organomet. Chem.*, **576**, 147–168.
8  Alonso, F., Beletskaya, I.P., and Yus, M. (2008) *Tetrahedron*, **64**, 3047–3101.
9  Zapf, A. (2004) Coupling of aryl and alkyl halides with organoboron reagents (Suzuki reaction), in *Transition Metals for Organic Synthesis: Building Blocks and Fine Chemicals*, 2nd edn (eds M. Beller and C. Bolm), Wiley-VCH Verlag GmbH, Weinheim, pp. 211–229.
10  Suzuki, A. (2005) Coupling reactions of areneboronic acids or esters with aromatic electrophiles, in *Boronic Acids: Preparation and Applications in Organic Synthesis and Medicine* (ed. D.G. Hall), Wiley-VCH Verlag GmbH, Weinheim.
11  Miura, M. (2004) *Angew. Chem. Int. Ed.*, **43**, 2201–2203.
12  Walker, S.D., Barder, T.E., Martinelli, J.R., and Buchwald, S.L. (2004) *Angew. Chem. Int. Ed.*, **43**, 1871–1876.
13  O'Brien, C.J., Kantchev, E.A.B., Chass, G.A., Hadei, N., Hopkinson, A.C., Organ, M.G., Setiadi, D.H., Tang, T.-H., and Fang, D.-C. (2005) *Tetrahedron*, **61**, 9723–9735.
14  Negishi, E.-i. (1978) *Aspects of Mechanism and Organometallic Chemistry* (ed. J.H. Brewster), Plenum, New York, p. 285.
15  Negishi, E.-I. (2002) *J. Organomet. Chem.*, **653**, 34–40.
16  Miyaura, N., Yamada, K., and Suzuki, A. (1979) *Tetrahedron Lett.*, **20**, 3437–3440.

17 Miyaura, N. and Suzuki, A. (1979) *Chem. Commun.*, 866–867.
18 Thomas, S.E. (1991) *Organic Synthesis: The Roles of Boron and Silicon*, vol. 1, Oxford University Press, Oxford.
19 de Meijere, A. and Diederich, F. (2004) *Metal-Catalyzed Cross-Coupling Reactions*, 2nd edn, vol. 1 (ed. N. Miyarua), Wiley-VCH Verlag GmbH, Weinheim, pp. 41–123.
20 Hall, D.G. (ed.) (2005) *Boronic Acids: Preparation and Applications in Organic Synthesis and Medicine*, Wiley-VCH Verlag GmbH, Weinheim.
21 Miyaura, N. (2002) *J. Organomet. Chem.*, **653**, 54–57.
22 Franzén, R. and Xu, Y. (2005) *Can. J. Chem.*, **83**, 266–272.
23 Aktoudianakis, E., Chan, E., Edward, A.R., Jarosz, I., Lee, V., Mui, L., Thatipamala, S.S., and Dicks, A.P. (2008) *J. Chem. Educ.*, **85**, 555–557.
24 Littke, A.F. and Fu, G.C. (2002) *Angew. Chem. Int. Ed.*, **41**, 4176–4211.
25 Grushin, V.V. and Alper, H. (1999) Activation of otherwise unreactive C-Cl bonds, in *Activation of Unreactive Bonds and Organic Synthesis* (ed. S. Murai), Springer, Berlin, pp. 193–226.
26 Netherton, M.R. and Fu, G.C. (2005) *Top. Organomet. Chem*, **14**, 85–108.
27 Netherton, M.R. and Fu, G.C. (2004) *Adv. Synth. Catal.*, **346**, 1525–1532.
28 Frisch, A.C. and Beller, M. (2005) *Angew. Chem. Int. Ed.*, **44**, 674–688.
29 Cárdenas, D.J. (1999) *Angew. Chem. Int. Ed.*, **38**, 3018–3020.
30 Grushin, V.V. and Alper, H. (1994) *Chem. Rev.*, **94**, 1047–1062.
31 Altenhoff, G., Goddard, R., Lehmann, C.W., and Glorius, F. (2003) *Angew. Chem. Int. Ed.*, **42**, 3690–3693.
32 Martin, R. and Buchwald, S.L. (2008) *Acc. Chem. Res.*, **41**, 1461–1473.
33 Fu, G.C. (2008) *Acc. Chem. Res.*, **41**, 1555–1564.
34 Surry, D.S. and Buchwald, S.L. (2008) *Angew. Chem. Int. Ed.*, **47**, 6338–6361.
35 Herrmann, W.A. (2002) *Angew. Chem. Int. Ed.*, **41**, 1290–1309.
36 Kantchev, E.A.B., O'Brien, C.J., and Organ, M.G. (2007) *Angew. Chem. Int. Ed.*, **46**, 2768–2813.
37 Kantchev, E.A.B., O'Brien, C.J., and Organ, M.G. (2006) *Aldrichim. Acta*, **39**, 97–111.
38 Organ, M.G., Chass, G.A., Fang, D.-C., Hopkinson, A.C., and Valente, C. (2008) *Synthesis*, **17**, 2776–2797.
39 Viciu, M.S. and Nolan, S.P. (2005) *Top. Organomet. Chem*, **14**, 241–278.
40 Herrmann, W.A., Öfele, K., Preysing, D.-v., and Schneider, S.K. (2003) *J. Organomet. Chem.*, **687**, 229–248.
41 Herrmann, W.A., Böhm, V.P.W., Gstöttmayr, C.W.K., Grosche, M., Reisinger, C.-P., and Weskamp, T. (2001) *J. Organomet. Chem.*, **617–618**, 616–628.
42 Wolfe, J.P., Singer, R.A., Yang, B.H., and Buchwald, S.L. (1999) *J. Am. Chem. Soc.*, **121**, 9550–9561.
43 Herrmann, W.A., Reisinger, C.-P., and Spiegler, M. (1998) *J. Organomet. Chem.*, **557**, 93–96.
44 Wilson, M.R., Woska, D.C., Prock, A., and Giering, W.P. (1993) *Organometallics*, **12**, 1742–1752.
45 Rahman, M.M., Liu, H.-Y., Eriks, K., Prock, A., and Giering, W.P. (1989) *Organometallics*, **8**, 1–7.
46 Birkholz, M.-N., Freixa, Z., and van Leeuwen, P.W.N.M. (2009) *Chem. Soc. Rev.*, **38**, 1099–1118.
47 Kranenburg, M., Kamer, P.C.J., and van Leeuwen, P.W.N.M. (1998) *Eur. J. Inorg. Chem.*, **2**, 155–157.
48 Brown, J.M. and Guiry, P.J. (1994) *Inorg. Chim. Acta*, **220**, 249–259.
49 Hayashi, T., Konishi, M., Kobori, Y., Kumada, M., Higuchi, T., and Hirotsu, K. (1984) *J. Am. Chem. Soc.*, **106**, 158–163.
50 Fihri, A., Meunier, P., and Hierso, J.-C. (2007) *Coord. Chem. Rev.*, **251**, 2017–2055.
51 Kohara, T., Yamamoto, T., and Yamamoto, A. (1980) *J. Organomet. Chem.*, **192**, 265–274.
52 Mann, G., Shelby, Q., Roy, A.H., and Hartwig, J.F. (2003) *Organometallics*, **22**, 2775–2789.
53 Gillie, A. and Stille, J.K. (1980) *J. Am. Chem. Soc.*, **102**, 4933–4941.
54 Culkin, D.A. and Hartwig, J.F. (2004) *Organometallics*, **23**, 3398–3416.

55 Negishi, E.-I., Takahashi, T., and Akiyoshi, K. (1987) *J. Organomet. Chem.*, **334**, 181–194.
56 Hartwig, J.F. (2007) *Inorg. Chem.*, **46**, 1936–1947.
57 Valentine, D.H.J. and Hillhouse, J.H. (2003) *Synthesis*, **16**, 2437–2460.
58 Dias, P.B., Piedade, M.E.M., and Simões, J.A.M. (1994) *Coord. Chem. Rev.*, **135/136**, 737–807.
59 Netherton, M.R. and Fu, G.C. (2001) *Org. Lett.*, **3**, 4295–4298.
60 Wanzlick, H.-W. (1962) *Angew. Chem. Int. Ed. Engl.*, **1**, 75–80.
61 Wanzlick, H.-W. and Schönherr, H.-J. (1968) *Angew. Chem. Int. Ed. Engl.*, **7**, 141–142.
62 Öfele, K. (1968) *J. Organomet. Chem.*, **12**, P42–P43.
63 Glorius, F. (2007) *Top. Organomet. Chem*, **21**, 1–20.
64 Herrmann, W.A., Schütz, J., Frey, G.D., and Herdtweck, E. (2006) *Organometallics*, **25**, 2437–2448.
65 Arduengo, A.J., III, Harlow, R.L., and Kline, M. (1991) *J. Am. Chem. Soc.*, **113**, 361–363.
66 Herrmann, W.A., Mihalios, D., Öfele, K., Kiprof, P., and Belmedjahed, F. (1992) *Chem. Ber.*, **125**, 1795–1799.
67 Öfele, K., Herrmann, W.A., Mihalios, D., Elison, M., Herdtweck, E., Scherer, W., and Mink, J. (1993) *J. Organomet. Chem.*, **459**, 177–184.
68 Herrmann, W.A., Öfele, K., Elison, M., Kuhn, F.E., and Roesky, P.W. (1994) *J. Organomet. Chem.*, **480**, C7–C9.
69 Perry, M.C. and Burgess, K. (2003) *Tetrahedron Asymmetry*, **14**, 951–961.
70 Marion, N., Díez-González, S., and Nolan, S.P. (2007) *Angew. Chem. Int. Ed.*, **46**, 2988–3000.
71 Nair, V., Bindu, S., and Sreekumar, V. (2004) *Angew. Chem. Int. Ed.*, **43**, 5130–5135.
72 Hahn, F.E. (2006) *Angew. Chem. Int. Ed.*, **45**, 1348–1352.
73 Cavallo, L., Correa, A., Costabile, C., and Jacobsen, H. (2005) *J. Organomet. Chem.*, **690**, 5407–5413.
74 Crudden, C.M. and Allen, D.P. (2004) *Coord. Chem. Rev.*, **248**, 2247–2273.
75 Diez-Gonzalez, S. and Nolan, S.P. (2007) *Coord. Chem. Rev.*, **251**, 874–883.
76 Nolan, S.P. and Scott, N.M. (2005) *Eur. J. Inorg. Chem.*, **10**, 1815–1828.
77 Crabtree, R.H. (2005) *J. Organomet. Chem.*, **690**, 5451–5457.
78 Garrison, J.C. and Youngs, W.J. (2005) *Chem. Rev.*, **105**, 3978–4008.
79 Dorta, R., Stevens, E.D., Hoff, C.D., and Nolan, S.P. (2003) *J. Am. Chem. Soc.*, **125**, 10490–10491.
80 Chianese, A.R., Li, X., Janzen, M.C., Faller, J.W., and Crabtree, R.H. (2003) *Organometallics*, **22**, 1663–1667.
81 Dorta, R., Stevens, E.D., Scott, N.M., Costabile, C., Cavallo, L., Hoff, C.D., and Nolan, S.P. (2005) *J. Am. Chem. Soc.*, **127**, 2485–2495.
82 Hadei, N., Kantchev, E.A.B., O'Brien, C.J., and Organ, M.G. (2005) *Org. Lett.*, **7**, 3805–3807.
83 Tolman, C.A. (1970) *J. Am. Chem. Soc.*, **92**, 2953–2956.
84 Tolman, C.A. (1970) *J. Am. Chem. Soc.*, **92**, 2956–2965.
85 Tolman, C.A. (1977) *Chem. Rev.*, **77**, 313–348.
86 Würtz, S. and Glorius, F. (2008) *Acc. Chem. Res.*, **41**, 1523–1533.
87 Fuller, A.A., Hester, H.R., Salo, E.V., and Stevens, E.P. (2003) *Tetrahedron Lett.*, **44**, 2935–2938.
88 Abraham, M.H. and Grellier, P.L. (1985) *The Chemistry of the Metal–Carbon Bond*, vol. 2 (eds F.R. Hartley, and S. Patai,), John Wiley & Sons, Inc., New York, p. 25.
89 Littke, A.F. and Fu, G.C. (1998) *Angew. Chem. Int. Ed.*, **37**, 3387–3388.
90 Old, D.W., Wolfe, J.P., and Buchwald, S.L. (1998) *J. Am. Chem. Soc.*, **120**, 9722–9723.
91 Littke, A.F., Dai, C., and Fu, G.C. (2000) *J. Am. Chem. Soc.*, **122**, 4020–4028.
92 Wolfe, J.P. and Buchwald, S.L. (1999) *Angew. Chem. Int. Ed.*, **38**, 2413–2416.
93 Yin, J., Rainka, M.P., Zhang, X.-X., and Buchwald, S.L. (2002) *J. Am. Chem. Soc.*, **124**, 1162–1163.
94 Barder, T.E., Walker, S.D., Martinelli, J.R., and Buchwald, S.L. (2005) *J. Am. Chem. Soc.*, **127**, 4685–4696.

95 Bonnet, V., Mongin, F., Trecourt, F., Breton, G., Marsais, F., Knochel, P., and Queguiner, G. (2002) *Synlett*, 1008–1010.
96 Billingsley, K.L. and Buchwald, S.L. (2007) *J. Am. Chem. Soc.*, **129**, 3358–3366.
97 Kudo, N., Perseghini, M., and Fu, G.C. (2006) *Angew. Chem. Int. Ed.*, **45**, 1282–1284.
98 Billingsley, K.L., Anderson, K.W., and Buchwald, S.L. (2006) *Angew. Chem. Int. Ed.*, **45**, 3484–3488.
99 Anderson, K.W. and Buchwald, S.L. (2005) *Angew. Chem. Int. Ed.*, **44**, 6173–6177.
100 Zapf, A., Jackstell, R., Rataboul, F., Riermeier, T., Monsees, A., Fuhrmann, C., Shaikh, N., Dingerdissen, U., and Beller, M. (2004) *Chem. Commun.*, 38–39.
101 Zapf, A., Ehrentraut, A., and Beller, M. (2000) *Angew. Chem. Int. Ed.*, **39**, 4153–4155.
102 Zhang, C., Huang, J., Trudell, M.L., and Nolan, S.P. (1999) *J. Org. Chem.*, **64**, 3804–3805.
103 Grasa, G.A., Viciu, M.S., Huang, J., Zhang, C., Trudell, M.L., and Nolan, S.P. (2002) *Organometallics*, **21**, 2866–2873.
104 Song, C., Ma, Y., Chai, Q., Ma, C., Jiang, W., and Andrus, M.B. (2005) *Tetrahedron*, **61**, 7438–7446.
105 Lavallo, V., Canac, Y., DeHope, A., Donnadieu, B., and Bertrand, G. (2005) *Angew. Chem. Int. Ed.*, **44**, 5705–5709.
106 Altenhoff, G., Goddard, R., Lehmann, C.W., and Glorius, F. (2004) *J. Am. Chem. Soc.*, **126**, 15195–15201.
107 O'Brien, C.J., Kantchev, E.A.B., Valente, C., Hadei, N., Chass, G.A., Lough, A., Hopkinson, A.C., and Organ, M.G. (2006) *Chem. Eur. J.*, **12**, 4743–4748.
108 Lebel, H., Janes, M.K., Charette, A.B., and Nolan, S.P. (2004) *J. Am. Chem. Soc.*, **126**, 5046–5047.
109 Marion, N. and Nolan, S.P. (2008) *Acc. Chem. Res.*, **41**, 1440–1449.
110 Gstöttmayr, C.W.K., Böhm, V.P.W., Herdtweck, E., Grosche, M., and Herrmann, W.A. (2002) *Angew. Chem. Int. Ed.*, **41**, 1363–1365.
111 Andreu, M.G., Zapf, A., and Beller, M. (2000) *Chem. Commun.*, 2475–2476.
112 Zapf, A. and Beller, M. (2005) *Chem. Commun.*, 431–440.
113 Selvakumar, K., Zapf, A., Spannenberg, A., and Beller, M. (2002) *Chem. Eur. J.*, **8**, 3901–3906.
114 Navarro, O., Kelly, R.A., and Nolan, S.P. (2003) *J. Am. Chem. Soc.*, **125**, 16194–16195.
115 Navarro, O., Kaur, H., Mahjoor, P., and Nolan, S.P. (2004) *J. Org. Chem.*, **69**, 3173–3180.
116 Navarro, O., Oonishi, Y., Kelly, R.A., Stevens, E.D., Briel, O., and Nolan, S.P. (2004) *J. Organomet. Chem.*, **689**, 3722–3727.
117 Organ, M.G., Çalimsiz, S., Sayah, M., Hoi, K.H., and Lough, A.J. (2009) *Angew. Chem. Int. Ed.*, **48**, 2383–2387.
118 Cesar, V., Bellemin-Laponnaz, S., and Gade, L.H. (2002) *Organometallics*, **21**, 5204–5208.
119 Palencia, H., Garcia-Jimenez, F., and Takacs, J.M. (2004) *Tetrahedron Lett.*, **45**, 3849–3853.
120 Schneider, S.K., Herrmann, W.A., and Herdtweck, E. (2006) *J. Mol. Catal.*, **245**, 248–254.
121 Marion, N., Navarro, O., Mei, J., Stevens, E.D., Scott, N.M., and Nolan, S.P. (2006) *J. Am. Chem. Soc.*, **128**, 4101–4111.
122 Navarro, O., Marion, N., Mei, J., and Nolan, S.P. (2006) *Chem. Eur. J.*, **12**, 5142–5148.
123 Terao, J. and Nobuaki, K. (2006) *Bull. Chem. Soc. Jpn.*, **79**, 633–672.
124 Brookhart, M., Green, M.L.H., and Parkin, G. (2007) *Proc. Natl. Acad. Sci. USA*, **104**, 6908–6914.
125 Kirchhoff, J.H., Netherton, M.R., Hills, I.D., and Fu, G.C. (2002) *J. Am. Chem. Soc.*, **124**, 13662–13663.
126 Netherton, M.R., Dai, C., Neuschutz, K., and Fu, G.C. (2001) *J. Am. Chem. Soc.*, **123**, 10099–10100.
127 Kirchhoff, J.H., Dai, C., and Fu, G.C. (2002) *Angew. Chem. Int. Ed.*, **41**, 1945–1947.
128 Netherton, M.R. and Fu, G.C. (2002) *Angew. Chem. Int. Ed.*, **41**, 3910–3912.
129 Brenstrum, T., Gerristma, D.A., Adjabeng, G.M., Frampton, C.S., Britten, J., Robertson, A.J., McNulty, J., and Capretta, A. (2004) *J. Org. Chem.*, **69**, 7635–7639.

130 Adjabeng, G., Brenstrum, T., Wilson, J., Frampton, C., Robertson, A., Hillhouse, J., McNulty, J., and Capretta, A. (2003) *Org. Lett.*, **5**, 953–955.

131 Zhou, J. and Fu, G.C. (2004) *J. Am. Chem. Soc.*, **126**, 1340–1341.

132 González-Bobes, F. and Fu, G.C. (2006) *J. Am. Chem. Soc.*, **128**, 5360–5361.

133 Singh, R., Viciu, M.S., Kramareva, N., Navarro, O., and Nolan, S.P. (2005) *Org. Lett.*, **7**, 1829–1832.

134 Saito, B. and Fu, G.C. (2007) *J. Am. Chem. Soc.*, **129**, 9602–9603.

135 Arp, F.O. and Fu, G.C. (2005) *J. Am. Chem. Soc.*, **127**, 10482–10483.

136 Rudolph, A. and Lautens, M. (2009) *Angew. Chem. Int. Ed.*, **48**, 2656–2670.

137 Jones, G.D., Martin, J.L., McFarland, C., Allen, O.R., Hall, R.E., Haley, A.D., Brandon, R.J., Konovalova, T., Desrochers, P.J., Pulay, P., and Vicic, D.A. (2006) *J. Am. Chem. Soc.*, **128**, 13175–13183.

138 Arentsen, K., Caddick, S., Cloke, F.G.N., Herring, A.P., and Hitchcock, P.B. (2004) *Tetrahedron Lett.*, **45**, 3511–3515.

139 Valente, C., Baglione, S., Candito, D., O'Brien, C.J., and Organ, M.G. (2008) *Chem. Commun.*, 735–737.

140 Baudoin, O. (2005) *Eur. J. Org. Chem.*, **20**, 4223–4229.

141 Bringmann, G., Mortimer, A.J.M., Keller, P.A., Gresser, M.J., Garner, J., and Breuning, M. (2005) *Angew. Chem. Int. Ed.*, **44**, 5384–5427.

142 Jensen, J.F. and Johannsen, M. (2003) *Org. Lett.*, **5**, 3025–3028.

143 Castanet, A.-S., Colobert, F., Broutin, P.-E., and Obringer, M. (2002) *Tetrahedron Asymmetry*, **13**, 659–665.

144 Kasák, P., Mereiter, K., and Widhalm, M. (2005) *Tetrahedron Asymmetry*, **16**, 3416–3426.

145 Genov, M., Almorín, A., and Espinet, P. (2007) *Tetrahedron Asymmetry*, **18**, 625–627.

146 Cammidge, A.N. and Crépy, K.V.L. (2004) *Tetrahedron*, **60**, 4377–4386.

147 Cammidge, A.N. and Crépy, K.V.L. (2000) *Chem. Commun.*, 1723–1724.

148 Genov, M., Almorín, A., and Espinet, P. (2006) *Chem. Eur. J.*, **12**, 9346–9352.

149 Mikami, K., Miyamoto, T., and Hatano, M. (2004) *Chem. Commun.*, 2082–2083.

150 Herrbach, A., Marinetti, A., Baudoin, O., Guénard, D., and Guéritte, F. (2003) *J. Org. Chem.*, **68**, 4897–4905.

151 Yin, J. and Buchwald, S.L. (2000) *J. Am. Chem. Soc.*, **122**, 12051–12052.

152 Bermejo, A., Ros, A., Fernández, R., and Lassaletta, J.M. (2008) *J. Am. Chem. Soc.*, **130**, 15798–15799.

153 Uozumi, Y., Matsuura, Y., Arakawa, T., and Yamada, Y.M.A. (2009) *Angew. Chem. Int. Ed.*, **48**, 2708–2710.

154 Sawai, K., Tatumi, R., Nakahodo, T., and Fujihara, H. (2008) *Angew. Chem. Int. Ed.*, **47**, 6917–6919.

155 Lipshutz, B.H. and Keith, J.M. (1999) *Angew. Chem. Int. Ed.*, **38**, 3530–3533.

156 Broutin, P.-E. and Colobert, F. (2003) *Org. Lett.*, **5**, 3281–3284.

157 Broutin, P.-E. and Colobert, F. (2005) *Eur. J. Org. Chem.*, **36**, 1113–1128.

158 Uemura, M., Nishimura, H., and Hayashi, T. (1993) *Tetrahedron Lett.*, **34**, 107–110.

159 Uemura, M. and Kamikawa, K. (1994) *Chem. Commun.*, 2697–2698.

160 Uemura, M., Nishimura, H., Kamikawa, K., Nakayama, K., and Hayashi, Y. (1994) *Tetrahedron Lett.*, **35**, 1909–1912.

161 Watanabe, T., Kamikawa, K., and Uemura, M. (1995) *Tetrahedron Lett.*, **36**, 6695–6698.

162 Kamikawa, K., Watanabe, T., and Uemura, M. (1996) *J. Org. Chem.*, **61**, 1375–1384.

163 Kamikawa, K. and Uemura, M. (2000) *Synlett*, 938–947.

164 Bromley, L.A., Davies, S.G., and Goodfellow, G.L. (1991) *Tetrahedron Asymmetry*, **2**, 139–156.

165 Jaouen, G. and Meyer, A. (1975) *J. Am. Chem. Soc.*, **97**, 4667–4672.

166 Mandelbaum, A., Zeuwirth, Z., and Cais, M. (1963) *Inorg. Chem.*, **2**, 902–903.

167 Solladie-Cavallo, A., Solladie, G., and Tsamo, E. (1979) *J. Org. Chem.*, **44**, 4189–4191.

168 Glorius, F. (2008) *Angew. Chem. Int. Ed.*, **47**, 8347–8349.

169 Saito, B. and Fu, G.C. (2008) *J. Am. Chem. Soc.*, **130**, 6694–6695.

170 Imao, D., Glasspoole, B.W., Laberge, V.S., and Crudden, C.M. (2009) *J. Am. Chem. Soc.*, **131**, 5024–5025.
171 Ishikawa, S. and Manabe, K. (2006) *Chem. Commun.*, 2589–2591.
172 Manabe, K. and Ishikawa, S. (2008) *Chem. Commun.*, 3829–3838.
173 Wang, C. and Glorius, F. (2009) *Angew. Chem. Int. Ed.*, **48**, 5240–5244.
174 Cheng, W. and Snieckus, V. (1987) *Tetrahedron Lett.*, **28**, 5097–5098.
175 Ernst, J.T., Kutzki, O., Debnath, A.K., Jiang, S., Lu, H., and Hamilton, A.D. (2002) *Angew. Chem. Int. Ed.*, **41**, 278–281.
176 Hensel, V. and Schlüter, A.-D. (1999) *Chem. Eur. J.*, **5**, 421–429.
177 Hensel, V., Lützow, K., Jacob, J., Gessler, K., Saenger, W., and Schlüter, A.-D. (1997) *Angew. Chem. Int. Ed.*, **36**, 2654–2656.
178 Read, M.W., Escobedo, J.O., Willis, D.M., Beck, P.A., and Strongin, R.M. (2000) *Org. Lett.*, **2**, 3201–3204.
179 Galda, P. and Rehahn, M. (1995) *Synthesis*, 614–615.
180 Blake, A.J., Cooke, P.A., Doyle, K.J., Gair, S., and Simpkins, N.S. (1998) *Tetrahedron Lett.*, **39**, 9093–9096.
181 Tobisu, M., Shimasaki, T., and Chatani, N. (2008) *Angew. Chem. Int. Ed.*, **47**, 4866–4869.
182 Nishikata, T. and Lipshutz, B.H. (2009) *J. Am. Chem. Soc.*, **131**, 12103–12105.
183 Guan, B.-T., Wang, Y., Li, B.-J., Yu, D.-G., and Shi, Z.-J. (2008) *J. Am. Chem. Soc.*, **130**, 14468–14470.
184 Quasdorf, K.W., Tian, X., and Garg, N.K. (2008) *J. Am. Chem. Soc.*, **130**, 14422–14423.
185 Quasdorf, K.W., Riener, M., Petrova, K.V., and Garg, N.K. (2009) *J. Am. Chem. Soc.*, **131**, 17748–17749.
186 Gooßen, L.J., Gooßen, K., and Stanciu, C. (2009) *Angew. Chem. Int. Ed.*, **48**, 3569–3571.
187 Macklin, T.K. and Snieckus, V. (2005) *Org. Lett.*, **7**, 2519–2522.
188 Snieckus, V. (1990) *Chem. Rev.*, **90**, 879–933.
189 Molander, G.A. and Sandrock, D.L. (2008) *J. Am. Chem. Soc.*, **130**, 15792–15793.
190 Iwadate, N. and Suginome, M. (2009) *J. Organomet. Chem.*, **694**, 1713–1717.
191 Iwadate, N. and Suginome, M. (2009) *Org. Lett.*, **11**, 1899–1902.
192 Noguchi, H., Hojo, K., and Suginome, M. (2007) *J. Am. Chem. Soc.*, **129**, 758–759.
193 Noguchi, H., Shioda, T., Chou, C.-M., and Suginome, M. (2008) *Org. Lett.*, **10**, 377–380.
194 Gillis, E.P. and Burke, M.D. (2007) *J. Am. Chem. Soc.*, **129**, 6716–6717.
195 Gillis, E.P. and Burke, M.D. (2008) *J. Am. Chem. Soc.*, **130**, 14084–10485.
196 Lee, S.J., Gray, K.C., Paek, J.S., and Burke, M.D. (2008) *J. Am. Chem. Soc.*, **130**, 466–468.
197 Uno, B.E., Gillis, E.P., and Burke, M.D. (2009) *Tetrahedron*, **65**, 3130–3138.
198 Gillis, E.P. and Burke, M.D. (2009) *Aldrichim. Acta*, **42**, 17–27.
199 Tobisu, M. and Chatani, N. (2009) *Angew. Chem. Int. Ed.*, **48**, 3565–3568.
200 Iwadate, N. and Suginome, M. (2010) *J. Am. Chem. Soc.*, **132**, 2548–2549.

# 5
# Rhodium- and Palladium-Catalyzed Asymmetric Conjugate Additions of Organoboronic Acids

*Guillaume Berthon-Gelloz and Tamio Hayashi*

## 5.1
## Introduction

Since the seminal report by Uemura [1] in 1995 for palladium and by Miyaura in 1997 for rhodium [2], late-transition metal-catalyzed conjugate addition of organoboron reagents to activated alkenes has emerged as one of the most functional group-tolerant and reliable carbon–carbon bond forming processes. This methodology has been applied in several large-scale processes and has been extensively used as a testing ground for new ligand design. In this chapter, we will provide an overview of the recent advances made in this type of transformation spanning from 2005 to 2009. This chapter is not comprehensive and will cover only selected examples. For more in-depth and comprehensive accounts, we direct the reader to refer to the previous edition of this book and to a number of excellent reviews on this subject [3–14].

## 5.2
## Rh-Catalyzed Enantioselective Conjugate Addition of Organoboron Reagents

In this section, we will describe the state of the art for the rhodium-catalyzed enantioselective conjugate addition of organoboron reagent to activated olefins. There will be a special emphasis on α,β-unsaturated ketones since this substrate class has received the most attention and has been thoroughly investigated with a plethora of different ligand systems. Many of the findings described for α,β-unsaturated ketones are also applicable to other olefin classes and other nucleophilic organometallic reagents.

*Boronic Acids: Preparation and Applications in Organic Synthesis, Medicine and Materials*, Second Edition.
Edited by Dennis G. Hall.
© 2011 Wiley-VCH Verlag GmbH & Co. KGaA. Published 2011 by Wiley-VCH Verlag GmbH & Co. KGaA.

## 5.2.1
### α,β-Unsaturated Unsaturated Ketones

#### 5.2.1.1 A Short History

The first example of conjugate addition of an arylboronic acid to an enone catalyzed by transition metal complexes can be traced back to a study from 1995 by Uemura and coworkers [1]. This reaction was carried out in the absence of ligand and with high catalyst loading (10 mol% of $Pd(OAc)_2$). The interest in this reaction remained limited until Miyaura described in 1997 that the $[Rh(acac)(CO)_2]$/dppb (acac = acetylacetonato; dppb = 1,4-bis(diphenylphosphino)butane) system efficiently catalyzes the conjugate addition of a wide range of aryl- and alkenylboronic acids to methylvinylketone in high yields and also to β-substituted enones including 2-cyclohexenone [2]. The hallmarks of this reaction are (1) no competitive uncatalyzed reaction of the organoboronic acids onto the enone; (2) no 1,2-addition of the organoboron reagent; (3) a large functional group tolerance that is in contrast to the addition of organolithium and Grignard reagents; and (4) mild reaction conditions. A real breakthrough in this methodology came in 1998 when Hayashi and Miyaura described the first example of a rhodium-catalyzed enantioselective conjugate addition [15]. For the first time, a wide range of aryl and alkenyl fragments could be added in high yields and with exquisite enantioselectivity to α,β-unsaturated ketones using (S)-binap (**L1**) as the chiral diphosphine ligand [15]. Since this initial study, there has also been great progress in the copper-catalyzed enantioselective conjugate addition using Grignard and organozinc reagents [16–19].

#### 5.2.1.2 Mechanism

In 2002, Hayashi and coworkers established the detailed mechanistic cycle for the rhodium-catalyzed enantioselective conjugate addition. The proposed catalytic cycle depicted in Scheme 5.1 goes through three identifiable intermediates, the hydroxyrhodium **A**, the phenylrhodium **B**, and the oxa-π-allylrhodium (rhodium-enolate) **C** complexes. These intermediates are related to the cycle as follows: the reaction is initiated through the transmetalation of a phenyl group from boron to hydroxyrhodium

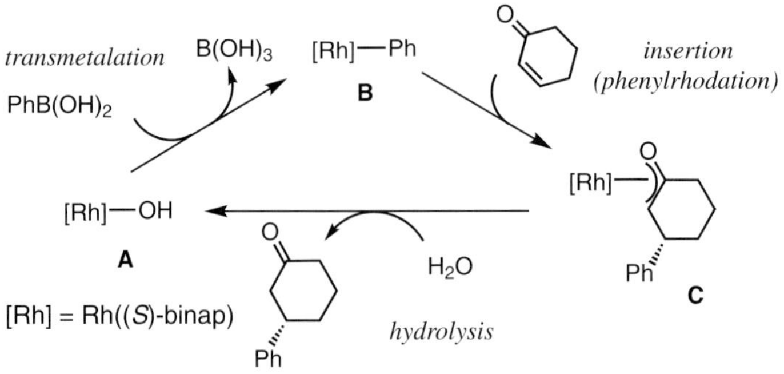

**Scheme 5.1** Accepted mechanism for the rhodium-catalyzed conjugate addition.

A to generate the phenylrhodium **B**. The enone will subsequently insert into Rh—Ph bond of **B** to form the oxa-π-allylrhodium **C**. The rhodium enolate **C** is unstable under protic conditions and will be readily hydrolyzed to regenerate **A** and liberate the enantioselective conjugate addition product. It is important to note that throughout the catalytic cycle, rhodium remains at a constant oxidation state of $+I$ [20]. This catalytic cycle was validated through the observation of the intermediates in stoichiometric NMR experiments [20].

Following this initial mechanistic study, Hayashi and coworkers performed a detailed kinetic study on the catalytic cycle of Rh-catalyzed enantioselective conjugate addition using the reaction calorimetry methodology and analysis developed by Blackmond [21, 22]. The key findings of this study are that the transmetalation step from boron to rhodium is rate determining and that most of the rhodium lies outside the catalytic cycle in the form of the [Rh(μ-OH)(R)-binap]$_2$ dimer. Using the same methodology, Hayashi and coworkers performed the kinetic analysis of the reaction catalyzed by [Rh(μ-OH)(cod)]$_2$ [23]. Under identical conditions, the rate of the reaction with [Rh(μ-OH)(cod)]$_2$ was 20 times faster than with [Rh(μ-OH)(R)-binap]$_2$. The remarkably large catalytic activity of rhodium–diene complexes can be attributed to both the higher rate of transmetalation with [Rh(μ-OH)(cod)] than with [Rh(μ-OH)(binap)] (**A**) and the fact that more of the Rh lies within the cycle. Other strongly π-accepting ligands, such as phosphoramidites, also have a beneficial effect on the reaction rate [24].

The rate-determining transmetalation from boron to rhodium or palladium under the conditions of conjugate addition is thought to occur through a metal hydroxy complex **A** that can coordinate to highly oxophilic organoboronic acid to give intermediate **D**, which can subsequently deliver the organic fragment to rhodium in an intramolecular fashion to furnish the aryl–rhodium species **B** (Scheme 5.2) [13, 25].

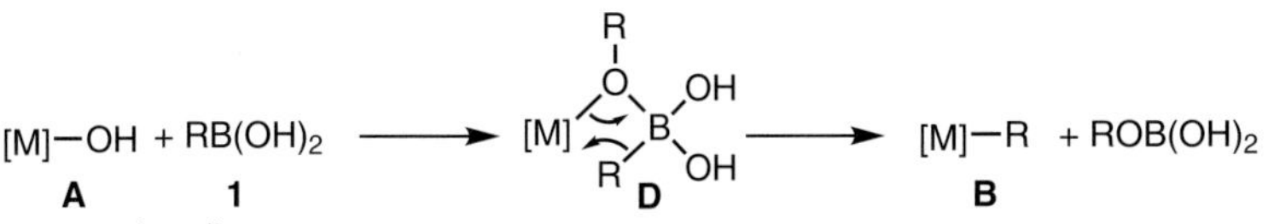

[M]—OH + RB(OH)$_2$ ⟶ [M]—D ⟶ [M]—R + ROB(OH)$_2$
**A**     **1**                                    **B**

[M] = Rh$^I$, Pd$^{II}$, R = H, Me

**Scheme 5.2** Proposed mechanism for the transmetalation of organoboronic acids.

Direct evidence of this mechanism was given by Hartwig and coworkers who showed that a boronic acid reacts cleanly with the hydroxy dimer **2**, **3** or the Rh-enolate **4** to give complex **5** [26]. Upon heating, **5** rearranges to form a Rh–aryl species **6** bond and extrudes an insoluble boroxine oligomer (Scheme 5.3) [26]. Although this process occurs under neutral conditions, it is greatly accelerated by the presence of stoichiometric amounts of base. This observation is rationalized by quaternization of the arylboronic acid, which facilitates the rupture of the B—C$_{sp2}$ bond [13, 27].

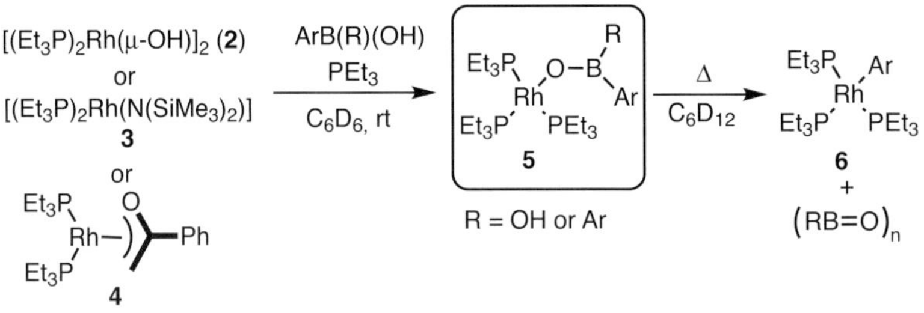

[(Et$_3$P)$_2$Rh(μ-OH)]$_2$ (**2**)
or
[(Et$_3$P)$_2$Rh(N(SiMe$_3$)$_2$)]
**3**
or

Et$_3$P–Rh–O–⟩–Ph  (Et$_3$P)  **4**

ArB(R)(OH) / PEt$_3$ / C$_6$D$_6$, rt →

Et$_3$P\\Rh/O–B(R)(Ar) /Et$_3$P/PEt$_3$  **5**

R = OH or Ar

—Δ→ C$_6$D$_{12}$

Et$_3$P\\Rh/Ar /Et$_3$P/PEt$_3$  **6**
+
(RB=O)$_n$

**Scheme 5.3** Direct observation of the transmetalation from boron to rhodium.

### 5.2.1.3 Model for Enantioselection

The proposed stereochemical pathway for the conjugate addition to *cis*- and *trans*-activated olefins catalyzed by rhodium complex coordinated with (*S*)-binap is depicted (Scheme 5.4) [15]. According to the highly skewed structure known for transition metal complexes coordinated with a binap ligand [28], the [Rh(Ar)(*S*)-binap)] intermediate has an open space at the lower part of the vacant coordination site, the upper part being blocked by one of the phenyl rings of the binap ligand. The *trans*-activated alkene coordinates to [Rh(Ar)(*S*)-binap)] to form **E**, while coordination of the *cis*-alkene leads to **E'**. In both cases, the coordination on the opposite enantiotopic face leads to unfavorable steric interactions. After insertion into the Rh–Ar bond (**E** and **E'**) and protonolysis, **F** and **F'** lead to opposite enantiomers. Hence, the geometry of the double bond plays a determining role in the stereochemical outcome and great care must be taken to avoid mixture of geometrical isomers in the starting alkenes. For cyclic enones, the observed stereochemical outcome is the same as for *cis*-alkenes.

This stereochemical model can be extended to a wide range of $C_2$ symmetric bidentate ligands by considering how the ligand, when it is coordinated to rhodium, is capable of bisecting the space around the rhodium into a quadrant and which enantiotopic face of the alkene will minimize steric interaction upon coordination to Rh (Figure 5.1) [29].

### 5.2.1.4 Organoboron Sources Other Than Boronic Acids

Although organoboronic acids (**1**) are the most practical and widespread source of organoboron reagents for Rh-catalyzed conjugate additions, other derivatives have proven to be equally effective (Figure 5.2). In organic solvents, boronic acids are in equilibrium with oligomeric species of various degrees of hydration. Complete dehydration of organoboronic acid leads to well-defined cyclic organoboroxine (**7**). Boroxines are readily hydrolyzed back to the corresponding boronic acid with one equivalent of water relative to boron under basic aqueous conditions [30]. Organoboroxines (**7**) have become the preferred reagents for Rh-catalyzed conjugate additions because they enable the use of a precise stoichiometry of organoboron reagent and are more stable toward protodeboronation than boronic acids especially at elevated temperatures (about 100 °C) [30]. Boronate esters (**8**) can be used in

## 5.2 Rh-Catalyzed Enantioselective Conjugate Addition of Organoboron Reagents | 267

$Y = C(O)R^2, C(O)OR^2, C(O)N(R^2)_2, P(O)(OR^2)_2, NO_2$

**Scheme 5.4** Stereocontrol model for the enantioselective conjugate addition of *trans*- and *cis*-olefins.

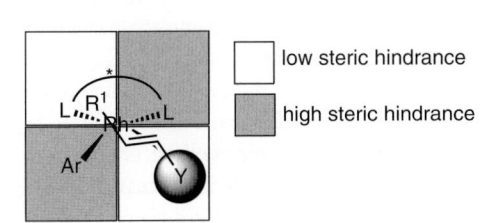

**Figure 5.1** Generalized model for the stereochemical outcome of an enantioselective conjugate addition.

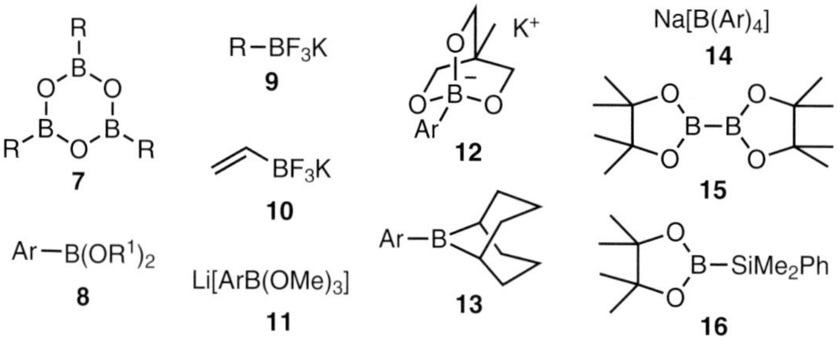

**Figure 5.2** Organoboron reagents competent in Rh-catalyzed enantioselective conjugate additions.

Rh-catalyzed enantioselective conjugate additions [31, 32]. The rate of the enantioselective conjugate addition reaction of these boronate esters is directly related to the ease of their hydrolysis back to the corresponding boronic acid [31]. This feature can be advantageous in the one-pot alkyne hydroboration with catecholborane followed by Rh-catalyzed enantioselective conjugate addition [32].

Potassium organotrifluoroborate salts (**9**) have become a very popular source of organoboron reagents [33–35] because they tend to be more stable than the corresponding boronic acids while still being reactive in Rh-catalyzed enantioselective conjugate additions [36]. One particularly useful reagent is the potassium vinyltrifluoroborate (**10**), which enables the introduction of a vinyl group in excellent yields with high enantioselectivities and does not readily polymerize [37]. It is important to note that potassium organotrifluoroborates (**9**) do not transmetalate directly to rhodium(I), but a monohydroxyborate (ArBF$_2$(OH)$^-$) is probably the boron species that effects the transmetalation step, with a mechanism akin to the one depicted in Scheme 5.2 [38–42].

Lithium trimethylarylborate (**11**) is also a very active reagent for the enantioselective conjugate addition, but it is relatively unstable and is best formed *in situ* [31, 43]. Cyclic aryl triolborates (**12**) are also a convenient and reactive class of reagents for Rh-catalyzed enantioselective conjugate additions [44, 45]. These reagents have the advantage of being very stable in air and water and being more soluble in organic solvents than related potassium organotrifluoroborates. The reactive ArB(9-BBN) (**13**) derivatives can be used in enantioselective conjugate addition reactions in aprotic solvent and in the absence of base to yield a stable rhodium enolate that can be further reacted with an electrophile to yield a ketone with a high diastereoselectivity [46]. Very recently, Shintani and Hayashi disclosed the use of sodium tetraarylborates (**14**) as aryl transfer reagents in the rhodium-catalyzed addition to β,β-disubstituted α,β-unsaturated ketones. This remarkable method enables the creation of very challenging chiral quaternary carbons, with high enantioselectivities [47]. Reagents such as bis(pinacolato)diboron (**15**) [48] and dimethylphenylsilylpinacolatoboron (**16**) [49, 50] have also been used to introduce a boron and silyl moiety in Rh-catalyzed addition reactions.

### 5.2.1.5 Rh Precatalysts

The nature of the Rh precatalyst used in the enantioselective conjugate addition reaction is of crucial importance. The precursor must enable rapid exchange of ligands with free chiral ligand to form quantitatively the enantioselective catalytic species. Thus, when [Rh(acac)(C$_2$H$_4$)$_2$] is used in conjunction with a chiral bidentate ligand, high enantioselectivities are observed, while when [Rh(acac)(CO)$_2$] is employed lower selectivities are obtained because of strong binding of CO [15]. Although acac containing precursors were initially favored, it was later observed that the free acac ligand generated can bind again to rhodium and inhibit the reaction. Therefore, acac-free [Rh(Cl)(C$_2$H$_4$)$_2$]$_2$ is the precatalyst of choice. Cod containing rhodium precursor precursors such as [Rh(Cl)(cod)]$_2$ and [Rh(μ-OH)(cod)]$_2$ should be avoided because they are more catalytic than Rh-phosphine complexes [27, 51]. Only trace amounts of these complexes (due to incomplete ligand exchange) will suffice to significantly lower the observed enantioselectivity of a conjugate addition. However, complexes bearing the 2,5-norbornadiene (nbd) as a diene ligand (e.g., [Rh(μ-OH)(nbd)]$_2$,) are poor catalysts for the reaction.

Chiral cationic rhodium complexes, formed *in situ* by reaction of [Rh(cod)$_2$]$^+$BF$_4^-$, [Rh(cod)(MeCN)$_2$]$^+$BF$_4^-$ [52–59], or [Rh(nbd)$_2$]$^+$BF$_4^-$ [27, 60] with a chiral ligand, are also active rhodium precatalysts [61]. Under the basic aqueous conditions used for Rh-catalyzed enantioselective conjugate additions, the cationic precursors are presumably converted *in situ* into the neutral Rh–OH species bearing chiral ligands. Because of the low intrinsic activity of [Rh(μ-OH)(nbd)]$_2$, the [Rh(nbd)$_2$]$^+$BF$_4^-$ complex is also a preferred rhodium precursor. One practical advantage of using cationic rhodium precursors is that they enable more robust reaction conditions and lead more consistently to higher enantioselectivities when the catalyst is generated *in situ* with a chiral ligand than with the corresponding neutral Rh precursor [62]. This is presumably due to the faster exchange of diene for chiral ligands on cationic rhodium compared to the neutral precursors [62]. Furthermore, with cationic rhodium precursors, Et$_3$N can be used instead of KOH as the activator, thus making the enantioselective conjugate addition protocol more functional group tolerant [63].

Hayashi and coworkers found that traces of phenol (0.05 ± 0.02 mol%) present in commercial phenylboronic acid can significantly deactivate chiral diene rhodium catalysts [64]. This deactivation pathway becomes prevalent under low-catalyst loading conditions (below 0.05 mol%). The phenol impurity can be removed by dehydration of the boronic acid to the boroxine (**7**) followed by washing with hexanes. These findings are probably applicable to other Rh systems for enantioselective conjugate additions that are used at low catalyst loading.

### 5.2.1.6 Ligand Systems

In this section, we will present an overview of the different ligand designs and concepts that have been applied to the rhodium-catalyzed enantioselective conjugate additions of organoboronic acids to α,β-unsaturated ketones. Very early on, the enantioselective conjugate addition of phenylboronic acid (**17**) to 2-cyclohexenone (**18**) was chosen as a model reaction for Rh-catalyzed enantioselective conjugate additions. The wealth of studies using this model reaction enables the direct comparison of a wide gamut of ligand structures. To facilitate the comparison, we have grouped the ligands by families (i.e., phosphorus-based bidentate, monodentate, mixed ligands, and others). When a family of ligands is prepared, only the best performing ligand would be discussed. One must keep in mind the limitation of a comparison on a fixed model reaction, as some ligands might be better suited for specific substrates. When possible, this will be highlighted.

### 5.2.1.6.1 Bidentate Phosphorus Ligands

Following the initial breakthrough for Rh-catalyzed enantioselective conjugate addition, using binap as a ligand, there has been a flurry of studies employing bidentate phosphorus ligands. Figures 5.3 and 5.4 summarize these results. Diop (**L2**) [65] and chiraphos (**L3**) [65] gave low enantioselectivities. The diphonane ligand **L4** bears an interesting backbone and gives good selectivities [66]. Binol-based bisphosphonites **L5** and **L6** performed well in Rh-catalyzed enantioselective conjugate addition; interestingly, depending on the carbon chain length separating the phosphonites in **L7a** and **L7b**, the enantioselectivity is reversed [67]. Similarly, binol-based bisphosphoramidite **L11** linked together in the 3-position gave excellent results [63, 68]. The water-soluble binap-based ligand **L8** catalyzed the enantioselective conjugate addition in aqueous media with a TON of 13 200 [69]. The observation that π-accepting ligands accelerate the rate-determining transmetalation in Rh-catalyzed enantioselective conjugate additions was confirmed by the use of π-accepting ligands **L10a** and **L10b** that displayed higher catalytic activity than the corresponding MeO-biphep [70]. In general, axially chiral ligands such as **L9** [71], (S)-MeO-biphep [70], **L10** [70], **L12** [72], **L13** [72], and polystyrene-supported binap **L14** [73] all produce excellent enantioselectivities on par with binap. Substitution in remote positions (not 3 and 3′) of the binaphthyl backbone of binap has little influence on the stereochemical outcome [74, 75]. The diphosphine ligand **L15** [76], **L16** [77], and **L17** [65] bearing planar chirality were investigated; however, only Re-based **L15** gave high enantiomeric excesses.

### 5.2.1.6.2 Monodentate Ligands

Although monodentate chiral ligands (P-stereogenic) were the first class of ligands used in asymmetric homogeneous catalysis, they have since been replaced by rigid bidentate ligands (Figure 5.5) [78]. Early attempts to use the monodentate (R)-MeO-mop (**L18**) as a ligand for Rh-catalyzed enantioselective conjugate addition proved to be disappointing with low conversion and enantioselectivities [65]. The discovery by de Vries and Feringa that binol-based phosphoramidites were highly active and enantioselective in Rh-catalyzed hydrogenation [79] has spurred a renewed interest in monodentate ligands in asymmetric catalysis. A great impetus for this growing trend in homogeneous catalysis is the cheap and rapid synthesis of monodentate ligands over bidentate ones. For the addition of phenylboronic acid to 2-cyclohexenone, $H_8$-binol-based phosphoramidite (S)-**L19** proved to be the most efficient [80]. Phosphoramidite ligands have also been used for the addition of potassium organotrifluoroborates [37]. In an early and elegant report of a chiral monodentate N-heterocyclic carbene (NHC), Andrus and coworkers demonstrated that the cyclophane-based

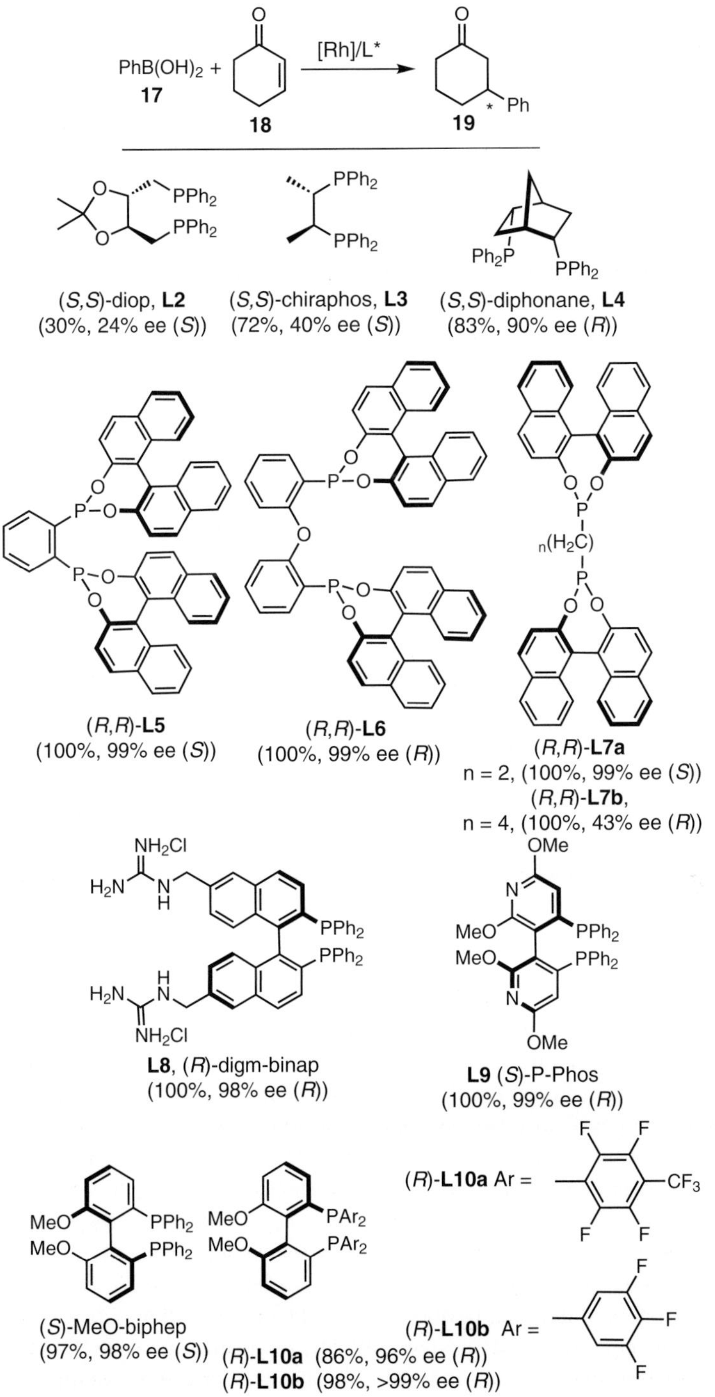

PhB(OH)$_2$ + [structure 18] $\xrightarrow{\text{[Rh]/L*}}$ [structure 19]
17

(S,S)-diop, **L2**
(30%, 24% ee (S))

(S,S)-chiraphos, **L3**
(72%, 40% ee (S))

(S,S)-diphonane, **L4**
(83%, 90% ee (R))

(R,R)-**L5**
(100%, 99% ee (S))

(R,R)-**L6**
(100%, 99% ee (R))

(R,R)-**L7a**
n = 2, (100%, 99% ee (S))
(R,R)-**L7b**,
n = 4, (100%, 43% ee (R))

**L8**, (R)-digm-binap
(100%, 98% ee (R))

**L9** (S)-P-Phos
(100%, 99% ee (R))

(R)-**L10a** Ar = [pentafluorophenyl-CF$_3$ group]

(S)-MeO-biphep
(97%, 98% ee (S))

(R)-**L10a** (86%, 96% ee (R))
(R)-**L10b** (98%, >99% ee (R))

(R)-**L10b** Ar = [trifluorophenyl group]

**Figure 5.3** Enantioselective conjugate addition catalyzed by C$_2$ symmetric bidentate ligand rhodium complexes.

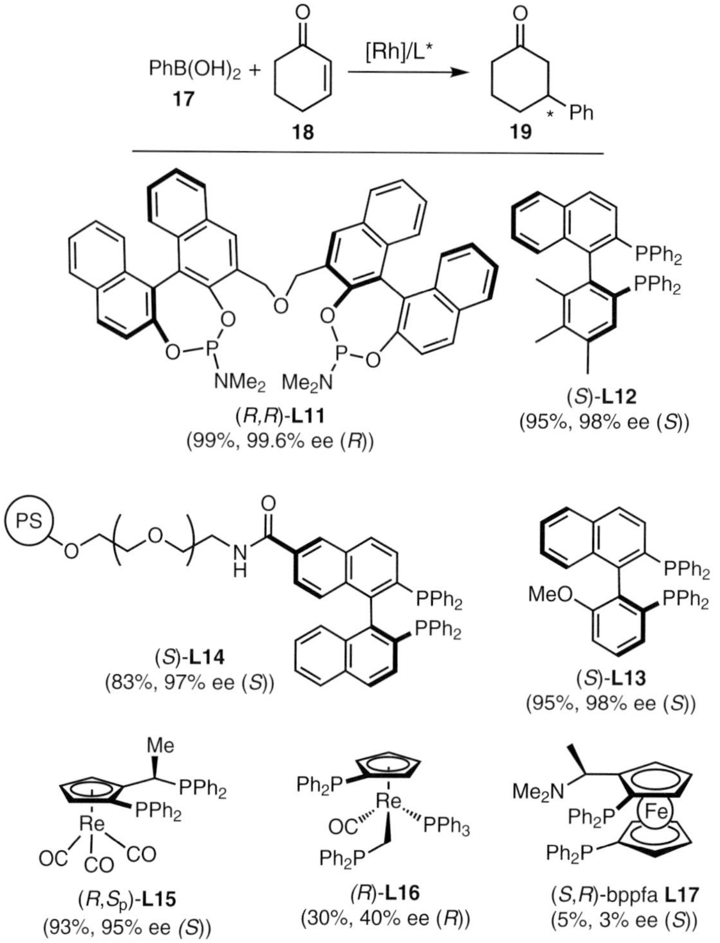

PhB(OH)$_2$ + **17**, **18** → **19** (cyclohexanone with Ph) [Rh]/L*

(R,R)-**L11** (99%, 99.6% ee (R))

(S)-**L12** (95%, 98% ee (S))

(S)-**L14** (83%, 97% ee (S))

(S)-**L13** (95%, 98% ee (S))

(R,S$_p$)-**L15** (93%, 95% ee (S))

(R)-**L16** (30%, 40% ee (R))

(S,R)-bppfa **L17** (5%, 3% ee (S))

**Figure 5.4** Enantioselective conjugate addition catalyzed by bidentate ligand rhodium complexes.

NHC (**L20**) was very effective for Rh-catalyzed enantioselective conjugate addition [81, 82]. Importantly, only one equivalent of chiral NHC relative to rhodium was necessary to obtain high selectivity; this is in stark contrast to other monodentate ligands, which require two equivalents per rhodium.

The use of inexpensive methyl deoxycholic ester **21** as the source of chirality in phosphite **L21** proved efficient [82, 83]. The deoxycholic moiety induces only one conformation in the *tropos* biphenyl backbone. This approach alleviates the need to use binol as the source of chirality. When the deoxycholic moiety is paired with each enantiomer of binol, only diastereoisomer **L22** was catalytically active, demonstrating the usefulness of having a flexible *tropos* backbone. In addition, depending on the molar ratio of phosphite **L21** relative to Rh, the reactivity and selectivity of the addition could be modulated. With only one equivalent of phosphite per Rh, the major product was **19**, while with two equivalents of phosphite **L21**, **19** underwent a

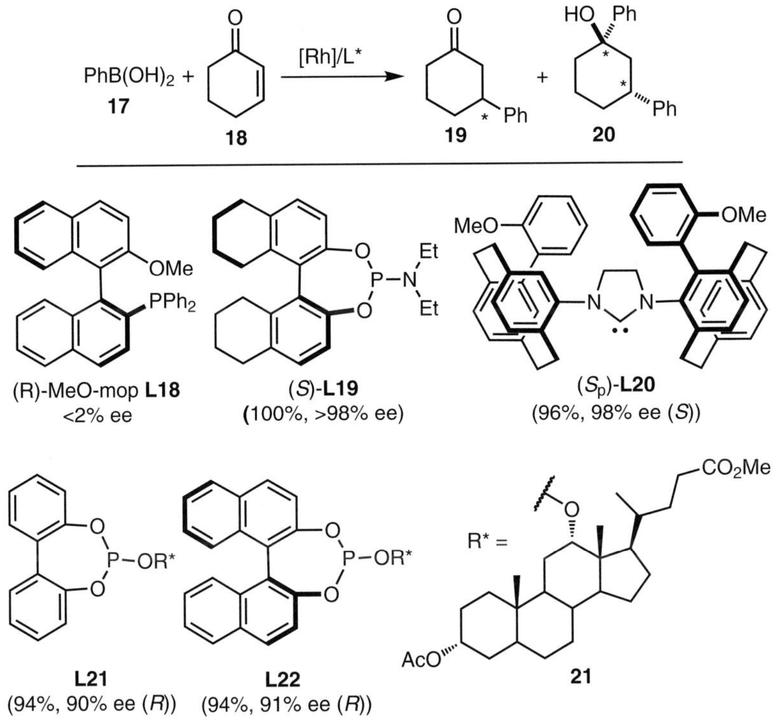

## 5.2 Rh-Catalyzed Enantioselective Conjugate Addition of Organoboron Reagents

PhB(OH)$_2$ + **17** + cyclohexenone **18** →[Rh]/L*→ **19** (ketone with *Ph) + **20** (HO, Ph cyclohexane)

**(R)-MeO-mop L18**
<2% ee

**(S)-L19**
(100%, >98% ee)

**(S$_p$)-L20**
(96%, 98% ee (S))

**L21**
(94%, 90% ee (R))

**L22**
(94%, 91% ee (R))

R* = **21** (steroid with CO$_2$Me and AcO substituents)

**Figure 5.5** Enantioselective conjugate addition catalyzed by monodentate ligand rhodium complexes.

diastereoselective 1,2-addition to furnish the bisphenylated **20** product as a single diastereoisomer [83].

The use of monodentate ligands offered the tantalizing possibility to mix different monodentate ligands together to quickly generate combinatorial libraries of complexes. This fascinating approach goes beyond the traditional parallel preparation of modular ligands. Thus, mixtures of monodentate ligands L$^a$ and L$^b$ can upon exposure to a transition metal [M] form not only the two homocombinations [M(L$^a$)$_2$] and [M(L$^b$)$_2$] but also the heterocombination [M(L$^a$)(L$^b$)]. For example, when a 1:1:1 mixture of L$^a$, L$^b$, and metal precursor is used (and no other interaction exists), a statistical mixture of [M(L$^a$)$_2$], [M(L$^b$)$_2$], and [M(L$^a$)(L$^b$)] in a 1:1:2 ratio will be obtained. If the heterocombination is more reactive and selective than the homocombinations, an improved catalyst system is formed without the need to synthesize new ligands. This approach, dubbed combinatorial transition metal catalysis, has been reviewed in an excellent article by Reetz [84]. The first example of this approach applied to Rh-catalyzed enantioselective conjugate addition was described by Feringa with mixtures of phosphoramidites [85]. Each homocombination of **L23** and **L24** performed significantly less well than the heterocombination generated *in situ* [85]. An interesting extension of this approach is to use mixtures of *tropos* monodentate phosphoramidite ligand. In this example as well, the combinatorial mixing of ligands led to the

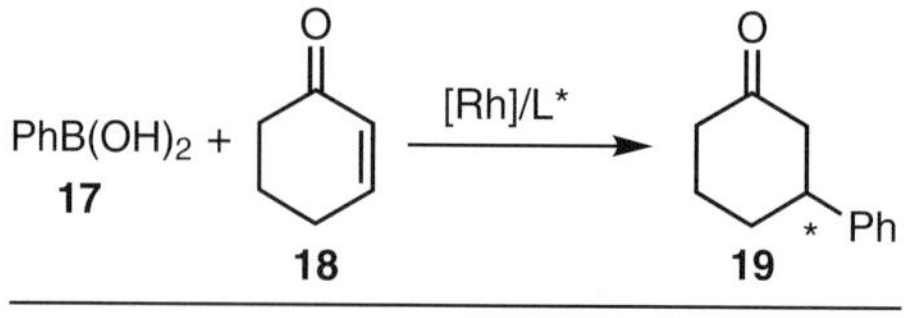

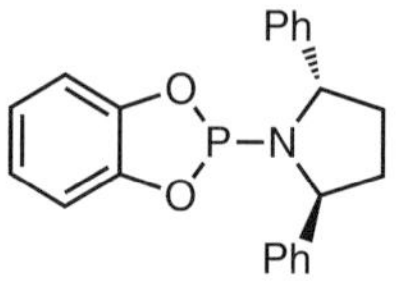

PhB(OH)₂ + **17** 　**18** → [Rh]/L* → **19** (cyclohexanone with *Ph)

(S)-**L23**
(26%, 33% ee (S))

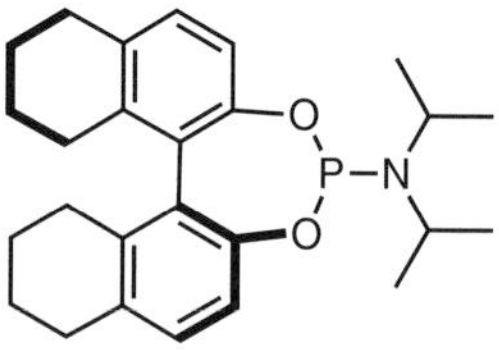

(S)-**L24** (22%, 27% ee (R))

(S)-**L23** + (S)-**L24** (93%, 75% ee (S))

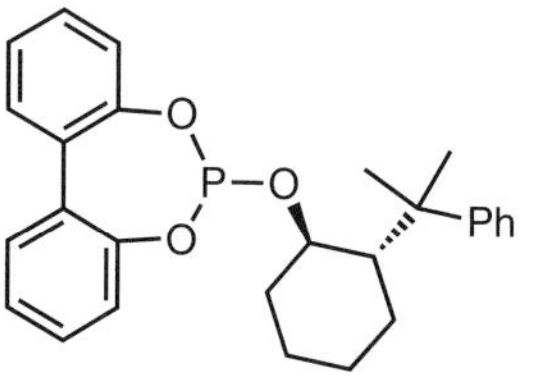

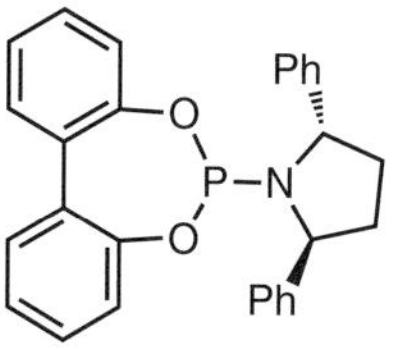

**L25** (100%, 70% ee (R))　　**L26** (100%, 36% ee (R))

**L25** + **L26** (100%, 95% ee (R))

**Figure 5.6** Rh-catalyzed enantioselective conjugate addition, using mixtures of chiral monodentate ligands.

identification of heterocombination **L25** and **L26** that performed significantly better than both homocombinations of ligands (Figure 5.6) [86].

### 5.2.1.6.3 Diene Ligands

The seminal observation by Miyaura and coworkers that [Rh(μ-OH)(cod)₂] is the most active catalyst, with TON of up to 375 000, for rhodium-catalyzed conjugate additions [27, 87] prompted the investigation of optically active dienes as ligands for this transformation. Since the first application of chiral diene as ligands in Rh-catalyzed conjugate addition (**L27**) by Hayashi and coworkers [88], a variety of bicyclic diene scaffolds [89–99] have been successfully applied for this transformation (Figure 5.7). Independently, Carreira and coworkers reported the application of a chiral diene for Ir-catalyzed allylic substitution [100]. These discoveries have spurred intense research efforts in homogeneous catalysis and have been compiled in 2008 in an excellent review [101].

A systematic exploration of different bicyclic scaffolds for chiral dienes revealed that the 2,5-disubstituted bicyclo[2.2.1]heptadiene (**L27**) [88, 102], bicyclo[2.2.2]octadienes (**L28, L29, L30, L31, L33**) [90, 100, 103–108], and bicyclo[3.3.0]octadiene (**L32**) [95, 109] gave high enantiomeric excess over a wide range of substrate. On the other hand, the first-generation 2,6-disubstituted bicyclo[3.3.1]nonadiene (**L34**) [110, 111] and bicyclo[3.3.2]decadiene (**L35**) [111] gave inferior results. Reexamination of the substitution pattern on these dienes revealed that 3,7-disubstituted bicyclo[3.3.1]

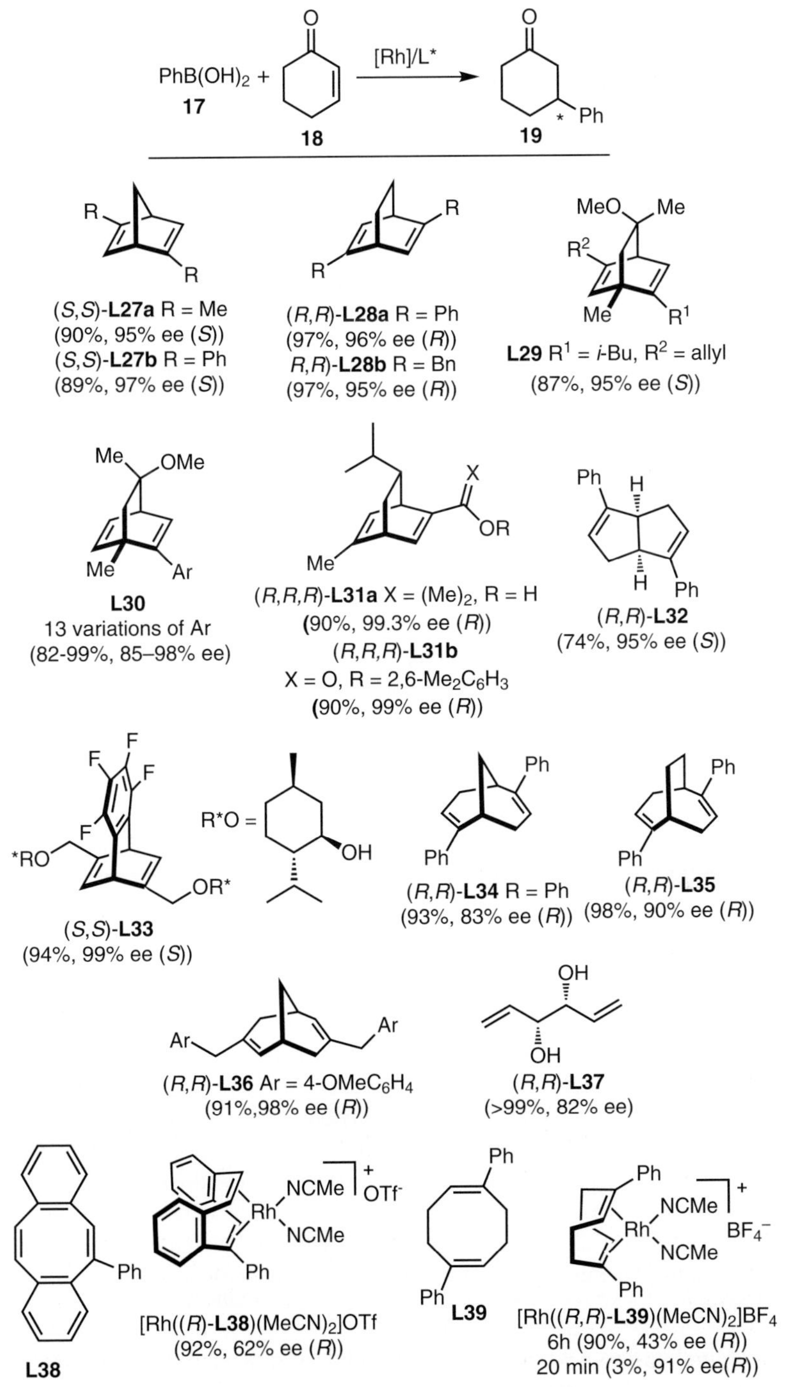

## 5.2 Rh-Catalyzed Enantioselective Conjugate Addition of Organoboron Reagents | 275

PhB(OH)$_2$ + [cyclohexenone] $\xrightarrow{[Rh]/L^*}$ [3-phenylcyclohexanone]
**17**    **18**    **19**

**(S,S)-L27a** R = Me
(90%, 95% ee (S))
**(S,S)-L27b** R = Ph
(89%, 97% ee (S))

**(R,R)-L28a** R = Ph
(97%, 96% ee (R))
**(R,R)-L28b** R = Bn
(97%, 95% ee (R))

**L29** R$^1$ = i-Bu, R$^2$ = allyl
(87%, 95% ee (S))

**L30**
13 variations of Ar
(82–99%, 85–98% ee)

**(R,R,R)-L31a** X = (Me)$_2$, R = H
(90%, 99.3% ee (R))
**(R,R,R)-L31b**
X = O, R = 2,6-Me$_2$C$_6$H$_3$
(90%, 99% ee (R))

**(R,R)-L32**
(74%, 95% ee (S))

**(S,S)-L33**
(94%, 99% ee (S))

**(R,R)-L34** R = Ph
(93%, 83% ee (R))

**(R,R)-L35**
(98%, 90% ee (R))

**(R,R)-L36** Ar = 4-OMeC$_6$H$_4$
(91%, 98% ee (R))

**(R,R)-L37**
(>99%, 82% ee)

**L38**

[Rh((R)-**L38**)(MeCN)$_2$]OTf
(92%, 62% ee (R))

**L39**

[Rh((R,R)-**L39**)(MeCN)$_2$]BF$_4$
6h (90%, 43% ee (R))
20 min (3%, 91% ee (R))

**Figure 5.7** Chiral diene ligands in the Rh-catalyzed addition of phenylboronic acid to 2-cyclohexenone.

nonadienes (**L36**) can, in fact, produce excellent results and greatly surpass previous chiral dienes as ligands for the enantioselective conjugate addition of alkenyl boronic acid to β-silyl α,β-unsaturated ketones [99]. Interestingly, with the chiral bicyclo[2:2:1] heptadiene scaffold, only two methyl groups in ligand **L27a** are sufficient to impart a high enantioselectivity (95% ee) in the enantioselective conjugate addition reaction [94]. It is only when going to a phenyl (**L27b**) substituent that a higher % ee is obtained. Interestingly, disubstituted dienes are not a prerequisite for high activity and enantioselectivity, and just monosubstitution of the bicyclo[2.2.2]octadiene framework with an aryl moiety (**L30**) is necessary to achieve these [42, 98]. A systematic investigation of the steric effects of *mono*substituted chiral dienes **L30** revealed that the *ortho* positions on the aromatic moiety (i.e., 2,6-Me$_2$C$_6$H$_3$) were found to be important to achieve highest enantioselectivities. It is worth noting that dienes **L29-L31** and **L33** are readily available in few synthetic steps. With monosubstituted bicyclo[2:2:1]heptadiene ligands, the order of addition of the acceptor and boronic acid is important because in the absence of an acceptor such strained ligands can undergo undesired carborhodation on the unsubstituted alkene leading to a direct loss of catalytic activity [112]. However, the less strained monosubstituted bicyclo[2:2:2]octadiene does not suffer from this side reaction.

Based on the observation that 1,5-hexadiene could effectively promote Rh-catalyzed conjugate additions, Du and coworkers reported that simple chiral chain dienes such as (3$R$,4$R$)-hexa-1,5-diene-3,4-diol (**L37**) could be efficient chiral ligands for this reaction leading to moderately high % ee of 82% [113]. This result is remarkable considering the flexibility of such ligands, the reactivity of terminal olefins, and the ease of access of such dienes. Ph-dbcot (**L38**) [93] and 1,5-Ph-cod (**L39**) [92] are achiral dienes, but when coordinated to Rh, they become conformationally locked, which leads to a pair of enantiomers that can be subsequently resolved and ultimately the enantiomerically pure cationic rhodium complexes [Rh($R$)-**L38**)(MeCN)$_2$](OTf) and [Rh($R,R$)-**L39**)(MeCN)$_2$](BF$_4$) can be obtained. Complex [Rh($R$)-**L38**)(MeCN)$_2$](OTf) yields a moderate 62% ee [92], while the complex [Rh($R,R$)-**L39**)(MeCN)$_2$](BF$_4$) gave high ee's at low conversion but the enantioselectivity eroded at higher conversion due to the conformational instability of ligand **L39** in the rhodium complex [93].

#### 5.2.1.6.4 Bis-Sulfoxides

Bis-sulfoxides are an emerging class of ligands in homogeneous catalysis [114, 115]. Dorta and coworkers reported the first chiral bis-sulfoxide ligands **L40** [116] and **L41** [117] and found them to be exceptional ligands for the Rh-catalyzed enantioselective conjugate addition of arylboronic acids, giving near-perfect enantioselectivities on a wide range of cyclic α,β-unsaturated ketones (Figure 5.8). The biphenyl bis-sulfoxide **L41** was found to be more active and selective than binaphthyl derivative **L40** [117]. A comparison of the X-ray crystal structures of [RhCl($R$)-binap)]$_2$, [RhCl($S,S$)-**L28a**)]$_2$ [118], and [RhCl($R,R$)-**L40**)]$_2$ [117] indicates that the ligating properties of bis-sulfoxides might lie somewhere between that of diene and bis-arylphosphine ligand.

#### 5.2.1.6.5 Mixed Ligands

A range of chiral ligands bearing a phosphorus center and another coordinating functionality have been investigated in the Rh-catalyzed enantioselective conjugate addition of boronic acid to α,β-unsaturated carbonyl compounds (Figure 5.9). Interestingly, ligand ($S$)-ip-phox family performed poorly

## 5.2 Rh-Catalyzed Enantioselective Conjugate Addition of Organoboron Reagents | 277

**Figure 5.8** Chiral bis-sulfoxide ligands for Rh-cat enantioselective conjugate addition.

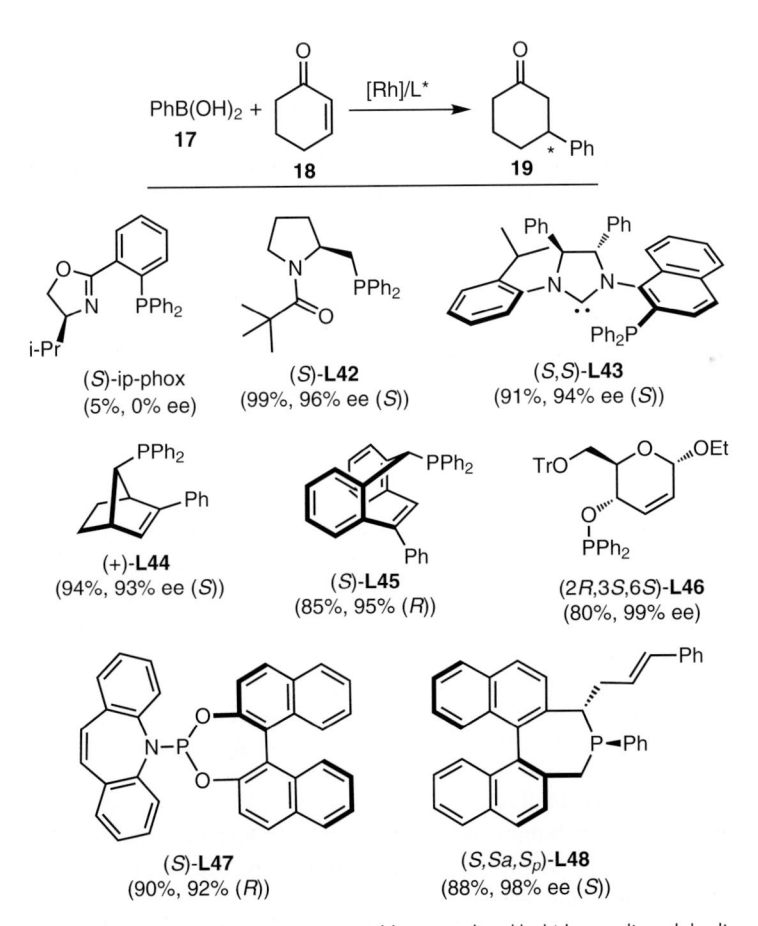

(P,R,R)-p-tol-BINASO **L40**
19 examples
(55–99%, 90–99% ee (S))

(M,S,S)-p-Tol-Me-bipheso **L41**
18 examples
(49–99%, 95–99% ee (R))

(S)-ip-phox
(5%, 0% ee)

(S)-**L42**
(99%, 96% ee (S))

(S,S)-**L43**
(91%, 94% ee (S))

(+)-**L44**
(94%, 93% ee (S))

(S)-**L45**
(85%, 95% (R))

(2R,3S,6S)-**L46**
(80%, 99% ee)

(S)-**L47**
(90%, 92% (R))

(S,Sa,Sp)-**L48**
(88%, 98% ee (S))

**Figure 5.9** Enantioselective conjugate addition catalyzed by bidentate ligand rhodium complexes.

in this transformation [65], while the L-proline-derived amido phosphines (S)-**L42** was found to be very active and selective [119, 120]. This ligand has also been applied in diastereoselective enantioselective conjugate addition processes [121]. The combination of N-heterocyclic carbene and a phosphine moiety **L43** was successful [122]. Based on the observation that phosphines coordinate more strongly to late-transition metals than alkenes, but that the latter provide an effective chiral environment, Shintani and Hayashi synthesized the chiral phosphine-alkene ligand **L44** that proved highly effective for both the enantioselective conjugate addition to enones and maleimides [123]. A kinetic study of the Rh/**L44** system in enantioselective conjugate addition revealed that the catalytic activity with **L44** is intermediate to that of diphosphines and dienes [124]. Good activities and enantioselectivities were obtained with phosphine-alkene ligands **L45** [125], D-glucose-derived **L46** [126], amidophosphine-alkene **L47** [127] (first synthesized by Carreira and coworkers [128]) and with the chiral phosphepine-olefin **L48** [129].

### 5.2.1.7 α,β-Unsaturated Aldehydes

Aldehydes are among the most versatile functional groups in organic chemistry; thus, an enantioselective conjugate addition protocol to generate chiral 3-arylpropanals is highly desirable and can be used in the synthesis of biologically active substances. However, enals represent an especially challenging class of substrates in Rh-catalyzed enantioselective conjugate additions [105, 130, 131]. This can be attributed to the high reactivity of aldehydes, which can undergo competitive 1,2-addition to yield alcohol **25** or after the 1,4-addition to give **24** (Scheme 5.5).

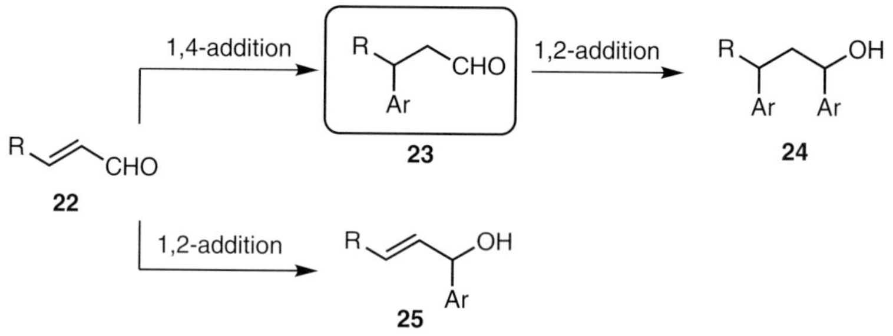

**Scheme 5.5** Reaction pathways in the Rh-catalyzed conjugate addition of arylboronic acids to enals.

The influence of the ligand on the selectivity of the transformation is depicted in Scheme 5.6. While the use of phosphine ligands in the addition of phenylboronic acid to cinnamaldehyde (**26**) leads selectively to the allylic alcohol **28**, Rh/diene-catalyzed processes result in the formation of the desired 1,4-adduct **27** [132].

The use of chiral dienes (**L30f**, **L27b**, and **L28b**) was optimal for the formation of a wide range of enantiomerically enriched 3,3-diarylpropanals and 3-arylalkanals (**23**) (Figure 5.10). Poorer results were obtained with conventional ligands such as (R)-binap (**L1**) or phosphoramidite **L49**.

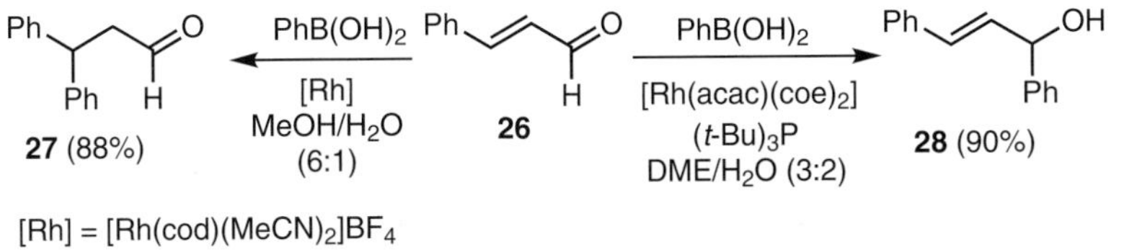

## 5.2 Rh-Catalyzed Enantioselective Conjugate Addition of Organoboron Reagents

**Scheme 5.6** Ligand control of the selectivity for 1,2- or 1,4-addition to enal **31a**.

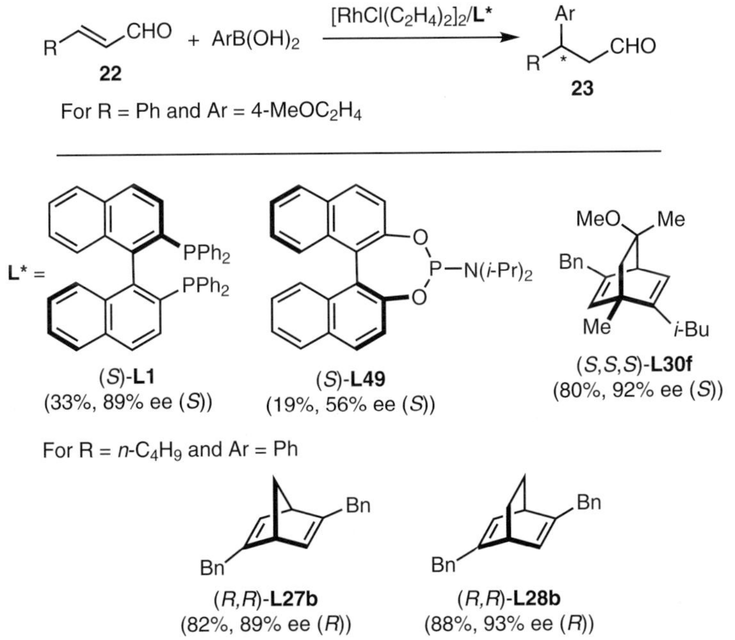

[Rh] = [Rh(cod)(MeCN)$_2$]BF$_4$

Reaction: Ph-CH=CH-CHO (**26**) with PhB(OH)$_2$ gives:
- Ph-CH(Ph)-CH$_2$-CHO **27** (88%) with [Rh], MeOH/H$_2$O (6:1)
- Ph-CH=CH-CH(OH)-Ph **28** (90%) with [Rh(acac)(coe)$_2$], (t-Bu)$_3$P, DME/H$_2$O (3:2)

R-CH=CH-CHO (**22**) + ArB(OH)$_2$ → [RhCl(C$_2$H$_4$)$_2$]$_2$/L* → Ar-CH(R*)-CH$_2$-CHO (**23**)

For R = Ph and Ar = 4-MeOC$_6$H$_4$

L* =

(S)-**L1** (BINAP, PPh$_2$/PPh$_2$)
(33%, 89% ee (S))

(S)-**L49** (binaphthyl phosphoramidite, P–N(i-Pr)$_2$)
(19%, 56% ee (S))

(S,S,S)-**L30f** (MeO, Me, Bn, Me, i-Bu)
(80%, 92% ee (S))

For R = n-C$_4$H$_9$ and Ar = Ph

(R,R)-**L27b** (Bn, Bn)
(82%, 89% ee (R))

(R,R)-**L28b** (Bn, Bn)
(88%, 93% ee (R))

**Figure 5.10** Rh-catalyzed enantioselective conjugate addition of arylboronic acids to enals.

### 5.2.2
### Enantioselective Addition to α,β-Unsaturated Esters and Amides

Cyclic α,β-unsaturated esters react well with arylboronic acids in the enantioselective conjugate addition catalyzed by the Rh/(S)-binap system [31, 61]. However, for linear enoates, the more reactive LiArB(OMe)$_3$ reagent (**11**, generated *in situ*) is necessary to obtain acceptable yields. In addition, the bulkier ester function (R = t-Bu) not only has a positive effect on the enantioselectivity on the process but also lowers the yield of the enantioselective conjugate addition.

The Rh-catalyzed enantioselective conjugate addition on α,β-unsaturated esters has been applied as the key step in the synthesis of a series of biologically active compounds such as baclofen [133] and tolterodine (**31**) (Scheme 5.7, Equation 5.1) [134]. Especially noteworthy is the first reported application of a Rh-catalyzed enantioselec-

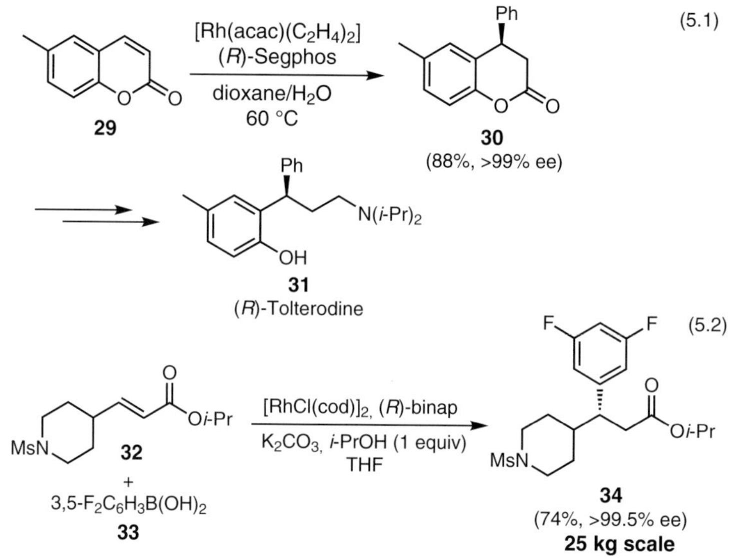

**Scheme 5.7** Applications of the Rh-catalyzed enantioselective conjugate addition of ester in the synthesis of APIs.

tive conjugate addition on a multikilogram scale on **34** (25 kg) (Scheme 5.7, Equation 5.2) [135]. A key result of the multikilogram scale-up of this reaction is the unexpected discovery that the use of a minimal quantity of a 2-propanol (1 equiv.), rather than water as the cosolvent, reduces the extent of rhodium-mediated protodeboronation of the boron species. In addition, potassium carbonate was found to be a useful base.

A cationic rhodium(I)–chiraphos (**L3**) system was developed by Miyaura for the enantioselective preparation of β-diaryl carbonyl compounds (**36**) via the enantioselective conjugate addition of arylboronic acids to β-aryl-α,β-unsaturated ketones or esters (**35**) (Scheme 5.8, Equation 5.3) [60]. As with other systems, the increase in the size of the ester group to $CO_2t$-Bu, led to a slight increase in enantioselectivity.

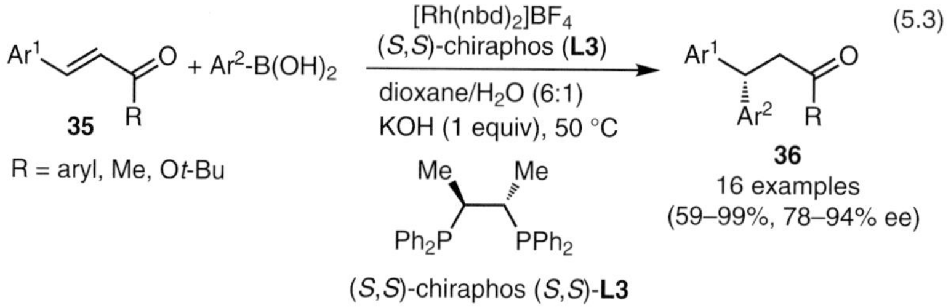

**Scheme 5.8** Rh-catalyzed enantioselective conjugate addition of arylboronic acid on β-aryl α,β-unsaturated esters.

The chiraphos ligand (**L3**) was quite effective for these substrates, while it was a poor ligand for the Rh-catalyzed enantioselective conjugate addition of cyclic enones (cf. Figure 5.3). This system proved quite versatile and functional group tolerant and was successfully applied as the key enantioselective step in the synthesis of two endothelin receptor antagonists **40** and **44** (Scheme 5.9) [136].

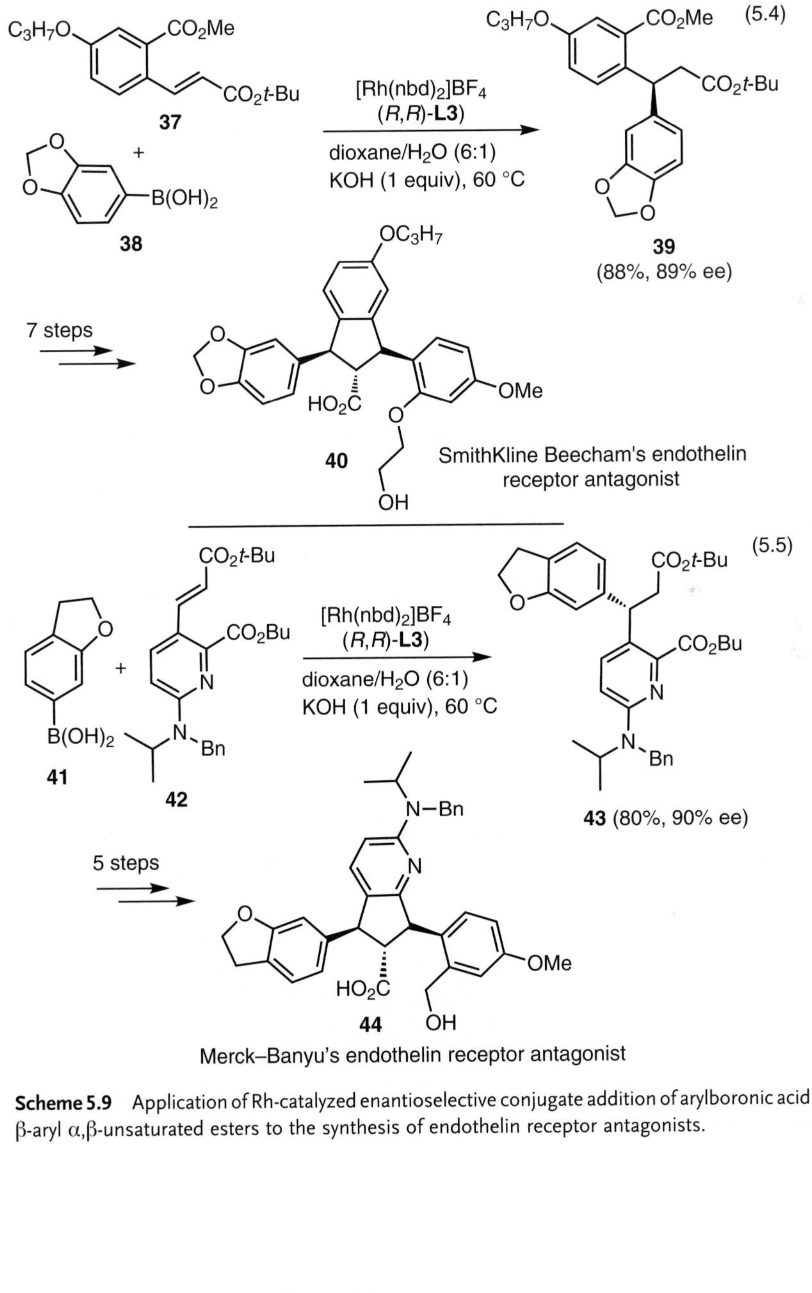

**Scheme 5.9** Application of Rh-catalyzed enantioselective conjugate addition of arylboronic acid on β-aryl α,β-unsaturated esters to the synthesis of endothelin receptor antagonists.

Although less reactive than enones and enoates, linear α,β-unsaturated amides perform similarly well under standard Rh-catalyzed enantioselective conjugate addition conditions [61]. The use of $K_2CO_3$ as a base significantly increased the overall reaction yield. The asymmetric addition to α,β-unsaturated amides has also been applied in total synthesis of hermitamides A and B [137]. The use of boroxines with the Rh/phosphoramidite (**L19**) catalytic system enabled the synthesis of 2-aryl-4-piperidones **46**, which are a useful framework found in a number of active pharmaceutical ingredients (Scheme 5.10) [138]. The use of aryl boroxines (**7**) and slow addition of water were necessary to minimize protodeboronation.

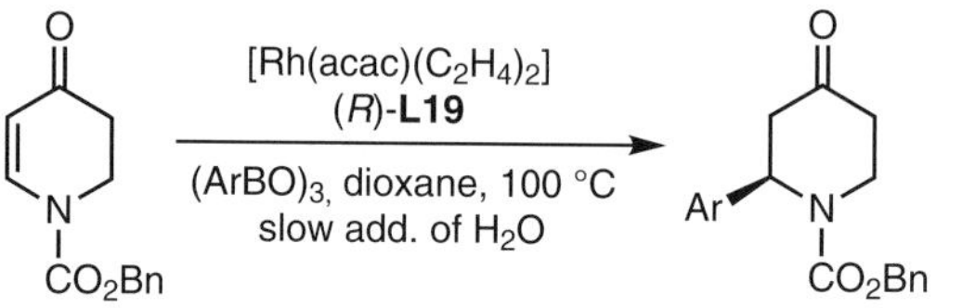

**Scheme 5.10**  Rh-catalyzed enantioselective conjugate addition to 2,3-dihydro-4-pyridones.

### 5.2.2.1 Diastereoselective Conjugate Addition

The presence of a stereogenic center in close proximity to the β-position of a cyclic α,β-unsaturated carbonyl compound can be used to control the diastereoselectivity of a Rh-catalyzed conjugate addition without the need for an extraneous chiral ligand on Rh. The first example of such diastereoselective reaction was reported for the synthesis of C-glycosides from the enantiomerically pure cyclic α,β-unsaturated esters [139]. A variant of this methodology has been applied as a key step in the total synthesis of C-aryl glycoside of natural product *diospongins B* (**49**) (Scheme 5.11, Equation 5.6) [140]. In this example, the conjugate addition to enantiomerically pure **47** yields **48** as one diastereoisomer. Another example is the Rh-catalyzed addition of aryl- and alkenylboronic acids to butenolide **50** leading to the products **51** with high diastereoselectivities (Scheme 5.11, Equation 5.7) [141]. In this regard, the presence of an unprotected hydroxyl group may also provide an enhancement of the diastereocontrol. The conjugate addition of arylboronic acids to unsaturated furano esters **52** also occurs with excellent diastereoselectivity and enables rapid access to trisubstituted furanolignans **53** (Scheme 5.11, Equation 5.8) [142]. In all these examples, the aromatic moiety is delivered on the least hindered face of the cyclic acceptor.

A similar approach was used for the asymmetric synthesis of functionalized pyrrolizidinones **55** (Scheme 5.12) [143]. In this case, chiral diene (*S,S,S*)-**L30f** was used to enhance the diastereoselectivity of the process leading to (*S,R*)-**55**. On the other hand, the use of the other enantiomer of the ligand (*R,R,R*)-**L30f** reversed the diastereoselectivity of the addition to afford (*S,S*)-**55** (Scheme 5.12). Therefore, the process is under ligand control.

## 5.2 Rh-Catalyzed Enantioselective Conjugate Addition of Organoboron Reagents

(5.6)

**47** + PhB(OH)$_2$ → [Rh(cod)$_2$]BF$_4$ (2.5 mol%), KOH (5 mol%), dioxane/H$_2$O, 100 °C, 2 h → **48** (98%, >99.9% de) → 7 steps → **49**

(5.7)

**50** (R = Aryl, Alkenyl) → RB(OH)$_2$, [Rh(Cl(cod)]$_2$, Ba(OH)$_2$, 1,4-dioxane/H$_2$O (10:1) → **51** 6 examples (70–95%, dr = 95:5)

(5.8)

**52** + ArB(OH)$_2$ → [Rh(Cl(cod)]$_2$ (5 mol%), dppb (5 mol%), Ba(OH)$_2$, dioxane/H$_2$O (10:1) → **53** (58%, dr >94:6)

**Scheme 5.11** Diastereoselective Rh-catalyzed conjugate addition.

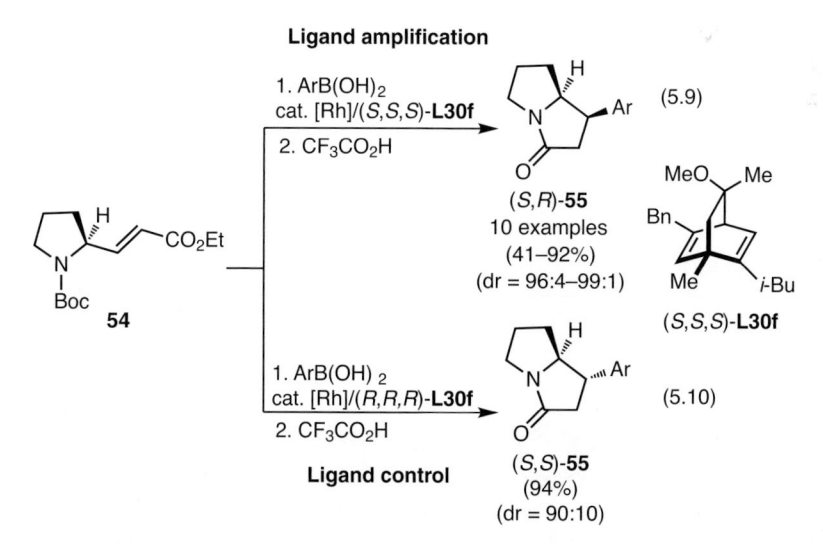

**Ligand amplification**

**54** → 1. ArB(OH)$_2$, cat. [Rh]/(S,S,S)-**L30f**; 2. CF$_3$CO$_2$H → (S,R)-**55** 10 examples (41–92%) (dr = 96:4–99:1) (5.9)

(S,S,S)-**L30f**

**Ligand control**

**54** → 1. ArB(OH)$_2$, cat. [Rh]/(R,R,R)-**L30f**; 2. CF$_3$CO$_2$H → (S,S)-**55** (94%) (dr = 90:10) (5.10)

**Scheme 5.12** Asymmetric synthesis of functionalized pyrrolizidinones using [Rh]/diene catalytic system.

### 5.2.2.2 Fumarate and Maleimides

The enantioselective conjugate addition products of fumarates and maleimides are synthetically useful 2-substituted 1,4-dicarbonyl compounds; however, they represent a difficult class of substrates because they are relatively unreactive. In the enantioselective conjugate addition of phenylboronic acid to di-*tert*-butyl fumarates (**56**), traditional diphosphine ligands (**L1**) and phosphoramidite ligands (**L19**) gave poor yields and enantioselectivities, while the bulky chiral diene **L27g** gave higher yields and synthetically useful enantiomeric excesses (Figure 5.11) [144].

Similarly, phosphine-based ligands such as (*R*)-binap (**L1**) only lead to moderate enantioselectivity for the conjugate addition of phenylboronic acid to benzylmaleimide (**58**) (Figure 5.12) [123]. First-generation chiral dienes such as **L27g** showed increased reactivity [123], but the enantioselectivity remained unsatisfactory. A breakthrough was achieved with the use of phosphorus–olefin hybrid ligands **L44** [123, 124] and **L45** [125] that gave excellent yields and enantioselectivities.

The efficient diastereoselective synthesis of axially chiral *N*-arylsuccinimides **61** has been achieved by using chiral diene **L28a** (Scheme 5.13) [145]. Diphosphine ligands gave lower diastereoselectivities and enantioselectivities. The axial chirality present in **61** can be efficiently used as a stereochemical relay for subsequent transformations [145].

The effective construction of chiral quaternary carbons is arguably one of the biggest challenges in asymmetric catalysis [146–148]. There are very few examples of enantioselective conjugate addition to β,β-disubstituted α,β-unsaturated carbonyl compounds [147]. During examination of the use of substituted maleimides **62**, Shintani and Hayashi discovered that the regioselectivity of the addition is a function of the ligand employed (Scheme 5.14) [149]. While Rh/(H$_8$-binap)-catalyzed processes preferably give rise to 1,4-adducts **63** with a quaternary stereogenic center, the Rh/(*R*,*R*)-**L28a**) catalyst leads to *cis/trans* mixtures of **64**.

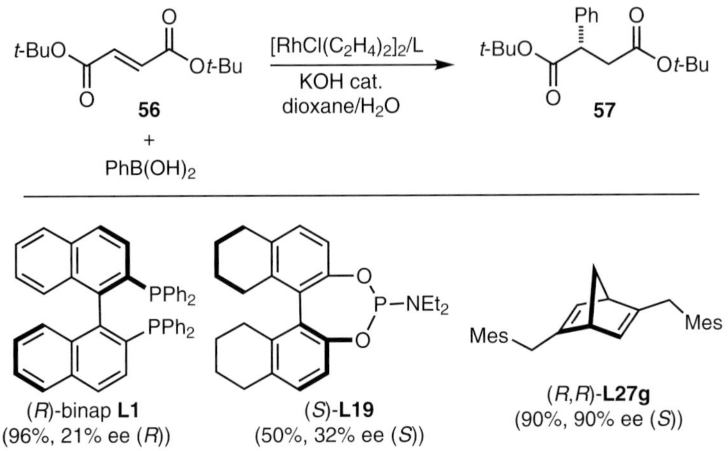

**Figure 5.11** Rh-catalyzed enantioselective conjugate addition of phenylboronic acid to di-*tert*-butyl fumarate.

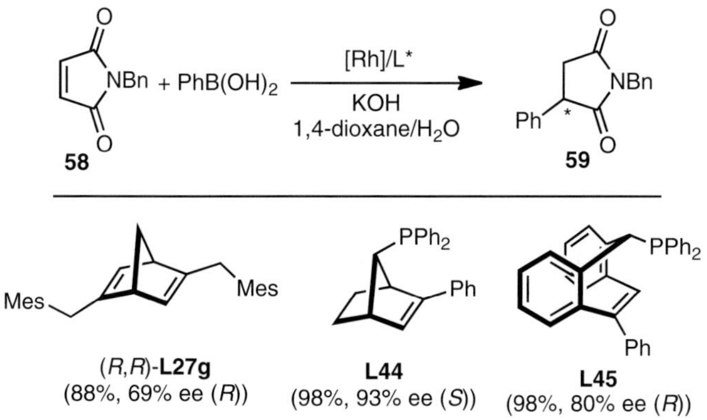

**Figure 5.12** Activity and selectivity of different ligands in the Rh-catalyzed enantioselective conjugate addition of PhB(OH)$_2$ to N-benzylmaleimide **58**.

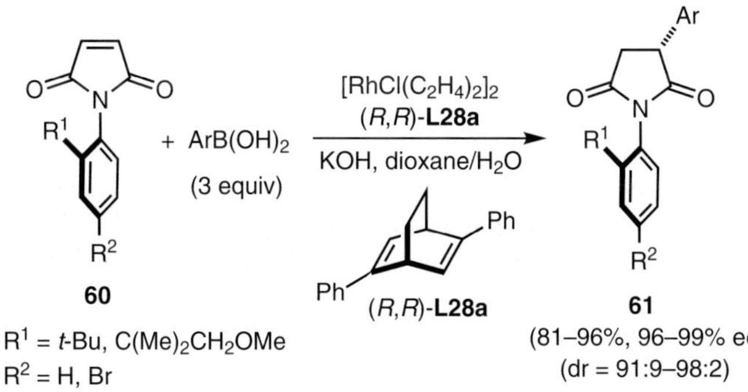

**Scheme 5.13** Diastereoselective synthesis of axially chiral arylsuccinimides.

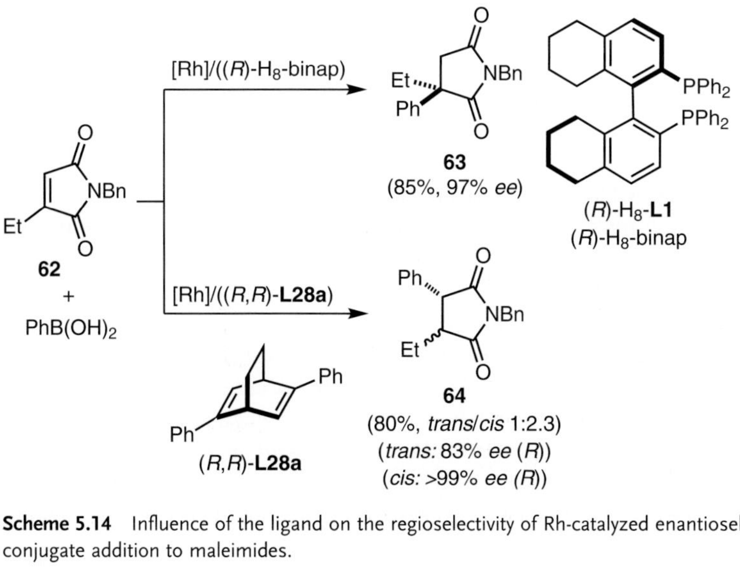

**Scheme 5.14** Influence of the ligand on the regioselectivity of Rh-catalyzed enantioselective conjugate addition to maleimides.

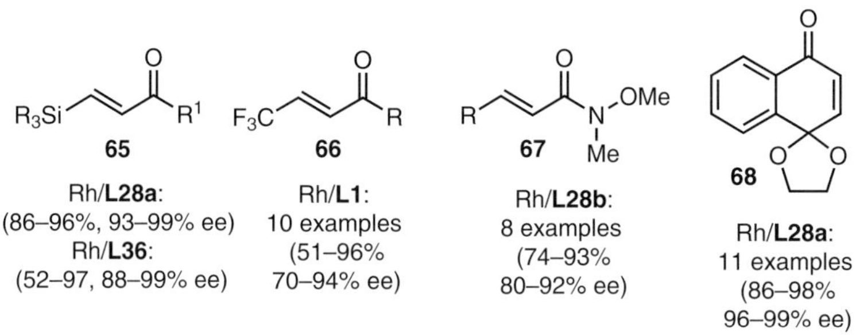

**Figure 5.13** Synthetically useful acceptors in Rh-catalyzed enantioselective conjugate additions.

**65** R$_3$Si-CH=CH-C(O)-R$^1$
Rh/L28a: (86–96%, 93–99% ee)
Rh/L36: (52–97, 88–99% ee)

**66** F$_3$C-CH=CH-C(O)-R
Rh/L1: 10 examples (51–96% 70–94% ee)

**67** R-CH=CH-C(O)-N(Me)-OMe
Rh/L28b: 8 examples (74–93% 80–92% ee)

**68** (tetralone quinone monoketal)
Rh/L28a: 11 examples (86–98% 96–99% ee)

### 5.2.2.3 Synthetically Useful Acceptors

There is now a range of synthetically useful acceptors that can be used in the Rh-catalyzed enantioselective conjugate addition (Figure 5.13). For example, the use of β-silyl-substituted α,β-unsaturated carbonyl compounds **65** as acceptors is of special interest since these compounds can be transformed to β-hydroxyketones by Tamao–Fleming oxidation [104]. In addition, the introduction of an alkenyl group to the β-silyl enone **65** leads to a chiral allylsilane that can further react with the ketone moiety in an intramolecular Sakurai [150] reaction [151]. The 3,7-disubstituted bicyclo[3.3.1]nonadiene **L36** performs particularly well with these substrates [99]. Another family of acceptors that enable straightforward modification of the resulting adducts are α,β-unsaturated esters **35** [90] and α,β-unsaturated Weinreb amides **67** [152]. The enantioselective conjugate addition product using β-trifluoromethyl-α,β-unsaturated ketones **66** are of particular interest in the medicinal, pharmaceutical, and agricultural fields [153]. Finally, α-arylated tetralones can be accessed in high yields and stereoselectivity by Rh/diene-catalyzed enantioselective conjugate addition of organoboron reagents to quinone monoketal **68** [107].

For example, the Rh-catalyzed enantioselective conjugate addition of alkenyltrifluoroborate (**70**) to quinone monoketals **69** was used by Corey and coworkers in the synthesis of chiral ketone **71**, a key intermediate for the synthesis of platensimycin (**72**), a potent anticancer agent (Scheme 5.15) [154, 155].

### 5.2.2.4 Conjugate Additions of Boryl and Silyl Groups

In addition to aryl and alkenyl groups, bis(pinacolato)diboron (**15**) and silyl boronic ester **16** can be used in Rh-catalyzed conjugate additions to transfer boron [48] and silyl groups, respectively. Oestreich and coworkers reported the rhodium-catalyzed enantioselective conjugate addition of the SiMe$_2$Ph group to linear and cyclic α,β-unsaturated carbonyl compounds to provide chiral β-silyl ester (**73**) and ketones (**74**) in good yields and with high enantioselectivity (Scheme 5.16) [49, 50].

To summarize the general trends in Rh-catalyzed enantioselective conjugate additions of organoboron reagents to α,β-unsaturated carbonyl compounds, the rate of the conjugate addition decreases with the reactivity of the acceptor and its steric bulk. Thus, acceptors can be classified into the following order of decreasing reactivities:

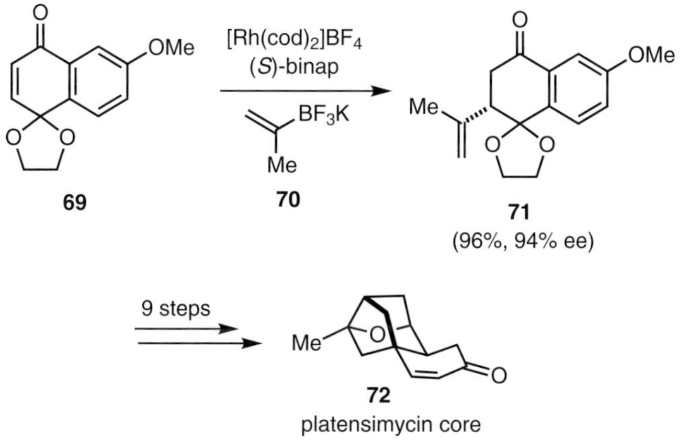

## 5.2 Rh-Catalyzed Enantioselective Conjugate Addition of Organoboron Reagents | 287

**Scheme 5.15** Application of Rh-catalyzed enantioselective conjugate addition as a key step in the synthesis of the platensimycin core.

$$(5.11)$$

X = CH$_2$ (45%, 96% ee)
X = O (58%, 96% ee)

$$(5.12)$$

74
(42–72%, 98–99% ee)

R$^1$ = OEt, SMe, —N⌐oxazolidinone
R$^2$ = alkyl, Ar

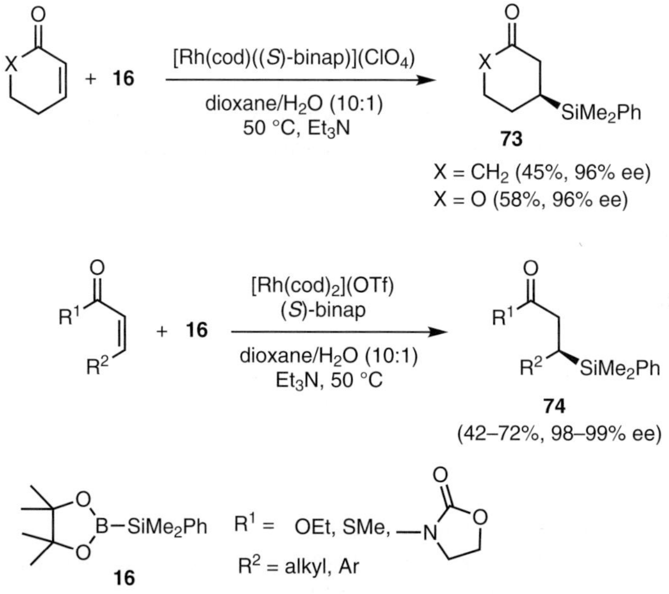

**Scheme 5.16** Rh-catalyzed ECA with (pin)B-SiMe$_2$Ph.

enals > enone > enoate > enamides > fumarates > maleimides. Because effective coordination of the acceptor to rhodium is crucial for the reaction, acceptors that bear steric bulk in proximity to the unsaturation will be less reactive than unhindered ones. In addition, the activity of Rh/ligand catalytic systems increases with increasing π-accepting properties of the ligands.

### 5.2.3
### Addition to Other Electron-Deficient Alkenes

#### 5.2.3.1 Arylmethylene Cyanoacetates

The enantioselective construction of stereogenic carbon centers substituted with two aryl groups and one alkyl group is a subject of importance because this structural motif is often found in pharmaceuticals and natural products. As discussed previously in Figures 5.10 and Scheme 5.8, their asymmetric synthesis has been reported using the chiral diene/rhodium-catalyzed asymmetric 1,4-addition of arylboronic acids to β-aryl-α,β-unsaturated aldehydes (22) and esters (35). An alternative approach to such chiral building blocks has been demonstrated by Hayashi and coworkers with the use of arylmethylene cyanoacetates 75 as substrates (Scheme 5.17) [108]. The chiral α-cyanoester 76 can be easily decarboxylated to give the corresponding enantiopure β,β′-diaryl nitrile. The best results were obtained with the chiral diene $(R,R)$-L28a that enabled consistently high enantioselectivities to be achieved.

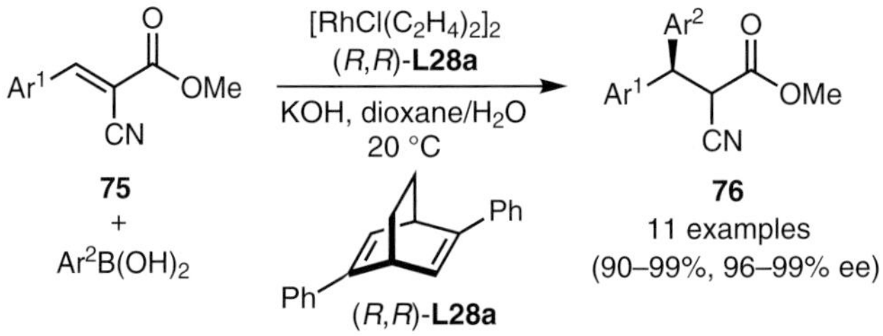

**Scheme 5.17** Rh-catalyzed enantioselective conjugate addition of arylboronic acids to arylmethylene cyanoacetates.

As shown in the Table 5.1, the presence of both cyano and ester groups at the α-position of the substrates is essential for the high reactivity and enantioselectivity in the present reaction. Other combinations gave either low yields or enantioselectivities or both (Table 5.1).

#### 5.2.3.2 Alkenylphosphonates

The Rh-catalyzed enantioselective conjugate addition to alkenylphosphonates was first reported by Hayashi and coworkers. This class of substrate is less reactive than α,β-unsaturated carbonyl compounds and the use of arylboroxine, instead of arylboronic acids, in the presence of one equivalent of water was necessary in order to obtain high yields [156]. In agreement with the stereochemical model depicted in Scheme 5.4, the *trans* and *cis* geometries of alkenylphosphonate give rise to opposite enantiomers.

#### 5.2.3.3 Nitroalkene

Hayashi first reported that nitroalkenes are good substrates for the rhodium-catalyzed enantioselective conjugate addition of organoboronic acids [157].

**Table 5.1** Rh-catalyzed enantioselective conjugate addition of arylboronic acids to arylmethylene cyanoacetates.

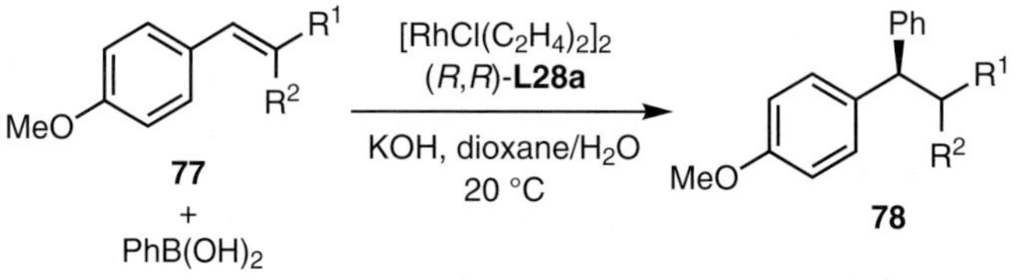

| Entry | Substrate | R$^1$ | R$^2$ | Product | Yield (%) | ee (%) |
|---|---|---|---|---|---|---|
| 1 | 77a | CO$_2$Me | CN | 78a | 99 | 99 (R) |
| 2 | 77b | CN | CN | 78b | 9 | n.d. |
| 3 | 77c | CO$_2$Me | CO$_2$Me | 78c | 11 | n.d. |
| 4 | 77d | CN | H | 78d | 74 | 52 (R) |
| 5 | 77e | CO$_2$Me | H | 78e | 99 | 57 (R) |

The nitroalkanes obtained can be readily converted into a wide variety of optically active compounds. For example, a Rh-catalyzed enantioselective conjugate addition of 2,3-difluorophenylboronic acid to nitroalkene **79** was the key step in the synthesis of nitroalkane **80**, a precursor to migraine headache treatment, on a 2 kg scale [158]. It was found that bicarbonate was a useful base in this transformation. Furthermore, in this acyclic system, the other stereocenter in **79** exerted only very modest diastereocontrol (41 : 59) on the reaction when (*rac*)-binap was employed (Scheme 5.18). This is in contrast to chiral cyclic acceptors with which good diastereocontrol is usually observed.

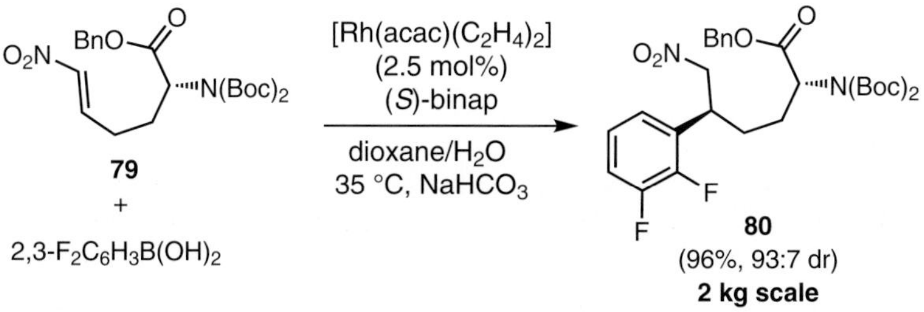

**Scheme 5.18** Application of Rh-catalyzed enantioselective conjugate addition of a nitroalkene on a kg scale.

### 5.2.3.4 Sulfones

Hayashi disclosed that α,β-unsaturated phenyl sulfones do not react with organoboron reagents under the usual rhodium-catalyzed conditions. When nucleophilic aryltitanium reagents were employed, elimination of the sulfonyl group occurred after the conjugate addition leading to desulfonylated alkenes as final products [159]. The key to this synthetic challenge was found through the use of a rhodium-coordinating α,β-unsaturated 2-pyridylsulfone (**81**). With these pyridyl sulfones, it was possible to obtain a general methodology for providing β-substituted sulfones in high yields and enantioselectivities ranging from 76 to 92% ee with (*S*,*S*)-chiraphos (**L3**) as chiral ligand (Scheme 5.19) [160, 161]. The corresponding 4-pyridyl sulfone analogues displayed no reactivity demonstrating the necessity of 2-pyridyl sulfone moieties to stabilize the rhodium intermediate **H**. The *cis*-alkenylsulfone (*cis*-**81**) gives the opposite enantiomer to the *trans*-alkenylsulfone (*trans*-**81**). This outcome is rationalized by the general stereochemical model depicted in Scheme 5.4 [160, 161]. The chiral β-substituted sulfones **82** readily participate in a Julia–Kociensky olefination [162] to provide a novel approach to the enantioselective synthesis of allylic substituted *trans*-alkenes **83** (Scheme 5.19). In addition, the chiral sulfones **82** can be alkylated and the sulfonyl group can be removed by Zn-mediated reduction [161]. This approach was extended to the addition of alkenylboronic acids to β-aryl-β-methyl-α,β-unsaturated pyridylsulfones, which enable the efficient stereoselective formation of quaternary centers with up to 99% ee [163].

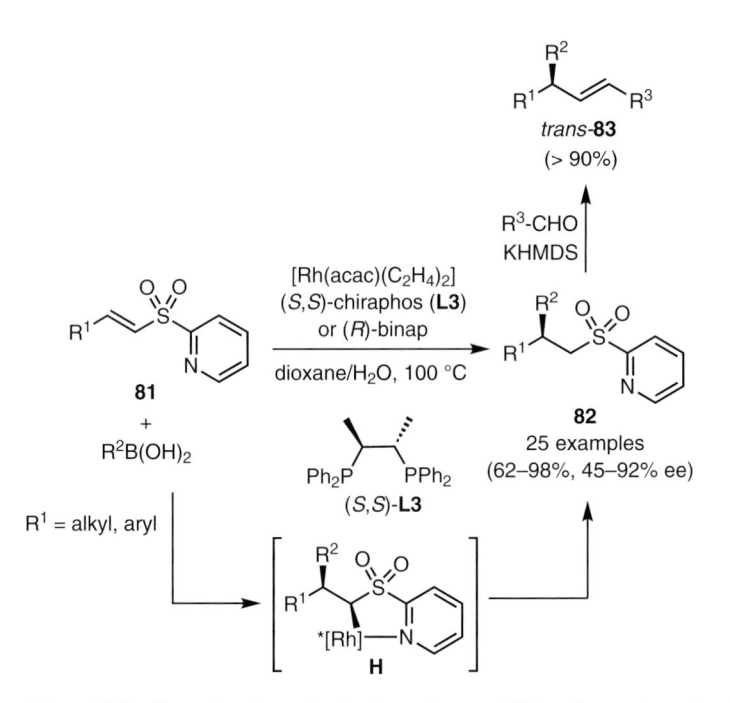

**Scheme 5.19** Rh-catalyzed enantioselective conjugate addition of organoboronic acid with alkenyl sulfones.

## 5.2.3.5 Addition to *cis*-Allylic Alcohols

A novel and original approach for the synthesis of enantiomerically enriched 2-arylbut-3-enols **86** uses *cis*-allylic diol **84** and arylboroxines **(7)** (Scheme 5.20) [164]. Under the reaction conditions, *cis*-diol **84** readily forms the cyclic arylboronic ester **85** that serves as an acceptor for a *syn*-1,2-carborhodation by [Rh(μ-OH)(**L30f**)] to give intermediate **I**. A subsequent β-oxygen elimination regenerates the active rhodium–hydroxide species **A** and releases the optically active alcohols **86**.

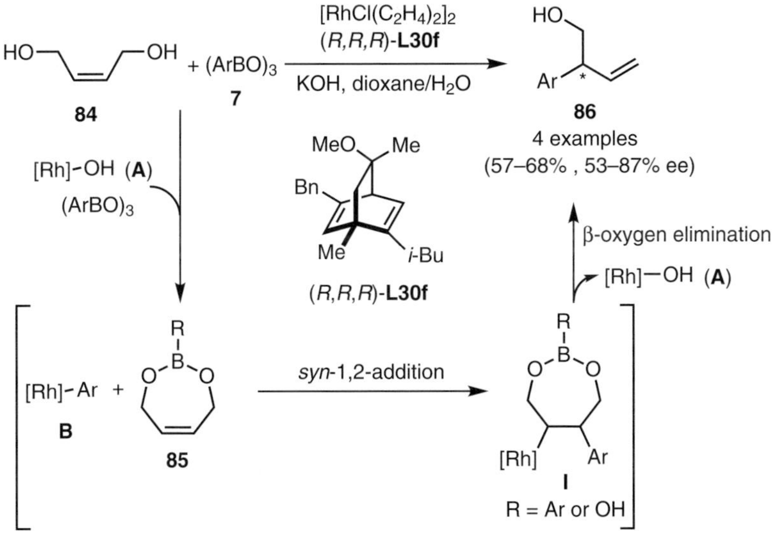

**Scheme 5.20** Rh-catalyzed substitutive arylation of a *cis*-allylic diol with arylboroxines.

### 5.2.3.6 1,4-Addition/Enantioselective Protonation

Thus far, the Rh-catalyzed enantioselective conjugate addition process with β-substituted α,β-unsaturated carbonyl compounds was reviewed. With this family of substrates, the enantio-determining step is the 1,2-insertion of the acceptor into the Rh–$C_{sp2}$ bond. However, when α-substituted β-unsubstituted α,β-unsaturated carbonyl compounds are employed, the enantio-determining step is protonation of the oxo-π-allylrhodium species. A general overview of enantioselective protonation processes has been recently published [165].

The asymmetric variant of this mechanistic manifold was first exploited by Reetz *et al.*, in the enantioselective addition of phenylboronic acid to α-acetamidoacrylic esters leading to phenylalanine derivatives **(88)** in up to 77% ee [67]. Similar asymmetric transformations have been reported by Frost and coworkers [166]. This reaction represents a convenient alternative to the synthesis of unnatural phenylalanine congeners **(88)**, which are commonly accessed through Rh-catalyzed asymmetric hydrogenation of β-aryl-α-acetamidoacrylic esters [167]. Recognizing the importance of the proton source in this process, Genêt and Darses investigated a wide range of phenols instead of water as the proton donor [36, 168]. Guaiacol (**89**) proved to be the best proton donor and enables satisfactory enantioselectivity with a range of potassium organotrifluoroborates (Scheme 5.21). Higher temperatures and the use of toluene or dioxane increased conversion and enantioselectivity. Importantly, the presence of water accelerates the reaction 10-fold compared to guaiacol, but drastically decreases the enantioselectivity to 16% ee. The use of organoboronic acids also leads to lower yields and enantioselectivities, presumably due to residual traces of water. As with the enantioselective conjugate addition of enoate, increasing the size of the ester substituent in **87** leads to a slight increase in enantiomeric excesses. The use of more π-acidic (S)-difluorophos (**L50**) yields to a faster and more selective transformation relative to binap (**L1**). In-depth mechanistic studies and DFT calculations of this process suggest that the actual mechanism goes through a sequential conjugate addition and β•hydride elimination to form an imine and a Rh–hydride species. The Rh-hydride subsequently reinserts into the imine enantioselectively, and the Rh-amino bond is then hydrolyzed to generate the reaction product [36].

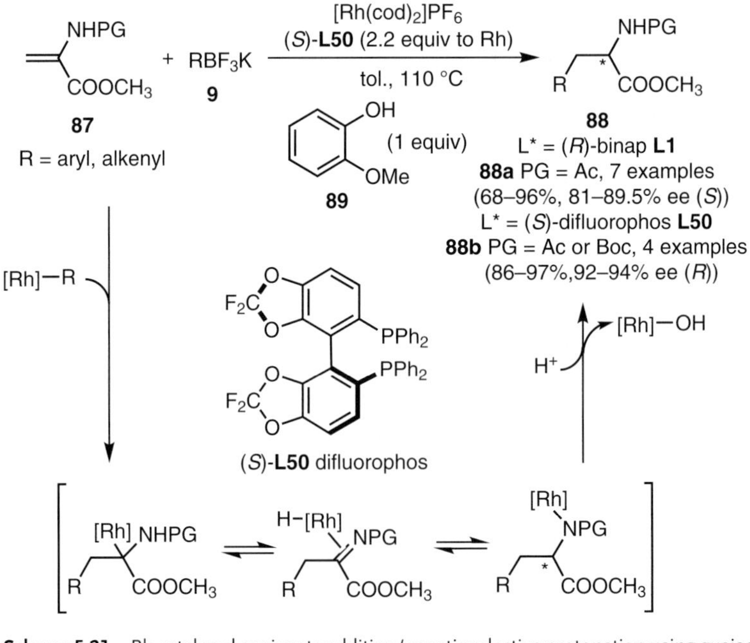

**Scheme 5.21** Rh-catalyzed conjugate addition/enantioselective protonation using guaiacol as the sole proton source.

The scope of substrates amenable to the enantioselective protonation was expanded to dimethyl itaconate (**90**) [169], α-benzyl acrylates (**93**) [170], and α-aminomethyl acrylates (**95**) [171] (Scheme 5.22) [166]. In all of these examples, the choice of the

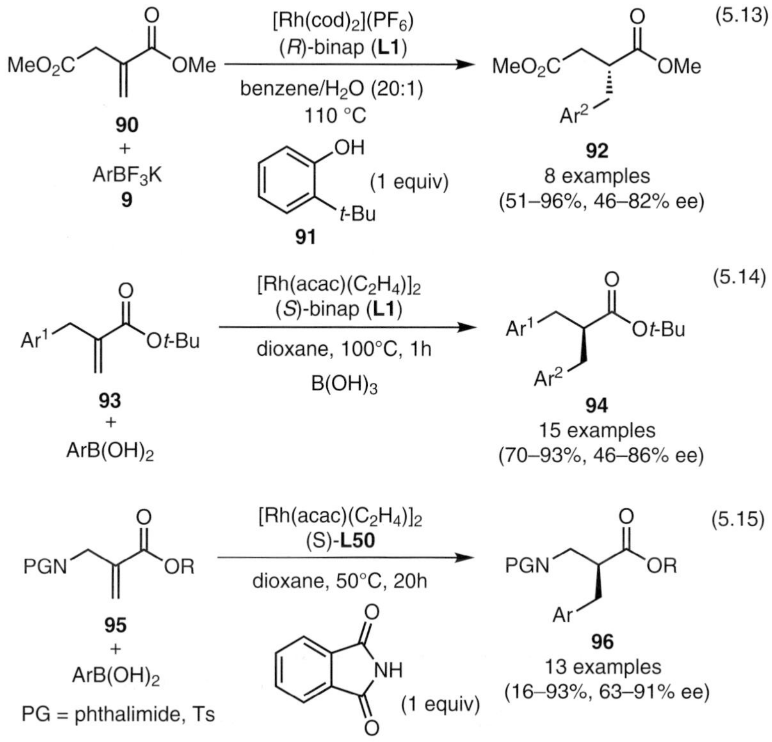

## 5.2 Rh-Catalyzed Enantioselective Conjugate Addition of Organoboron Reagents | 293

**Scheme 5.22** Applications of Rh-catalyzed addition/enantioselective protonation.

additional proton source (phenol **91**, boric acid (B(OH)$_3$) or phthalimide) was critical to obtain high yields and good selectivities.

An interesting application of the Rh-catalyzed enantioselective protonation is the peptide modification through site-selective residue interconversion, following an elimination and conjugate addition sequence (Scheme 5.23). Thus, a serine or cysteine (**97**) can be selectively eliminated from a peptide chain to form a dehydroalanine fragment (**98**) that can subsequently undergo a Rh-catalyzed enantioselective conjugate addition to yield peptide with modified Ph-alanine fragment (**99**). Although the diastereoisomeric excesses are modest, this transformation can be applied in a range of di- and tripeptides [172].

The enantioselective protonation was also applied to the hydroarylation of diphenylphosphinylallenes [173]. Unlike the oxo-π-allylrhodium species in previous examples, the π-allylrhodium intermediate formed in this transformation can be protonated α and γ to the phosphorus center to give regioisomers **101a** and **101b**, respectively (Scheme 5.24). THF was found to minimize the amount of **101b** formed. Furthermore, bulkier R groups (*t*-Bu) in allene **100** lead to **101b** because protonation at the least hindered position is favored. Under these conditions, the boronic acid acts as the proton source.

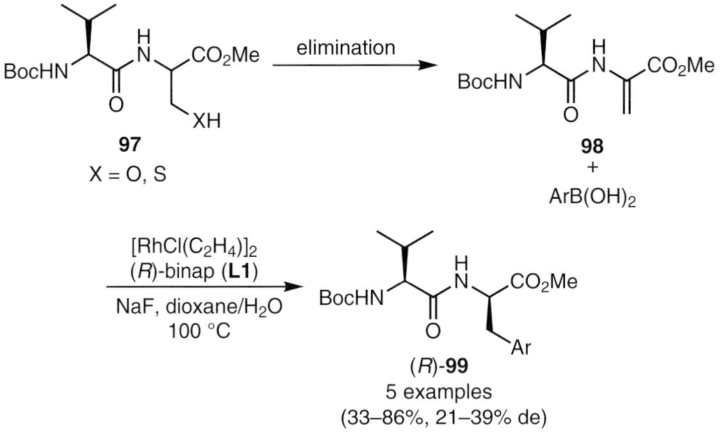

**Scheme 5.23** Application of the addition/enantioselective protonation to the synthesis of the site interconversion peptides.

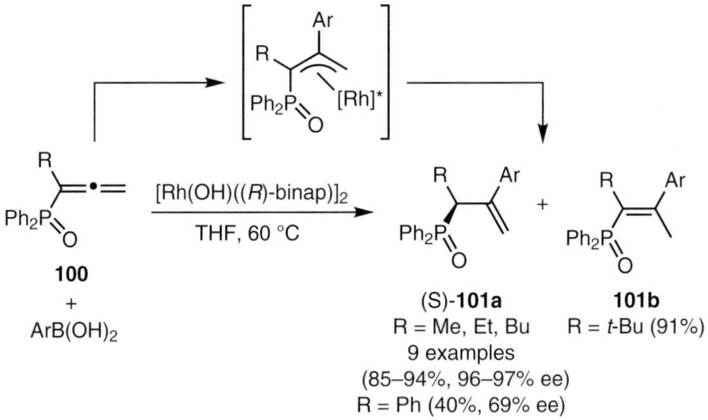

**Scheme 5.24** Rh-catalyzed enantioselective addition to diphenylphosphinylallenes.

## 5.2.4
### 1,6-Conjugate Additions

From this review, it is apparent that Rh-catalyzed enantioselective conjugate addition chemistry is now a well understood process. Arguably, one of the new frontiers in Rh-catalyzed enantioselective conjugate addition methodology is now the control of the regio- and enantioselectivity in 1,6-addition processes. 1,6-Additions to $\alpha,\beta,\gamma,\delta$-diunsaturated carbonyl compounds are particularly challenging because of the multitude of possible reaction pathways that are under substrate control. This issue is clearly illustrated in Scheme 5.25 [174], where three competitive reaction pathways coexist, depending on the substitution pattern of the $\alpha,\beta,\gamma,\delta$-diunsaturated esters **102**

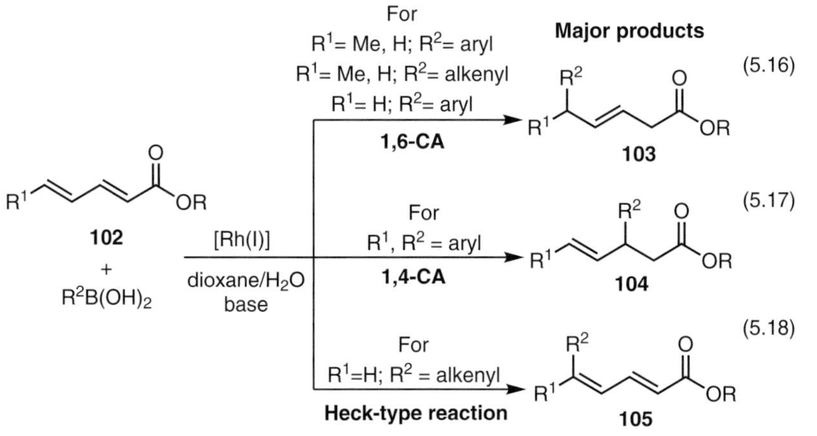

## 5.2 Rh-Catalyzed Enantioselective Conjugate Addition of Organoboron Reagents | 295

**Scheme 5.25** Rh-catalyzed 1,6-addition of organoboronic acid to α,β,γ,δ-diunsaturated esters.

and the nature of the organoboron reagent. For unhindered dienoates **102** ($R^1$ = H or Me), the 1,6-addition product **103** is favored. When $R^1$ and $R^2$ are aromatic, the 1,4-addition (**104**) becomes predominant, whereas when $R^2$ is an alkenyl group the fully conjugated Heck-type product **105** is observed.

The regioselectivity of the 1,6-addition of β-substituted dienoates can be controlled by using reactive arylzinc reagents [175].

In 2006, Hayashi and coworkers reported a breakthrough in transition metal-catalyzed 1,6-addition methodology with organoboronic acids, through the use of [Ir(μ-OH)(cod)]$_2$ the catalyst (Scheme 5.26) [176]. Importantly, this is the first study of an iridium-catalyzed addition of organoboronic acid to an electron-deficient alkene or diene. The 1,6-addition of an arylboronic acid to **106** leads to products **107** as the *cis* isomer. Compound **107** was hydrogenated to **108** to facilitate analysis. The high 1,6-selectivity obtained with the iridium catalyst is in stark contrast to that observed with

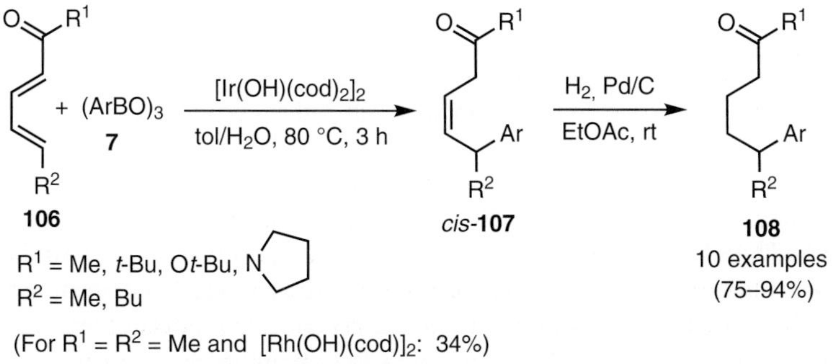

**Scheme 5.26** Ir-catalyzed 1,6-conjugate addition with aryboroxines.

the parent [Rh(μ-OH)(cod)]$_2$ complex as a catalyst under the same conditions (Scheme 5.26). Thus, rhodium-catalyzed reaction gave the 1,4-adduct as the main isomer (55% yield) and a minor amount (34% yield) of 1,6-adducts **107** (as a mixture of geometrical isomers).

Competition experiments revealed that the iridium catalyst has a much stronger reactivity toward the dienone than the enone, while the opposite reactivity is observed for [Rh(μ-OH)(cod)]$_2$. On the basis of the high reactivity toward the diene moiety and the high *cis* selectivity in the 1,6-addition product, the catalytic cycle that was surmised is depicted in Scheme 5.27. Transmetalation of an aryl group from the boron to iridium-hydroxide **I** forms an aryl–iridium species **J** [177]. The coordination of the dienone to the aryl–iridium complex with a cisoid diene moiety results in the formation of a (η$^4$-diene)–iridium complex **K**. Insertion of the diene into the aryl–iridium bond leads to π-allyl–iridium moiety **L**. This step is followed by a selective hydrolysis of **L** at the α-position to the carbonyl with the assistance of phenylboronic acid or boric acid gives the 1,6-addition product *cis*-**107** and regenerates the hydroxo–iridium species **I**.

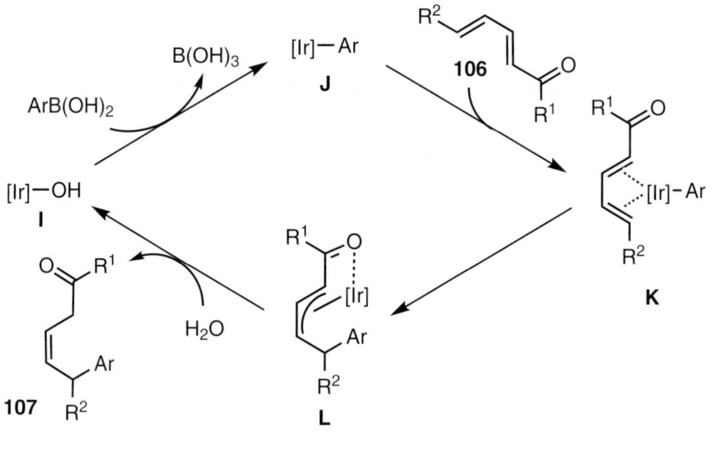

**Scheme 5.27** Proposed catalytic cycle to the Ir-catalyzed 1,6-addition with arylboronic acids.

### 5.2.5
### Rh-Catalyzed Enantioselective Conjugate Addition with Other Organometallic Reagents

Apart from boron, several other organometallic reagents such as Ti [159, 178, 179], Zn [92, 131, 175, 180], Si [57, 151, 181–185], Zr [58, 186, 187], Bi [188], Pb [189], Sn [53, 56, 190], and indium [191] have been used successfully in Rh-catalyzed enantioselective conjugate additions. It appears that the main criteria for the effectiveness of an organometallic reagent in rhodium-catalyzed conjugate additions are that it can smoothly and efficiently transmetalate to the rhodium center and that it is stable under the conditions necessary to hydrolyze the rhodium-enolate or that the reagent can transmetalate directly with the Rh-enolate.

## 5.2.6
## Rh-Catalyzed Tandem Processes

In the previous sections, we have described rhodium-catalyzed asymmetric conjugate additions of organometallic reagents to a wide range of acceptors. The more electron-deficient the unsaturated substrate is, the more readily it will react with the organorhodium species generated by transmetalation from boron to rhodium. Thus, it is possible to program carborhodation cascades by assembling inter- or intramolecular acceptors of differing reactivity. The cascade is initiated by carborhodation onto the most reactive acceptors, which subsequently reacts with the second most reactive acceptor until the sequence is terminated by protonolysis of the organorhodium intermediate. Such cascade sequences that consist of multiple carbometalation steps provide powerful methods for the construction of structurally complex molecules in an efficient and atom-economical manner [192]. These transformations have been the object of several reviews [193–195].

### 5.2.6.1 Tandem Enantioselective Conjugate Addition/Aldol Reaction
An elegant three-component Rh-catalyzed tandem enantioselective conjugate addition/aldol reaction was developed by Hayashi and coworkers [196]. The reaction of ArB(9-BBN), methyl vinyl ketone, and propanal catalyzed by [Rh($\mu$-OH)($S$)-binap)]$_2$ as a catalyst gave optically active products, *syn*-aldol products in 41% ee and *anti*-aldol products in 94% ee, though the *syn/anti* selectivity is only 0.8/1.0. The formation of the enantiomerically enriched products demonstrated that the reaction proceeds through a chiral (oxa-$\pi$-allyl)rhodium complex and that this intermediate undergoes an aldol-type reaction. A boron alkoxide as an intermediate would lead to a racemic aldol product. An analogous zirconium-based three-component coupling reaction was reported by Nicolaou and coworkers [187].

Krische and coworkers reported the intramolecular version of this sequence using acceptors bearing a pendant electrophile that can react with the incipient (chiral) rhodium enolate [197]. This method was elegantly applied to the desymmetrization of symmetrical diketoenones, which resulted in the stereoselective formation of bicycle with four contiguous stereogenic centers including a tandem conjugate addition/1,2-addition. In a similar fashion, nitrile containing enoate **109** can undergo the conjugate addition of arylboron reagents to form five- and six-membered β-enamino esters **110** (Scheme 5.28) [198]. In this process, an (oxa-$\pi$-allyl)rhodium intermediate (**M**), generated by the initial conjugate addition of an arylrhodium species, undergoes a facile intramolecular addition to the cyano group, leading to **N**, followed by sequential transmetalation with Ar-B(9-BBN) (**13**) to form **O**, which after hydrolysis generates **110**. The use of chiral (R)-H$_8$-binap for this transformation leads to enantioselectivities for **110** in up to 94% ee.

**Scheme 5.28** Tandem Rh-catalyzed enantioselective conjugate addition/1,2-addition to a cyano group.

### 5.2.6.2 Tandem Carborhodation/Conjugate Addition

The electron-deficient Rh/chiral-diene (**L28b**) catalyst can efficiently promote the chemo- and enantioselective arylative cyclization of alkyne-tethered electron-deficient olefin **111** to chiral cyclopentene **112** (Scheme 5.29) [199]. Such selectivities are not observed with the more electron-rich rhodium/diphosphine catalytic systems. This chemoselectivity, which involves initial carborhodation of the triple bond of **111** instead of the conjugated double bond, is in accord with the observation that a Rh/

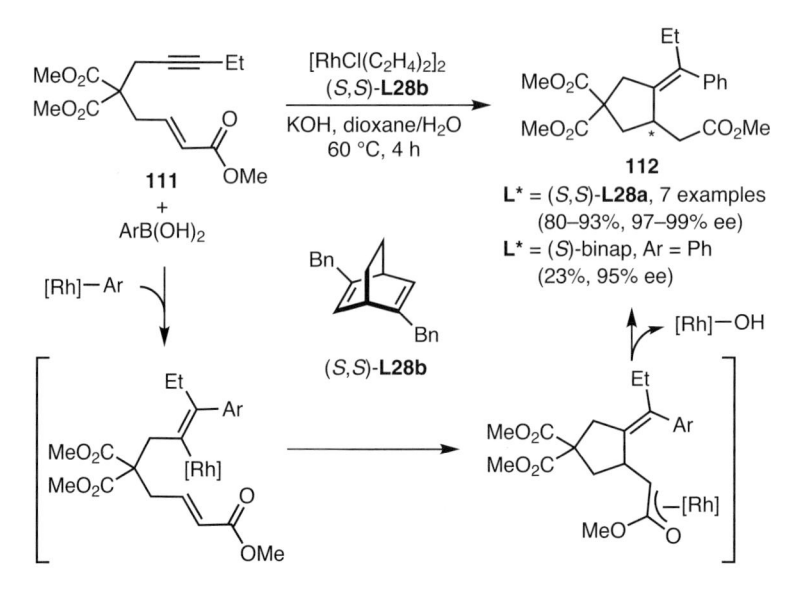

**Scheme 5.29** Rh/diene-catalyzed arylative cyclization of alkyne-tethered electron-deficient olefins.

diene catalyst displays higher activity in the arylation of alkynes than in the 1,4-addition to α,β-unsaturated esters, whereas a Rh/diphosphine catalyst behaves in the opposite manner. This behavior could be due to the more electrophilic nature of a Rh–diene complex relative to a Rh/phosphine center. A related addition/cyclization, involving the initial carborhodation of an alkyne followed by an enantioselective 1,2-addition to a tethered aldehyde was reported by Hayashi and coworkers [200]. For this reaction, chiral dienes **L27b** and **L28b** also proved instrumental in accessing high activity and enantioselectivity.

*Syn*-1,2-addition is the most common pathway for the insertion of an alkyne into a Rh–carbon bond. However, recently, a 1,1-carborhodation pathway was observed with **113**, in which an *endo*-olefin cyclic product **114** is formed (Scheme 5.30) [201]. This novel addition/cyclization reaction occurred through an alkylidenerhodium-mediated 1,1-carborhodation process. Following the formation of vinylidenerhodium **P**, α-migration of the $R^2$ group from the Rh center to the vinylidene ligand provides alkenylrhodium intermediate **Q**, which can then undergo addition to the pendant enone to give rhodium enolate **R**. Finally, protonation of **R** produces cyclopentene derivatives (**114**) and regenerates the methoxyrhodium species.

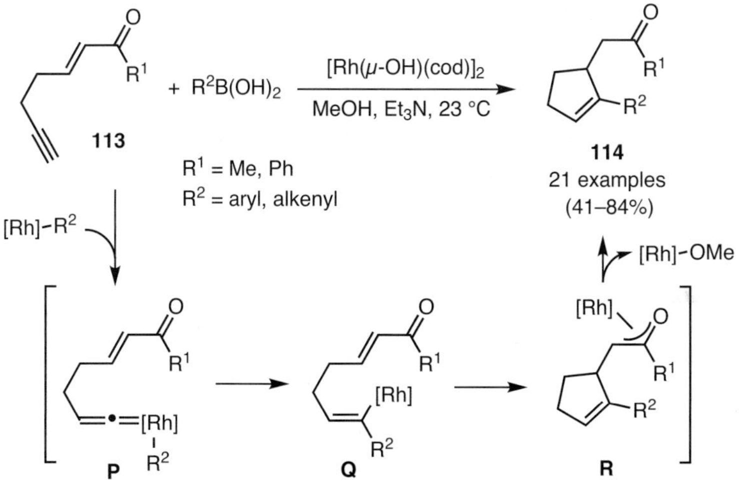

**Scheme 5.30** Tandem 1,1-carborhodation/conjugate addition.

## 5.3
## Pd-Catalyzed Enantioselective Conjugate Addition of Organoboron Reagents

### 5.3.1
### Introduction

As described in the previous sections of this chapter, the rhodium-catalyzed enantioselective conjugate addition provides a very robust and flexible approach to introduce aryl and alkenyl groups in high yields with excellent enantioselectivities and chemoselectivities. However, from an economic point of view, rhodium remains the most onerous metal and its price has been the subject of intense speculation over the past decade. At its peak in 2008, Rh at $10 000/ounce was 23 times more expensive than Pd at $471/ounce. Such high and fluctuating prices hamper the use of Rh-catalyzed enantioselective conjugate addition on a large industrial scale and creates a need for alternative methodologies based on cheaper transition metals such as palladium. There are several reviews on Pd-catalyzed enantioselective conjugate additions [12, 14, 202–205]. In this section, we will cover the recent advances made in palladium-catalyzed conjugate addition of boronic acids to Michael acceptors with an emphasis on enantioselective transformations.

### 5.3.2
### Addition to α,β-Unsaturated Ketones

The first use of palladium as a catalyst for conjugate addition of an organometallic reagent to α,β-unsaturated ketones can be traced back to the seminal work of Cacchi who employed organotin [206] and organomercury [207] reagents in an acidic biphasic system. In 1995, Uemura and coworkers reported [1] that organoboronic acids could be used in the Pd-catalyzed addition to enones. This reaction was performed under acidic conditions and with a high catalyst loading. Although these early studies showed the potential of the palladium-catalyzed conjugate addition, this methodology has lagged behind the rhodium-catalyzed reaction. Probably the main reason for the underdevelopment of Pd-catalyzed conjugate addition is the propensity of neutral Pd-enolates, which are carbon centered rather than oxygen centered (i.e., Rh-enolates), to undergo competitive β-hydride elimination rather than hydrolysis (Scheme 5.31). This generates Heck-type coupling with concomitant Pd-black formation.

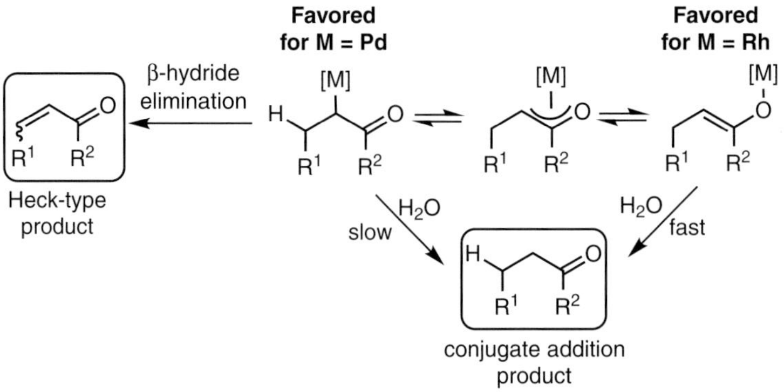

**Scheme 5.31** Difference in reactivity between neutral Pd- and Rh-enolates.

## 5.3 Pd-Catalyzed Enantioselective Conjugate Addition of Organoboron Reagents

Based on the findings that cationic palladium-enolates are much more susceptible to hydrolytic Pd-carbon bond cleavage than are neutral species [208] and that alkenes can insert readily into cationic organopalladium, Miyaura and coworkers developed a highly efficient Pd-catalyzed conjugate addition using dicationic [Pd(dppe)(MeCN)$_2$] (SbF$_6$)$_2$ complex as the catalyst [209, 210]. The use of a cationic palladium source in the presence of water effectively shuts down the β-hydride elimination from the Pd-enolate by increasing its hydrolysis rate. The cationic Pd source is very active, promoting the conjugate addition at room temperature. The use of a bidentate phosphine ligands with a two-carbon spacer, such as dppe (1,2-bis(diphenylphosphino)ethane), proved essential to obtain reactivity. Diphosphine ligands with larger bite angles [211] such as 1,3-bis(diphenylphosphino)propane and binap were ineffective. Although the presence of a base such as K$_2$CO$_3$ accelerates the reaction rate, it also promotes β-hydride elimination. Under these conditions, addition to β-arylenals **22** proved completely chemoselective, with only the 1,4-addition product being observed with this catalytic system. Enoates were sluggish substrates and α,β-unsaturated amides were unreactive. A Pd(OAc)$_2$/2,2'-bipyridine catalytic system is highly effective for the addition of arylboronic acids to enones, enals, nitroalkenes, and very interestingly to cinnamates and acrylates. The last substrates have posed long-standing problems [212]. The reaction can also be run in water in the presence of anionic surfactants [213].

In 2004, Miyaura and coworkers reported the first Pd-catalyzed enantioselective conjugate addition to α,β-unsaturated ketones using triarylbismuth reagents [214]. In this seminal study, the combination of [Pd(MeCN)$_2$](SbF$_6$)$_2$ with (S,S)-dipamp (**L51**) or (S,S)-chiraphos (**L3**), as chiral ligands, with Cu(BF$_4$)$_2$ as substoichiometric additive in a MeO/H$_2$O (6:1) mixture proved to be the most reactive and enantioselective system (Scheme 5.32). The organobismuth reagents were especially reactive and gave full conversion at just −5 °C. Using the same conditions, the scope of the reaction was expanded to potassium aryltrifluoroborates and aryltrifluorosilanes [215]. The yields

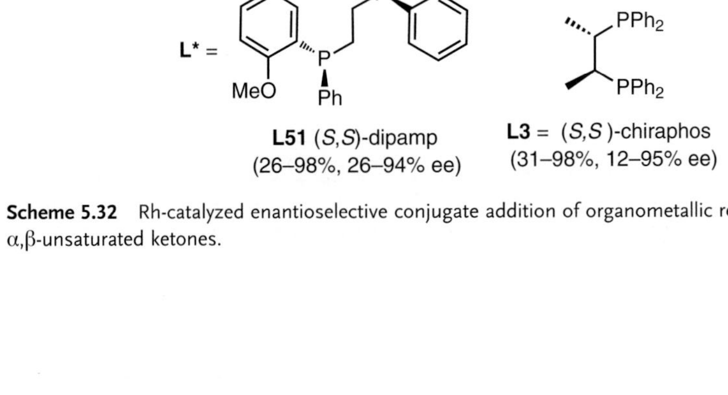

**Scheme 5.32** Rh-catalyzed enantioselective conjugate addition of organometallic reagents to α,β-unsaturated ketones.

were both high with the triarylbismuth and with ArBF$_3$K reagents, while they were slightly lower with ArSiF$_3$. Importantly, the enantioselectivities observed did not vary with the nature of the nucleophilic reagent (BiAr$_3$, ArSiF$_3$, or ArBF$_3$K).

Miyaura and coworkers further developed the methodology to realize the use of more readily available arylboronic acids. Under optimized conditions, the use of acetone was found to be a more suitable solvent than MeOH. Addition of 5–10 mol% of AgBF$_4$ or AgSbF$_6$ was found to improve the catalytic activity and stability of the dicationic Pd(II) catalyst by presumably oxidizing the inactive Pd(0) complexes back to the dicationic active species [210]. With these additives, the catalyst loading can be lowered to 0.01 mol% [216]. Under theses conditions, β-arylenones (**35**) underwent smoothly the enantioselective conjugate addition with organoboronic acid at 0 °C or 25 °C with a low catalyst loading (Scheme 5.33). An interesting application of this methodology is the efficient synthesis of optically active chromenes **116** through a enantioselective conjugate addition/dehydration sequence of β-arylenones **115** (Scheme 5.33) [216].

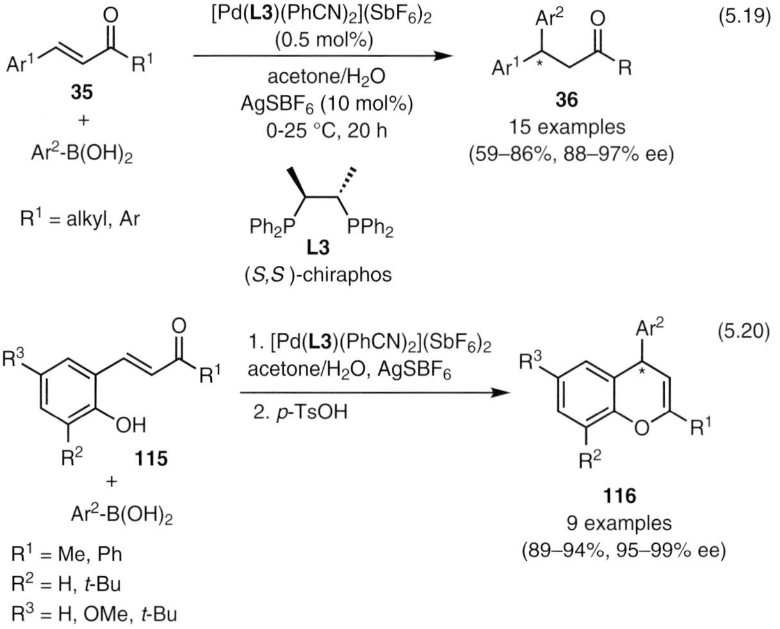

**Scheme 5.33** Pd-catalyzed enantioselective conjugate addition of arylboronic acids to β-arylenones.

Cationic palladium(II) complexes (*R*)-**117** bearing a chelating chiral bidentate N-heterocyclic carbene was reported to be active in the asymmetric conjugate addition of arylboronic acids to cyclic enones (Scheme 5.34) [217].

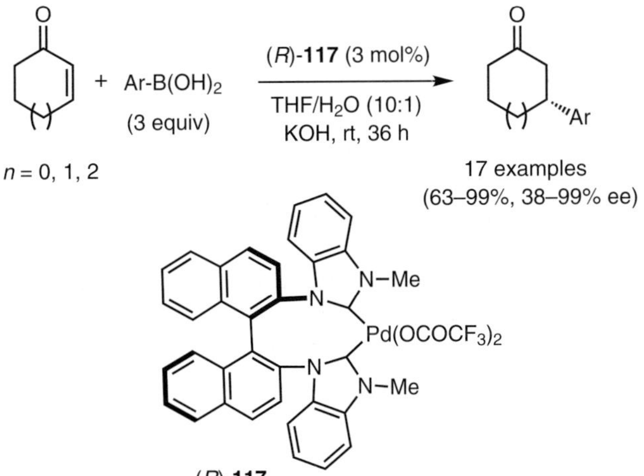


(*R*)-**117**

**Scheme 5.34** Pd-catalyzed enantioselective conjugate addition of arylboronic acids to α,β-unsaturated carbonyls.

In 2007, Hu and coworkers demonstrated that palladacycle **118** is a highly efficient catalyst for the addition of organoboronic acids to enones (Figure 5.14) [218]. The reaction is performed in toluene at room temperature using $K_3PO_4$ as the base to activate the boronic acid. In contrast to dicationic Pd catalysts, Lewis acidic activator such as $AgSbF_6$ or $HBF_4$ is not necessary. It is worth noting that palladacycle **118** can also promote the 1,2-addition to α-ketoesters [218]. Similarly, palladacycle **119** derived from the inexpensive and π-acidic tris(2,4-di-*tert*-butylphenyl)phosphite, a plasticizer, was also found to promote the conjugate addition (Figure 5.14) [219, 220]. The enantiopure palladacycle **120**, generated *in situ* from optically active ferrocenyl phosphine ligand and $Pd(dba)_2$, was applied to the conjugate addition of arylboronic

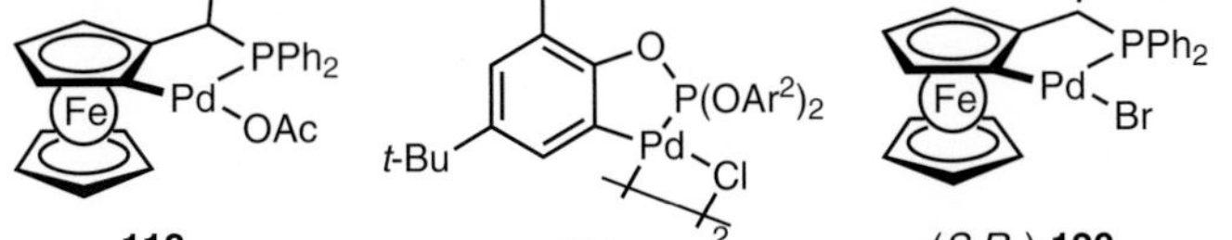


**118**
11 examples
(81–98%)

**119**
$Ar^2$ = 2,4-($t$-Bu)$_2C_6H_3$
14 examples
(47–99%)

($S$,$R_p$)-**120**
(formed in situ)
(82–90%, 42–71% ee)

**Figure 5.14** Palladacycles as catalysts for the conjugate addition of organoboronic acids.

### 5.3.3
### Addition to α,β-Unsaturated Esters, Amides, and Aldehydes

The Pd-catalyzed enantioselective conjugate addition with linear α,β-unsaturated esters and amides has proved problematic due to the competing formation of Heck-type products. An alternative entry into this substrate class is the use of α,β-unsaturated aryl esters **121** [222] and α,β-unsaturated N-benzoylamides **123** [223], which can afford the enantioselective conjugate addition product **122** and **124** in high yields and enantioselectivities without concomitant formation of Heck products (Scheme 5.35). The effectiveness of the imide unit in **123** resides in the coordinating ability of the two carbonyls that shifts the carbon-centered Pd-enolate to an O-centered one. Similarly, maleimides **125** [223] proved competent substrates in Pd-catalyzed enantioselective conjugate addition; however, the enantioselectivity strongly depends on the N-substituent (Scheme 5.35).

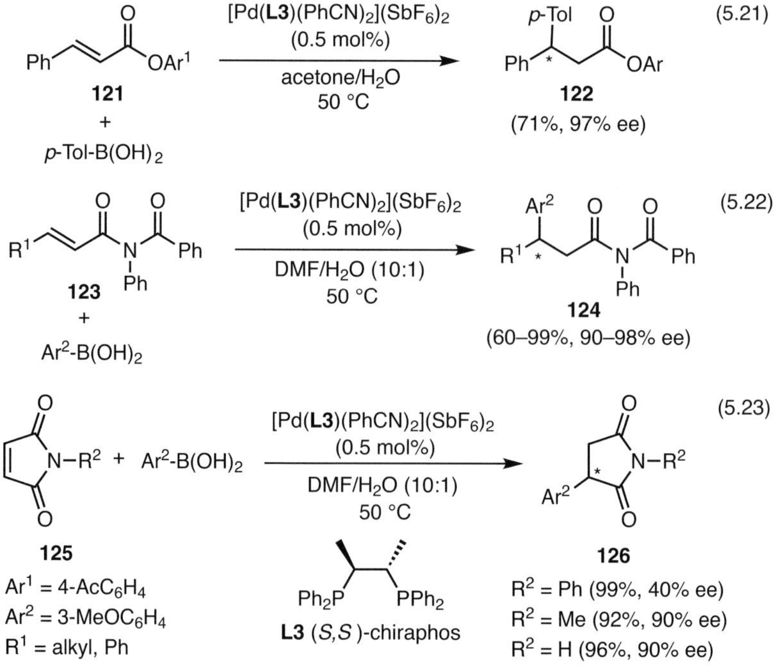

$$\text{Ph}\diagup\!\!\!\diagup\text{C(O)OAr}^1 + p\text{-Tol-B(OH)}_2 \xrightarrow[\text{acetone/H}_2\text{O},\ 50\ ^\circ\text{C}]{[\text{Pd}(\mathbf{L3})(\text{PhCN})_2](\text{SbF}_6)_2\ (0.5\ \text{mol\%})} \text{Ph-CH}(p\text{-Tol})\text{-CH}_2\text{-C(O)OAr} \quad (5.21)$$

**121** → **122** (71%, 97% ee)

$$\text{R}^1\diagup\!\!\!\diagup\text{C(O)N(Ph)C(O)Ph} + \text{Ar}^2\text{-B(OH)}_2 \xrightarrow[\text{DMF/H}_2\text{O (10:1)},\ 50\ ^\circ\text{C}]{[\text{Pd}(\mathbf{L3})(\text{PhCN})_2](\text{SbF}_6)_2\ (0.5\ \text{mol\%})} \text{R}^1\text{-CH}(\text{Ar}^2)\text{-CH}_2\text{-C(O)N(Ph)C(O)Ph} \quad (5.22)$$

**123** → **124** (60–99%, 90–98% ee)

$$\text{maleimide N-R}^2 + \text{Ar}^2\text{-B(OH)}_2 \xrightarrow[\text{DMF/H}_2\text{O (10:1)},\ 50\ ^\circ\text{C}]{[\text{Pd}(\mathbf{L3})(\text{PhCN})_2](\text{SbF}_6)_2\ (0.5\ \text{mol\%})} \text{succinimide-Ar}^2,\text{N-R}^2 \quad (5.23)$$

**125** → **126**

Ar¹ = 4-AcC₆H₄
Ar² = 3-MeOC₆H₄
R¹ = alkyl, Ph

L3 (S,S)-chiraphos (Ph₂P PPh₂)

R² = Ph (99%, 40% ee)
R² = Me (92%, 90% ee)
R² = H (96%, 90% ee)

**Scheme 5.35** Pd-catalyzed enantioselective conjugate addition of organoboronic acids to α,β-unsaturated esters and amides.

This methodology was extended to β-aryl enals (**22**), and again it was found that addition of HBF$_4$ and AgBF$_4$ dramatically increases the reaction rate. It was postulated that an additional role of HBF$_4$ is to accelerate the rate of exchange between aldehydes and their corresponding hydrates, which is the favored form in aqueous solvents (Scheme 5.36) [224].

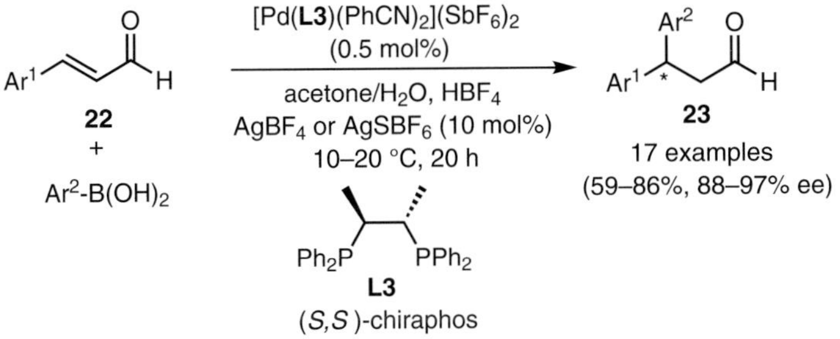

**Scheme 5.36** Pd-catalyzed enantioselective conjugate addition of organoboronic acids to β-aryl enals.

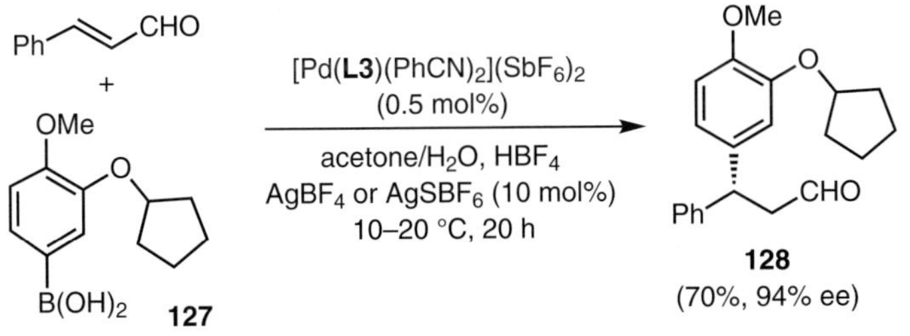

Minnaard and coworkers reported that the Pd-catalyzed enantioselective conjugate addition is efficiently catalyzed by the combination of Pd(OCOCF$_3$)$_2$ and (*R*,*R*)-Me-duphos (**L52**) [225]. The reaction did not take place if Pd(OAc)$_2$ was used as precatalyst, requiring additional activation with triflic acid (CF$_3$SO$_3$H). The yields and enantioselectivities were high for cyclic α,β-unsaturated ketone and esters; however, the % ee was much lower for linear enals (**129**) and enones (**130**). For linear enoate **131**, the formation of the Heck-type product became predominant (Scheme 5.37).

### 5.3.4
**Palladium-Catalyzed Tandem Processes**

There are several reports of palladium-catalyzed cascade reactions, such as the addition of an aryl boronic acid to enone **132**, which leads to the enantioselective formation of 1-aryl-1*H*-indenes **133** via a tandem Pd-catalyzed enantioselective conjugate addition/aldol condensation of substrate **132** depicted in Scheme 5.38 [226].

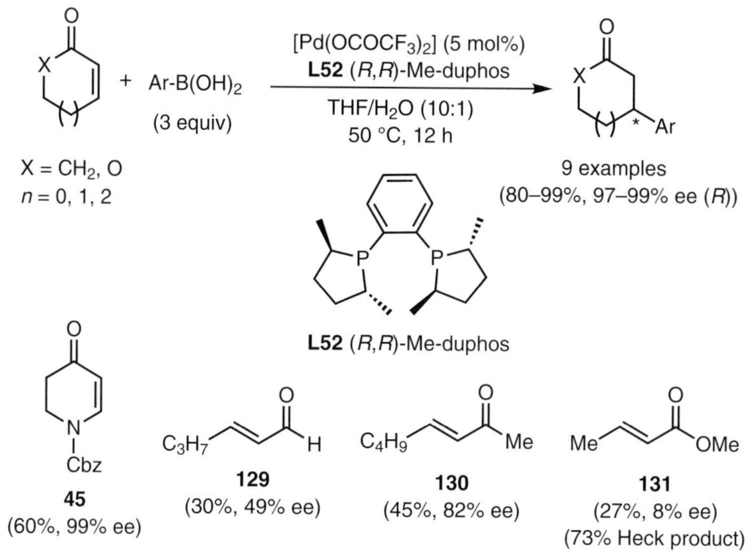

[Pd(OCOCF$_3$)$_2$] (5 mol%)
L52 (R,R)-Me-duphos

THF/H$_2$O (10:1)
50 °C, 12 h

X = CH$_2$, O
n = 0, 1, 2

Ar-B(OH)$_2$ (3 equiv)

9 examples
(80–99%, 97–99% ee (R))

**L52** (R,R)-Me-duphos

**45**
(60%, 99% ee)

**129**
(30%, 49% ee)

**130**
(45%, 82% ee)

**131**
(27%, 8% ee)
(73% Heck product)

**Scheme 5.37** Pd-catalyzed enantioselective conjugate addition of arylboronic acids to α,β-unsaturated carbonyls.

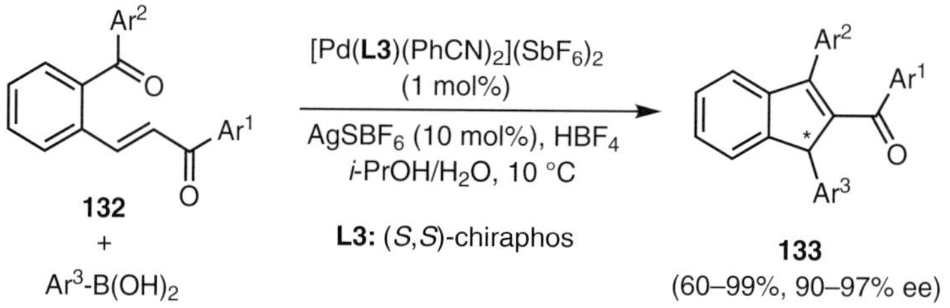

**132**
+
Ar$^3$-B(OH)$_2$

[Pd(**L3**)(PhCN)$_2$](SbF$_6$)$_2$
(1 mol%)

AgSBF$_6$ (10 mol%), HBF$_4$
*i*-PrOH/H$_2$O, 10 °C

**L3**: (S,S)-chiraphos

**133**
(60–99%, 90–97% ee)

**Scheme 5.38** Tandem Pd-catalyzed enantioselective conjugate addition/aldol condensation.

In analogy to the tandem carborhodation/ECA processes developed for rhodium by Lautens and Marquardt [227], Lu and coworkers developed a Pd-catalyzed variation of this transformation using **134** and internal alkynes (**135**) to form chiral indenes **136** (Scheme 5.39) [228]. Interestingly, binap-type ligand (S)-**L53** proved the best chiral ligand, while it proved ineffective in other Pd-catalyzed conjugate addition.

## 5.4
## Conclusions

Since the first edition of *Boronic Acids* (2005), great progress has been made in the field of Rh-catalyzed enantioselective conjugate addition of organoboron reagents. Synthetic organic chemists have now at their disposal a large toolbox of conditions, ligands, and substrate classes to choose from to confidently tackle enantioselective

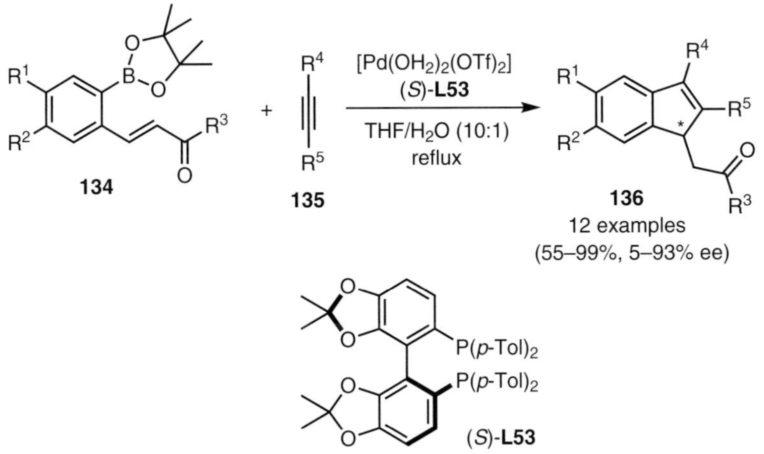

**Scheme 5.39** Pd-catalyzed carborhodation/ECA.

conjugate additions in a complex setting and on a large scale. However, there are still some unsolved issues with rhodium-catalyzed enantioselective conjugate additions, such as the addition of an alkylboronic to an activated alkene. In this respect, the Rh- and Pd-based methodologies are complementary to the copper-catalyzed processes that are well suited for the enantioselective conjugate addition of more reactive organometallic reagents such as alkyl, zinc, or aluminum reagents to α,β-unsaturated carbonyl compounds.

## References

1 Cho, C.S., Motofusa, S., Ohe, K., and Uemura, S. (1995) *J. Org. Chem.*, **60**, 883–888.
2 Sakai, M., Hayashi, T., and Miyaura, N. (1997) *Organometallics*, **16**, 4229–4231.
3 Hayashi, T. (2001) *Synlett*, 879–887.
4 Hayashi, T. (2003) *Russ. Chem. Bull.*, **52**, 2595–2605.
5 Hayashi, T. and Yamasaki, K. (2003) *Chem. Rev.*, **103**, 2829–2844.
6 Lautens, M., Fagnou, K., and Hiebert, S. (2003) *Acc. Chem. Res.*, **36**, 48–58.
7 Fagnou, K. and Lautens, M. (2003) *Chem. Rev.*, **103**, 169–196.
8 Hayashi, T. (2004) *Bull. Chem. Soc. Jpn.*, **77**, 13–21.
9 Hayashi, T. (2004) *Pure Appl. Chem.*, **76**, 465–475.
10 Yoshida, K. and Hayashi, T. (2005) *Boronic Acids* (ed. D.G. Hall), Wiley-VCH Verlag GmbH, Weinheim, pp. 171–204.
11 Yoshida, K. and Hayashi, T. (2005) *Modern Rhodium-Catalyzed Organic Reactions* (ed. P.A. Evans), Wiley-VCH Verlag GmbH, Weinheim, pp. 55–78.
12 Yamamoto, Y., Nishikata, T., and Miyaura, N. (2008) *Pure Appl. Chem.*, **80**, 807–817.
13 Miyaura, N. (2008) *Bull. Chem. Soc. Jpn.*, **81**, 1535–1553.
14 Miyaura, N. (2009) *Synlett*, 2039–2050.
15 Takaya, Y., Ogasawara, M., Hayashi, T., Sakai, M., and Miyaura, N. (1998) *J. Am. Chem. Soc.*, **120**, 5579–5580.
16 Krause, N. (1998) *Angew. Chem., Int. Ed.*, **37**, 283–285.
17 Alexakis, A. (2002) *Pure Appl. Chem.*, **74**, 37–42.

18. Lopez, F., Minnaard, A.J., and Feringa, B.L. (2007) *Acc. Chem. Res.*, **40**, 179–188.
19. Jerphagnon, T., Pizzuti, M.G., Minnaard, A.J., and Feringa, B.L. (2009) *Chem. Soc. Rev.*, **38**, 1039–1075.
20. Hayashi, T., Takahashi, M., Takaya, Y., and Ogasawara, M. (2002) *J. Am. Chem. Soc.*, **124**, 5052–5058.
21. Blackmond, D.G. (2005) *Angew. Chem. Int. Ed.*, **44**, 4302–4320.
22. Kina, A., Iwamura, H., and Hayashi, T. (2006) *J. Am. Chem. Soc.*, **128**, 3904–3905.
23. Kina, A., Yasuhara, Y., Nishimura, T., Iwamura, H., and Hayashi, T. (2006) *Chem. Asian J.*, **1**, 707–711.
24. Martina, S.L.X., Minnaard, A.J., Hessen, B., and Feringa, B.L. (2005) *Tetrahedron Lett.*, **46**, 7159–7163.
25. Matos, K. and Soderquist, J.A. (1998) *J. Org. Chem.*, **63**, 461–470.
26. Zhao, P.J., Incarvito, C.D., and Hartwig, J.F. (2007) *J. Am. Chem. Soc.*, **129**, 1876–1877.
27. Itooka, R., Iguchi, Y., and Miyaura, N. (2003) *J. Org. Chem.*, **68**, 6000–6004.
28. Ozawa, F., Kubo, A., Matsumoto, Y., Hayashi, T., Nishioka, E., Yanagi, K., and Moriguchi, K. (1993) *Organometallics*, **12**, 4188–4196.
29. Hayashi, T., Takahashi, M., Takaya, Y., and Ogasawara, M. (2002) *J. Am. Chem. Soc.*, **124**, 5052–5058.
30. Senda, T., Ogasawara, M., and Hayashi, T. (2001) *J. Org. Chem.*, **66**, 6852–6856.
31. Takaya, Y., Senda, T., Kurushima, H., Ogasawara, M., and Hayashi, T. (1999) *Tetrahedron Asymmetry*, **10**, 4047–4056.
32. Takaya, Y., Ogasawara, M., and Hayashi, T. (1998) *Tetrahedron Lett.*, **39**, 8479–8482.
33. Darses, S. and Genêt, J.-P. (2003) *Eur. J. Org. Chem.*, 4313–4327.
34. Molander, G.A. and Figueroa, R. (2005) *Aldrichim. Acta*, **38**, 49–56.
35. Darses, S. and Genet, J.P. (2008) *Chem. Rev.*, **108**, 288–325.
36. Navarre, L., Martinez, R., Genêt, J.P., and Darses, S. (2008) *J. Am. Chem. Soc.*, **130**, 6159–6169.
37. Duursma, A., Boiteau, J.G., Lefort, L., Boogers, J.A., De Vries, A.H., De Vries, J., Minnaard, A.J., and Feringa, B.L. (2004) *J. Org. Chem.*, **69**, 8045–8052.
38. Batey, R.A. and Quach, T.D. (2001) *Tetrahedron Lett.*, **42**, 9099–9103.
39. Molander, G.A. and Biolatto, B. (2002) *Org. Lett.*, **4**, 1867–1870.
40. Molander, G.A. and Biolatto, B. (2003) *J. Org. Chem.*, **68**, 4302–4314.
41. Yuen, A.K.L. and Hutton, C.A. (2005) *Tetrahedron Lett.*, **46**, 7899–7903.
42. Gendrineau, T., Genêt, J.P., and Darses, S. (2009) *Org. Lett.*, **11**, 3486–3489.
43. Takaya, Y., Ogasawara, M., and Hayashi, T. (1999) *Tetrahedron Lett.*, **40**, 6957–6961.
44. Yamamoto, Y., Takizawa, M., Yu, X., and Miyaura, N. (2008) *Angew. Chem. Int. Ed.*, **47**, 928–931.
45. Yu, X., Yamamoto, Y., and Miyaura, N. (2009) *Synlett*, 994–998.
46. Yoshida, K., Ogasawara, M., and Hayashi, T. (2003) *J. Org. Chem.*, **68**, 1901–1905.
47. Shintani, R., Tsutsumi, Y., Nagaosa, M., Nishimura, T., and Hayashi, T. (2009) *J. Am. Chem. Soc.*, **131**, 13588–13589.
48. Kabalka, G.W., Das, B.C., and Das, S. (2002) *Tetrahedron Lett.*, **43**, 2323–2325.
49. Walter, C., Auer, G., and Oestreich, M. (2006) *Angew. Chem. Int. Ed.*, **45**, 5675–5677.
50. Walter, C. and Oestreich, M. (2008) *Angew. Chem. Int. Ed.*, **47**, 3818–3820.
51. Sakuma, S. and Miyaura, N. (2001) *J. Org. Chem.*, **66**, 8944–8946.
52. Oi, S., Moro, M., and Inoue, Y. (1997) *Chem. Commun.*, 1621–1622.
53. Oi, S., Moro, M., Ono, S., and Inoue, Y. (1998) *Chem. Lett.*, 83–84.
54. Oi, S., Moro, M., and Inoue, Y. (2001) *Organometallics*, **20**, 1036–1037.
55. Oi, S., Honma, Y., and Inoue, Y. (2002) *Org. Lett.*, **4**, 667–669.
56. Oi, S., Moro, M., Ito, H., Honma, Y., Miyano, S., and Inoue, Y. (2002) *Tetrahedron*, **58**, 91–97.

57 Oi, S., Taira, A., Honma, Y., and Inoue, Y. (2003) *Org. Lett.*, **5**, 97–99.
58 Oi, S., Sato, T., and Inoue, Y. (2004) *Tetrahedron Lett.*, **45**, 5051–5055.
59 Oi, S., Taira, A., Honma, Y., Sato, T., and Inoue, Y. (2006) *Tetrahedron Asymmetry*, **17**, 598–602.
60 Itoh, T., Mase, T., Nishikata, T., Iyama, T., Tachikawa, H., Kobayashi, Y., Yarnamoto, Y., and Miyaura, N. (2006) *Tetrahedron*, **62**, 9610–9621.
61 Sakuma, S., Sakai, M., Itooka, R., and Miyaura, N. (2000) *J. Org. Chem.*, **65**, 5951–5955.
62 Lukin, K., Zhang, Q.Y., and Leanna, M.R. (2009) *J. Org. Chem.*, **74**, 929–931.
63 Yamamoto, Y., Kurihara, K., Sugishita, N., Oshita, K., Piao, D.G., and Miyaura, N. (2005) *Chem. Lett.*, **34**, 1224–1225.
64 Chen, F.-X., Kina, A., and Hayashi, T. (2006) *Org. Lett.*, **8**, 341–344.
65 Takaya, Y., Ogasawara, M., and Hayashi, T. (2000) *Chirality*, **12**, 469–471.
66 Vandyck, K., Matthys, B., Willen, M., Robeyns, K., Van Meervelt, L., and Van Der Eycken, J. (2006) *Org. Lett.*, **8**, 363–366.
67 Reetz, M.T., Moulin, D., and Gosberg, A. (2001) *Org. Lett.*, **3**, 4083–4085.
68 Kurihara, K., Sugishita, N., Oshita, K., Piao, D., Yamamoto, Y., and Miyaura, N. (2007) *J. Organomet. Chem.*, **692**, 428–435.
69 Amengual, R., Michelet, V., and Genêt, J.P. (2002) *Synlett*, 1791–1794.
70 Korenaga, T., Osaki, K., Maenishi, R., and Sakai, T. (2009) *Org. Lett.*, **11**, 2325–2328.
71 Shi, Q., Xu, L.J., Li, X.S., Jia, X., Wang, R.H., Au-Yeung, T.T.L., Chan, A.S.C., Hayashi, T., Cao, R., and Hong, M.C. (2003) *Tetrahedron Lett.*, **44**, 6505–6508.
72 Madec, J., Michaud, G., Genêt, J.P., and Marinetti, A. (2004) *Tetrahedron Asymmetry*, **15**, 2253–2261.
73 Otomaru, Y., Senda, T., and Hayashi, T. (2004) *Org. Lett.*, **6**, 3357–3359.
74 Shimada, T., Suda, M., Nagano, T., and Kakiuchi, K. (2005) *J. Org. Chem.*, **70**, 10178–10181.
75 Yuan, W., Cun, L., Mi, A., Jiang, Y., and Gong, L. (2009) *Tetrahedron*, **65**, 4130–4141.
76 Stemmler, R.T. and Bolm, C. (2005) *J. Org. Chem.*, **70**, 9925–9931.
77 Kromm, K., Eichenseher, S., Prommesberger, M., Hampel, F., and Gladysz, J.A. (2005) *Eur. J. Org. Chem.*, 2983–2998.
78 Lagasse, F. and Kagan, H.B. (2000) *Chem. Pharm. Bull.*, **48**, 315–324.
79 Van Den Berg, M., Minnaard, A.J., Schudde, E.P., Van Esch, J., De Vries, A.H.M., De Vries, J.G., and Feringa, B.L. (2000) *J. Am. Chem. Soc.*, **122**, 11539–11540.
80 Boiteau, J.G., Minnaard, A.J., and Feringa, B.L. (2003) *J. Org. Chem.*, **68**, 9481–9484.
81 Ma, Y.D., Song, C., Ma, C.Q., Sun, Z.J., Chai, Q., and Andrus, M.B. (2003) *Angew. Chem. Int. Ed.*, **42**, 5871–5874.
82 Facchetti, S., Cavallini, I., Funaioli, T., Marchetti, F., and Iuliano, A. (2009) *Organometallics*, **28**, 4150–4158.
83 Iuliano, A., Facchetti, S., and Funaioli, T. (2009) *Chem. Commun.*, 457–459.
84 Reetz, M.T. (2008) *Angew. Chem. Int. Ed.*, **47**, 2556–2588.
85 Duursma, A., Pena, D., Minnaard, A.J., and Feringa, B.L. (2005) *Tetrahedron Asymmetry*, **16**, 1901–1904.
86 Monti, C., Gennari, C., and Piarulli, U. (2007) *Chem. Eur. J.*, **13**, 1547–1558.
87 Itooka, R., Iguchi, Y., and Miyaura, N. (2001) *Chem. Lett.*, **30**, 722–723.
88 Hayashi, T., Ueyama, K., Tokunaga, N., and Yoshida, H. (2003) *J. Am. Chem. Soc.*, **125**, 11508–11509.
89 Defieber, C., Paquin, J.-F., Serna, S., and Carreira, E.M. (2004) *Org. Lett.*, **6**, 3873–3876.
90 Paquin, J.-F., Stephenson, C.R.J., Defieber, C., and Carreira, E.M. (2005) *Org. Lett.*, **7**, 3821–3824.
91 Otomaru, Y., Okamoto, K., Shintani, R., and Hayashi, T. (2005) *J. Org. Chem.*, **70**, 2503–2508.
92 Kina, A., Ueyama, K., and Hayashi, T. (2005) *Org. Lett.*, **7**, 5889–5892.

93 Lang, F., Breher, F., Stein, D., and Grützmacher, H. (2005) *Organometallics*, **24**, 2997–3007.

94 Berthon-Gelloz, G. and Hayashi, T. (2006) *J. Org. Chem.*, **71**, 8957–8960.

95 Helbig, S., Sauer, S., Cramer, N., Laschat, S., Baro, A., and Frey, W. (2007) *Adv. Synth. Catal.*, **349**, 2331–2237.

96 Nishimura, T., Nagaosa, M., and Hayashi, T. (2008) *Chem. Lett.*, **37**, 860–861.

97 Okamoto, K., Hayashi, T., and Rawal, V.H. (2008) *Org. Lett.*, **10**, 4387–4389.

98 Gendrineau, T., Chuzel, O., Eijsberg, H., Genêt, J.P., and Darses, S. (2008) *Angew. Chem. Int. Ed.*, **47**, 7669–7672.

99 Shintani, R., Ichikawa, Y., Takatsu, K., Chen, F.X., and Hayashi, T. (2009) *J. Org. Chem.*, **74**, 869–873.

100 Fischer, C., Defieber, C., Takeyuki, S., and Carreira, E.M. (2004) *J. Am. Chem. Soc.*, **126**, 1628–1629.

101 Defieber, C., Grützmacher, H., and Carreira, E.M. (2008) *Angew. Chem. Int. Ed.*, **47**, 4482–4502.

102 Shintani, R., Ueyama, K., Yamada, I., and Hayashi, T. (2004) *Org. Lett.*, **6**, 3425–3427.

103 Hayashi, T., Tokunaga, N., Okamoto, K., and Shintani, R. (2005) *Chem. Lett.*, **34**, 1480–1481.

104 Shintani, R., Okamoto, K., and Hayashi, T. (2005) *Org. Lett.*, **7**, 4757–4759.

105 Paquin, J.-F., Defieber, C., Stephenson, C.R.J., and Carreira, E.M. (2005) *J. Am. Chem. Soc.*, **127**, 10850–10851.

106 Chen, F., Kina, A., and Hayashi, T. (2006) *Org. Lett.*, **8**, 341–344.

107 Tokunaga, N. and Hayashi, T. (2007) *Adv. Synth. Catal.*, **349**, 513–516.

108 Soergel, S., Tokunaga, N., Sasaki, K., Okamoto, K., and Hayashi, T. (2008) *Org. Lett.*, **10**, 589–592.

109 Wang, Z.Q., Feng, C.G., Xu, M.H., and Lin, G.Q. (2007) *J. Am. Chem. Soc.*, **129**, 5336–5337.

110 Otomaru, Y., Tokunaga, N., Shintani, R., and Hayashi, T. (2005) *Org. Lett.*, **7**, 307–310.

111 Otomaru, Y., Kina, A., Shintani, R., and Hayashi, T. (2005) *Tetrahedron Asymmetry*, **16**, 1673–1679.

112 Brown, M.K. and Corey, E.J. (2010) *Org. Lett.*, **12**, 172–175.

113 Hu, X., Zhuang, M., Cao, Z., and Du, H. (2009) *Org. Lett.*, **11**, 4744–4747.

114 Chen, M.S., Prabagaran, N., Labenz, N.A., and White, C.M. (2005) *J. Am. Chem. Soc.*, **127**, 6970–6971.

115 Chen, M.S. and White, C.M. (2004) *J. Am. Chem. Soc.*, **126**, 1346–1347.

116 Mariz, R., Luan, X., Gatti, M., Linden, A., and Dorta, R. (2008) *J. Am. Chem. Soc.*, **130**, 2172–2173.

117 Bürgi, J.J., Mariz, R., Gatti, M., Drinkel, E., Luan, X., Blumentritt, S., Linden, A., and Dorta, R. (2009) *Angew. Chem. Int. Ed.*, **48**, 2768–2771.

118 Tokunaga, N., Otomaru, Y., Okamoto, K., Ueyama, K., Shintani, R., and Hayashi, T. (2004) *J. Am. Chem. Soc.*, **126**, 13584–13585.

119 Kuriyama, M. and Tomioka, K. (2001) *Tetrahedron Lett.*, **42**, 921–923.

120 Kuriyama, M., Nagai, K., Yamada, K., Miwa, Y., Taga, T., and Tomioka, K. (2002) *J. Am. Chem. Soc.*, **124**, 8932–8939.

121 Chen, Q., Soeta, T., Kuriyama, M., Yamada, K.I., and Tomioka, K. (2006) *Adv. Synth. Catal.*, **348**, 2604–2608.

122 Becht, J.M., Bappert, E., and Helmchen, G. (2005) *Adv. Synth. Catal.*, **347**, 1495–1498.

123 Shintani, R., Duan, W.L., Nagano, T., Okada, A., and Hayashi, T. (2005) *Angew. Chem. Int. Ed.*, **44**, 4611–4614.

124 Duan, W.L., Iwamura, H., Shintani, R., and Hayashi, T. (2007) *J. Am. Chem. Soc.*, **129**, 2130–2138.

125 Piras, E., Lang, F., Ruegger, H., Stein, D., Worle, M., and Grützmacher, H. (2006) *Chem. Eur. J.*, **12**, 5849–5858.

126 Minuth, T. and Boysen, M.M.K. (2009) *Org. Lett.*, **11**, 4212–4215.

127 Mariz, R., Briceno, A., Dorta, R., and Dorta, R. (2008) *Organometallics*, **27**, 6605–6613.

128 Defieber, C., Ariger, M.A., Moriel, P., and Carreira, E.M. (2007) *Angew. Chem. Int. Ed.*, **46**, 3139–3143.

129 Kasak, P., Arion, V.B., and Widhalm, M. (2006) *Tetrahedron Asymmetry*, **17**, 3084–3090.

130 Hayashi, T., Tokunaga, N., Okamoto, K., and Shintani, R. (2005) *Chem. Lett.*, **34**, 1480–1481.
131 Tokunaga, N. and Hayashi, T. (2006) *Tetrahedron Asymmetry*, **17**, 607–613.
132 Ueda, M. and Miyaura, N. (2000) *J. Org. Chem.*, **65**, 4450–4452.
133 Meyer, O., Becht, J.M., and Helmchen, G. (2003) *Synlett*, 1539–1541.
134 Chen, G., Tokunaga, N., and Hayashi, T. (2005) *Org. Lett.*, **7**, 2285–2288.
135 Brock, S., Hose, D.R.J., Moseley, J.D., Parker, A.J., Patel, I., and Williams, A.J. (2008) *Org. Process. Res. Dev.*, **12**, 496–502.
136 Song, Z.G.J., Zhao, M.Z., Desmond, R., Devine, P., Tschaen, D.M., Tillyer, R., Frey, L., Heid, R., Xu, F., Foster, B., Li, J., Reamer, R., Volante, R., Grabowski, E.J.J., Dolling, U.H., Reider, P.J., Okada, S., Kato, Y., and Mano, E. (1999) *J. Org. Chem.*, **64**, 9658–9667.
137 Frost, C.G., Penrose, S.D., and Gleave, R. (2008) *Org. Biomol. Chem.*, **6**, 4340–4347.
138 Jagt, R.B.C., De Vries, J.G., Feringa, B.L., and Minnaard, A.J. (2005) *Org. Lett.*, **7**, 2433–2435.
139 Ramnauth, J., Poulin, O., Bratovanov, S.S., Rakhit, S., and Maddaford, S.P. (2001) *Org. Lett.*, **3**, 2571–2573.
140 Kumaraswamy, G., Ramakrishna, G., Naresh, P., Jagadeesh, B., and Sridhar, B. (2009) *J. Org. Chem.*, **74**, 8468–8471.
141 Navarro, C., Moreno, A., and Csákÿ, A.G. (2009) *J. Org. Chem.*, **74**, 466–469.
142 Mondière, A., Pousse, G., Bouyssi, D., and Balme, G. (2009) *Eur. J. Org. Chem.*, 4225–4229.
143 Zoute, L., Kociok-Köhn, G., and Frost, C.G. (2009) *Org. Lett.*, **11**, 2491–2494.
144 Shintani, R., Ueyama, K., Yamada, I., and Hayashi, T. (2004) *Org. Lett.*, **6**, 3425–3427.
145 Duan, W.L., Imazaki, Y., Shintani, R., and Hayashi, T. (2007) *Tetrahedron*, **63**, 8529–8536.
146 Trost, B.M. and Jiang, C. (2006) *Synthesis*, 369–396.
147 Riant, O. and Hannedouche, J. (2007) *Org. Biomol. Chem.*, **5**, 873–888.
148 Cozzi, P.G., Hilgraf, R., and Zimmermann, N. (2007) *Eur. J. Org. Chem.*, 5969–5994.
149 Shintani, R., Duan, W.L., and Hayashi, T. (2006) *J. Am. Chem. Soc.*, **128**, 5628–5629.
150 Jacques, T., Markó, I.E., and Pospísil, J. (2005) *Multicomponent Reactions* (eds J. Zhu and H. Bienaymé), Wiley-VCH Verlag GmbH, Weinheim, pp. 398–452.
151 Shintani, R., Ichikawa, Y., Hayashi, T., Chen, J., Nakao, Y., and Hiyama, T. (2007) *Org. Lett.*, **9**, 4643–4645.
152 Shintani, R., Kimura, T., and Hayashi, T. (2005) *Chem. Commun.*, 3213–3214.
153 Konno, T., Tanaka, T., Miyabe, T., Morigaki, A., and Ishihara, T. (2008) *Tetrahedron Lett.*, **49**, 2106–2110.
154 Lalic, G. and Corey, E.J. (2007) *Org. Lett.*, **9**, 4921–4923.
155 Lalic, G. and Corey, E.J. (2008) *Tetrahedron Lett.*, **49**, 4894–4896.
156 Hayashi, T., Senda, T., Takaya, Y., and Ogasawara, M. (1999) *J. Am. Chem. Soc.*, **121**, 11591–11592.
157 Hayashi, T., Senda, T., and Ogasawara, M. (2000) *J. Am. Chem. Soc.*, **122**, 10716–10717.
158 Burgey, C.S., Paone, D.V., Shaw, A.W., Deng, J.Z., Nguyen, D.N., Potteiger, C.M., Graham, S.L., Vacca, J.P., and Williams, T.M. (2008) *Org. Lett.*, **10**, 3235–3238.
159 Yoshida, K. and Hayashi, T. (2003) *J. Am. Chem. Soc.*, **125**, 2872–2873.
160 Mauleon, P. and Carretero, J.C. (2004) *Org. Lett.*, **6**, 3195–3198.
161 Mauleon, P., Alonso, I., Rivero, M.R., and Carretero, J.C. (2007) *J. Org. Chem.*, **72**, 9924–9935.
162 Plesniak, K., Zarecki, A., and Wicha, J. (2007) *Top. Curr. Chem.*, **275**, 163–250.
163 Mauleón, P. and Carretero, J. (2005) *Chem. Commun.*, 4961–4963.
164 Miura, T., Takahashi, Y., and Murakami, M. (2007) *Chem. Commun.*, 595–597.
165 Mohr, J., Hong, A., and Stoltz, B. (2009) *Nature Chem.*, **1**, 359–369.
166 Chapman, C.J., Wadsworth, K.J., and Frost, C.G. (2003) *J. Organomet. Chem.*, **680**, 206–211.
167 Au-Yeung, T.T.-L., Chan, S.-S., and Chan, A.S.C. (2004) *Transition Metals for Organic Synthesis*, vol. 2

(eds M., Beller and C. Bolm), Wiley-VCH Verlag GmbH, Weinheim, pp. 14–28.
168 Navarre, L., Darses, S., and Genêt, J.P. (2004) *Angew. Chem. Int. Ed.*, **43**, 719–723.
169 Moss, R.J., Wadsworth, K.J., Chapman, C.J., and Frost, C.G. (2004) *Chem. Commun.*, 1984–1985.
170 Frost, C.G., Penrose, S.D., Lambshead, K., Raithby, P.R., Warren, J.E., and Gleave, R. (2007) *Org. Lett.*, **9**, 2119–2122.
171 Sibi, M.P., Tatamidani, H., and Patil, K. (2005) *Org. Lett.*, **7**, 2571–2573.
172 Chapman, C.J., Hargrave, J.D., Bish, G., and Frost, C.G. (2008) *Tetrahedron*, **64**, 9528–9539.
173 Nishimura, T., Hirabayashi, S., Yasuhara, Y., and Hayashi, T. (2006) *J. Am. Chem. Soc.*, **128**, 2556–2557.
174 Herrán, G.D.L., Murcia, C., and CsáKÿ, A.G. (2005) *Org. Lett.*, **7**, 5629–5632.
175 Hayashi, T., Yamamoto, S., and Tokunaga, N. (2005) *Angew. Chem. Int. Ed.*, **44**, 4224–4227.
176 Nishimura, T., Yasuhara, Y., and Hayashi, T. (2006) *Angew. Chem. Int. Ed.*, **45**, 5164–5166.
177 Koike, T., Du, X.L., Sanada, T., Danda, Y., and Mori, A. (2003) *Angew. Chem. Int. Ed.*, **42**, 89–92.
178 Hayashi, T., Tokunaga, N., Yoshida, K., and Han, J.W. (2002) *J. Am. Chem. Soc.*, **124**, 12102–12103.
179 Hayashi, T., Kawai, M., and Tokunaga, N. (2004) *Angew. Chem. Int. Ed.*, **43**, 6125–6128.
180 Shintani, R., Tokunaga, N., Doi, H., and Hayashi, T. (2004) *J. Am. Chem. Soc.*, **126**, 6240–6241.
181 Mori, A., Danda, Y., Fujii, T., Hirabayashi, S., and Osakada, K. (2001) *J. Am. Chem. Soc.*, **123**, 10774–10775.
182 Huang, T. and Li, C. (2001) *Chem. Commun.*, 2348–2349.
183 Nakao, Y., Imanaka, H., Sahoo, A.K., Yada, A., and Hiyama, T. (2005) *J. Am. Chem. Soc.*, **127**, 6952–6953.
184 Nakao, Y., Chen, J., Imanaka, H., Hiyama, T., Ichikawa, Y., Duan, W.L., Shintani, R., and Hayashi, T. (2007) *J. Am. Chem. Soc.*, **129**, 9137–9143.
185 Nakao, Y., Imanaka, H., Chen, J., Yada, A., and Hiyama, T. (2007) *J. Organomet. Chem.*, **692**, 585–603.
186 Kakuuchi, A., Taguchi, T., and Hanzawa, Y. (2004) *Tetrahedron*, **60**, 1293–1299.
187 Nicolaou, K.C., Tang, W.J., Dagneau, P., and Faraoni, R. (2005) *Angew. Chem. Int. Ed.*, **44**, 3874–3879.
188 Venkatraman, S. and Li, C.J. (2001) *Tetrahedron Lett.*, **42**, 781–784.
189 Ding, R., Chen, Y.J., Wang, D., and Li, C.J. (2001) *Synlett*, 1470–1472.
190 Venkatraman, S., Meng, Y., and Li, C.J. (2001) *Tetrahedron Lett.*, **42**, 4459–4462.
191 Miura, T. and Murakami, M. (2005) *Chem. Commun.*, 5676–5677.
192 Trost, B.M. (1991) *Science*, **254**, 1471–1477.
193 Guo, H.C. and Ma, J.A. (2006) *Angew. Chem. Int. Ed.*, **45**, 354–366.
194 Miura, T. and Murakami, M. (2007) *Chem. Commun.*, 217–224.
195 Youn, S. (2009) *Eur. J. Org. Chem.*, 2597–2605.
196 Yoshida, K., Ogasawara, M., and Hayashi, T. (2002) *J. Am. Chem. Soc.*, **124**, 10984–10985.
197 Cauble, D.F., Gipson, J.D., and Krische, M.J. (2003) *J. Am. Chem. Soc.*, **125**, 1110–1111.
198 Miura, T., Harumashi, T., and Murakami, M. (2007) *Org. Lett.*, **9**, 741–743.
199 Shintani, R., Tsurusaki, A., Okamoto, K., and Hayashi, T. (2005) *Angew. Chem. Int. Ed.*, **44**, 3909–3912.
200 Shintani, R., Okamoto, K., Otomaru, Y., Ueyama, K., and Hayashi, T. (2005) *J. Am. Chem. Soc.*, **127**, 54–55.
201 Chen, Y. and Lee, C. (2006) *J. Am. Chem. Soc.*, **128**, 15598–15599.
202 Yamamoto, Y., Nishikata, T., and Miyaura, N. (2006) *J. Synth. Org. Chem. Jpn.*, **64**, 1112–1121.
203 Gutnov, A. (2008) *Eur. J. Org. Chem.*, 4547–4554.
204 Kobayashi, K., Nishikata, T., Yamamoto, Y., and Miyaura, N. (2008) *Bull. Chem. Soc. Jpn.*, **81**, 1019–1025.

205 Yamamoto, Y. and Miyaura, N. (2008) *J. Synth. Org. Chem. Jpn.*, **66**, 194–204.
206 Cacchi, S., Misiti, D., and Palmieri, G. (1981) *Tetrahedron*, **37**, 2941–2946.
207 Cacchi, S., Latorre, F., and Misiti, D. (1979) *Tetrahedron Lett.*, **20**, 4591–4594.
208 Albeniz, A.C., Catalina, N.M., Espinet, P., and Redon, R. (1999) *Organometallics*, **18**, 5571–5576.
209 Nishikata, T., Yamamoto, Y., and Miyaura, N. (2003) *Angew. Chem. Int. Ed.*, **42**, 2768–2770.
210 Nishikata, T., Yamamoto, Y., and Miyaura, N. (2004) *Organometallics*, **23**, 4317–4324.
211 Kamer, P.C.J., Van Leeuwen, P.W.N., and Reek, J.N.H. (2001) *Acc. Chem. Res.*, **34**, 895–904.
212 Lu, X.Y. and Lin, S.H. (2005) *J. Org. Chem.*, **70**, 9651–9653.
213 Lin, S. and Lu, X. (2006) *Tetrahedron Lett.*, **47**, 7167–7170.
214 Nishikata, T., Yamamoto, Y., and Miyaura, N. (2004) *Chem. Commun.*, 1822–1823.
215 Nishikata, T., Yamamoto, Y., Gridnev, I.D., and Miyaura, N. (2005) *Organometallics*, **24**, 5025–5032.
216 Nishikata, T., Yamamoto, Y., and Miyaura, N. (2007) *Adv. Synth. Catal.*, **349**, 1759–1764.
217 Zhang, T. and Shi, M. (2008) *Chem. Eur. J.*, **14**, 3759–3764.
218 He, P., Lu, Y., Dong, C., and Hu, Q. (2007) *Org. Lett.*, **9**, 343–346.
219 Bedford, R., Betham, M., Charmant, J.P.H., Haddow, M.F., Orpen, A., Pilarski, L.T., Coles, S., and Hursthouse, M. (2007) *Organometallics*, **26**, 6346–6353.
220 Bedford, R.B., Dumycz, H., Haddow, M.F., Pilarski, L.T., Orpen, A.G., Pringle, P.G., and Wingad, R.L. (2009) *Dalton Trans.*, 7796–7804.
221 Suzuma, Y., Yamamoto, T., Ohta, T., and Ito, Y. (2007) *Chem. Lett.*, **36**, 470–471.
222 Nishikata, T., Kiyomura, S., Yamamoto, Y., and Miyaura, N. (2008) *Synlett*, 2487–2490.
223 Nishikata, T., Yamamoto, Y., and Miyaura, N. (2007) *Chem. Lett.*, **36**, 1442–1443.
224 Nishikata, T., Yamamoto, Y., and Miyaura, N. (2007) *Tetrahedron Lett.*, **48**, 4007–4010.
225 Gini, F., Hessen, B., and Minnaard, A.J. (2005) *Org. Lett.*, **7**, 5309–5312.
226 Nishikata, T., Kobayashi, Y., Kobayshi, K., Yamamoto, Y., and Miyaura, N. (2007) *Synlett*, 3055–3057.
227 Lautens, M. and Marquardt, T. (2004) *J. Org. Chem.*, **69**, 4607–4614.
228 Zhou, F., Yang, M., and Lu, X. (2009) *Org. Lett.*, **11**, 1405–1408.

# 6
# Recent Advances in Chan–Lam Coupling Reaction: Copper-Promoted C–Heteroatom Bond Cross-Coupling Reactions with Boronic Acids and Derivatives

*Jennifer X. Qiao and Patrick Y.S. Lam*

## 6.1
### General Introduction

The palladium-catalyzed *Suzuki–Miyaura coupling* between a boronic acid and an aryl halide is one of the most powerful and effective methods for carbon–carbon (C–C) bond formation [1]. On the other hand, the corresponding aryl carbon–heteroatom (C–X, where X = O, N, S) bond cross-coupling was less well established prior to the discovery of *Buchwald–Hartwig coupling reaction* [2, 3] with Pd and *Chan–Lam coupling reaction* with Cu. Like the C–C bond formation, the C–heteroatom transformation is equally essential because aryl ethers, anilines, and thioethers are ubiquitous moieties in a wide range of molecules with many important applications, especially in the areas of pharmaceutical, agricultural, and material science research. The classic copper-mediated *Ullmann–Goldberg reaction* [4] to generate aryl amines and aryl ethers with aryl halides involves harsh conditions, such as very high temperatures and strong bases. Recent modification with addition of ligands functions at a lower temperature [5, 6]. Similarly, Pd-catalyzed *Buchwald–Hartwig coupling reaction*, a very useful reaction, has the drawback of high temperature, strong base, and need for expensive Pd (Pd, ~$1000 per ounce; Cu, ~$0.1 per ounce).

The recent development of copper(II)-promoted *O*- and *N*-arylation with boronic acids is a major breakthrough in the C–heteroatom transformation. The prototype of this transformation is illustrated in Equation 6.1. The Cu(II)-mediated C–N, C–O, and C–S bond formation between O-, N-, or S-containing nucleophilic substrates and aryl- and alkenylboronic acids to form the *O*-arylated, *N*-arylated, or *S*-arylated products is now referred as the *Chan–Lam coupling reaction* [7]. One reason for its popularity is the mild reaction conditions needed, for example, room temperature, weak base, and ambient atmosphere ("open-flask" chemistry). This approach also takes advantage of the ready availability of the boronic acids and the chemistry developed in the Suzuki–Miyaura coupling arena.

---

*Boronic Acids: Preparation and Applications in Organic Synthesis, Medicine and Materials*, Second Edition.
Edited by Dennis G. Hall.
© 2011 Wiley-VCH Verlag GmbH & Co. KGaA. Published 2011 by Wiley-VCH Verlag GmbH & Co. KGaA.

$$R^1{-}X{-}H \;+\; HO{-}B(OH){-}\underset{R^2}{C_6H_4} \;\xrightarrow[\text{room temperature}]{\substack{\text{Cu(OAc)}_2 \\ \text{amine base} \\ \text{CH}_2\text{Cl}_2}}\; R^1{-}X{-}\underset{R^2}{C_6H_4} \qquad (6.1)$$

$$X = N, O, S$$

In the previous edition of *Boronic Acids* edited by Hall, Chan, and Lam, the original discovery of Chan–Lam C–N and C–O cross-coupling reactions is described in detail [8]. In 1998, the cumulated research efforts in the laboratories of Chan, Lam, and Evans were disclosed in three back-to-back publications [9–11]. In the following years, the research teams from Chan and Lam and other research groups made considerable progress in expanding this copper-mediated cross-coupling methodology. Over the years, this methodology has proven to be mild, versatile, and robust. In addition, the different aspects of this chemistry were surveyed by several authors [7, 12–17]. The excellent review by Thomas and Ley [14] covers literatures up to late 2003, while the book chapter by Chan and Lam [8] covers literature from 2003 until mid-2005.

This chapter focuses primarily on recent work since mid-2005. Owing to the significance, Chan's original discovery and Chan and Lam's initial studies are included. Unpublished works from Lam's laboratory are also included. In the past several years, this area has continued to attract attention and further refinement, as corroborated by the increasing number of research groups around the world adopting and expanding the scope of this methodology. A literature search was performed on Cu(OAc)$_2$-promoted cross-coupling with boronic acids/esters in June 2009. There are about 400 references including more than 150 patent applications using and studying the methodology of Chan–Lam coupling reaction since its discovery. Recent developments include further expansion of the scope of the substrates and the boron reagents, fine-tuning of the reaction with various solvents and additives, and the expansion of the type of boron reagents as well as other organometalloids. The main focus of this chapter will be the examination of various types of C–X cross-coupling using boronic acids and their derivatives, as well as mechanistic considerations. Readers are encouraged to peruse the aforementioned reviews and book chapters for a more complete survey of the earlier development of this area.

## 6.2
## C–O Cross-Coupling with Arylboronic Acids

### 6.2.1
### Intermolecular C–O Cross-Coupling

Chan et al. reported the original discovery for C(aryl)–O bond formation in 1998 [9]. The process involved simply stirring a phenol with an arylboronic acid (2–3 equiv), stoichiometric copper(II) acetate (1–2 equiv), and Et$_3$N (2–3 equiv) as the base in methylene chloride at room temperature for 1–2 days to give the diaryl ethers in good yields (Equation 6.2). Electron-rich boronic acids **2b** and **2c** gave

6.2 C–O Cross-Coupling with Arylboronic Acids | **317**

higher yields (73%) than phenylboronic acid **2a** (40%). The reaction also worked for *ortho*-substituted phenol **4** and electron-deficient boronic acid **5**, giving biaryl ether **6** (Equation 6.3).

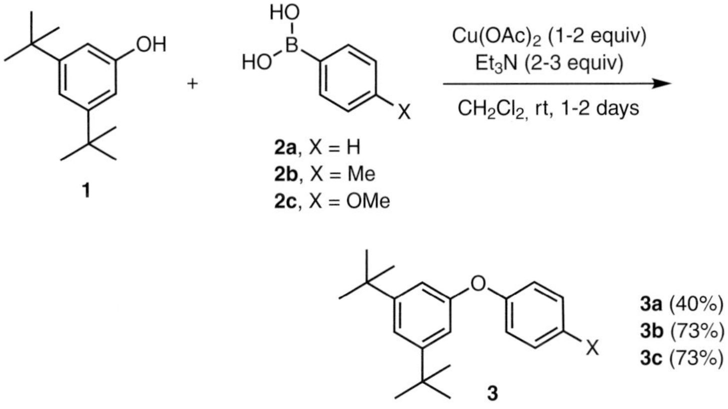

(6.2)

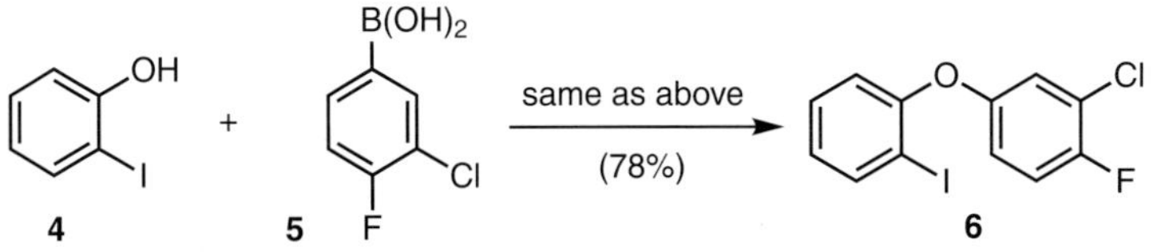

(6.3)

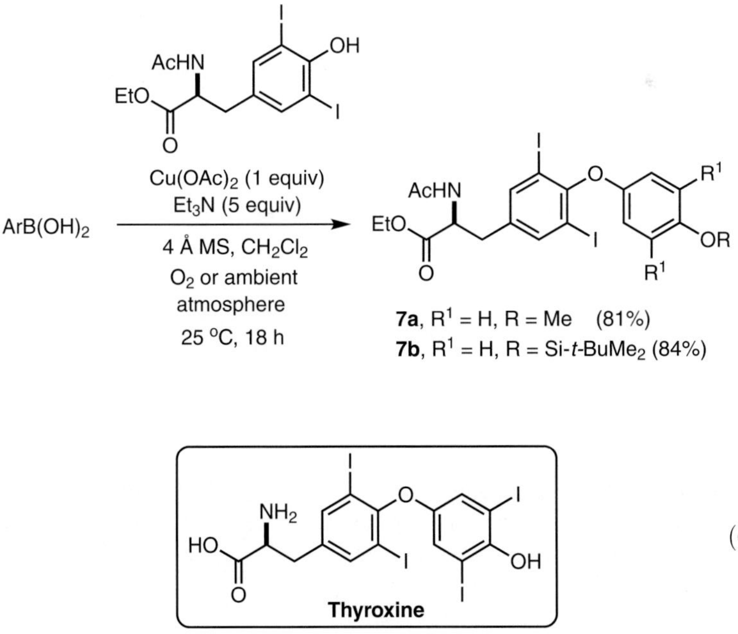

(6.4)

In the same year, Evans et al. quickly optimized the above O-arylation reaction in the expedient synthesis of thyroxine (Equation 6.4) [10]. The optimization process was summarized in the previous book chapter [8]. The optimized procedure for C(aryl)–O bond formation developed by Evans et al. is as follows: to the heterogeneous mixture of phenol (1 equiv), Cu(OAc)$_2$ (1 equiv), arylboronic acid (1–3 equiv), and powdered 4 Å molecular sieves (4 Å MS) in dichloromethane (about 0.1 M in phenol) was added the amine base (Et$_3$N, 5 equiv, or pyridine, 2 equiv, in the synthesis of thyroxine). The mixture was stirred at room temperature for 18 h under ambient atmosphere and the diaryl ether product was isolated in good yields.

Many groups then utilized the above boronic acid O-arylation methodology in the synthesis of biologically active molecules and natural products. For example, an efficient convergent synthesis of (S,S)-isodityrosine **8** was developed from two natural aromatic amino acids, L-tyrosine and L-phenylalanine, by Jung and Lazarova (Scheme 6.1) [18]. Because of the mildness of the copper-mediated C–O bond formation reaction, all the stereocenters of the molecule were preserved.

Besides phenols, other hydroxyl-containing nucleophilic reaction partners were identified [19–23]. For example, coupling of N-hydroxyimides with boronic acids using pyridine as the base under ambient air in 1,2-dichloroethane followed by treatment with hydrazine provided N-aryl hydroxylamines [22]. Lam et al. devised an O-arylation of N-hydroxybenzotriazole to provide the corresponding O-phenylated product **9** (Equation 6.5) [20].

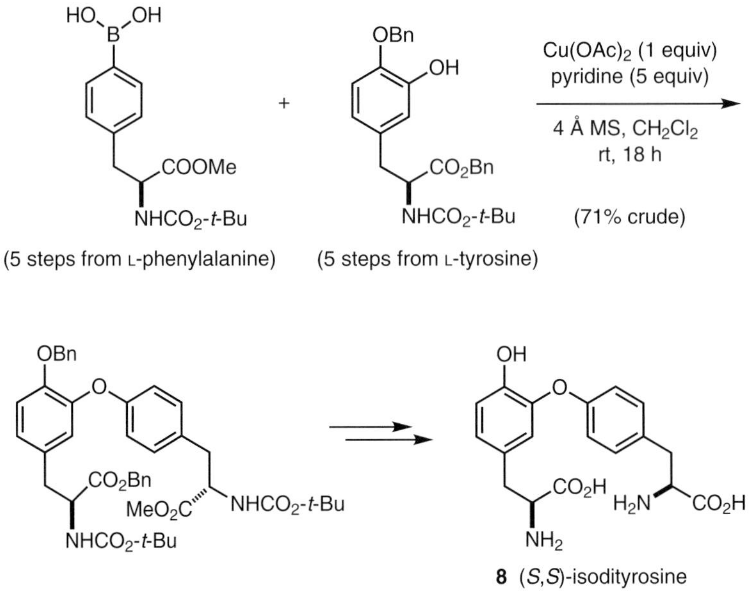

**Scheme 6.1** Synthesis of (S,S)-isodityrosine **8**.

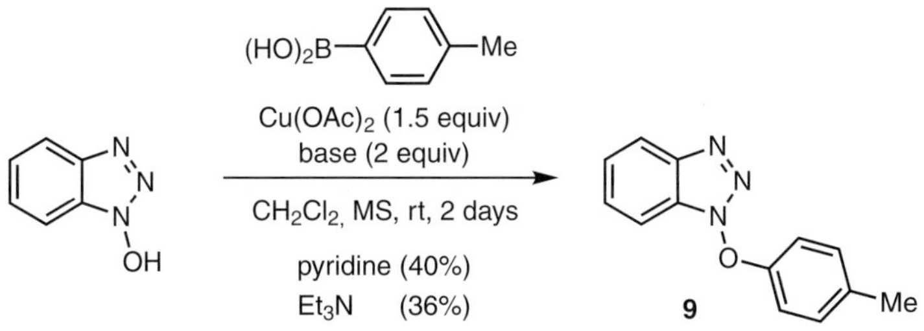

$$(HO)_2B-\underset{}{\bigcirc}-Me$$

$$\underset{\underset{OH}{|}}{\text{benzotriazole}} \xrightarrow[\text{CH}_2\text{Cl}_2, \text{MS, rt, 2 days}]{\text{Cu(OAc)}_2 \ (1.5 \ \text{equiv}) \atop \text{base (2 equiv)}} \mathbf{9} \qquad (6.5)$$

pyridine (40%)
Et$_3$N (36%)

Similarly, O-arylation of hydroxyquinolin-4(1H)-one gave the corresponding phenoxyquinolin-4(1H)-one **10** (Equation 6.6) [23].

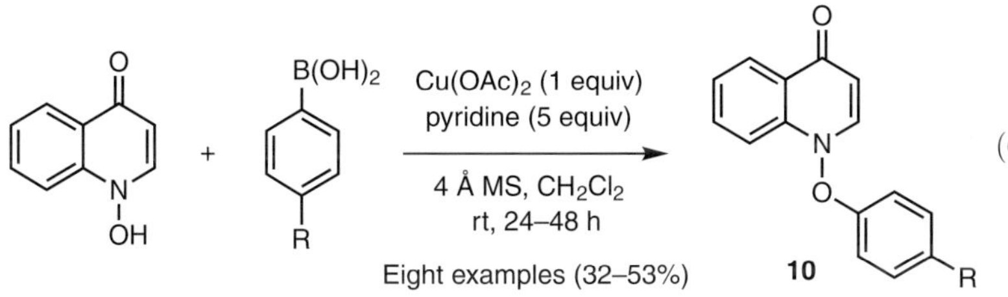

$$\text{hydroxyquinolinone} + \text{ArB(OH)}_2 \xrightarrow[\text{4 Å MS, CH}_2\text{Cl}_2, \text{rt, 24–48 h}]{\text{Cu(OAc)}_2 \ (1 \ \text{equiv}) \atop \text{pyridine (5 equiv)}} \mathbf{10} \qquad (6.6)$$

Eight examples (32–53%)

Hartwig and coworkers developed a methodology for 3,5-disubstituted aryl ethers **11** from arenes by a sequential iridium-catalyzed C—H borylation of **12**, oxidative hydrolysis of the boronic esters **13** with aqueous NaIO$_4$, followed by copper-mediated coupling of the corresponding crude boronic acids **14** with phenols (Scheme 6.2) [24]. 3,5-Disubstituted boronic esters **13** were generated by reaction of bispinacolatodiboron (B$_2$pin$_2$), 0.025 mol% of [(COD)Ir(OMe)]$_2$, and 0.05 mol% of 4,4′-di-*tert*-butyl-2,2′-bipyridine (dtbpy) in cyclohexane.

In the aforementioned optimization process for stoichiometric Cu(II)-mediated C(aryl)—O bond formation reactions, Evans and coworkers observed that when only 10% Cu(OAc)$_2$ was used, the reaction gave 30% yield under O$_2$ but only 9% yield

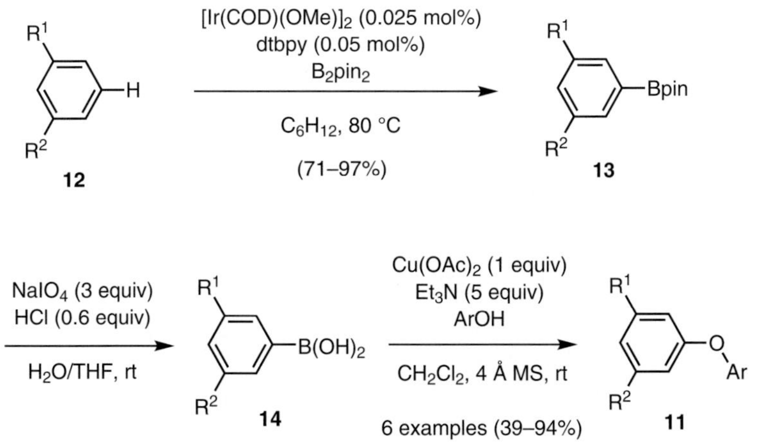

**12** $\xrightarrow[\text{C}_6\text{H}_{12}, 80\ °\text{C}]{\substack{[\text{Ir(COD)(OMe)}]_2 \ (0.025 \ \text{mol\%}) \\ \text{dtbpy (0.05 mol\%)} \\ \text{B}_2\text{pin}_2}}$ **13** (71–97%)

$\xrightarrow[\text{H}_2\text{O/THF, rt}]{\substack{\text{NaIO}_4 \ (3 \ \text{equiv}) \\ \text{HCl (0.6 equiv)}}}$ **14** $\xrightarrow[\text{CH}_2\text{Cl}_2, \text{4 Å MS, rt}]{\substack{\text{Cu(OAc)}_2 \ (1 \ \text{equiv}) \\ \text{Et}_3\text{N (5 equiv)} \\ \text{ArOH}}}$ **11**

6 examples (39–94%)

**Scheme 6.2** Synthesis of 3,5-disubstituted aryl ethers **11** from 3,5-disubstituted arenes.

under Ar, suggesting a catalytic process might be developed. In 2001, Lam et al. reported catalytic Cu(II)-mediated C(aryl)–O formation by studying different co-oxidants or oxygen (Equation 6.7) [25]. The best yield (75%) for **15** was obtained when oxygen was used in DMF using pyridine as the base.

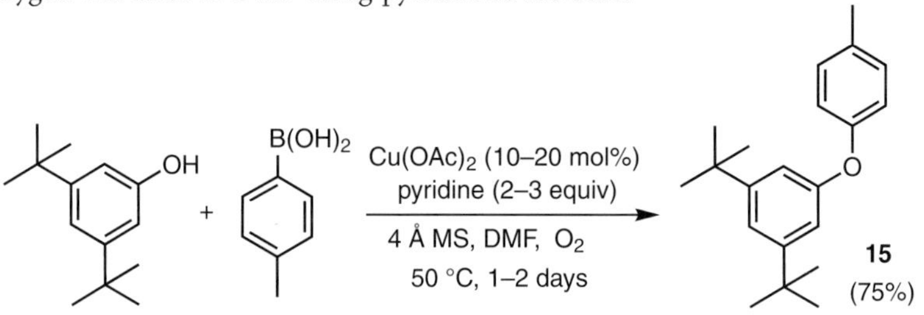

$$\text{(6.7)}$$

## 6.2.2
### Intramolecular C–O Cross-Coupling

Decicco's laboratory demonstrated the first successful extension of the boronic acid intermolecular C(aryl)–O cross-coupling to an intramolecular system **16** for the synthesis of metalloprotease inhibitors (Equation 6.8) [26]. Other classical C–O formation methods failed to provide the macrocycle. Since then, multiple macrocycles, for example, chloropeptin I [27] and cycloisodityrosines [28] (compounds **17** and **18**), were synthesized using the intramolecular Cu(II)-mediated arylation process.

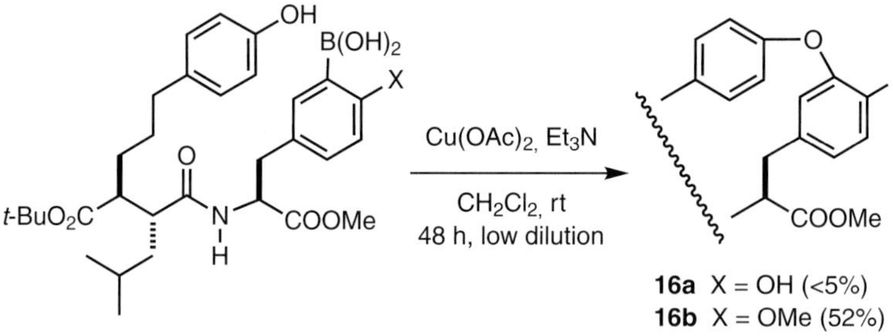

$$\text{(6.8)}$$

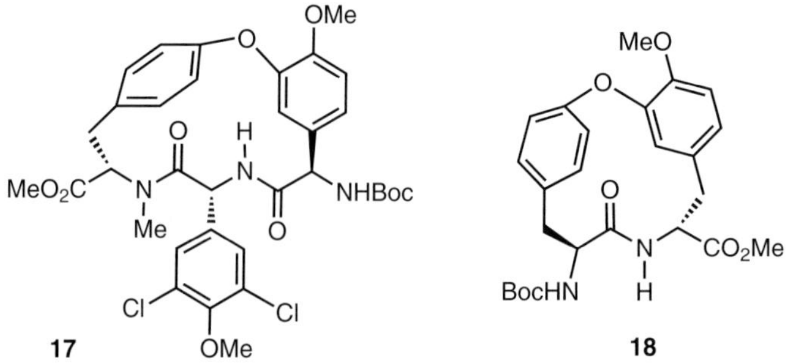

**16a** X = OH (<5%)
**16b** X = OMe (52%)

In general, copper-catalyzed O-arylation of phenols using arylboronic acids is a robust reaction.

## 6.3
## C—N Cross-Coupling with Arylboronic Acids

### 6.3.1
### C—N (Nonheteroarene NH) Cross-Coupling

The advantage of copper-mediated boronic acid C(aryl)—N bond formation reaction is its high tolerability of a wide range of functional groups and its high success rate on a broad spectrum of substrates. In the first preliminary communication, the Chan group demonstrated that a wide range of the NH-containing nucleophilic partners **19**, including amines, amides, imides, ureas, carbamates, and sulfonamides, underwent stoichiometric copper-mediated C—N bond formation reaction with *p*-tolylboronic acid to afford compounds **20** as shown in Equation 6.9 [9].

Besides electron-rich anilines illustrated in the first publication by Chan, Lam *et al.* reported that a variety of heterocyclic amines (**21**, $H_2N$-Het) reacted with arylboronic acids to afford the corresponding anilines **22** in variable yields (Equation 6.10) (Lam, P.Y.S., Boone, D., and Clark, C.G., unpublished work, Lam, P.Y.S., Vincent, G., and Clark, C.G., unpublished work). The electron-rich heterocyclic amines and substrates containing more nucleophilic amines gave higher yields, although the presence of chelating nitrogen atoms could influence the chemistry. For instance, 3-amino pyridine gave 70% of the *N*-arylated product, while 2-aminopyridine led to threefold lower yield and 4-aminopyridine failed to offer any *N*-arylated product.

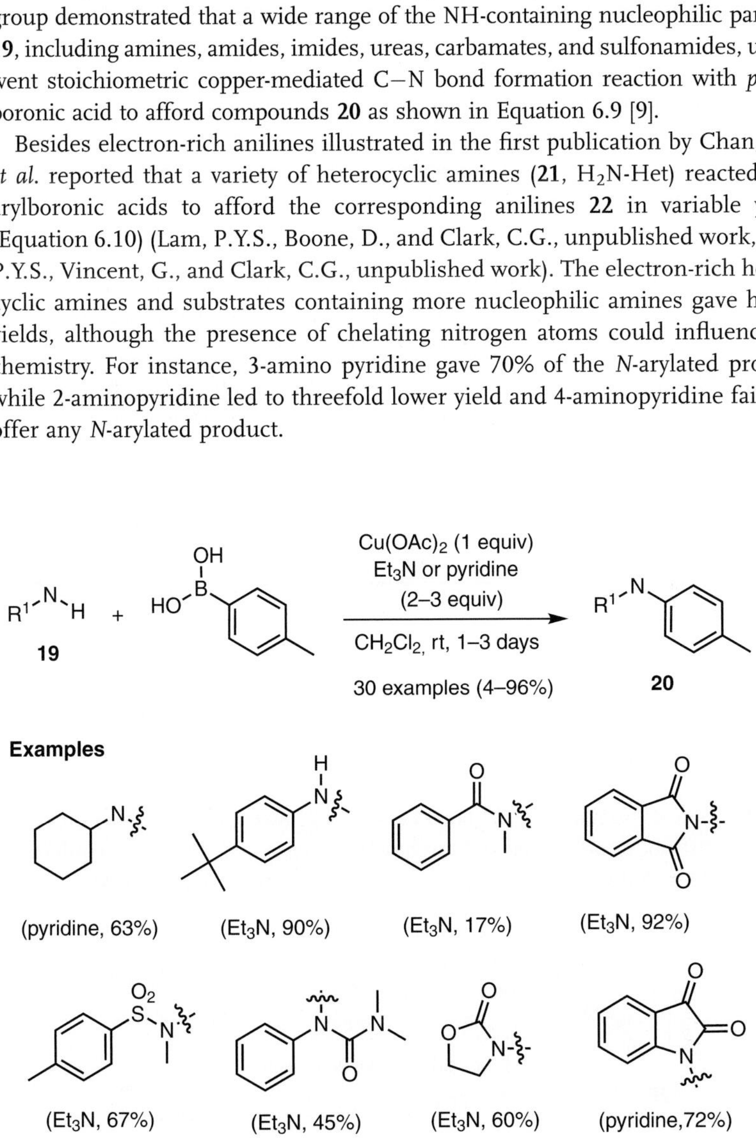

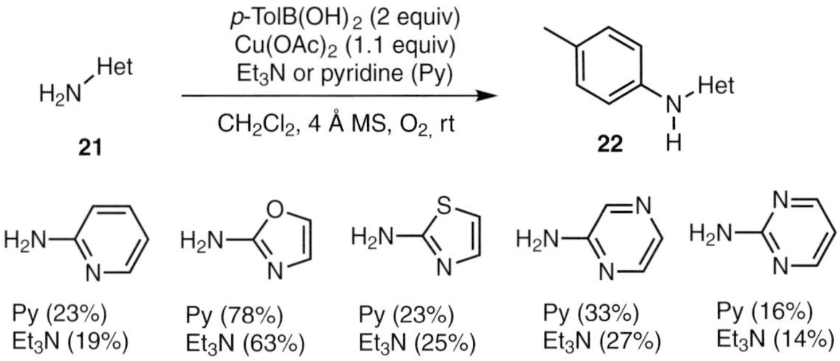

$$\text{H}_2\text{N}{-}\text{Het} \quad \xrightarrow[\text{CH}_2\text{Cl}_2,\ 4\ \text{Å MS, O}_2,\ \text{rt}]{\substack{p\text{-TolB(OH)}_2\ (2\ \text{equiv}) \\ \text{Cu(OAc)}_2\ (1.1\ \text{equiv}) \\ \text{Et}_3\text{N or pyridine (Py)}}} \quad \text{Ar-N(H)-Het}$$

**21** → **22** (6.10)

Py (23%), Et$_3$N (19%) — 2-aminopyridine
Py (78%), Et$_3$N (63%) — 2-aminooxazole
Py (23%), Et$_3$N (25%) — 2-aminothiazole
Py (33%), Et$_3$N (27%) — 2-aminopyrazine
Py (16%), Et$_3$N (14%) — 2-aminopyrimidine

In the past 5 years, there have been numerous publications regarding *N*-arylation using copper-mediated Chan–Lam couplings, and some of which will be discussed herein. For instance, Cu–Al hydrotalcite in refluxing methanol was used in the *N*-arylation of imides with arylboronic acids with continuing air bubbling under base-free, solvent-free conditions [29].

One of the best examples to showcase the mild conditions of this chemistry is the copper-catalyzed *N*-arylation of labile azetidinone esters **23** with a variety of arylboronic acids by Devasthale and coworkers in BMS (Equation 6.11) [30]. The *N*-arylated azetidinone esters **24** were obtained in nearly quantitative yields, remarkably, with no racemization. The original *N*-arylation conditions reported by the Lam laboratory were used without modification. The alpha-carbonyl activating effect is critical for the excellent yield, since the corresponding alpha-acetal group gave much lower yields. The corresponding azetidinone carboxylate acids were potent dual PPARα/γ agonists as antidiabetic agents.

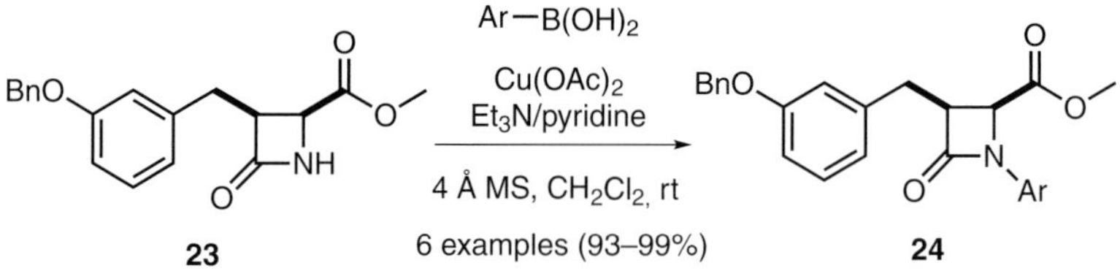

**23** → **24**, conditions: Ar–B(OH)$_2$, Cu(OAc)$_2$, Et$_3$N/pyridine, 4 Å MS, CH$_2$Cl$_2$, rt; 6 examples (93–99%) (6.11)

*N*-Arylation of 2-azabicyclo[2,2,1]hept-5-en-3-one **25** was developed by using catalytic Cu(OAc)$_2$ (10 mol%), pulverized KOH (5 equiv) as the base, and Me$_3$NO (1.1 equiv) as the oxidant in MeCN at 80 °C under microwave irradiation (Equation 6.12) [31].

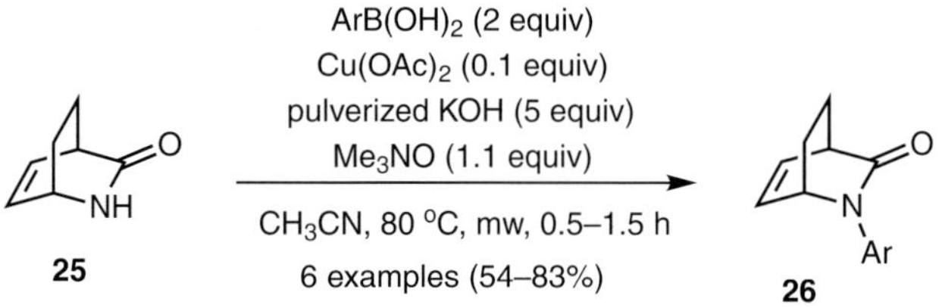

**25** → **26**, conditions: ArB(OH)$_2$ (2 equiv), Cu(OAc)$_2$ (0.1 equiv), pulverized KOH (5 equiv), Me$_3$NO (1.1 equiv), CH$_3$CN, 80 °C, mw, 0.5–1.5 h; 6 examples (54–83%) (6.12)

Arylation at the N1-position of 2(1H)-pyrazinone scaffold **27** was achieved using 2 equiv of $Cu(OAc)_2$ in mixed bases of $Et_3N$ and pyridine (1 : 2) under microwave irradiation at the maximum power of 300 W while simultaneously cooling at 0 °C (Equation 6.13) [32]. While the reaction tolerated both electron-donating and electron-withdrawing groups in the arylboronic acids, *ortho*-substituted arylboronic acid and heteroarylboronic acids failed. The use of maximum power of irradiation with simultaneously cooling slowed down the decomposition of the desired product under microwave irradiation condition.

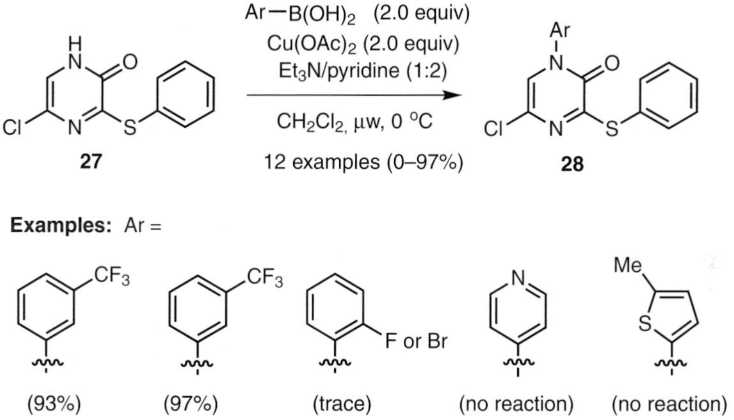

$$\text{27} \xrightarrow[\text{CH}_2\text{Cl}_2,\ \mu w,\ 0\ °\text{C}]{\substack{\text{Ar-B(OH)}_2\ (2.0\ \text{equiv}) \\ \text{Cu(OAc)}_2\ (2.0\ \text{equiv}) \\ \text{Et}_3\text{N/pyridine}\ (1:2)}} \text{28}$$

12 examples (0–97%)

**Examples:** Ar =

- 3-CF$_3$-C$_6$H$_4$ (93%)
- 3-CF$_3$-C$_6$H$_4$ (97%)
- 2-F or 2-Br-C$_6$H$_4$ (trace)
- 4-pyridyl (no reaction)
- 5-methyl-2-thienyl (no reaction)

(6.13)

The hydantoin substrate **29** was cross-coupled with 4-CN-3-CF$_3$-phenylboronic acid **30** using stoichiometric amount of $Cu(OAc)_2$ to afford the *N*-arylated compound **31** in 79% yield (Equation 6.14). On the other hand, S$_N$Ar reaction of **29** with the corresponding aryl chloride or fluoride (NaH, 18-crown-6, DMF, 140 °C, 60 h) afforded **31** in only 55% yield [33].

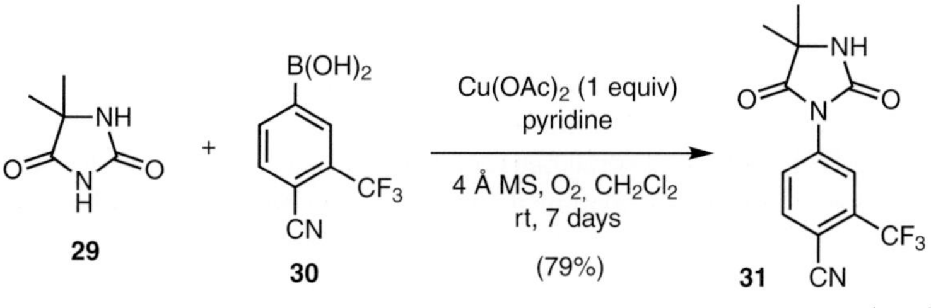

$$\text{29} + \text{30} \xrightarrow[\substack{4\ \text{Å MS, O}_2,\ \text{CH}_2\text{Cl}_2 \\ \text{rt, 7 days}}]{\substack{\text{Cu(OAc)}_2\ (1\ \text{equiv}) \\ \text{pyridine}}} \text{31}\ (79\%)$$

(6.14)

Microwave-assisted *N*-arylation of amines with arylboronic acids was reported using 2 equiv of $Cu(OAc)_2$ and 2 equiv of DBU as the base in DMSO under microwave irradiation at 100 °C (Equation 6.15) [34].

$$R^1\text{-}N(R^2)H + HO\text{-}B(OH)\text{-}C_6H_4\text{-}R \xrightarrow[\text{DMSO, μw,}\\ \text{100 °C, 30 min}]{\text{Cu(OAc)}_2 \text{ (2 equiv)} \\ \text{DBU (2 equiv)}} R^1\text{-}N(R^2)\text{-}C_6H_4\text{-}R \quad (6.15)$$

22 examples (29–96%)

#### 6.3.1.1 Application of Chan–Lam Cross-Coupling in Solid-Phase Synthesis

The copper-mediated arylation with boronic acids has been applied in solid-phase synthesis and combinatorial chemistry. Two approaches were reported: solid-supported catalyst and resin-supported substrates. The supported catalyst has the added advantage of easy recovery and recycling. Besides the previously reported methacrylic resin solid-supported copper catalyst (MPI-Cu) [35] and solid-phase copper catalytic system with a β-ketoester linkage [36], heterogeneous basic copper-exchanged fluorapatite (CuFAP) was used to catalyze the N-arylation of imidazoles and amines under base-free conditions [37]. In addition, cellulose-supported copper(0) [38] and polymer-supported copper(II) acetate [39], as well as silica-supported copper(II) catalyst [40], were also employed in the Chan–Lam C−N and C−O bond formation reactions.

Combs et al. pioneered the use of resin-supported substrates, reporting efficient cross-coupling reactions with boronic acids. Examples of the resin-supported substrates were solid-supported sulfonamides [41], primary and secondary aliphatic amines [42], and heteroarenes such as benzimidazoles, imidazoles, pyrazoles, and benzotriazoles [43].

### 6.3.2
### C−N (Heteroarene) Cross-Coupling

While working on the design and discovery of pyrazole-based factor Xa inhibitors as novel anticoagulants, Lam et al. searched for mild reaction conditions for the N-arylation of azoles [44]. Prior to the advent of the Buchwald–Hartwig coupling, C−N (heteroarene) cross-coupling chemistry was underexplored in academics, presumably because no natural product containing the C(arene)−N (heteroarene) bond was known.[1] The state-of-the-art Buchwald–Hartwig palladium-catalyzed N-arylation chemistry with aryl halides did not work well for the N-arylation of many azoles such as imidazoles and some heterocycles such as quinazolinediones [45].

Lam et al. expanded the original scope of copper-mediated C−N bond formation and discovered that a variety of aromatic heterocycles **32**, such as imidazole, pyrazole, triazoles, tetrazole, benzimidazole, and indazole, can also be used as N-containing nucleophiles [11]. For examples, imidazoles and pyrazoles underwent N-arylation with 2.0 equiv of p-tolylboronic acid under typical conditions (Equation 6.16). Electron-poor azoles such as triazole and tetrazole gave low yields. Although pyrazoles and imidazoles gave good yields, the parent pyrrole and indole gave very poor yields.

---

1) Search of natural product database in 1997.

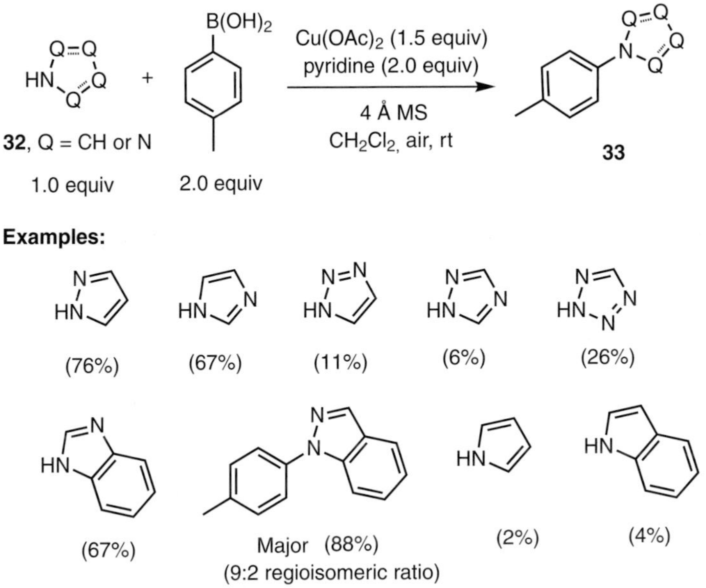

## 6.3 C—N Cross-Coupling with Arylboronic Acids

$$\text{32, Q = CH or N} + \text{B(OH)}_2\text{-Ar} \xrightarrow[\substack{4\text{ Å MS} \\ \text{CH}_2\text{Cl}_2,\ \text{air, rt}}]{\substack{\text{Cu(OAc)}_2\ (1.5\ \text{equiv}) \\ \text{pyridine (2.0 equiv)}}} \text{33} \quad (6.16)$$

1.0 equiv + 2.0 equiv

**Examples:**

- pyrazole (76%)
- imidazole (67%)
- 1,2,3-triazole (11%)
- 1,2,4-triazole (6%)
- tetrazole (26%)
- benzimidazole (67%)
- indazole — Major (88%) (9:2 regioisomeric ratio)
- pyrrole (2%)
- indole (4%)

Later on, N-arylation proceeded in good yields with pyrroles and indoles containing a chelating aldehyde, ketone, or ester alpha to the NH group [46, 47].

Recently, N-arylation of electron-deficient pyrroles and indoles having no carbonyl group at the C2-position (compounds **34**) was developed in good to excellent yields using diisopropylethylamine as the base (Equation 6.17) [48]. Triethylamine or pyridine did not give the desired product under these conditions.

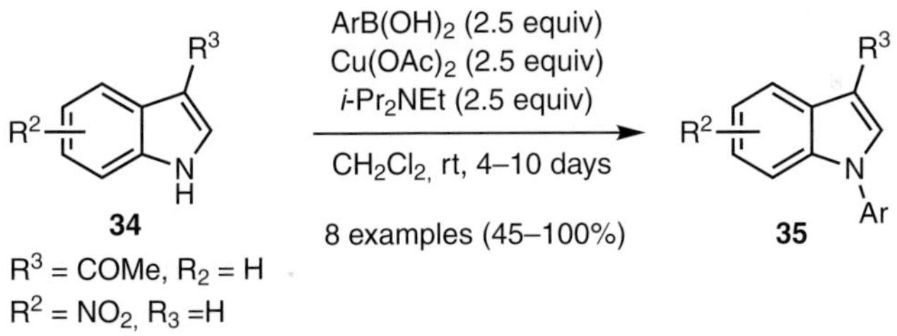

$$\text{34} \xrightarrow[\substack{\text{CH}_2\text{Cl}_2,\ \text{rt, 4–10 days}}]{\substack{\text{ArB(OH)}_2\ (2.5\ \text{equiv}) \\ \text{Cu(OAc)}_2\ (2.5\ \text{equiv}) \\ i\text{-Pr}_2\text{NEt (2.5 equiv)}}} \text{35} \quad (6.17)$$

$R^3 = \text{COMe}, R^2 = H$
$R^2 = NO_2, R^3 = H$

8 examples (45–100%)

On the other hand, N-arylation of 3-trimethylsilyl indazoles **36** with arylboronic acid proceeded regioselectively and the resulting 1-aryl-3-trimethylsilylimidazoles **37** were easily converted to 1-arylindazoles (Equation 6.18). Moreover, the trimethylsilyl group at the 3-position of indazoles accelerated the reaction rate of N-arylation at the 1-position [49].

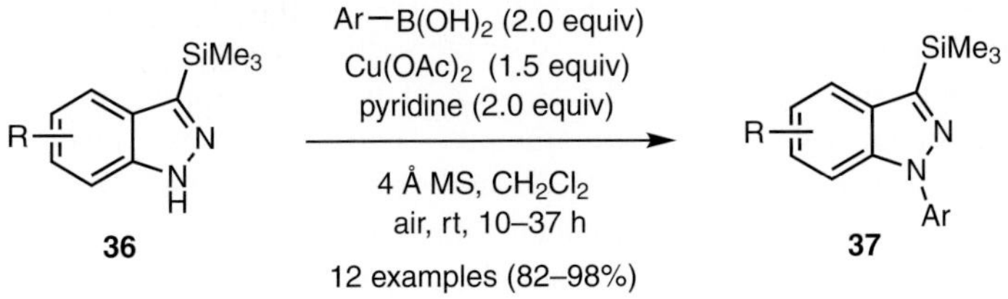

$$\text{36} \xrightarrow[\substack{4\text{ Å MS, CH}_2\text{Cl}_2 \\ \text{air, rt, 10–37 h}}]{\substack{\text{Ar–B(OH)}_2\ (2.0\ \text{equiv}) \\ \text{Cu(OAc)}_2\ (1.5\ \text{equiv}) \\ \text{pyridine (2.0 equiv)}}} \text{37} \quad (6.18)$$

12 examples (82–98%)

#### 6.3.2.1 Factor Xa Inhibitors

Lam and Clark studied the regioselectivity in the *N*-arylation of pyrazoles (Equation 6.19) (Lam, P.Y.S. and Clark, C.G., unpublished work). For 3-CF$_3$-5-methylpyrazole **38a** and 3-CF$_3$-pyrazole **38b**, *N*-arylation occurred at the least hindered nitrogen. For 3-methyl-5-ethoxycarbonylpyrazole **38c**, 15% of the minor regioisomer of **39c** was also obtained. In this case, the α-activating ester is the directing group, even though it is bigger than the methyl group. In general, for many substrates besides pyrazoles, the reaction is very sensitive to steric effects.

$$\underset{\textbf{38}}{\text{pyrazole-X,Y}} + \underset{F_3C}{\text{ArB(OH)}_2} \xrightarrow[\text{CH}_2\text{Cl}_2, \text{ rt, 4d}]{\text{Cu(OAc)}_2, \text{ pyridine}} \underset{\textbf{39a,b,c}}{\text{product}} \quad (6.19)$$

**38a** X = CF$_3$, Y = CH$_3$ (70%)
**38b** X = CF$_3$, Y = H (72%)
**38c** X = CH$_3$, Y = COOEt (50%); minor isomer X = COOEt, Y = CH$_3$ (15%)

Subsequently, as demonstrated in the previous book chapter [8], *N*-arylated pyrazoles can also be transferred to substituted pyrazoles as factor Xa inhibitors. Scheme 6.3 illustrates the preparation of FXa inhibitor **40** starting from commercially available 3-propyl-1*H*-purine-2,6(3*H*,7*H*)-dione **41** [50]. The synthesis involved three consecutive coupling reactions (Chan–Lam C–N, Suzuki–Miyaura C–C, and Chan–Lam C–N) from Boc-protected **42** to construct the requisite skeleton **43** in a concise manner.

#### 6.3.2.2 Purines

Purines are important building blocks of biologically active molecules. One of the applications of Chan–Lam copper-mediated coupling reaction is the synthesis of *N*-arylated purines. Gray and coworkers first published the reaction of 2,6-dichloropurine with boronic acids in the presence of Cu(OAc)$_2$ and triethylamine, which led to the desired N9-arylated products as the major regioisomer (>9:1). The N7 regioisomer was produced in much lower yield [51].

The analogous *N*-arylation with 6-chloropurines was performed using phenanthroline as the base [52]. Only the N9-arylated regioisomer was formed, even in the presence of a 2-amino functional group. N1 arylation of nucleosides was also realized using Lam's catalytic copper conditions with pyridine *N*-oxide [25]. In addition, various nucleoside bases such as inosine **44**, hypoxanthine **45**, and xanthine **46** were *N*-arylated as shown in Figure 6.1 [50].

Recently, Yue and coworkers developed an efficient and mild method for the direct *N*-arylation of nucleosides **47–50** with arylboronic acids catalyzed by Cu(OAc)$_2$·xH$_2$O (Equation 6.20) [53, 54]. The presence of H$_2$O in the reaction was important. Replacing Cu(OAc)$_2$ with Cu(OAc)$_2$·H$_2$O in the absence of molecular sieves significantly increased the yield (90% versus trace amount). The mixed solvents MeOH:H$_2$O (4:1) were optimal. In addition, only TMEDA as the base gave good yields of products **51–54**; pyridine, TEA, DMAP, and 1,10-phenanthroline gave only trace amount of the desired *N*-arylated products. Several simple copper salts, such as Cu(OAc)$_2$,

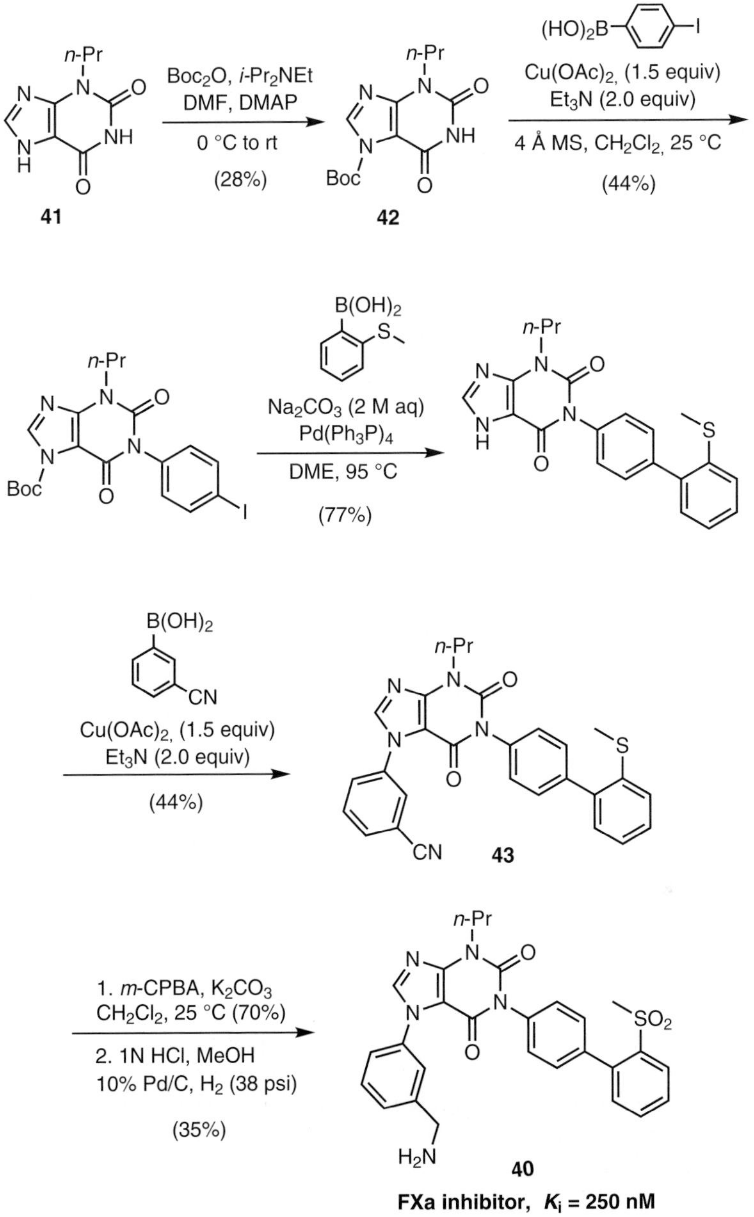

**Scheme 6.3** Synthesis of FXa inhibitor **40**.

CuSO$_4$·H$_2$O, CuCl, CuBr, and Cu(OTf)$_2$, gave good to high yields. The reaction was chemoselective for the N-arylation of cytosine (at N1) or adenine (at N9). Boronic acids with both electron-donating and electron-withdrawing groups gave moderate to high yields. The nucleobases showed the following order in terms of yield and ease of the reaction: cytosine > uracil and thymine > adenine.

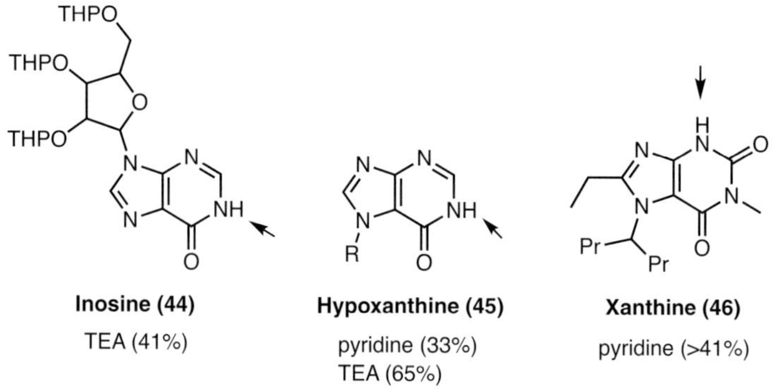
THPO
THPO
THPO
Inosine (44)
TEA (41%)
Hypoxanthine (45)
pyridine (33%)
TEA (65%)
Xanthine (46)
pyridine (>41%)
Pr
Pr

**Figure 6.1** Inosine, hypoxanthine, and xanthine.

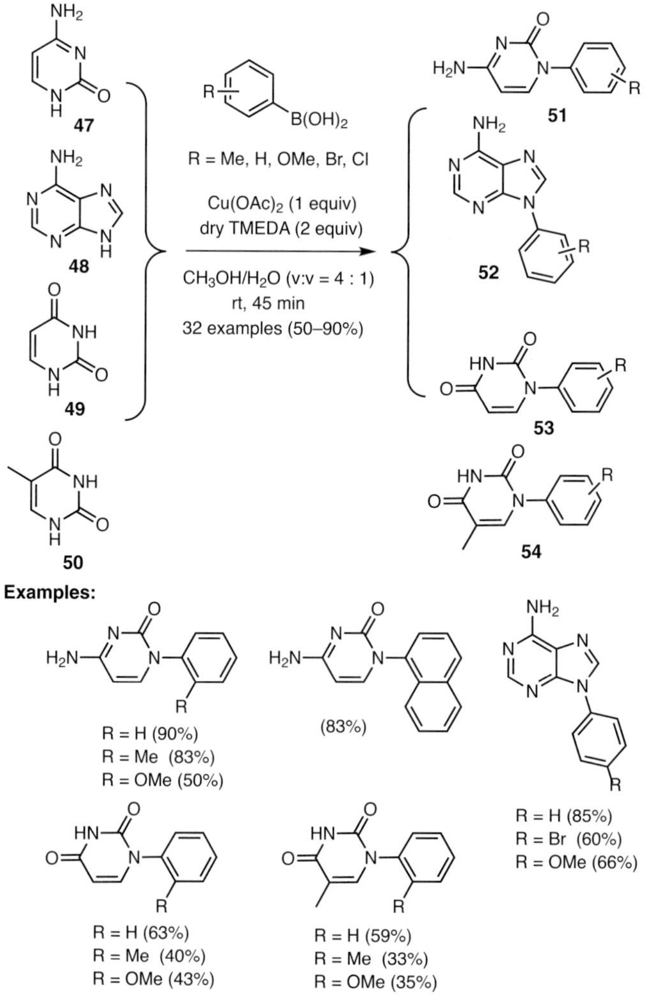
47
48
49
50
R = Me, H, OMe, Br, Cl
Cu(OAc)2 (1 equiv)
dry TMEDA (2 equiv)
CH3OH/H2O (v:v = 4 : 1)
rt, 45 min
32 examples (50–90%)
51
52
53
54
Examples:
R = H (90%)
R = Me (83%)
R = OMe (50%)
(83%)
R = H (85%)
R = Br (60%)
R = OMe (66%)
R = H (63%)
R = Me (40%)
R = OMe (43%)
R = H (59%)
R = Me (33%)
R = OMe (35%)

(6.20)

Gothelf and coworkers reported a general approach to the N-arylation and N-alkenylation of all five properly protected or masked natural nucleobases to form the corresponding products **55–58** (Equation 6.21) [55]. The reactions tolerated various substitution patterns on the arylboronic acids including *ortho*-substitution and electron-donating groups. The products **55–58** were obtained in good to high yields and could subsequently be readily converted to the corresponding deprotected or unmasked N-arylated or N-alkenylated nucleobases **59–63**. In the case of the thymine-related nucleosides, it was found that substrates with benzoyl protecting group on the N3-position gave better yields than protecting groups such as Boc and 4-*t*-Bu-Bn.

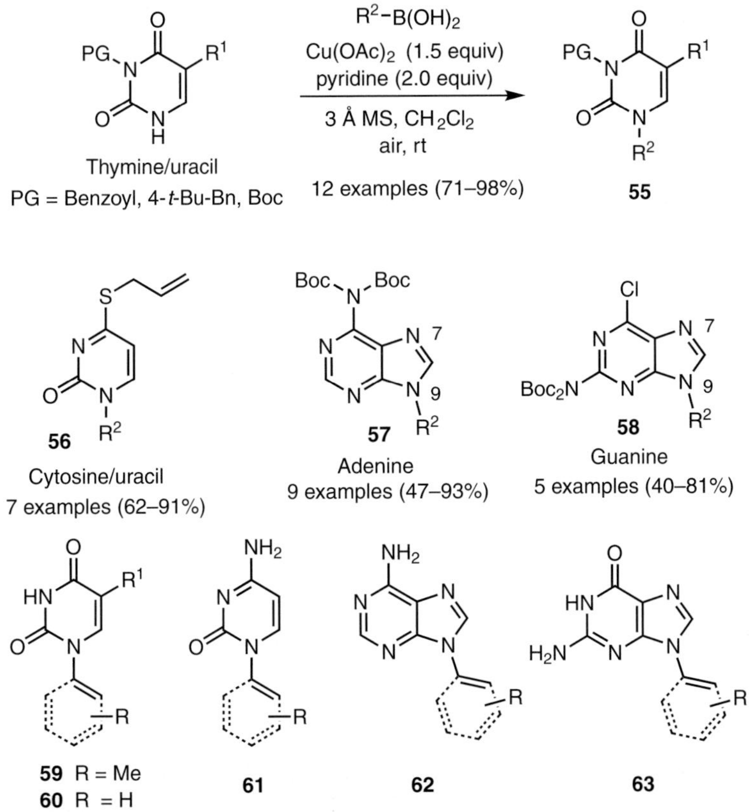

(6.21)

With these discoveries, one can envision making oligonucleotides, where the ribose ring is replaced by a properly substituted aryl/heteroaryl ring, with the right topology. This may have implications in antisense oligonucleotide (ASO) and SiRNA nucleic acid therapeutics.

### 6.3.2.3 Heteroarene–Heteroarene Cross-Coupling

C—N cross-coupling between two heteroarenes is an important process in medicinal, crop protection, and material science chemistry. Lam and coworkers explored the cross-coupling between 3-pyridylboronic acid and benzimidazole and obtained only 22% yield [56]. However, changing the boron reagent to the corresponding propylene glycol boronic ester resulted in a higher yield (54%) (Equation 6.22). The C–N cross-coupling between two heteroarenes using Chan–Lam coupling reaction has been widely used, in particular in pharmaceutical research for the synthesis of drug-like small molecules.

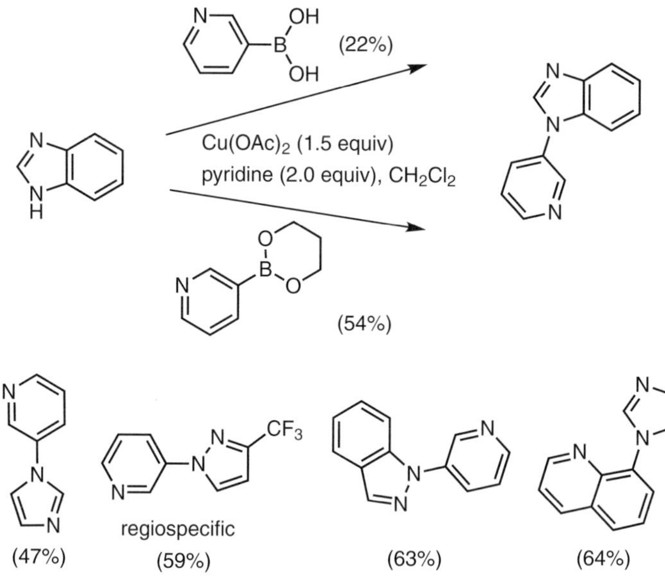

(6.22)

### 6.3.3
### Intramolecular C–N Cross-Coupling

In a patent application [57], an intramolecular Chan–Lam C–N coupling reaction was reported by a group of Novartis researchers for the preparation of fused 1,2,4-thiadiazine derivatives **64** and **65**. Thus, the thiophene boronic acid was cross-coupled with the guanidine NH group on the thiophene ring intramolecularly in **66** and **67** (Equation 6.23) using stoichiometric Cu(OAc)$_2$, pyridine as the base and N-methyl-2-pyrrolidinone as the solvent at room temperature. The fused 1,2,4-thiadiazines **64** and **65** were made in 44–80% yields, which were higher than those under Ullmann conditions reported previously [14, 17].

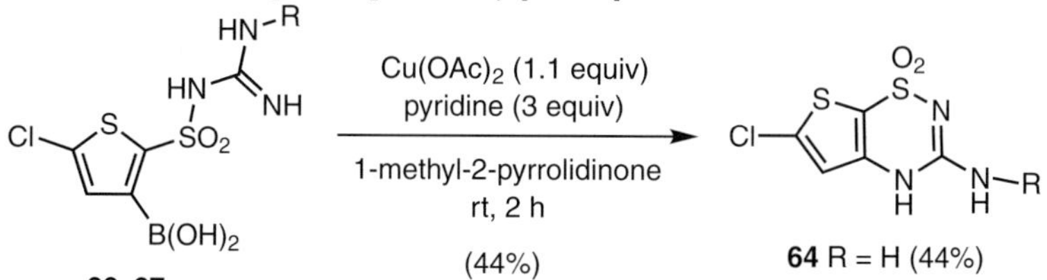

(6.23)

## 6.3.4
### Catalytic Copper-Mediated C—N Cross-Coupling

Collman and coworkers [58, 59] first introduced catalytic C—N coupling by using [Cu(OH)·TMEDA]$_2$Cl$_2$ (10 mol%) in the presence of O$_2$. The reaction also occurs in water, albeit in lower yield. N-arylation of imidazole is faster than O-arylation of bulk water. The significance of running the reaction in water is the possibility of N-arylation of histidine residues on proteins. Among all the bidentate ligands screened, TMEDA offered the best yield.

As mentioned in Section 6.2, Lam *et al.* also developed a catalytic pathway using Cu(OAc)$_2$ (10 mol%) and a co-oxidant additive, such as pyridine N-oxide [25]. Depending on the substrates, pyridine N-oxide may or may not improve the yield. For about half of the general substrates, catalytic copper in the presence of air with no added oxidant works just fine.

Xie and coworkers reported the N-arylation of imidazole in refluxing methanol in the presence of catalytic CuCl or Cu(OAc)$_2$ in air [60]. Yields were generally higher than both [Cu(OH)·TMEDA]$_2$Cl$_2$ and Cu(OAc)$_2$ (10 mol%)/pyridine N-oxide conditions. Conversely, Xie's conditions gave lower yield than Lam's conditions for aniline and sulfonamide substrates. On the other hand, Batey's ligandless and base-free conditions [61] were the best for basic amines.

Arylboronic acids can also react with azoles, such as imidazoles, benzimidazoles, and pyrazoles, and amines with heterogeneous Cu$_2$O as the catalyst in methanol at room temperature under base-free condition to give good to excellent yields [62]. Recently, an effective and simple catalyst system, Cu(NO$_3$)$_2$–TMEDA, was developed to form hindered C—N biaryls **68** in good to excellent yield, as illustrated in the cross-coupling of imidazoles **69** and arylboronic acids **70** (Equation 6.24) [63]. This is a significant development since it sets up the stage to make chiral C—N atropisomers such as C—N-linked biaryls, where steric bulk is critical for maintaining chiral integrity.

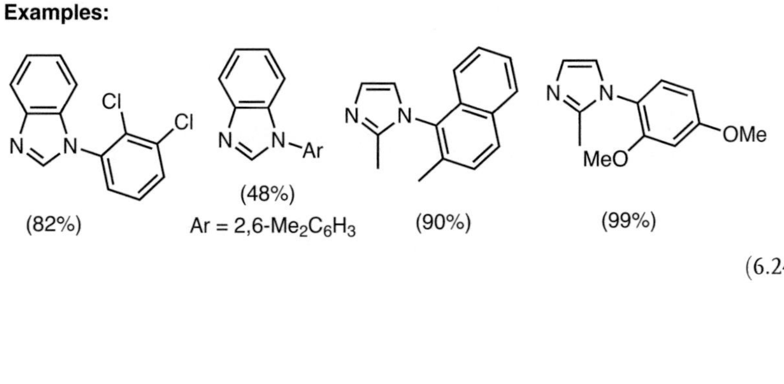

(6.24)

### 6.3.5
### Additional N-Containing Substrates in Chan–Lam Cross-Coupling

In the past several years, the scope of the nitrogen-containing nucleophiles has been broadened significantly. Examples of additional substrates for Chan–Lam C—N cross-coupling reaction are diazodicarboxylate, N-Boc-aryl hydrazines, sulfoximines, sodium azide, aqueous ammonia, oxime O-carboxylates, and O-acetyl hydroxamic acids. The products of these reactions are hydrazines (common synthetic precursors for amines), N-arylated sulfoximines (effective chiral ligands in catalytic asymmetric reactions), aryl azides and 1,2,3-triazoles, anilines, N-arylated or N-alkenylated imines and highly substituted pyridines, and amides.

Arylboronic acids were added to the N=N bond in diazodicarboxylate **71** in the presence of catalytic amount of $Cu(OAc)_2$ in THF or DMF for 20 h at room temperature under argon to afford aryl- or vinyl-substituted hydrazines **72** (Equation 6.25) [64]. The reaction was not significantly affected by the electronic nature of the functional groups; however, sterically hindered boronic acids such as 2,6-dimethylphenyl and 2-methoxylphenylboronic acids gave less than 5% yield.

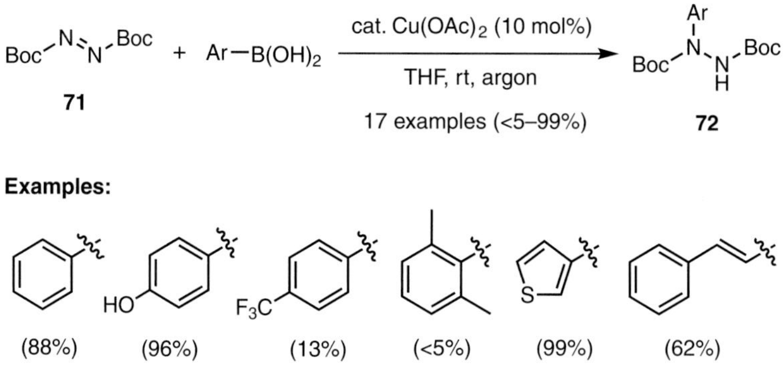

$$Boc\text{-}N\text{=}N\text{-}Boc + Ar\text{-}B(OH)_2 \xrightarrow[\text{THF, rt, argon}]{\text{cat. } Cu(OAc)_2 \text{ (10 mol\%)}} Boc\text{-}N(Ar)\text{-}N(H)\text{-}Boc$$

**71**  17 examples (<5–99%)  **72**

**Examples:**

(88%)   (96%)   (13%)   (<5%)   (99%)   (62%)

(6.25)

A copper-mediated cross-coupling reaction of N-Boc-aryl hydrazines **73** with arylboronic acids was developed for the synthesis of N,N′-diaryl hydrazines **74** in good yields under typical Chan–Lam coupling conditions (Equation 6.26) [65].

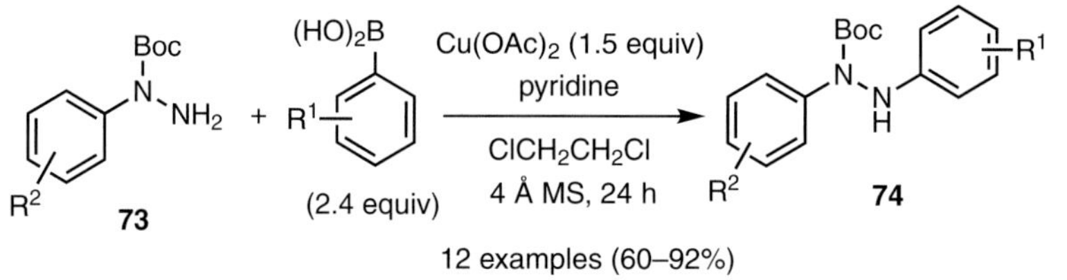

$$\text{73} + R^1\text{-}Ar\text{-}B(OH)_2 \xrightarrow[\text{ClCH}_2\text{CH}_2\text{Cl, 4 Å MS, 24 h}]{\text{Cu(OAc)}_2 \text{ (1.5 equiv), pyridine}} \text{74}$$

(2.4 equiv)

12 examples (60–92%)

(6.26)

Sulfoximines **75** were also used as substrate in the Chan–Lam cross-coupling reaction. They were reacted with arylboronic acids in the presence of a catalytic amount of $Cu(OAc)_2$ using MeOH as the solvent under base-free conditions to give N-arylated sulfoximines **76** (Equation 6.27) [66]. Methanol was a superior solvent compared to CH$_2$Cl$_2$, DMSO, and EtOH. No homocoupling of arylboronic acids was observed. However, in some cases, less than 10% of the arylmethyl ether side product was detected. Although the reaction tolerated some steric hindrance, 2,4,6-t-butyl-phenylboronic acid did not give any desired product.

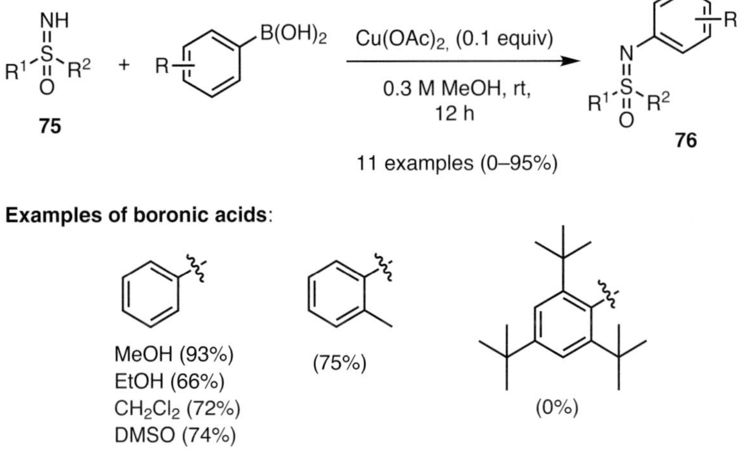

$$\underset{75}{R^1\overset{NH}{\underset{O}{\overset{\parallel}{S}}}R^2} + R\text{—}\langle\rangle\text{—}B(OH)_2 \xrightarrow[\text{0.3 M MeOH, rt,}]{Cu(OAc)_2, (0.1 \text{ equiv})} \underset{76}{R^1\overset{N=\langle\rangle\text{—}R}{\underset{O}{\overset{\parallel}{S}}}R^2}$$

11 examples (0–95%)

**Examples of boronic acids:**

- Phenyl: MeOH (93%), EtOH (66%), CH$_2$Cl$_2$ (72%), DMSO (74%)
- o-tolyl: (75%)
- 2,4,6-tri-t-butylphenyl: (0%)

(6.27)

The inorganic salt sodium azide was also used as the nucleophile in Chan–Lam cross-coupling reactions using a catalytic amount of copper salt (CuSO$_4$, 10 mol%) to cross-couple aryl- or vinylboronic acids in MeOH at room temperature under air (Equation 6.28) [67]. In contrast, traditional azidonation of aryl halides is usually performed at elevated temperature under inert atmosphere. Both electron-rich and electron-poor boronic acids gave moderate to excellent yields of the aryl or vinyl azides. 2-Substituted arylboronic acids also gave 90% yields. This method appears to be a safer way of preparing sensitive aryl azides. The azide formed was sequentially reacted with terminal alkynes to form 1-aryl-1,2,3-triazoles **77** (Equation 6.28). Thus, the one-pot synthesis of 1-aryl-1,2,3-triazole used 1 equiv of boronic acids, 1.1 equiv of terminal alkyne, 1.1 equiv of NaN$_3$, and 0.1 equiv of CuSO$_4$ in MeOH at room temperature to afford the triazoles in high yields.

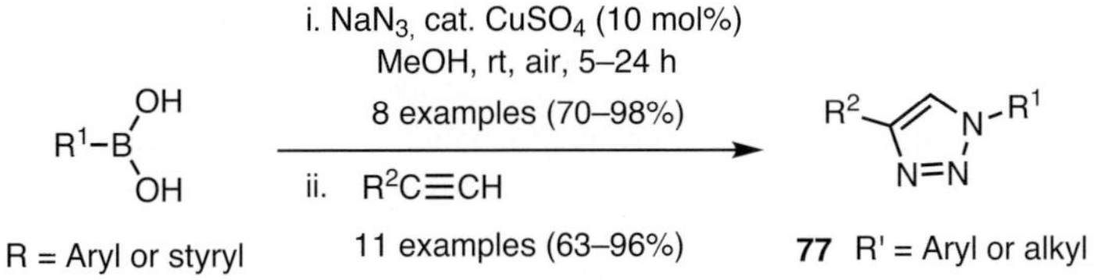

$$R^1\text{-B}(OH)_2 \xrightarrow[\text{ii. } R^2C\equiv CH]{\text{i. NaN}_3, \text{ cat. CuSO}_4 \text{ (10 mol\%)}, \text{ MeOH, rt, air, 5–24 h}} \underset{\substack{77 \\ R' = \text{Aryl or alkyl}}}{R^2\text{—}\overset{N-R^1}{\underset{N=N}{\langle\rangle}}}$$

R = Aryl or styryl

8 examples (70–98%)

11 examples (63–96%)

(6.28)

Recently, an important advancement was discovered by Zhao and coworkers, who developed a simple and efficient way for the synthesis of primary aromatic amines **78** using aqueous ammonia as the nucleophilic substrate in Chan–Lam coupling reactions (Equation 6.29) [68a]; however, monoarylation is feasible only when saturated ammonia is used [68b,c].

The optimal condition reported was as follows: 10 mol% of Cu$_2$O as the catalyst, aqueous ammonia (NH$_3$·H$_2$O, 25% aqueous solution) as the amine source, and methanol as the solvent. The reaction was carried out in air at room temperature. A base, ligand, or additive was not necessary. All the boronic acids including sterically hindered 2,4,6-trimethylphenyl boronic acid (yield: 92%) as well as electron-rich and electron-deficient boronic acids gave good to excellent yields. Boronic esters also reacted with aqueous ammonia to give good yields of the corresponding amines **78**.

$$NH_3 \cdot H_2O + Ar-B(OH)_2 \xrightarrow[\text{MeOH, rt, air}]{\text{Cu}_2\text{O (10 mol\%)}} Ar-NH_2 \quad \textbf{78}$$

21 examples (65–93%)

(6.29)

**Examples:**

(87%)  (72%)  (92%)  (65%)  (84%)

With all these recent successes, one wonders if other simple off-the-shelf inorganic reagents, such as KF, $Na_2SO_4$, and $NaNO_3$, and so on, could be arylated.

Liebeskind and coworkers reported a copper-catalyzed N-imination of aryl- or alkenylboronic acids and organostannanes using oxime O-carboxylates **79** as iminating agents and Cu(I) thiophene-2-carboxylate (CuTC) or Cu(OAc)$_2$ as the catalyst under nonbasic and nonoxidizing conditions to afford the N-arylated or N-alkenylated imines **80** (Equation 6.30) [69]. The idea was to increase the efficiency by using a higher oxidation state of the imine substrates such as oxime O-carboxylates. Transitioning from O-acetyl- and O-benzoyl- to O-pentafluorobenzoyl-derived oximes completely suppressed the boronic acid homocoupling side reaction. Among the oxime O-carboxylates, the O-pentafluorobenzoyl oximes had greater reactivity. Using DMF as solvent instead of THF, toluene and dioxane minimized the formation of competitive hydrolysis of the product imines by water generated *in situ* from the boronic acid–boroxine equilibrium. Many different copper sources (CuTC, CuCl, CuBr, CuI, Cu(OAc)$_2$, CuBr$_2$) were effective catalysts. Interestingly, by using boronic acids as the aryl donors, the reaction can be carried out in the presence of air; however, Cu(I) salt and an inert atmosphere were required when organostannanes were employed. Both electronic-rich and neutral boronic acids gave very good yields, and electron-withdrawing boronic acids were also tolerated. The reaction conditions were mild; however, they were not stereospecific and gave a mixture of E/Z imine isomers.

$$\underset{\textbf{79}}{\overset{R^1}{\underset{R^2}{\diagdown}}}C=N-OR' + \underset{\text{or}}{R^3B(OH)_2} \underset{R^3Sn(n\text{-Bu})_3}{\xrightarrow[\text{DMF, Ar or air, 50–70 °C}]{\text{cat. CuTC or Cu(OAc)}_2 \text{ (10–20 mol\%)}}} \underset{\textbf{80}}{\overset{R^1}{\underset{R^2}{\diagdown}}}C=N-R^3$$

21 examples (52–98%)

(6.30)

R' = Ac, $COC_6F_5$
$R^1, R^2$ = aryl, heteroaryl, alkyl
$R^3$ = aryl and alkenyl for B; aryl, heteroaryl, and alkenyl for Sn

Subsequently, the N-alkenylated α,β-unsaturated ketoxime O-pentafluorobenzoates, obtained from the cross-coupling reaction of **81** and **82**, were precursors in a cascade reaction for the synthesis of highly substituted pyridines **83** in moderate to excellent isolated yields (Equation 6.31) [70].

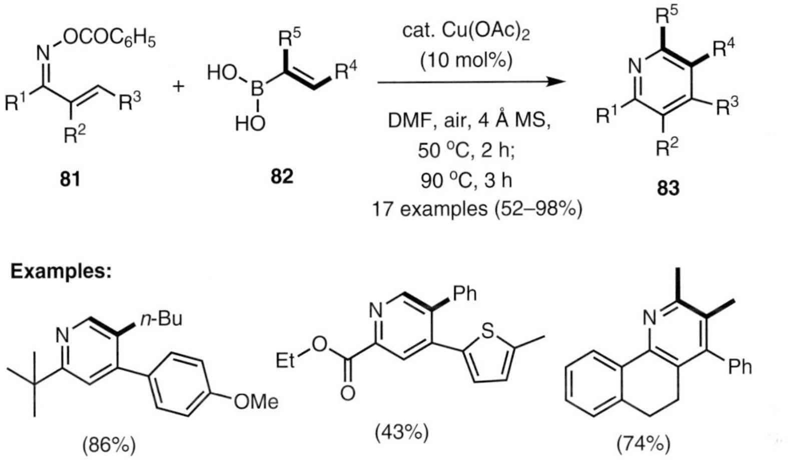

$$\text{81} + \text{82} \xrightarrow[\substack{\text{DMF, air, 4 Å MS,} \\ 50\,°C,\,2\,h; \\ 90\,°C,\,3\,h \\ 17\text{ examples (52–98\%)}}]{\text{cat. Cu(OAc)}_2\ (10\ \text{mol\%})} \text{83}$$

**Examples:**

(86%)  (43%)  (74%)

(6.31)

In addition, Liebeskind and coworkers developed a general nonoxidative N-amidation of organostannanes and boronic acids by coupling a wider variety of aryl, alkenyl, and heteroaryl organostannanes and boronic acids with O-acetyl hydroxamic acids **84** in the presence of Cu(I) sources (Equation 6.32) [71]. The reaction occurred with either CuTC (1 equiv) in THF or hexanes/THF at 60 °C or Cu(I) diphenylphosphinate (CuDPP, 2 equiv) in DMF at 60 °C for boronic acids.

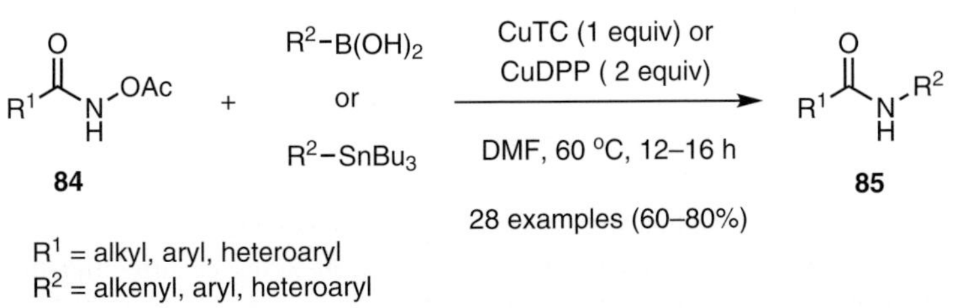

$$\text{84} + R^2\text{-B(OH)}_2 \text{ or } R^2\text{-SnBu}_3 \xrightarrow[\substack{\text{DMF, 60\,°C, 12–16 h} \\ 28\text{ examples (60–80\%)}}]{\text{CuTC (1 equiv) or CuDPP (2 equiv)}} \text{85}$$

(6.32)

$R^1$ = alkyl, aryl, heteroaryl
$R^2$ = alkenyl, aryl, heteroaryl

## 6.4
### Substrate Selectivity and Reactivity in Chan–Lam Cross-Coupling Reaction

Lam et al. studied the competition between O- and N-arylation in a 1 : 1 mixture of 4-*t*-butylaniline **86** and 3,5-di-*t*-butylphenol **87** with *p*-tolylboronic acid (Equation 6.33) (Lam, P.Y.S., Boone, D., and Clark, C.G., unpublished work, Lam, P.Y.S., Vincent, G., and Clark, C.G., unpublished work). N-Arylation was nine times faster than O-arylation (**88** versus **89**). Conversely, in the synthesis of tumor necrosis factor-α converting enzyme (TACE) inhibitors, O-arylation occurred predominately in the presence of the more hindered secondary aniline [72].

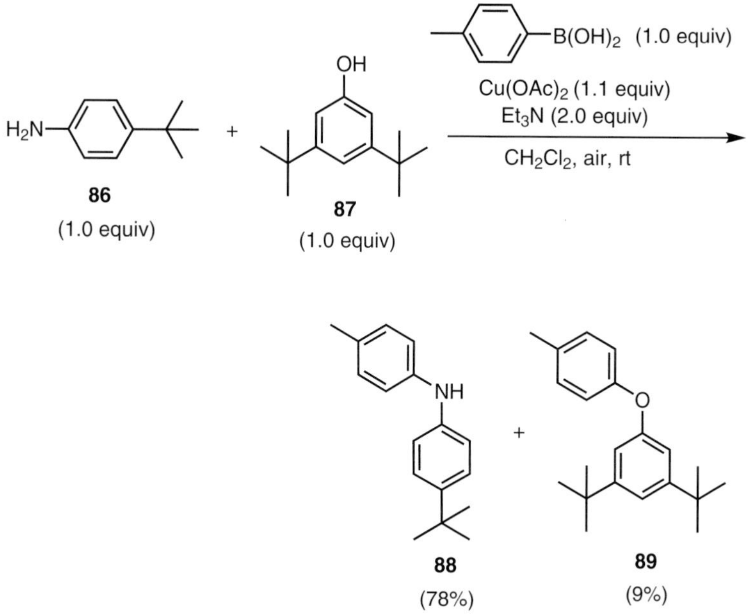

$$(6.33)$$

In general, N-arylation is faster than O-arylation. However, the reaction is very sensitive to steric effects, and it may be possible to selectively O-arylate in the presence of sterically hindered NH-containing substrates. Among the "amide-like" substrates, Stahl and Huffman found that compounds with more acidic NH group reacted faster than the one with a less acidic group as illustrated in Figure 6.2 [73].

## 6.5
## C—N and C—O Cross-Coupling with Alkenylboronic Acids

N-Vinyl groups can be used as protecting groups and easily cleaved using acidic hydrolysis or ozonolysis and as precursors for Grubbs' ring-closure metathesis reaction, as well as can be transformed to other functional groups such as cyclopropyl. Thus, the development of mild N- and O-vinylation using Chan–Lam chemistry could greatly facilitate these applications.

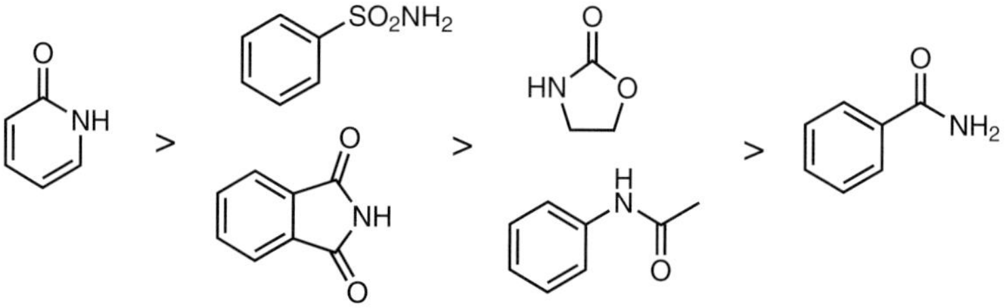

**Figure 6.2** Scale of reactivity in Chan–Lam cross-coupling reaction.

Lam et al. originally discovered that alkenylboronic acids **90** underwent efficient copper-promoted alkenylation of N—H or O—H substrates **91** (Equation 6.34) [21]. The reaction can also be run under catalytic conditions, although yields were lower. The unstable vinylboronic acid was then replaced with trivinylboroxine/pyridine complex for O-vinylation by McKinley and O'Shea [74].

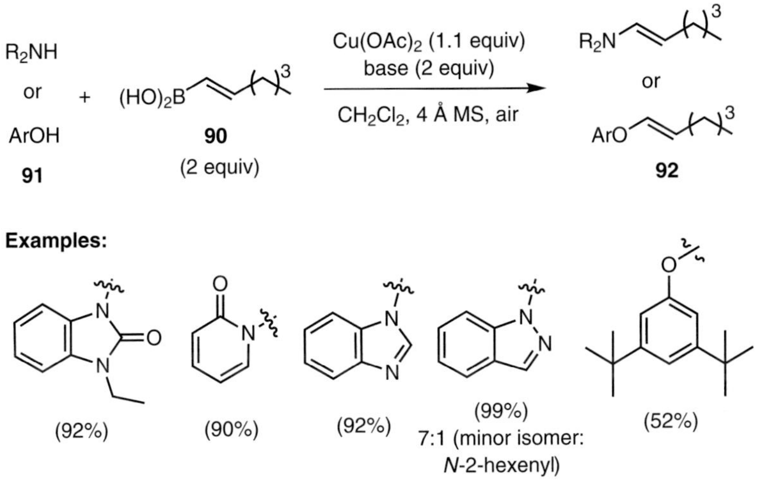

$$R_2NH \text{ or } ArOH \quad + \quad (HO)_2B\diagdown\diagup^3 \quad \xrightarrow[CH_2Cl_2, 4 \text{ Å MS, air}]{Cu(OAc)_2 \text{ (1.1 equiv)} \atop \text{base (2 equiv)}} \quad R_2N\diagdown\diagup^3 \text{ or } ArO\diagdown\diagup^3$$

**91**     **90** (2 equiv)     **92**

**Examples:**

(92%)    (90%)    (92%)    (99%) 7:1 (minor isomer: N-2-hexenyl)    (52%)

(6.34)

Recently, Batey and Bolshan reported an efficient synthesis of enamides **93** obtained from amides **94** and potassium alkenyltrifluoroborate salts **95** (Equation 6.35) [75]. Using N-methylimidazole as the ligand and catalytic $Cu(OAc)_2$, phthalimide reacted with alkenyltrifluoroborate salt **95** in quantitative yield; however, 2-pyrrolidinone and benzamide gave 22% and 0% yield, respectively, presumably due to the poor coordination ability and lower NH acidity, as well as poor solubility in $CH_2Cl_2$. Nevertheless, under ligandless condition using catalytic $Cu(OAc)_2$ in 1:1 $CH_2Cl_2$/DMSO, benzamides and pyrrolidones were N-alkenylated in good yields. Interestingly, electron-withdrawing benzamides afforded the corresponding enamides in higher yields than the electron-rich benzamides in contrast to observations made in the Buchwald–Hartwig chemistry [76]. In all cases, the reactions were highly stereoselective and only the *trans* enamides **93** were observed.

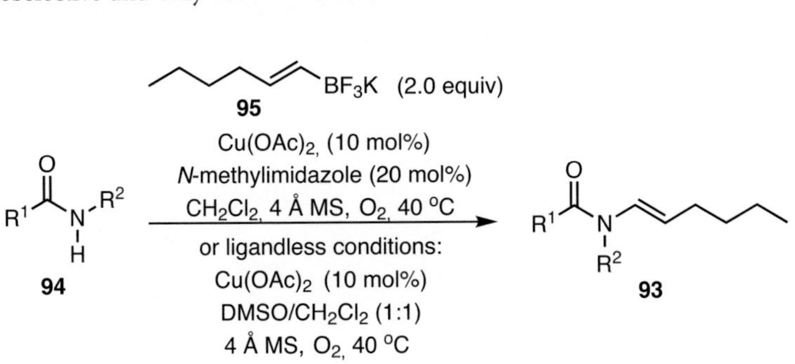

$$R^1\underset{H}{\overset{O}{\underset{\|}{C}}}N-R^2 \quad + \quad \diagup\diagdown\diagup BF_3K \text{ (2.0 equiv)} \quad \xrightarrow[\text{or ligandless conditions:} \atop Cu(OAc)_2 \text{ (10 mol\%)} \atop DMSO/CH_2Cl_2 \text{ (1:1)} \atop 4\text{ Å MS, } O_2, 40\,°C]{Cu(OAc)_2 \text{ (10 mol\%)} \atop N\text{-methylimidazole (20 mol\%)} \atop CH_2Cl_2, 4\text{ Å MS, } O_2, 40\,°C} \quad R^1\underset{R^2}{\overset{O}{\underset{\|}{C}}}N\diagup\diagdown\diagup$$

**94**    **95**    **93**

**Examples:**

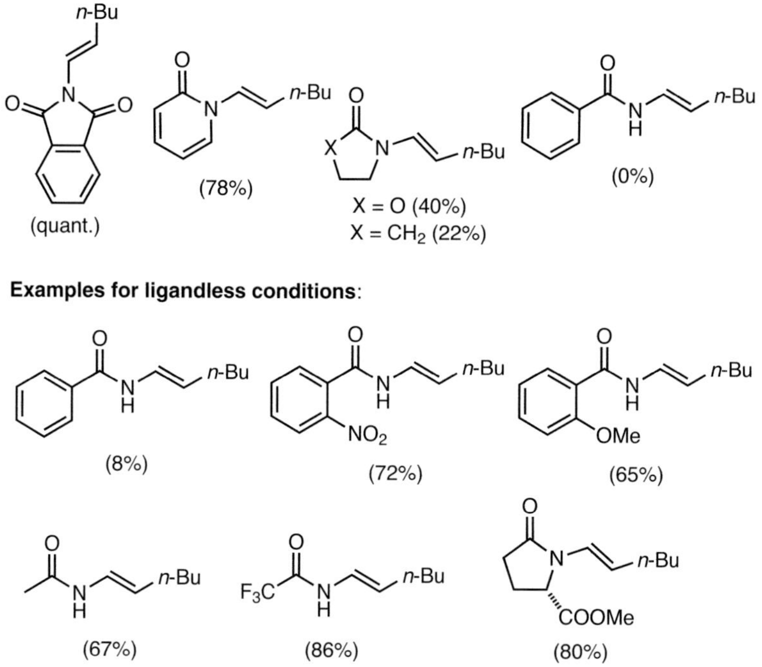

(quant.) (78%) X = O (40%) X = CH₂ (22%) (0%)

**Examples for ligandless conditions:**

(8%) (72%) (65%)

(67%) (86%) (80%)

(6.35)

## 6.6
## C—N and C—O Cross-Coupling with Boronic Acid Derivatives

### 6.6.1
### Boroxines, Boronic Esters, and Trifluoroborate Salts

As reported in the previous edition [8], Chan and coworkers demonstrated that arylboronic esters can be used in place of arylboronic acids in both O- and N-arylations (Figure 6.3) [56]. Boronic esters **96–98** and triphenylboroxine **99** were more efficient than phenylboronic acid **100**, while the corresponding catechol ester and sterically bulky pinacolate were not efficient. On the other hand, some researchers achieved some success with the sterically bulky pinacolates [45, 77]. Recently, a facile

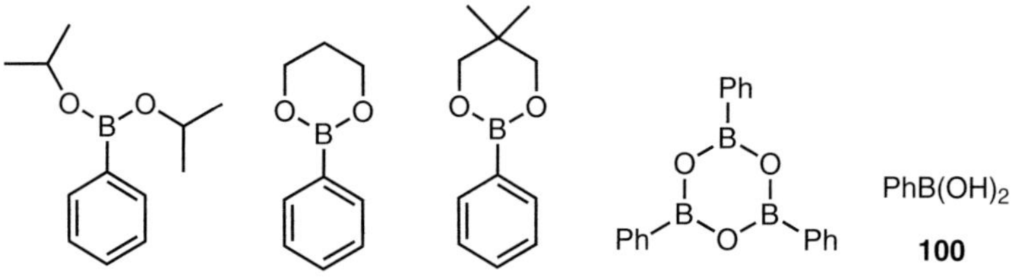

**96**    **97**    **98**    **99** (0.33 equiv)    **100** PhB(OH)₂

**Figure 6.3** Boronic esters **96–98**, triphenylboroxine **99**, and phenylboronic acid **100**.

route to the preparation of aryl boronates such as aryl pinacolates was developed by Marder and coworkers using copper-catalyzed borylation of aryl halides with alkoxy diboron reagents such as pinacolborane [78]. The reaction occurs with both electron-rich and sterically hindered aryl bromides using CuI (0.1 equiv) as the catalyst and n-Bu$_3$P as the ligand in THF at room temperature.

Recently, N-vinylation of azoles with alkoxydienyl boronates **101** and **102** and alkoxystyryl boronate **103** was developed by using 1 equiv of the boronate, 0.3 mol% of Cu(OAc)$_2$, 1 equiv of base (such as t-BuOK, Cs$_2$CO$_3$, pyridine, or Et$_3$N), and 1 equiv of CsF in CH$_2$Cl$_2$ at room temperature (Equation 6.36) [79]. It was speculated that the presence of fluoride in CsF increased the rate of transmetalation, as indicated by the increasing yield with t-BuOK/CsF (88%) compared to the reaction using just t-BuOK as the base (62%).

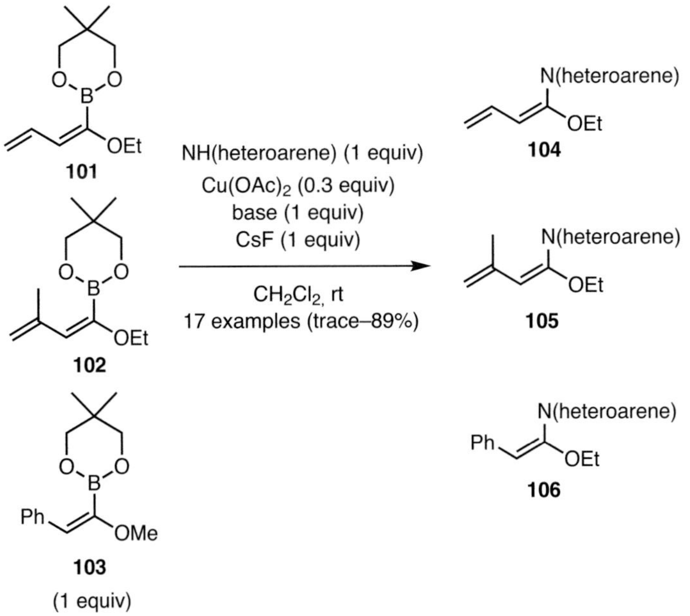

(6.36)

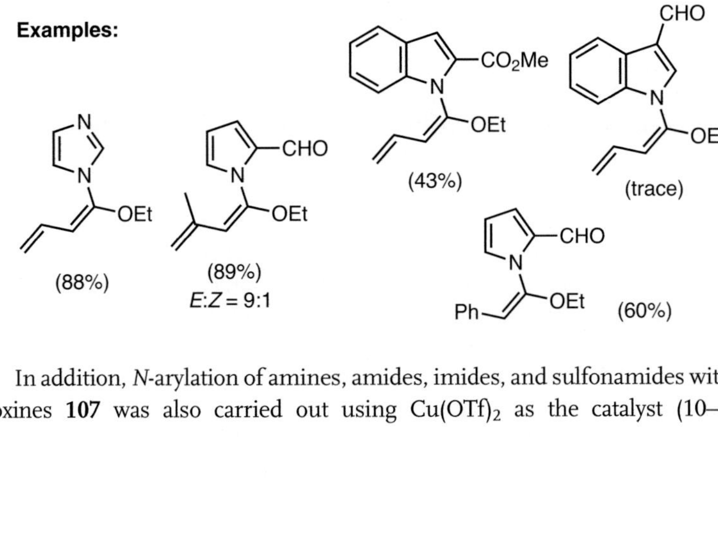

In addition, N-arylation of amines, amides, imides, and sulfonamides with arylboroxines **107** was also carried out using Cu(OTf)$_2$ as the catalyst (10–20 mol%)

in EtOH in the absence of base or additive (Equation 6.37) [80]. The corresponding N-arylation products **108** were obtained in moderate to excellent yield. The use of Cu(OTf)$_2$ gave higher yield (98%) and shorter reaction time (6 h) compared to Cu(OAc)$_2$ (48 h, 86% yield).

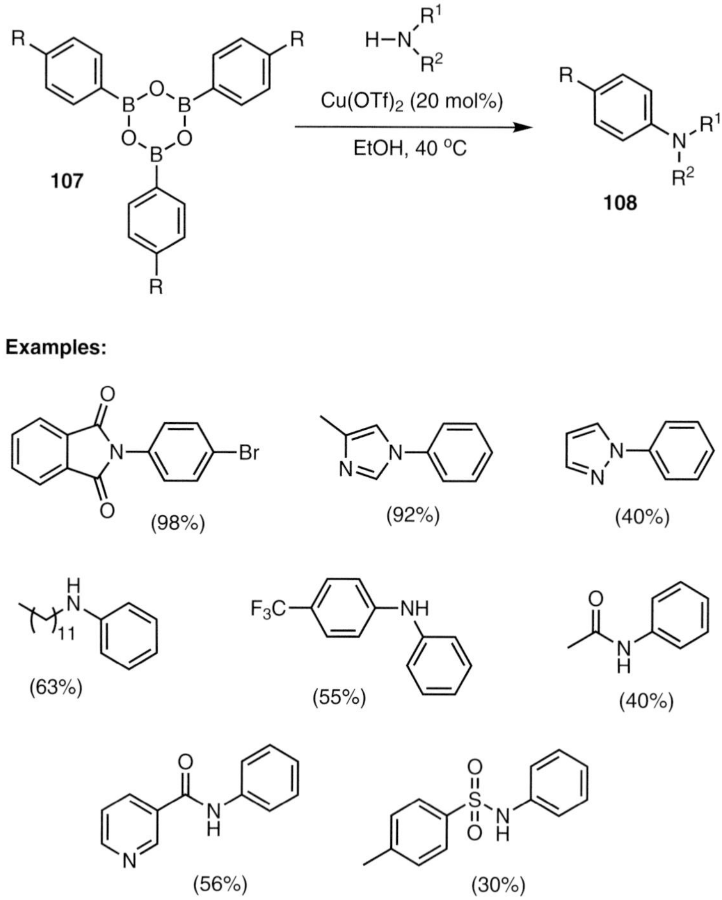

$$\underset{\textbf{107}}{\text{Ar-boroxine}} + \text{H-N}\underset{R^2}{\overset{R^1}{\diagdown}} \xrightarrow[\text{EtOH, 40 °C}]{\text{Cu(OTf)}_2 \text{ (20 mol\%)}} \underset{\textbf{108}}{\text{Ar-N}R^1R^2} \quad (6.37)$$

**Examples:**

- Phthalimide-C$_6$H$_4$-Br (98%)
- 2-methylimidazole-N-Ph (92%)
- pyrazole-N-Ph (40%)
- H-N(C$_{11}$H$_{23}$)-Ph (63%)
- F$_3$C-C$_6$H$_4$-NH-Ph (55%)
- CH$_3$C(O)NH-Ph (40%)
- 3-pyridyl-C(O)NH-Ph (56%)
- Tol-S(O)(=NH)-NH-Ph (30%)

Batey and Quach investigated alkenyl and aryl trifluoroborate salts **109** as coupling agents. Their earlier publication [81] disclosed a C–O cross-coupling protocol involving catalytic Cu(OAc)$_2$ and DMAP in the presence of oxygen and molecular sieves (Equation 6.38). The fluoroborate salts gave better yields than the corresponding boronic acids. The reaction did not require additional base other than the 20 mol% DMAP used. More important, the chemistry worked for aliphatic primary and secondary alcohols. This was the first reported case of alcohols participating in Chan–Lam cross-coupling reactions. Batey and coworkers also developed base-free N-arylation [61] and N-alkenylation reactions with trifluoroarylborates and potassium alkenyltrifluoroborate salts (Equation 6.35) [75].

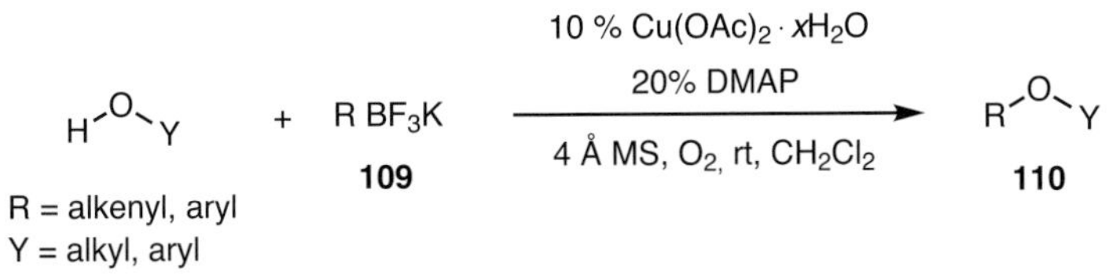

### 6.6 C—N and C—O Cross-Coupling with Boronic Acid Derivatives | 341

$$\text{H}^{\diagdown\text{O}\diagdown}\text{Y} \;+\; \text{R BF}_3\text{K} \quad \xrightarrow[\text{4 Å MS, O}_2,\text{ rt, CH}_2\text{Cl}_2]{\substack{10\%\ \text{Cu(OAc)}_2\cdot x\text{H}_2\text{O} \\ 20\%\ \text{DMAP}}} \quad \text{R}^{\diagdown\text{O}\diagdown}\text{Y} \quad (6.38)$$

**109** → **110**

R = alkenyl, aryl
Y = alkyl, aryl

Miyaura and coworkers developed the novel and easily prepared potassium aryl triolborates **111** to replace boronic acids in the Chan–Lam cross-coupling reactions. These cyclic aryl triolborates are air stable and water stable ate complexes of organoboronic acids and possesses high nucleophilicity, even compared to trifluoroborates (Equation 6.39) [82, 83].

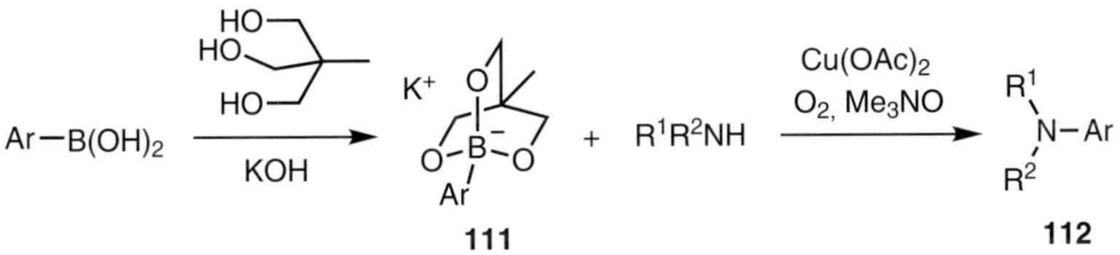

$$\text{Ar-B(OH)}_2 \xrightarrow{\substack{\text{HO} \\ \text{HO} \\ \text{HO} \\ \text{KOH}}} \underset{\textbf{111}}{\text{K}^+ \text{[triolborate-Ar]}^-} + \text{R}^1\text{R}^2\text{NH} \xrightarrow{\substack{\text{Cu(OAc)}_2 \\ \text{O}_2,\ \text{Me}_3\text{NO}}} \underset{\textbf{112}}{\text{R}^1\text{R}^2\text{N-Ar}} \quad (6.39)$$

*N*-arylation of aliphatic amines

$$\text{R}^1\text{R}^2\text{NH} + \underset{\substack{\textbf{111} \\ (1.5\ \text{equiv})}}{\text{K}^+\text{[triolborate-Ar]}^-} \xrightarrow[\substack{\text{4 Å MS, toluene, O}_2 \\ \text{rt – 60 °C, 20 h}}]{\substack{\text{Cu(OAc)}_2\ (10\ \text{mol\%}) \\ \text{O}_2\ (1\ \text{atm})}} \text{R}^1\text{R}^2\text{N-Ar}$$

20 examples (60–96%)

**Examples:**

n-Oct-NH₂ (90%)

t-Bu-CH₂-NH₂ (60%)

MeO-(CH₂)₃-NH₂ (88%)

geranyl-NH₂ (93%)

Cl-(CH₂)₂-NH₂·HCl (88%)

PhC(O)NH₂ (90%)

EtO₂C-CH₂-NH₂·HCl (89%)

morpholine (96%)

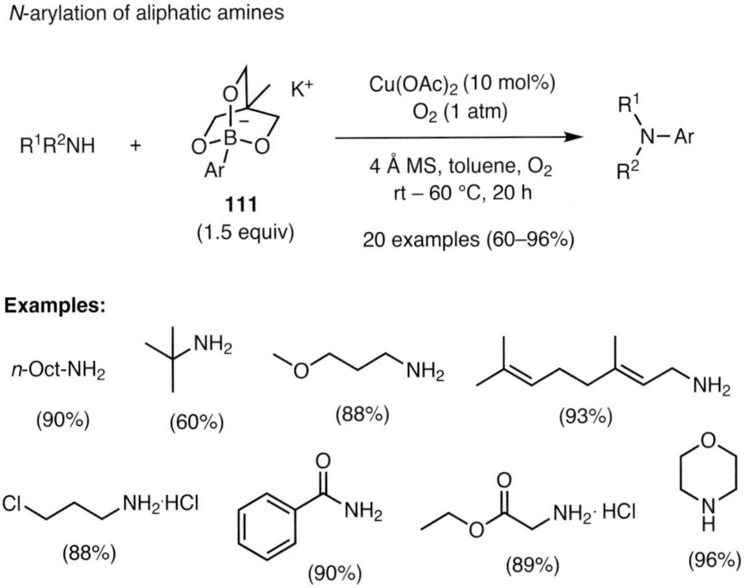

(6.40)

The *N*-arylation of primary and secondary aliphatic amines (Equations 6.40 and 41), anilines, and imidazoles (Equation 6.42) with potassium aryl triolborates **111** was carried out in the presence of a reoxidant (e.g., O₂ and trimethylamine *N*-oxide) and catalytic Cu(OAc)₂ (10 mol%). Aryl triolborates were found to be better aryl donors than the corresponding boronic acids (three times slower rate than triolborates) or potassium aryl trifluoroborates (very low reaction yields), presumably because of the easier transmetalation of the negatively charged complex in the trialkoxyborate relative to the neutral boronic acid compound. On the other hand, the analogous complex $PhBF_3K$ resulted in a very slow reaction owing to its low nucleophilicity and poor solubility in toluene.

The cross-coupling reactions of 2-pyridinylboronic acid and 3-pyridinylboronic acid are known to be challenging. Notably, the corresponding potassium 2-pyridine triolborate **112a** was cross-coupled with piperidine to afford the desired product **113a** in 70% yield (Equation 6.41). Similarly, potassium 3-pyridine triolborate **112b** gave **113b** in 85% yield.

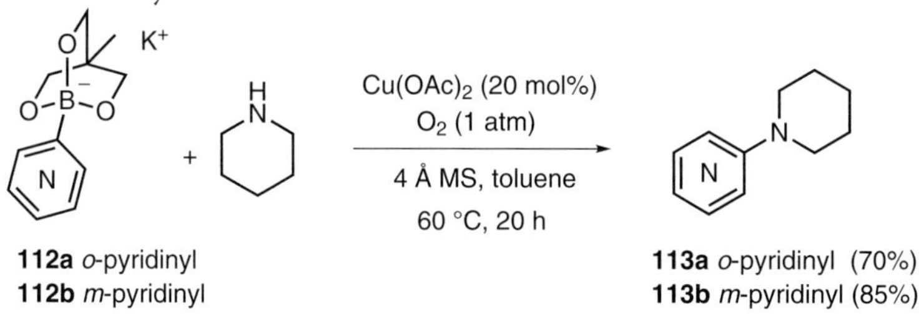
K+
Cu(OAc)2 (20 mol%)
O2 (1 atm)
4 Å MS, toluene
60 °C, 20 h
112a o-pyridinyl
112b m-pyridinyl
113a o-pyridinyl (70%)
113b m-pyridinyl (85%)

(6.41)

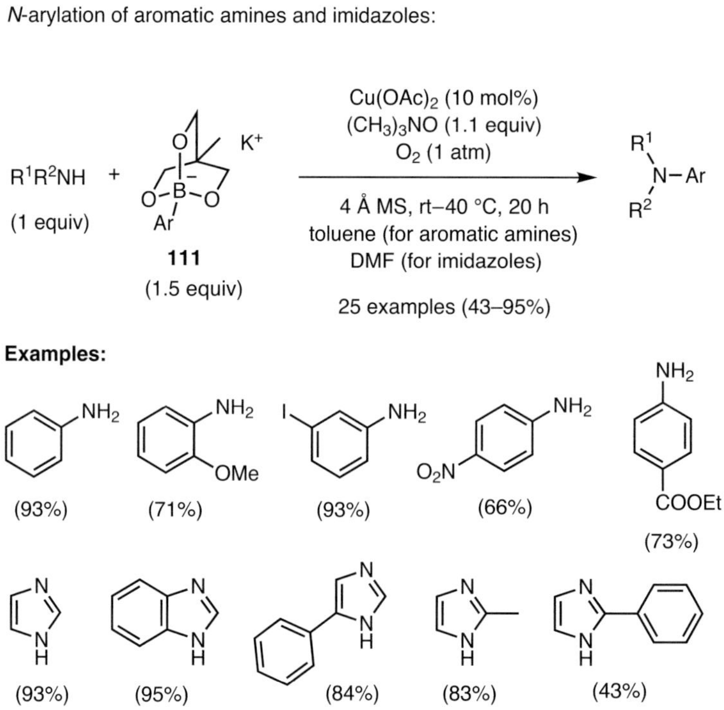
N-arylation of aromatic amines and imidazoles:
R1R2NH +
(1 equiv)
K+
111
(1.5 equiv)
Cu(OAc)2 (10 mol%)
(CH3)3NO (1.1 equiv)
O2 (1 atm)
4 Å MS, rt–40 °C, 20 h
toluene (for aromatic amines)
DMF (for imidazoles)
25 examples (43–95%)
N−Ar
Examples:
(93%)
OMe
(71%)
(93%)
O2N
(66%)
COOEt
(73%)
(93%)
(95%)
(84%)
(83%)
(43%)

(6.42)

Toluene was the solvent of choice for the cross-coupling of primary and secondary aliphatic amines and anilines with aryl triolborates since the formation of phenol or biaryl ether side products were observed in other solvent systems, such as $CH_2Cl_2$, DMF, THF, or ether. Trimethylamine *N*-oxide was found to be the best oxidant for anilines and imidazole substrates as shown in Equation 6.42.

## 6.6.2
## Alkylboronic Acids

Lam et al. (Lam, P.Y.S., Boone, D., and Clark, C.G., unpublished work) initially investigated the use of alkylboronic acids in the Chan–Lam cross-coupling reactions. For instance, cyclohexylboronic acid reacted with *t*-butylaniline **114** in low yields in dichloroethane at 70 °C for 2 days under the standard conditions (Equation 6.43).

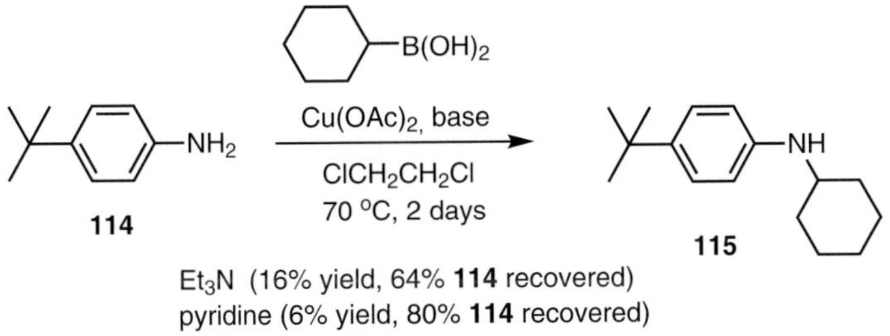

$$\underset{\mathbf{114}}{t\text{-Bu-C}_6\text{H}_4\text{-NH}_2} + \text{C}_6\text{H}_{11}\text{-B(OH)}_2 \xrightarrow[\text{ClCH}_2\text{CH}_2\text{Cl}]{\text{Cu(OAc)}_2,\text{ base}} \underset{\mathbf{115}}{t\text{-Bu-C}_6\text{H}_4\text{-NH-C}_6\text{H}_{11}} \quad (6.43)$$

Et$_3$N (16% yield, 64% **114** recovered)
pyridine (6% yield, 80% **114** recovered)

One of the recent significant expansions of the Chan–Lam coupling is the *N*-cyclopropanation on a variety of substrates with cyclopropylboronic acid at elevated temperature, as reported by two research groups independently [84, 85]. Tsuritani et al. reported a general *N*-cyclopropanation reaction of indoles **116** (Equation 6.44) and cyclic amides **117** (Equation 6.45) [84] using Cu(OAc)$_2$ as the catalyst, pyridine or DMAP as the base, and NaHMDS as the additive in hot toluene at 95 °C. Both the choice of base and the use of molecular oxygen were important for the rate of the reaction. For electron-deficient indole systems, a stoichiometric amount of Cu(OAc)$_2$ significantly improved the yield, while for electron-rich indoles, using trialkoxyborates in the absence of NaHMDS offered the desired products **118** in good yield. In addition, cyclic amides and benzamides **117** also reacted with cyclopropylboronic acid to give the corresponding cyclopropylamides **119** in good to excellent yields (49–93%) (Equation 6.45). The reaction tolerated a variety of functional groups such as Cl, ester, ketone, nitrile, and nitro. Acetanilide did not afford the desired product, presumably due to the preference of Z-configuration and the cleavage of the amide bond at 95 °C under the reaction conditions.

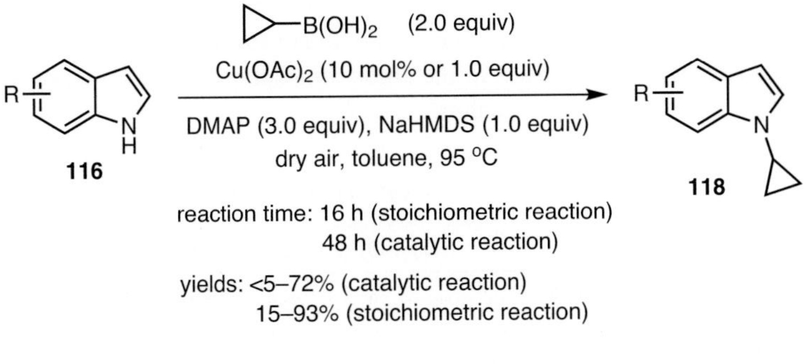

$$\underset{\mathbf{116}}{\text{indole-NH}} \xrightarrow[\substack{\text{DMAP (3.0 equiv), NaHMDS (1.0 equiv)} \\ \text{dry air, toluene, 95 °C}}]{\text{cyclopropyl-B(OH)}_2 \text{ (2.0 equiv)} \\ \text{Cu(OAc)}_2 \text{ (10 mol\% or 1.0 equiv)}} \underset{\mathbf{118}}{\text{N-cyclopropylindole}} \quad (6.44)$$

reaction time: 16 h (stoichiometric reaction)
48 h (catalytic reaction)

yields: <5–72% (catalytic reaction)
15–93% (stoichiometric reaction)

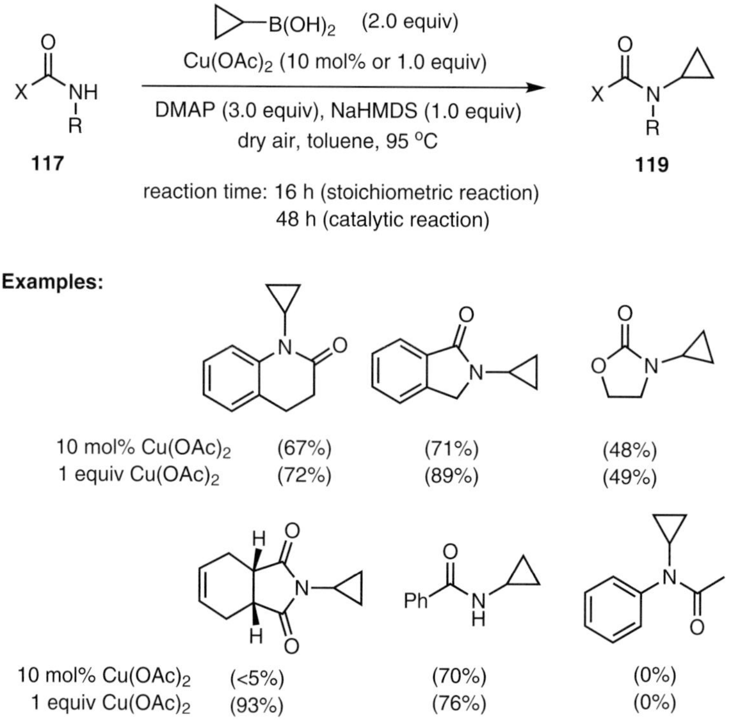

$$\underset{117}{\overset{\overset{O}{\|}}{X \diagdown \underset{R}{N}H}} \xrightarrow[\substack{\text{DMAP (3.0 equiv), NaHMDS (1.0 equiv)} \\ \text{dry air, toluene, 95 °C}}]{\triangleright\text{–B(OH)}_2 \quad (2.0 \text{ equiv}) \\ \text{Cu(OAc)}_2 \text{ (10 mol% or 1.0 equiv)}} \underset{119}{\overset{\overset{O}{\|}}{X \diagdown \underset{R}{N} \diagdown \triangleleft}} \qquad (6.45)$$

reaction time: 16 h (stoichiometric reaction)
48 h (catalytic reaction)

**Examples:**

| | | | |
|---|---|---|---|
| 10 mol% Cu(OAc)$_2$ | (67%) | (71%) | (48%) |
| 1 equiv Cu(OAc)$_2$ | (72%) | (89%) | (49%) |

| | | | |
|---|---|---|---|
| 10 mol% Cu(OAc)$_2$ | (<5%) | (70%) | (0%) |
| 1 equiv Cu(OAc)$_2$ | (93%) | (76%) | (0%) |

Zhu and coworkers reported the successful N-cyclopropanation of NH-containing azoles, amides, and sulfonamides **120** (Equation 6.46) [85]. The best conditions for N-cyclopropanation of phthalimide was established as follows: Cu(OAc)$_2$ (0.2 equiv), 2,2'-bipyridine (0.2 equiv), Na$_2$CO$_3$ (2 equiv), air in dichloroethane at 70 °C for 6 h to give the desired product in 80% yield, or using Cu(OAc)$_2$ (1 equiv), 2,2'-bipyridine (1.0 equiv), Na$_2$CO$_3$ (2 equiv), air in dichloroethane at 70 °C for 3 h to give the product quantitatively. Various N-heterocycles such as imidazole, benzimidazole, 2-acetyl-pyrrole, benzotriazole, carbazole and oxazolidinone, thymine, 6-methyluracil, indoles, amides, and sulfonamides can be N-cyclopropanated under the above reaction conditions. Acetanilide did not give the desired product similar to that observed by Tsuritani et al. [84]. The reaction tolerated functional groups such as halogen atoms, ester and carbamate, ether, ketone, and nitro groups. Interestingly, N-cyclopropylation can also take place in the argon atmosphere in the presence of a stoichiometric amount of copper catalyst.

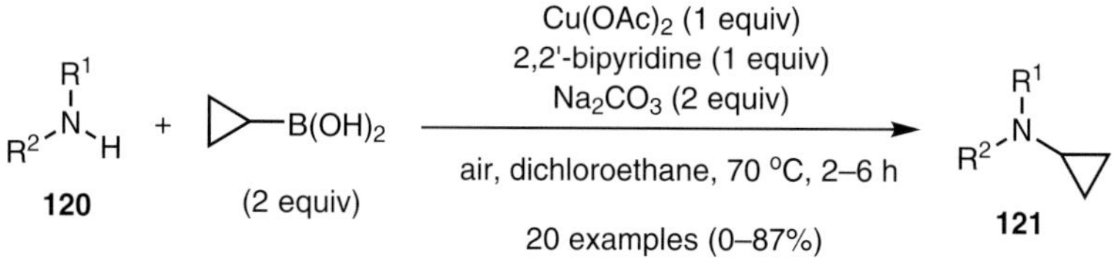

$$\underset{120}{\overset{R^1}{\underset{R^2 \diagdown N \diagdown H}{|}}} + \underset{(2 \text{ equiv})}{\triangleright\text{–B(OH)}_2} \xrightarrow[\substack{\text{Cu(OAc)}_2 \text{ (1 equiv)} \\ \text{2,2'-bipyridine (1 equiv)} \\ \text{Na}_2\text{CO}_3 \text{ (2 equiv)} \\ \text{air, dichloroethane, 70 °C, 2–6 h} \\ \text{20 examples (0–87%)}}]{} \underset{121}{\overset{R^1}{\underset{R^2 \diagdown N \diagdown \triangleleft}{|}}}$$

**Examples:**

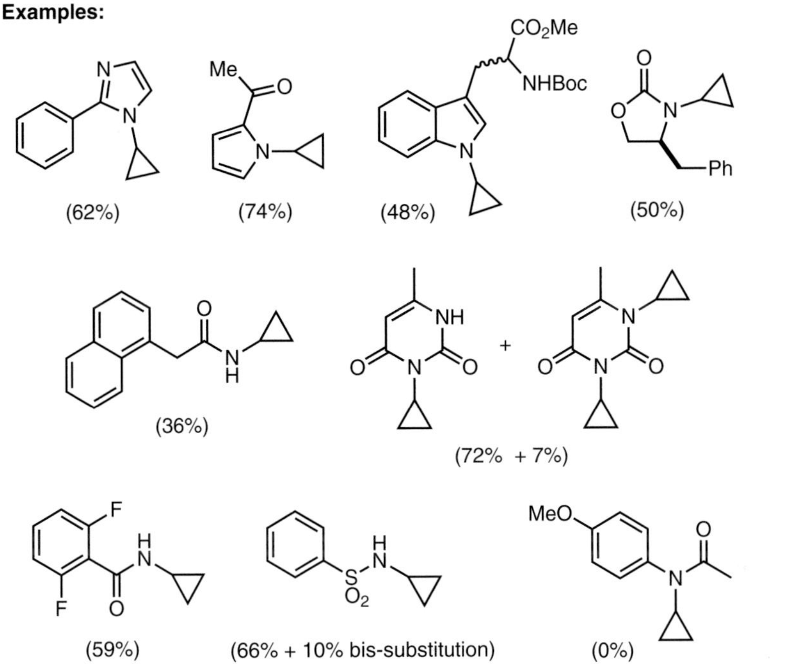

(62%)    (74%)    (48%)    (50%)

(36%)    (72% + 7%)

(59%)    (66% + 10% bis-substitution)    (0%)

(6.46)

Besides N-cyclopropanation, Cruces and coworkers recently reported the first successful Chan–Lam coupling reaction of anilines **122** with methylboronic acid for the selective synthesis of monomethylated anilines **123** in good to excellent yields (Equation 6.47) [86]. The order of addition of reagents, that is, preincubation of copper catalyst 10–15 min prior to the addition of MeB(OH)$_2$, and the use of 2.5 equiv of Cu(OAc)$_2$ and 2.5 equiv of MeB(OH)$_2$ were crucial for improving the yields. The optimized conditions are as follows: copper(II) acetate (2.5 equiv) was added to a solution of aniline (1 equiv) and pyridine (3–5 equiv) in dioxane. The mixture was stirred for 15 min, methylboronic acid (2.5 equiv) was added, and the reaction was refluxed until aniline was totally consumed (1.5–18 h).

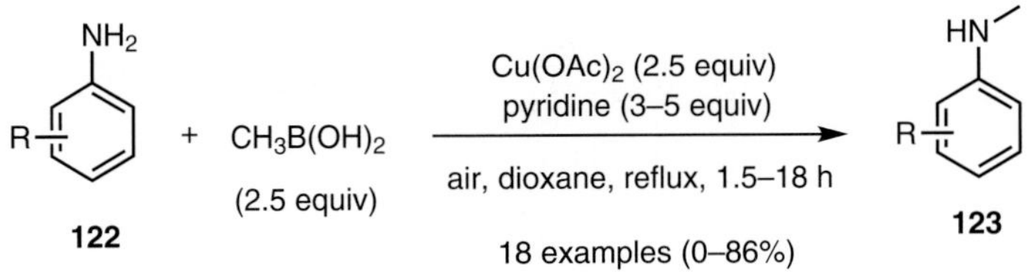

18 examples (0–86%)

**Examples:**

(86%)    (75%)    (0%)    (30%)    (53%)

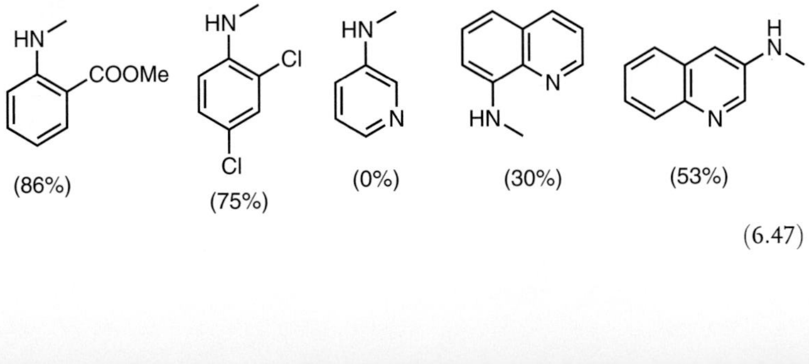

(6.47)

## 6.7
## C–S and C–Se/C–Te Cross-Coupling

Guy and coworkers reported the first C–S cross-coupling reaction of electron-rich alkyl thiols [87]. Refluxing DMF and argon atmosphere were required to suppress disulfide formation. The reaction may actually involve a Cu(I)-mediated coupling of the boronic acid with the corresponding disulfide since thiols are expected to be rapidly oxidized by Cu(OAc)$_2$ [88]. A mild, nonbasic synthesis of thioethers was then developed using a Cu(I)-catalyzed C–S cross-coupling of boronic acids with N-thioalkyl-, aryl-, and heteroarylimides, where the imide moiety serves as a sulfide surrogate. Later on, arylation of a cyclic thiourea at the sulfur was developed using phenylboronic acid, Cu(OAc)$_2$, and 1,10-phenanthroline [89]. The reaction time was shortened considerably by using microwave irradiation and dichloroethane as solvent, affording similar yields. Furthermore, cyclic thioureas/thioamides with latent free thiol functionalities can react with arylboronic acids to form desulfitative carbon cross-coupling (catalyzed by Pd(PPh$_3$)$_4$ and Cu(I)) or carbon–sulfur cross-coupling (Pd-free, catalyzed by Cu(I) or Cu(II) salt) products depending on the catalytic system being used (Equations 6.48 and 6.49) [90, 91].

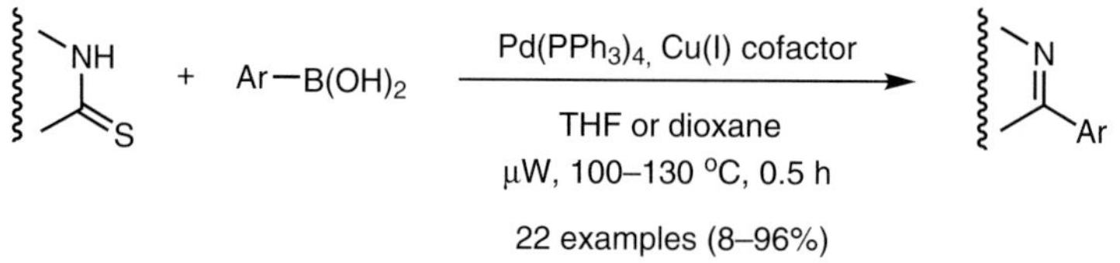

$$\text{NH-C(=S)-} + \text{Ar-B(OH)}_2 \xrightarrow[\substack{\text{THF or dioxane} \\ \mu W,\ 100-130\ ^\circ C,\ 0.5\ h}]{\text{Pd(PPh}_3)_4,\ \text{Cu(I) cofactor}} \text{N=C(-Ar)-} \quad (6.48)$$

22 examples (8–96%)

The protocol for C–S cross-coupling to obtain complete conversion was developed: a mixture of thioamide **124** or **126**, phenylboronic acid (4 equiv), Cu(OAc)$_2$ (1 equiv), and 1,10-phenanthroline (2 equiv) in dichloroethane was stirred under air for 15 min to form the disulfide that was subsequently heated at 110 °C for 30 min (2–3 bar) under microwave irradiation to deliver the desired sulfide **125** or **127** in 81% and 79% isolated yields, respectively (Equation 6.49).

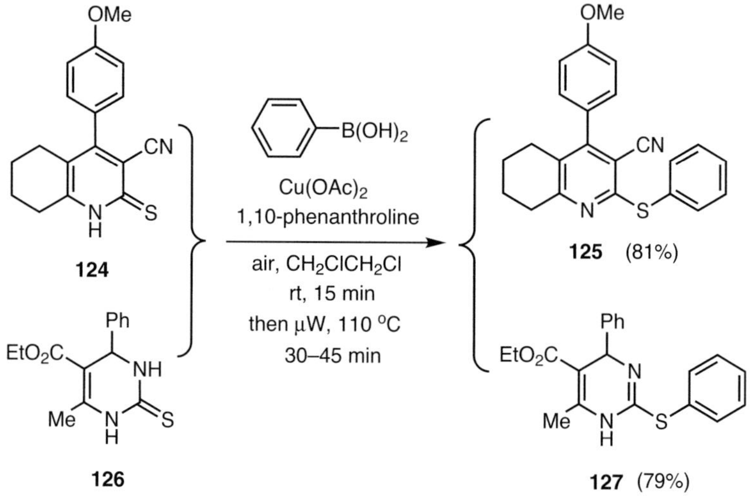

**124** / **126** + Ph–B(OH)$_2$ →[Cu(OAc)$_2$, 1,10-phenanthroline, air, CH$_2$ClCH$_2$Cl, rt, 15 min then μW, 110 °C, 30–45 min] **125** (81%) / **127** (79%)

(6.49)

Another major application in C–S cross-coupling chemistry is the synthesis of aryl and vinyl sulfones from aryl- or vinylboronic acids and sodium sulfinates. Since the report of stoichiometric conditions for aryl sulfone formation by Evans and coworkers in 2004 [92], two research groups independently published examples of catalytic copper-mediated sulfonylation of arylboronic acids in 2007 [93, 94].

Batey and Huang reported a mild synthesis of aryl and vinyl sulfones **128** via a ligand-accelerated protocol by utilizing catalytic Cu(OAc)$_2$-catalyzed cross-coupling between aryl- or vinylboronic acids **129** with sodium sulfinates **130** (10 mol% Cu(OAc)$_2$·H$_2$O, 1,10-phenanthroline as the ligand (2:1 ligand/copper stoichiometry) and a CH$_2$Cl$_2$/DMSO (15:1) cosolvent mixture in the absence of an external base under an atmosphere of O$_2$ for 72 h (Equation 6.50) [93]. Arylboronic acids bearing electron-donating groups gave higher yields. While *ortho* substituents were tolerated, bis-*ortho* substitution caused a drastic drop of yields. Potassium trifluorophenyl borate surprisingly gave low yields. Sodium phenylsulfinate or sodium methylsulfinate gave similar yields. While alkenylboronic acids afforded the product in good yields, no reaction was observed for alkylboronic acids.

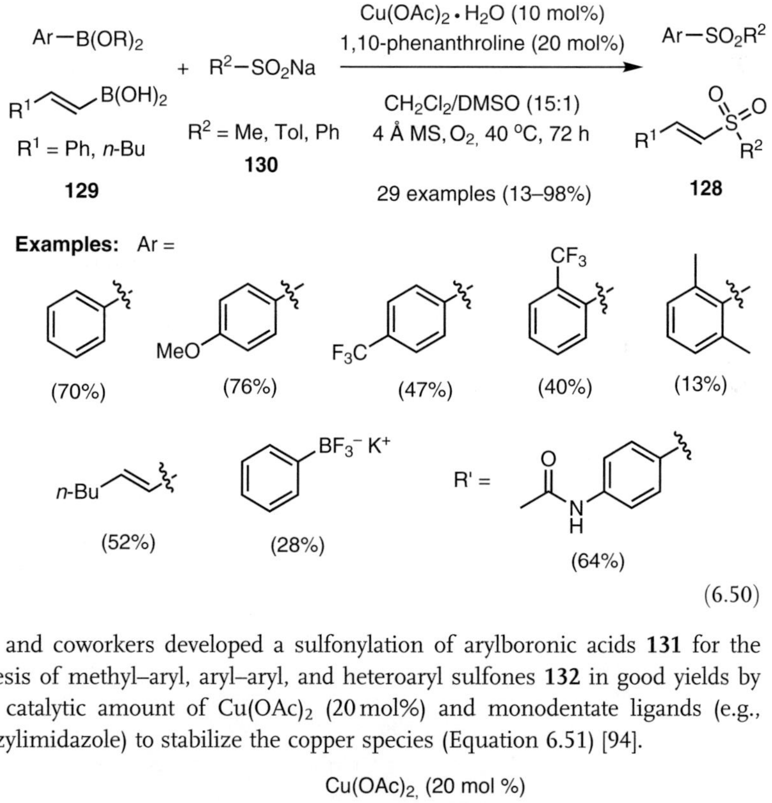

(6.50)

Tse and coworkers developed a sulfonylation of arylboronic acids **131** for the synthesis of methyl–aryl, aryl–aryl, and heteroaryl sulfones **132** in good yields by using catalytic amount of Cu(OAc)$_2$ (20 mol%) and monodentate ligands (e.g., 1-benzylimidazole) to stabilize the copper species (Equation 6.51) [94].

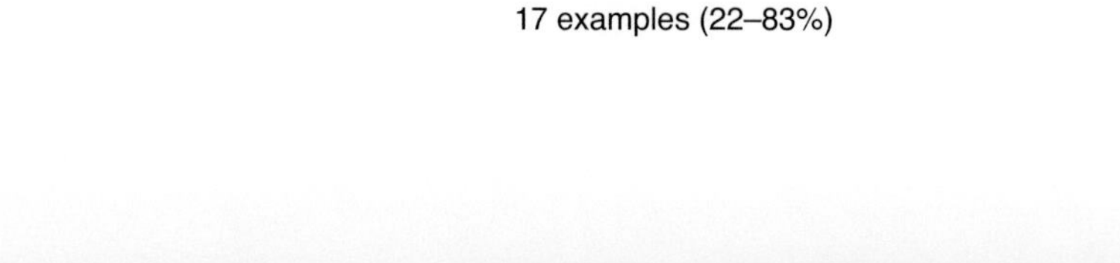

**Examples:**

Ar =

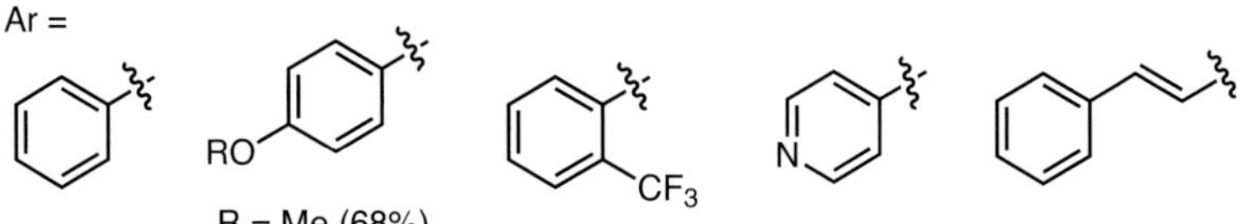

(64%)   R = Me (68%), R = H (27%)   (43%)   (22%)   (65%)

(6.51)

It was also reported that $Cu(OAc)_2$ in ionic liquids (such as 1-*n*-butyl-3-methylimidazolium triflate (BmimOTf)) catalyzed the cross-coupling of arylboronic acids with sulfinic acid salts (sodium salt of *p*-tolyl sulfinic acid and methyl sulfinic acid) to afford aryl sulfones in good yields (68–82%, 11 examples) under ambient conditions [95].

Besides the aforementioned C–S copper-mediated cross-coupling of boronic acids, Wang *et al.* reported the synthesis of unsymmetrical diaryl selenides and tellurides **133** via the reaction of potassium aryltrifluoroborates **134** or aryl- or vinylboronic acids **135** with diaryl diselenide or diphenyl ditelluride **136** in the presence of a copper catalyst (such as CuI or $Cu(OAc)_2$) (Equations 6.52 and 6.53) [96]. The reaction was carried out at 100 °C in DMSO or ionic liquid 1-*n*-butyl-3-methylimidazolium tetrafluoroborate ($BmimBF_4$). The reactions proceeded in high yields without any base. The selection of solvent was important. While DMSO gave high yields, other solvents such as $CH_2Cl_2$ and THF gave no reaction. Furthermore, ionic liquid and copper salt can be recycled without a significant decrease in yields.

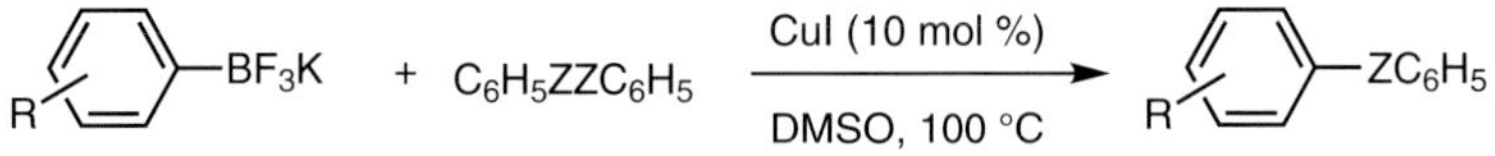

R–Ar–$BF_3K$ + $C_6H_5ZZC_6H_5$ →[CuI (10 mol %)][DMSO, 100 °C] R–Ar–$ZC_6H_5$

**134**   **136**   **133**

R = *p*-$CH_3$, Z = Se (88%)
R = *p*-$OCH_3$, Z = Se (92%)
R = *p*-$CH_3$, Z = Te (67%)

(6.52)

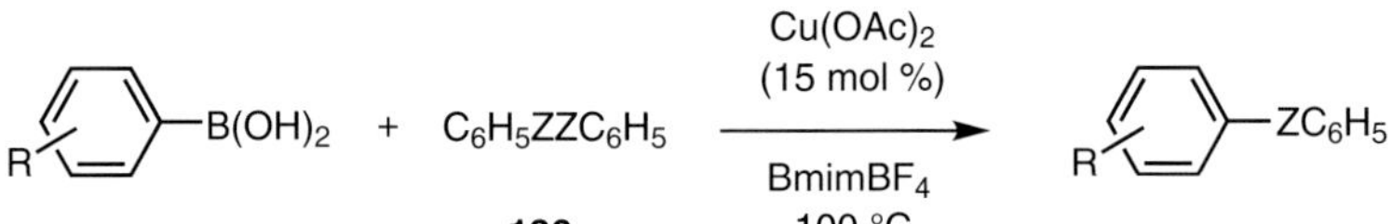

R–Ar–$B(OH)_2$ + $C_6H_5ZZC_6H_5$ →[$Cu(OAc)_2$ (15 mol %)][$BmimBF_4$, 100 °C] R–Ar–$ZC_6H_5$

**135**   **136**   **133**

R = *p*-$CH_3$, Z = Se (92%)
R = *p*-$OCH_3$, Z = Se (78%)
R = *p*-$CH_3$, Z = Te (82%)

(6.53)

## 6.8
## Mechanistic Considerations

The mechanism of Cu-catalyzed oxidative N/O-arylation reactions is yet fully understood, and various observations by Stahl and Lam are compiled herein to shed some light on this cross-coupling reaction. Empirical studies of the influence of substrate electronic properties, solvent identity, and ligand/base effects have been performed, but the results are difficult to interpret in the absence of a more complete mechanistic picture. Analysis of reaction side products has provided some useful insights into the reaction mechanism. The first detailed mechanistic study of one of these reactions was described recently by King and Stahl, who investigated the Cu-catalyzed methoxylation of tolylboronic acid in methanol [97].

### 6.8.1
### Empirical Observations

Electronic effects of the N-arylation of phthalimides with arylboronic acids were reported by Galemmo et al. (Galemmo, R.A., Chang, R.K., and Lam, P.Y.S., unpublished work). Phthalimide was chosen for study since it is one of the best substrates for N-arylation. The data in Figure 6.4 reveal that the two electron-rich phthalimides performed better than the nitro-substituted derivative, but the arylboronic acid substrate exhibited little electronic effect. This general trend was also observed for

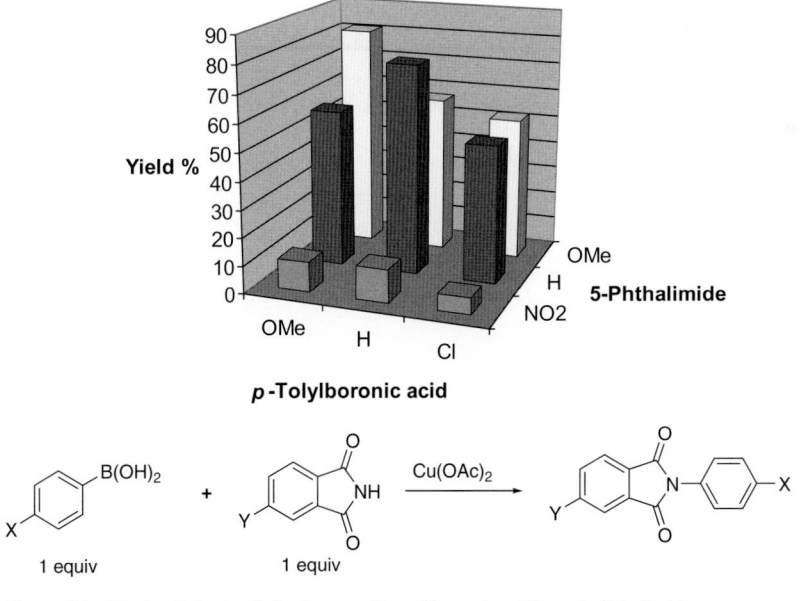

**Figure 6.4** Study of electronic in the coupling of boronic acids and phthalimides.

the N-arylation of sulfonamides (Galemmo, R.A., Chang, R.K., and Lam, P.Y.S., unpublished work) and azoles [11].

A solvent study, performed by Combs et al., used the N-arylation of morpholine as a prototype (Scheme 6.4) (Combs, A.P., Saubern, S., and Lam, P.Y.S., unpublished work). It was found that methylene chloride and 1,4-dioxane are good solvents. Lower yields were obtained with other solvents, including those with both higher and lower polarities (DMF, DMSO, and toluene). The preferred solvent appears to depend on the substrate. For example, dimethylformamide was identified as the best solvent with benzimidazole. Methanol (10 equiv) used as an additive or in refluxing methanol has been used by Snapper, Hoveyda [27], and Xie [60, 98] to improve the yields of other coupling reactions.

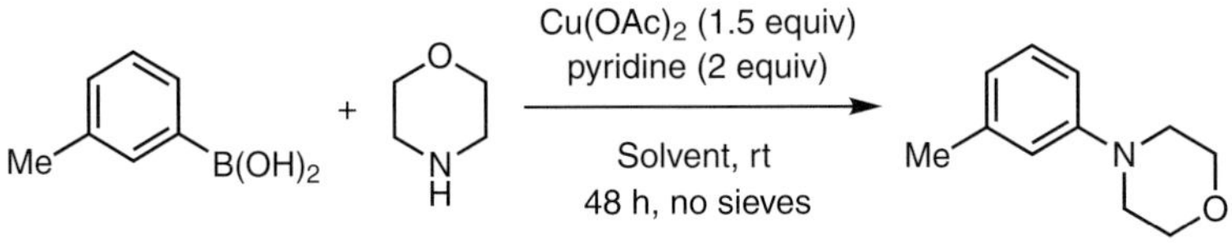

| Solvent | $CH_2Cl_2$ | 1,4-dioxane | DMF | EtOAc | THF | toluene | DMSO |
|---|---|---|---|---|---|---|---|
| Yield | 62 % | 73 % | 43 % | 26 % | 23 % | 18 % | 16 % |

**Scheme 6.4** Solvent study of N-arylation of morpholine.

Another important component of the reaction is the added base/ligand. It has been reported that large excess of triethylamine inhibits the reaction by coordinating to the copper center. In the absence of triethylamine, however, the substrate reacts with arylboronic acid to form the arylboronic acid monoamide adduct, which is less active or inactive [35]. Lam et al. (Lam, P.Y.S., Boone, D., and Clark, C.G., unpublished work) briefly examined the role of amine additives (Scheme 6.5). For an O-arylation

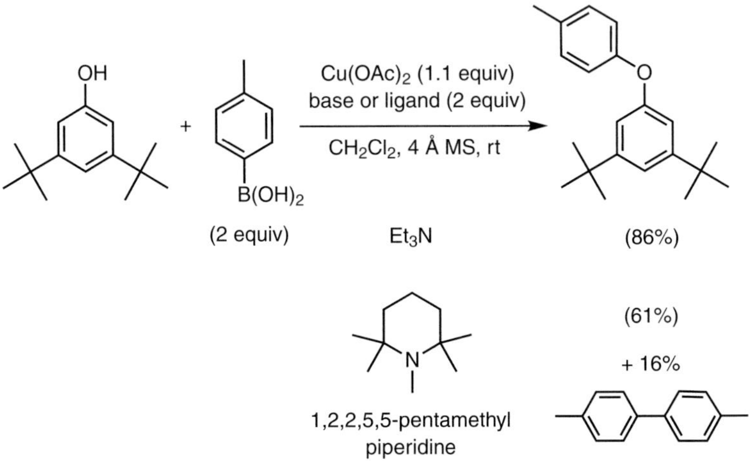

**Scheme 6.5** Study on the role of base/ligand.

reaction, the more sterically hindered 1,2,2,5,5-pentamethylpiperidine gave 25% lower yield compared to triethylamine, and 4,4′-dimethylbiphenyl was obtained as a side product in 16% yield with the former hindered amine. As discussed in Section 6.2.2, the protiodeboronation side reaction can be suppressed with the addition of 5 equiv of DMAP [28]. For N-arylation, the substrates can sometimes serve as base/ligand, and no external base/ligand needs to be added in many cases [60]. On the other hand, for O-arylation, base/ligand often must be used, the reactions of trifluoroborate salts being an important exception [9, 10, 71].

The empirical studies outlined above and related observations reported in the literature provide useful insights into the catalytic reactions, but the mechanistic origin of the observations depends on a number of factors, including the turnover-limiting step in the catalytic cycle, the identity of the catalyst resting state (e.g., the oxidation state; the number, identity, and coordination mode of ligands bound to the metal center; the nuclearity of the catalyst), the solubility of the catalyst and reagents under the reaction conditions, and the influence of the nucleophile and reaction conditions on the identity of the arylboron species that participates in the reaction. In most cases, these parameters have not been defined under the catalytic conditions.

## 6.8.2
### General Mechanistic Observations

First, this reaction does not appear to proceed by a free radical mechanism. For example, the addition of the radical trap 1,1-diphenylethylene has no effect on the yield of the N-arylation of benzimidazole to give p-tolylbenzimidazole (Lam, P.Y.S., Boone, D., and Clark, C.G., unpublished work). Triplet oxygen itself is a radical trap and is unlikely to be compatible with a radical-based coupling mechanism.

The reactions typically employ excess arylboronic acid (1.5–2.0 equiv) because this substrate partner undergoes two major side reactions: protiodeboronation and conversion to phenol [10, 99]. Lam et al. found that in the absence of substrates, 4-biphenylboronic acid has a half-life of $\leq 30$ min, with protiodeboronation being the major side reaction (Figure 6.5) (Lam, P.Y.S., Boone, D., and Clark, C.G., unpublished work). Phenol formation has two possible origins. Evans et al. [10] originally proposed that O-arylation of adventitious water (formed in the arylboronic acid/triarylboroxine equilibrium) can account for the phenol side product. On the other hand, it is also possible that arylboronic acid is oxidized to phenol via an oxocopper(III) intermediate or by hydrogen peroxide formed via reduction of $O_2$. To differentiate between these two mechanisms, a study was performed with labeled $O_2$ and $H_2O$ in the absence of substrates (Scheme 6.6) [99]. No $^{18}O$ incorporation was observed in the isolated phenol using $^{18}O_2$, and when $H_2^{18}O$ was used, $^{18}O$ incorporation into phenol was observed. These results support the hypothesis of Evans et al. [10] that the phenol side product comes from adventitious water. These observations may account for the beneficial effect of molecular sieves sometimes observed in catalytic reactions; however, molecular sieves are not always beneficial. The precise role of molecular sieves in catalytic reactions can be quite unexpected [100].

**Decomposition of *p*-biphenylboronic acid in presence of copper(II) acetate and triethylamine**

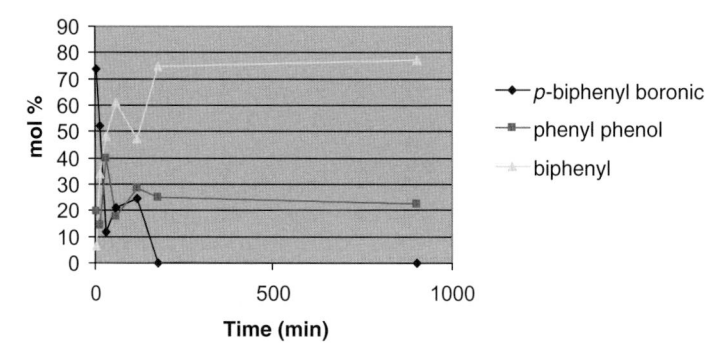

[Graph: mol % vs Time (min), showing three curves: *p*-biphenyl boronic, phenyl phenol, biphenyl]

**Figure 6.5** Side reactions of biphenylboronic acid in the absence of substrates.

[Scheme showing biphenyl-B($O^{16}H$)$_2$ reacting under Cu(OAc)$_2$, CH$_2$Cl$_2$, Et$_3$N, rt with $O_2^{18}$ to give biphenyl-$O^{16}$H (12%) + biphenyl (~50%)]

[Scheme showing biphenyl-B($O^{16}H$)$_2$ reacting under Cu(OAc)$_2$, CH$_2$Cl$_2$, Et$_3$N, rt with 10 equiv. H$_2O^{18}$/$O_2^{16}$ to give biphenyl-$O^{18}$H + biphenyl-$O^{16}$H (~22% (3:2)) + biphenyl (~25%)]

**Scheme 6.6** Studies of the source of phenol by-product.

## 6.8.3
### Mechanistic Study of the Catalytic Reaction

The first systematic mechanistic study of a Cu-catalyzed N/O-arylation of this type was reported by Stahl and coworkers in 2009 for the methoxylation of tolylboronic

**Scheme 6.7** Cu-catalyzed methoxylation of tolylboronic ester.

p-tolB(OMe)$_2$ + MeOH $\xrightarrow[\text{O}_2 \text{ (1 atm), MeOH}]{\text{5\% Cu(OAc)}_2}$ p-tolOMe + B(OMe)$_3$

ester (Scheme 6.7) [97]. Kinetic and EPR spectroscopic studies revealed that the catalyst resting state consists of a Cu(II) species with weak anionic ligands, such as acetate or methoxide, and the turnover-limiting step is transmetalation of the aryl group from boron to the Cu center. Analysis of the reaction stoichiometry demonstrated that C−O reductive elimination does not occur from Cu(II), but rather proceeds via an aryl–Cu(III) intermediate, proposed to form via oxidation of a transient aryl–Cu(II) intermediate by another equivalent of Cu(II). The latter disproportionation reaction, which yields aryl–Cu(III) and Cu(I), has precedent from the work of Ribas et al., who observed formation of a well-defined aryl–Cu(III) complex **137** via Cu(II)-mediated C−H activation (Scheme 6.8) [101]. Stahl and Huffman demonstrated that the aryl–Cu(III) complex **137** undergoes very facile C−N coupling in the presence of various amide-type nitrogen nucleophiles, even in the absence of exogenous base [73]. The Cu(I), which forms in the disproportionation reaction and following C−O/N reductive elimination, is rapidly oxidized by O$_2$ in a 4:1 Cu(I):O$_2$ stoichiometry [97].

**Scheme 6.8** Formation of a well-defined aryl–copper(III) species via C−H activation and disproportionation of Cu(II).

The mechanistic observations noted above are summarized in the catalytic cycle shown in Scheme 6.9. This mechanism not only provides a valuable framework for understanding the reaction pathway, but also draws attention to some important unanswered questions. Stahl and coworkers speculate that transmetalation is the turnover-limiting step for other reactions, for example, those carried out in other solvents, with other types of catalysts (e.g., those with coordinating ligands), and with other nucleophiles; however, the transmetalation step is poorly understood. This step will be influenced by the identity of the Cu(II) species present in solution, which can change, depending on the nucleophile used in the reaction. For example,

$$(HO)B(OMe)_2 + Cu^{II}X_2 \quad\quad Cu^{II}X_2$$

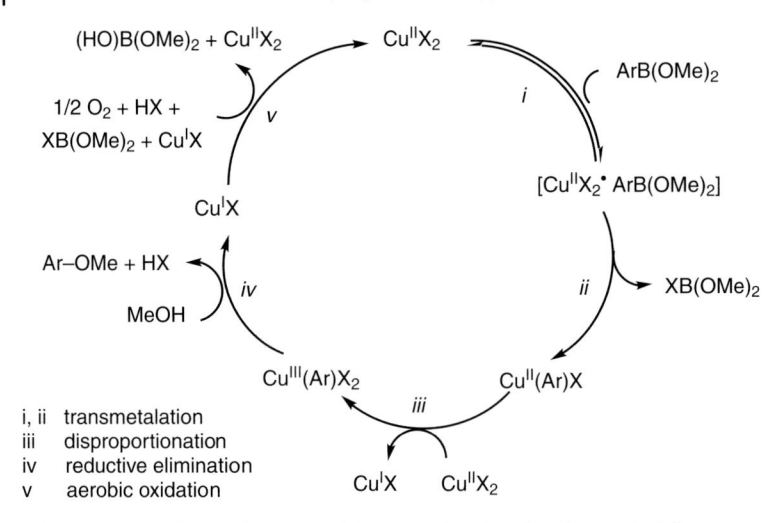

$$1/2\, O_2 + HX + XB(OMe)_2 + Cu^I X$$

$$v$$

$$Cu^I X$$

$$Ar\text{-}OMe + HX$$
$$MeOH$$
$$iv$$

$$Cu^{III}(Ar)X_2 \xrightarrow{iii} Cu^{II}(Ar)X$$

$$Cu^I X \quad Cu^{II}X_2$$

$$ArB(OMe)_2$$
$$i$$
$$[Cu^{II}X_2 \cdot ArB(OMe)_2]$$
$$ii \rightarrow XB(OMe)_2$$

i, ii  transmetalation
iii    disproportionation
iv    reductive elimination
v     aerobic oxidation

**Scheme 6.9** Catalytic cycle proposed for Cu-catalyzed methoxylation of tolylboronic ester (see Scheme 6.7) on the basis of mechanistic studies by Stahl and coworkers.

cupric acetate is completely insoluble in $CH_2Cl_2$, the preferred solvent, and the suspension is colorless. Addition of imidazole results in instantaneous formation of a deep blue solution, suggesting that the coordination/ligand exchange and dissolution of cupric acetate is rapid. Future insights into the influence of different nucleophiles and ligands on the identity of the Cu(II) catalyst and the effectiveness of the transmetalation step should play an important role in expanding the scope of this chemistry.

## 6.9
### Other Organometalloids

Because of their high efficiency and broad commercial availability, arylboronic reagents are the most versatile organometalloids used in the Chan–Lam cross-coupling reactions. Other organometalloids can also participate in this copper-promoted cross-coupling reaction. They are aryltrialkylsiloxanes [25, 102, 103], aryltrimethylstannanes [104, 105], organobismuth reagents (e.g., triarylbismuths [106, 107] and $Ph_3Bi(OAc)_2$) [108, 109]), aryl leads [110], diaryliodonium salts [111], dialkyl- and diarylzinc [112, 113], dialkylaluminum chlorides [114], alkenylstannanes [115], alkynes [116], and organomagnesium reagents [117]. For example, arylstannanes were used complementarily in the Chan–Lam C–O coupling reactions as demonstrated by the synthesis of biaryl ethers **138** with phenols **140** and arylstannanes **139** (Equation 6.54) [105]. DMAP was used as the base and acetonitrile as the solvent. The reactions gave higher yields with electron-rich stannanes, and a small amount of water did not retard the reaction progress.

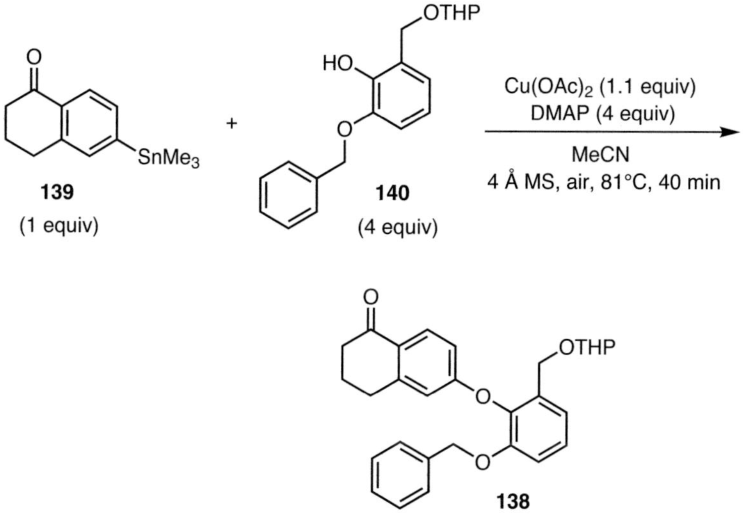

$$\text{139 (1 equiv)} + \text{140 (4 equiv)} \xrightarrow[\text{MeCN} \\ \text{4 Å MS, air, 81°C, 40 min}]{\text{Cu(OAc)}_2 \text{ (1.1 equiv)} \\ \text{DMAP (4 equiv)}} \text{138} \quad (6.54)$$

In summary, the potential scope of copper-promoted C–heteroatom cross-coupling reactions is illustrated in Table 6.1. The set of suitable organometalloids is similar to that of Pd-catalyzed C–C bond coupling reaction. However, in terms of nucleophiles, the Chan–Lam chemistry has much broader scope. This is also true when compared with Pd-catalyzed Buchwald–Hartwig C–heteroatom cross-coupling reaction (Table 6.2).

## 6.10
## Conclusion

Copper-promoted Chan–Lam cross-coupling chemistry has come a long way since its invention more than 10 years ago. As presented in this chapter, significant progress has been made in expanding the scope and the applications, as well as understanding the mechanism of this reaction. Studies on improving yields for certain classes of substrates and fine-tuning the reaction conditions are continually being conducted. Despite these challenges, the copper-promoted C–heteroatom cross-coupling reaction is now a powerful new synthetic tool, particularly for analoging programs in pharmaceutical, crop protection, and material science areas where expedient methodologies are in high demand. The copper-promoted Chan–Lam cross-coupling reaction could potentially become as important and useful for C–heteroatom bonds as the palladium-catalyzed Suzuki–Miyaura C–C cross-coupling reaction.

## 6.11
## Note Added in Proof

In 2010, Cheng and coworkers reported carboxylic acids as new HO-containing substrates in the copper(II) triflate-mediated Chan–Lam reaction for the synthesis of

Table 6.1 Potential scope of copper-promoted C–heteroatom cross-coupling reactions.

$$\text{R-M} + \text{H-}X\text{R'} \xrightarrow[\text{CH}_2\text{Cl}_2,\text{ air}]{\text{Cu(OAc)}_2,\text{ base}} \text{R-}X\text{R'} + \text{M-H}$$

(organometalloid) (Heteroatom nucleophile) (C-heteroatom cross-coupled product)

| Heteroatom Nucleophile H-XR′ Organometalloids R-M | O-H | N-H >15 distinctive classes | S-H RTe-H | RSe-H | ArSO$_2^-$ | (RO)$_2$P(O)-H | Ph$_2$P-H |
|---|---|---|---|---|---|---|---|
| ArB(OH)$_2$ | ✓ | ✓ | ✓ | ✓ | ✓ | ✓ | ✓ |
| (Alkenyl)B(OH)$_2$ | ✓ | ✓ | | | ✓ | | |
| ArSnMe$_3$ | ✓ | ✓ | × | | | | |
| (Alkenyl)SnMe$_3$ | ✓ | ✓ | | | | | |
| ArSi(OR)$_3$/F$^-$ | ✓ | ✓ | | | | | |
| (Alkenyl)Si(OR)$_3$/F$^-$ | | ✓ | | | | | |
| Ar$_3$Bi | ✓ | ✓ | | | | | |
| ArPb(OAc)$_3$ | | ✓ | | | | | |
| Ar$_2$I$^+$X$^-$ | | ✓ | | | | | |
| Alkyne | | ✓ | | | | | |
| ArN$_2^+$X$^-$ | | × | | | | | |
| ArNMe$_3^+$ | | × | | | | | |
| ArZnI | | ✓ (=N–OAc) | | | | | |
| Et$_2$Zn | | ✓ | | | | | |
| ArMgBr | | ✓ (=N–OAc) | | | | | |
| R(C=O)CH$_2$R | | × | | | | | |

Experimentals: ✓, successful; ×, unsuccessful.

Table 6.2 C–heteroatom cross-coupling methodology comparison.

| | Buchwald–Hartwig Pd chemistry | Chan–Lam Cu chemistry |
|---|---|---|
| Reaction condition | ~100 °C or t-BuONa as base<br>Nitrogen atmosphere | rt and weak base<br>Air |
| Substrate | Alcohols, amides, amines, anilines, carbamates, sulfonamides, phenols, thiols | Alcohols, amides, amines, anilines, azides, hydantoins, hydrazines, imides, imines, nitroso, phenols, pyrazinones, pyridones, purines, pyrimidines, sulfonamides, sulfinates, sulfoximines, thiols, thiourea, ureas |
| Reaction yield | Good to excellent yields | Good to excellent yields |
| Aryl donor | Cheap aryl halides and alkenyl halides | More expensive aryl/alkenylboronic acids, siloxanes = stannanes; aryl iodides |
| Catalyst | Catalytic<br>Pd expensive | In general catalytic<br>Cu very cheap |

phenolic esters from carboxylic acids and arylboronic acids [118]. Liu and coworkers also reported similar methodology [119]. Another significant advancement in the C−O cross-coupling is the synthesis of allyl vinyl ethers using copper-promoted coupling of vinyl boronates with neat aliphatic or allylic alcohols in the presence of anhydrous copper(II) acetate by Merlic and coworkers [120a]. The reaction is mild and stereospecific and tolerates functional groups sensitive to acidic, basic, nucleophilic, oxidative, and radical conditions. Optimization, namely, finding a way to not use the substrate as solvent, is ongoing [120b]. New development is also seen in the area of copper-mediated C−C coupling reactions. Chu and Qing reported the first copper-mediated oxidative trifluoromethylation of aryl- and alkenylboronic acids with (trifluoromethyl)trimethylsilane [121]. Hartwig and coworkers reported the first copper-mediated cyanation of arylboronate esters at 80–100 °C [122]. Jiang and coworkers, on the other hand, performed cyanation of boronic acids at a lower temperature (60 °C) [123]. In addition, N-alkylation with alkylboronic acid under Chan–Lam coupling conditions is now feasible by simply heating up in refluxing dioxane with a large excess of the reagents [124]. Copper-catalyzed alkynylation of amides with potassium alkynyltrifluoroborates was also reported under mild and base-free conditions [125]. Besides the mechanistic studies from Stahl's laboratory, Tromp *et al.* reported the mechanistic study of the N-arylation of imidazole using a multitechnique approach [126].

## Acknowledgment

We would like to thank Professor Shannon Stahl for providing Section 6.8.

## References

1 Suzuki, A. (2002) *Modern Arene Chemistry* (ed. D. Astruc), Wiley-VCH Verlag GmbH, Weinheim, pp. 53–106.
2 Jiang, L. and Buchwald, S.L. (2004) Palladium-Catalyzed Aromatic Carbon-Nitrogen Bond Formation, in *Metal-Catalyzed Cross-Coupling Reactions* (eds A. de Meijere and F. Diederich), Wiley-VCH Verlag GmbH, Weinheim, 699–760.
3 Hartwig, J.F. (1998) *Angew. Chem., Int. Ed.*, **37**, 2046–2067.
4 Lindley, J. (1984) *Tetrahedron*, **40**, 1433–1456.
5 Altman, R.A., Koval, E.D., and Buchwald, S.L. (2007) *J. Org. Chem.*, **72**, 6190–6199.
6 Martin, R. and Buchwald, S.L. (2008) *Acc. Chem. Res.*, **41**, 1461–1473.
7 Li, J.J. (2009) *Name Reactions: A Collection of Detailed Reaction Mechanisms*, 4th edn, Springer, pp. 102–104.
8 Chan, D.M.T. and Lam, P.Y.S. (2005) Chapter 5, in *Boronic Acids* (ed. D.G. Hall), Wiley-VCH Verlag GmbH, Weinheim, pp. 205–240.
9 Chan, D.M.T., Monaco, K.L., Wang, R.-P., and Winters, M.P. (1998) *Tetrahedron Lett.*, **39**, 2933–2936.
10 Evans, D.A., Katz, J.L., and West, T.R. (1998) *Tetrahedron Lett.*, **39**, 2937–2940 (Professor Evan's group found out about the discovery of copper-mediated O-arylation reaction on a National Organic Symposium poster of Dr. Chan and became interested because of the importance of novel biaryl ether synthesis for vancomycin total synthesis).

11 Lam, P.Y.S., Clark, C.G., Saubern, S., Adams, J., Winters, M.P., Chan, D.M.T., and Combs, A. (1998) *Tetrahedron Lett.*, **39**, 2941–2944.

12 Finet, J.-P., Fedorov, A.Y., Combes, S., and Boyer, G. (2002) *Curr. Org. Chem.*, **6**, 597–626.

13 Kunz, K., Scholz, U., and Ganzer, D. (2003) *Synlett*, 2428–2439.

14 Ley, S.V. and Thomas, A.W. (2003) *Angew. Chem., Int. Ed.*, **42**, 5400–5449.

15 Scott Sawyer, J. (2000) *Tetrahedron*, **56**, 5045–5065.

16 Theil, F. (2002) *Organic Synthesis Highlights V* (eds H. Schmalz and T. Wirth), Wiley-VCH Verlag GmbH, Weinheim, pp. 15–21.

17 Thomas, A.W. and Ley, S.V. (2008) *Modern Arylation Methods* (ed L. Ackermann), Wiley-VCH Verlag GmbH, Weinheim, pp. 121–154.

18 Jung, M.E. and Lazarova, T.I. (1999) *J. Org. Chem.*, **64**, 2976–2977.

19 Dehli, J.R., Legros, J., and Bolm, C. (2005) *Chem. Commun.*, 973–986.

20 Lam, P.Y.S., Clark, C.G., Saubern, S., Adams, J., Averill, K.M., Chan, D.M.T., and Combs, A. (2000) *Synlett*, 674–676.

21 Lam, P.Y.S., Vincent, G., Bonne, D., and Clark, C.G. (2003) *Tetrahedron Lett.*, **44**, 4927–4931.

22 Petrassi, H.M., Sharpless, K.B., and Kelly, J.W. (2001) *Org. Lett.*, **3**, 139–142.

23 Wang, Z. and Zhang, J. (2005) *Tetrahedron Lett.*, **46**, 4997–4999.

24 Tzschucke, C.C., Murphy, J.M., and Hartwig, J.F. (2007) *Org. Lett.*, **9**, 761–764.

25 Lam, P.Y.S., Deudon, S., Hauptman, E., and Clark, C.G. (2001) *Tetrahedron Lett.*, **42**, 2427–2429.

26 Decicco, C.P., Song, Y., and Evans, D.A. (2001) *Org. Lett.*, **3**, 1029–1032.

27 Deng, H., Jung, J.-K., Liu, T., Kuntz, K.W., Snapper, M.L., and Hoveyda, A.H. (2003) *J. Am. Chem. Soc.*, **125**, 9032–9034.

28 Hitotsuyanagi, Y., Ishikawa, H., Naito, S., and Takeya, K. (2003) *Tetrahedron Lett.*, **44**, 5901–5903.

29 Kantam, M.L., Prakash, B.V., and Reddy, C.V. (2005) *J. Mol. Catal. A*, **241**, 162–165.

30 Wang, W., Devasthale, P., Farrelly, D., Gu, L., Harrity, T., Cap, M., Chu, C., Kunselman, L., Morgan, N., Ponticiello, R., Zebo, R., Zhang, L., Locke, K., Lippy, J., O'Malley, K., Hosagrahara, V., Zhang, L., Kadiyala, P., Chang, C., Muckelbauer, J., Doweyko, A.M., Zahler, R., Ryono, D., Hariharan, N., and Cheng, P.T.W. (2008) *Bioorg. Med. Chem. Lett.*, **18**, 1939–1944.

31 Abe, T., Takeda, H., Yamada, K., and Ishikura, M. (2008) *Heterocycles*, **76**, 133–136.

32 Singh, B.K., Appukkuttan, P., Claerhout, S., Parmar, V.S., and Van der Eycken, E. (2006) *Org. Lett.*, **8**, 1863–1866.

33 Hugel, H.M., Rix, C.J., and Fleck, K. (2006) *Synlett*, 2290–2292.

34 Chen, S., Huang, H., Liu, X., Shen, J., Jiang, H., and Liu, H. (2008) *J. Comb. Chem.*, **10**, 358–360.

35 Biffis, A., Filippi, F., Palma, G., Lora, S., Macca, C., and Corain, B. (2003) *J. Mol. Catal. A*, **203**, 213–220.

36 Chiang, G.C.H. and Olsson, T. (2004) *Org. Lett.*, **6**, 3079–3082.

37 Kantam, M.L., Venkanna, G.T., Sridhar, C., Sreedhar, B., and Choudary, B.M. (2006) *J. Org. Chem.*, **71**, 9522–9524.

38 Reddy, K.R., Kumar, N.S., Sreedhar, B., and Kantam, M.L. (2006) *J. Mol. Catal. A*, **252**, 136–141.

39 Biffis, A., Gardan, M., and Corain, B. (2006) *J. Mol. Catal. A*, **250**, 1–5.

40 Zhang, L.-Y. and Wang, L. (2006) *Chin. J. Chem.*, **24**, 1605–1608.

41 Combs, A.P. and Rafalski, M. (2000) *J. Comb. Chem.*, **2**, 29–32.

42 Combs, A.P., Tadesse, S., Rafalski, M., Haque, T.S., and Lam, P.Y.S. (2002) *J. Comb. Chem.*, **4**, 179–182.

43 Combs, A.P., Saubern, S., Rafalski, M., and Lam, P.Y.S. (1999) *Tetrahedron Lett.*, **40**, 1623–1626.

44 Lam, P.Y.S., Clark, C.G., Li, R., Pinto, D.J.P., Orwat, M.J., Galemmo, R.A., Fevig, J.M., Teleha, C.A., Alexander, R.S., Smallwood, A.M., Rossi, K.A., Wright, M.R., Bai, S.A., He, K., Luettgen, J.M., Wong, P.C., Knabb, R.M., and Wexler, R.R. (2003) *J. Med. Chem.*, **46**, 4405–4418.

45 (a) Guy, C.S. and Jones, T.C. (2009) *Synlett*, 2253–2256; (b) Sreeramamurthy, K., Ashok, E., Mahendar, V., Santoshkumar, G., and Das, P. (2010) *Synlett*, 721–724.

46 Mederski, W.W.K.R., Lefort, M., Germann, M., and Kux, D. (1999) *Tetrahedron*, **55**, 12757–12770.

47 Yu, S., Saenz, J., and Srirangam, J.K. (2002) *J. Org. Chem.*, **67**, 1699–1702.

48 Bekolo, H. (2007) *Can. J. Chem.*, **85**, 42–46.

49 Hari, Y., Shoji, Y., and Aoyama, T. (2005) *Tetrahedron Lett.*, **46**, 3771–3774.

50 Lam, P.Y., Clark, C.G., Han, Q., and Richardson, T.E. (2003) US Patent 6673810B2.

51 Ding, S., Gray, N.S., Ding, Q., and Schultz, P.G. (2001) *Tetrahedron Lett.*, **42**, 8751–8755.

52 Bakkestuen, A.K. and Gundersen, L.-L. (2003) *Tetrahedron Lett.*, **44**, 3359–3362.

53 Tao, L., Yue, Y., Zhang, J., Chen, S.-Y., and Yu, X.-Q. (2008) *Helv. Chim. Acta*, **91**, 1008–1014.

54 Yue, Y., Zheng, Z.-G., Wu, B., Xia, C.-Q., and Yu, X.-Q. (2005) *Eur. J. Org. Chem.*, 5154–5157.

55 Jacobsen, M.F., Knudsen, M.M., and Gothelf, K.V. (2006) *J. Org. Chem.*, **71**, 9183–9190.

56 Chan, D.M.T., Monaco, K.L., Li, R., Bonne, D., Clark, C.G., and Lam, P.Y.S. (2003) *Tetrahedron Lett.*, **44**, 3863–3865.

57 Nielsen, F.E., Korno, H.T., and Rasmussen, K.G. (2002) WO 2002050085.

58 Collman, J.P. and Zhong, M. (2000) *Org. Lett.*, **2**, 1233–1236.

59 Collman, J.P., Zhong, M., Zhang, C., and Costanzo, S. (2001) *J. Org. Chem.*, **66**, 7892–7897.

60 Lan, J.-B., Zhang, G.-L., Yu, X.-Q., You, J.-S., Chen, L., Yan, M., and Xie, R.-G. (2004) *Synlett*, 1095–1097.

61 Quach, T.D. and Batey, R.A. (2003) *Org. Lett.*, **5**, 4397–4400.

62 Sreedhar, B., Venkanna, G.T., Kumar, S., Balaji, K., and Balasubrahmanyam, V. (2008) *Synthesis*, 795–799.

63 Wentzel, M.T., Hewgley, J.B., Kamble, R.M., Wall, P.D., and Kozlowski, M.C. (2009) *Adv. Synth. Catal.*, **351**, 931–937.

64 Uemura, T. and Chatani, N. (2005) *J. Org. Chem.*, **70**, 8631–8634.

65 Wang, Y.F., Zhou, Y., Wang, J.R., Liu, L., and Guo, Q.X. (2007) *Chin. Chem. Lett.*, **18**, 499–501.

66 Moessner, C. and Bolm, C. (2005) *Org. Lett.*, **7**, 2667–2669.

67 Tao, C.-Z., Cui, X., Li, J., Liu, A.-X., Liu, L., and Guo, Q.-X. (2007) *Tetrahedron Lett.*, **48**, 3525–3529.

68 (a) Rao, H., Fu, H., Jiang, Y., and Zhao, Y. (2009) *Angew. Chem., Int. Ed.*, **48**, 1114–1116. (b) Jiang, Z., Wu, Z., Wang, L., Wu, D., and Zhou, X. (2010) *Can. J. Chem.*, **88**, 964–968. (c) Zhou, C., Chen, F., Yang, D., Jiam, X., Zhang, L., and Cheng, J. (2009) *Chem. Lett.*, **38**, 708–709.

69 Liu, S., Yu, Y., and Liebeskind, L.S. (2007) *Org. Lett.*, **9**, 1947–1950.

70 Liu, S. and Liebeskind, L.S. (2008) *J. Am. Chem. Soc.*, **130**, 6918–6919.

71 Zhang, Z., Yu, Y., and Liebeskind, L.S. (2008) *Org. Lett.*, **10**, 3005–3008.

72 Cherney, R.J., Duan, J.J.W., Voss, M.E., Chen, L., Wang, L., Meyer, D.T., Wasserman, Z.R., Hardman, K.D., Liu, R.-Q., Covington, M.B., Qian, M., Mandlekar, S., Christ, D.D., Trzaskos, J.M., Newton, R.C., Magolda, R.L., Wexler, R.R., and Decicco, C.P. (2003) *J. Med. Chem.*, **46**, 1811–1823.

73 Huffman, L.M. and Stahl, S.S. (2008) *J. Am. Chem. Soc.*, **130**, 9196–9197.

74 McKinley, N.F. and O'Shea, D.F. (2004) *J. Org. Chem.*, **69**, 5087–5092.

75 Bolshan, Y. and Batey, R.A. (2008) *Angew. Chem., Int. Ed.*, **47**, 2109–2112.

76 Altman, R.A. and Buchwald, S.L. (2007) *Org. Lett.*, **9**, 643–646.

77 Voisin, A.S., Bouillon, A., Lancelot, J.-C., Lesnard, A., and Rault, S. (2006) *Tetrahedron*, **62**, 6000–6005.

78 Kleeberg, C., Dang, L., Lin, Z., and Marder, T.B. (2009) *Angew. Chem., Int. Ed.*, **48**, 5350–5354.

79 Deagostino, A., Prandi, C., Zavattaro, C., and Venturello, P. (2007) *Eur. J. Org. Chem.*, 1318–1323.

80 Zheng, Z.-G., Wen, J., Wang, N., Wu, B., and Yu, X.-Q. (2008) *Beilstein J. Org. Chem.*, **4** (40).
81 Quach, T.D. and Batey, R.A. (2003) *Org. Lett.*, **5**, 1381–1384.
82 Yamamoto, Y., Takizawa, M., Yu, X.-Q., and Miyaura, N. (2008) *Angew. Chem., Int. Ed.*, **47**, 928–931.
83 Yu, X.-Q., Yamamoto, Y., and Miyaura, N. (2008) *Chem. Asian J.*, **3**, 1517–1522.
84 Tsuritani, T., Strotman, N.A., Yamamoto, Y., Kawasaki, M., Yasuda, N., and Mase, T. (2008) *Org. Lett.*, **10**, 1653–1655.
85 Benard, S., Neuville, L., and Zhu, J. (2008) *J. Org. Chem.*, **73**, 6441–6444.
86 Gonzalez, I., Mosquera, J., Guerrero, C., Rodriguez, R., and Cruces, J. (2009) *Org. Lett.*, **11**, 1677–1680.
87 Herradura, P.S., Pendola, K.A., and Guy, R.K. (2000) *Org. Lett.*, **2**, 2019–2022.
88 Savarin, C., Srogl, J., and Liebeskind, L.S. (2002) *Org. Lett.*, **4**, 4309–4312.
89 Lengar, A. and Kappe, C.O. (2004) *Org. Lett.*, **6**, 771–774.
90 Prokopcova, H. and Kappe, C.O. (2007) *Adv. Synth. Catal.*, **349**, 448–452.
91 Prokopcova, H. and Kappe, C.O. (2007) *J. Org. Chem.*, **72**, 4440–4448.
92 Beaulieu, C., Guay, D., Wang, Z., and Evans, D.A. (2004) *Tetrahedron Lett.*, **45**, 3233–3236.
93 Huang, F. and Batey, R.A. (2007) *Tetrahedron*, **63**, 7667–7672.
94 Kar, A., Sayyed, I.A., Lo, W.F., Kaiser, H.M., Beller, M., and Tse, M.K. (2007) *Org. Lett.*, **9**, 3405–3408.
95 Kantam, M.L., Neelima, B., Sreedhar, B., and Chakravarti, R. (2008) *Synlett*, 1918.
96 Wang, L., Wang, M., and Huang, F. (2005) *Synlett*, 2007–2010.
97 (a) King, A.E., Brunold, T.C., and Stahl, S.S. (2009) *J. Am. Chem. Soc.*, **131**, 5044–5045 (part of the mechanistic work was funded by and accomplished in collaboration with BMS); (b) King, A.E., Huffman, L.M., Casitas, A., Costas, M., Ribas, X., and Stahl, S.S. (2010) *J. Am. Chem. Soc.*, **132**, 12068–12073.
98 Lan, J.-B., Chen, L., Yu, X.-Q., You, J.-S., and Xie, R.-G. (2004) *Chem. Commun.*, 188–189.
99 Lam, P.Y.S., Bonne, D., Vincent, G., Clark, C.G., and Combs, A.P. (2003) *Tetrahedron Lett.*, **44**, 1691–1694.
100 Steinhoff, B.A., King, A.E., and Stahl, S.S. (2006) *J. Org. Chem.*, **71**, 1861–1868.
101 Ribas, X., Jackson, D.A., Donnadieu, B., Mahia, J., Parella, T., Xifra, R., Hedman, B., Hodgson, K.O., Llobet, A., and Stack, T.D.P. (2002) *Angew. Chem., Int. Ed.*, **41**, 2991–2994.
102 Lam, P.Y.S., Deudon, S., Averill, K.M., Li, R., He, M.Y., DeShong, P., and Clark, C.G. (2000) *J. Am. Chem. Soc.*, **122**, 7600–7601.
103 (a) Song, R.-J., Deng, C.-L., Xie, Y.-X., and Li, J.-H. (2007) *Tetrahedron Lett.*, **48**, 7845–7848; (b) Lin, B., Liu, M., Ye, Z., Ding, J., Wu, H., and Cheng, J. (2009) *Org. Biomol. Chem.*, **7**, 869–873; (c) Pan, C., Cheng, J., Wu, H., Ding, J., and Liu, M. (2009) *Synth. Commun.*, **39**, 2082–2092; (d) Luo, P.-S., Yu, M., Tang, R.-Y., Zhong, P., and Li, J.-H. (2009) *Tetrahedron Lett.*, **50**, 1066–1070; (e) Luo, F., Pan, C., Qian, P., and Cheng, J. (2010) *Synthesis*, 2005–2010; (f) Luo, F., Pan, C., Qian, P., and Cheng, J. (2010) *J. Org. Chem.*, **75** 5379–5381.
104 Lam, P.Y.S., Vincent, G., Bonne, D., and Clark, C.G. (2002) *Tetrahedron Lett.*, **43**, 3091–3094.
105 Vakalopoulos, A., Kavazoudi, X., and Schoof, J. (2006) *Tetrahedron Lett.*, **47**, 8607–8610.
106 Arnauld, T., Barton, D.H.R., and Doris, E. (1997) *Tetrahedron*, **53**, 4137–4144.
107 Chan, D.M.T. (1996) *Tetrahedron Lett.*, **37**, 9013–9016.
108 Fedorov, A.Y. and Finet, J.-P. (1999) *Tetrahedron Lett.*, **40**, 2747–2748.
109 Tsubrik, O., Kisseljova, K., and Maeorg, U. (2006) *Synlett*, 2391–2394.
110 Lopez-Alvarado, P., Avendano, C., and Menendez, J.C. (1995) *J. Org. Chem.*, **60**, 5678–5682.
111 Kang, S.-K., Lee, S.-H., and Lee, D. (2000) *Synlett*, 1022–1024.
112 Berman, A.M. and Johnson, J.S. (2006) *J. Org. Chem.*, **71**, 219–224.
113 Brielles, C., Harnett, J.J., and Doris, E. (2001) *Tetrahedron Lett.*, **42**, 8301–8302.

114 Barton, D.H.R. and Doris, E. (1996) *Tetrahedron Lett.*, **37**, 3295–3298.
115 van Otterlo, W.A.L., Ngidi, E.L., Kuzvidza, S., Morgans, G.L., Moleele, S.S., and de Koning, C.B. (2005) *Tetrahedron*, **61**, 9996–10006.
116 Hamada, T., Ye, X., and Stahl, S.S. (2008) *J. Am. Chem. Soc.*, **130**, 833–835.
117 Campbell, M.J. and Johnson, J.S. (2007) *Org. Lett.*, **9**, 1521–1524.
118 Zhang, L., Zhang, G., Zhang, M., and Cheng, J. (2010) *J. Org. Chem.*, **75**, 7472–7474.
119 Dai, J.-J., Liu, J.-H., Luo, D.-F., and Liu, L. (2011) *Chem. Commun.*, **47**, 677–679.
120 (a) Shade, R.E., Alan, M., Hyde, A.M., Olsen, J.-C., and Merlic, C.A. (2010) *J. Am. Chem. Soc.*, **132**, 1202–1203; (b) Winternheimer, D.J. and Merlic, C.A. (2010) *Org. Lett.*, **12**, 2508–2510.
121 Chu, L. and Qing, F.-L. (2010) *Org. Lett.*, **12**, 5060–5063.
122 Liskey, C.W., Liao, C., and Hartwig, J.F. (2010) *J. Am. Chem. Soc.*, **132**, 11389–11391.
123 Zhang, G., Zhang, L., Hu, M., and Cheng, J. (2011) *Adv. Synth. Catal.* doi: 10.1002/adsc.201000747.
124 Larrosa, M., Guerrero, C., Rodríguez, R., and Cruces, J. (2010) *Synlett*, 2101–2105.
125 Jouvin, K., Couty, F., and Evano, G. (2010) *Org. Lett.*, **12**, 3272–3275.
126 Tromp, M., van Strijdonck, G.P.F., van Berkel, S.S., van den Hoogenband, A., Feiters, M.C., de Bruin, B., Fiddy, S.G., van der Eerden, A.M.J., van Bokhoven, J.A., van Leeuwen, P.W.N.M., and Koningsberger, D.C. (2010) *Organometallics*, **29**, 3085–3097.